Nanoparticles in Catalysis

Nanoparticles in Catalysis

Advances in Synthesis and Applications

Edited by
Karine Philippot
Alain Roucoux

WILEY-VCH

Editors

Dr. Karine Philippot
Laboratoire de Chimie de Coordination
du CNRS UPR 8241
205, route de Narbonne
BP44099
31077 Toulouse Cedex 04
France

Prof. Alain Roucoux
ENS Chimie de Rennes
UMR CNRS 6226
11, allée de Beaulieu
CS 50837
35708 Rennes Cedex 7
France

Library of Congress Card No.:
applied for

British Library Cataloguing-in-Publication Data
A catalogue record for this book is available from the British Library.

Bibliographic information published by the Deutsche Nationalbibliothek
The Deutsche Nationalbibliothek lists this publication in the Deutsche Nationalbibliografie; detailed bibliographic data are available on the Internet at <http://dnb.d-nb.de>.

Print ISBN: 978-3-527-34607-3
ePDF ISBN: 978-3-527-82174-7
ePub ISBN: 978-3-527-82175-4
oBook ISBN: 978-3-527-82176-1

Typesetting SPi Global, Chennai, India
Printing and binding CPI Group (UK) Ltd, Croydon, CR0 4YY

Printed on acid-free paper

C092935_040321

Contents

Foreword

In the recent decades, nanosciences and nanotechnologies have revolutionized the approaches of the researchers and the view of the public, and nanocatalysis stands as one of their most prominent aspects. Catalysis is involved in a majority of biological processes as well as in the transformation of chemicals and energy sources indispensable in today society.

A number of crucial industrial catalytic processes discovered in the beginning of the twentieth century have allowed the transformation of raw materials into fine chemicals using heterogeneous catalysts, whereas homogeneous catalysis involving a good control of kinetics, mechanisms, and consequently of high selectivities was only developed in the last quarter of the twentieth century. Neither heterogeneous nor homogeneously catalysis has produced perfect catalysts with the key desired properties of efficiency, selectivity, recyclability, and green chemistry like those encountered in enzymes. It is the merging of these two communities of catalysis that is at the origin of the of bottom-up concept of nanosized catalysts bringing a molecular dimension to heterogeneous catalysis.

Perhaps the understanding of the advantage of conducting catalysis with the smallest particle was realized by Paul Sabatier more than hundred years ago. Nanocatalysis itself was seldom addressed before the 1980's, however, when instruments such as the scanning tunneling microscope permitting the adequate characterization of nanoparticles were invented and later also applied to the fabrication of useful nanomaterials.

Our modern Society urgently requires meeting the fundamental needs involving sustainable development, clean energy, and advanced medicine, all calling for novel catalytic tools with which synergies, optimized nanocatalysts, and designed interfaces between nanoparticles and supports must be developed.

An immense body of research advances in nanocatalysis has already been achieved in the past 40 years and particularly since the 2000's. Therefore, this new book "Nanoparticles in Catalysis: Advances in Synthesis and Applications," edited by Dr. Karine Philippot (LCC, Toulouse) and Professor Alain Roucoux (ENSCR, Rennes), distinguished by their high standing and in particular their reputation in the field defined by the book title, is timely and most welcome. The editors have themselves authored a useful introductory chapter showing the new trends in the design of metal nanoparticles and derived nanomaterials for catalysis. They have also gathered among the best experts who have authored 15 excellent book chapters.

Altogether, this book, judiciously divided into four parts: nanoparticles in solution, supported nanoparticles, application, and activation and theory, covers the whole area of modern nanocatalysis and lays the foundation for further research challenges, developments, and applications.

Bordeaux, April 2020 *Didier Astruc*

1

New Trends in the Design of Metal Nanoparticles and Derived Nanomaterials for Catalysis

Alain Roucoux[1] and Karine Philippot[2]

[1] *Université de Toulouse, UPS, INPT, CNRS, LCC (Laboratoire de Chimie de Coordination), UPR 8241, F-31077, Toulouse Cedex 4, France*
[2] *Univ Rennes, Ecole Nationale Supérieure de Chimie de Rennes, CNRS, ISCR – UMR 6226, 11, allée de Beaulieu, F-35000, Rennes, France*

1.1 Nanocatalysis: Position, Interests, and Perspectives

Being part of heterogeneous catalysts, metal nanoparticles (NPs) have been known for a long time (see for instance the pioneering works of P. Sabatier [1] and L.D. Rampino and F.F. Nord [2]), but a renewed interest has emerged in the past three decades for the design of better-defined systems [3]. Interestingly, a great part of the former heterogeneous catalysis community is now merging into the nanoparticle one [4]. Thus, as it can be observed in the literature through the overincreasing number of papers and patents published by both academic and industrial institutes, huge research efforts are devoted to the synthesis of more precisely defined metal nanospecies and even more recently at an atomic precision level [5–7], as well as the study of their characteristics. This keen interest for nanoscale metal-based species derives from their particular matter state (finely divided metals) and their electronic parameters, thus influencing the physical and chemical properties that these entities possess in comparison with bulk metals and molecular complexes. Small nanoparticles are usually called nanoclusters, but there is a continuum of situations from molecules to solid state between small clusters defined by molecular orbitals and larger nanoparticles defined by energy band structures. These various types differ by the number of metal atoms, the nature of the stabilizing ligands, and the dispersity. Rigorously, the term cluster or nanocluster only concerns molecularly polymetallic assemblies with ligands for which the X-ray crystal structure is known, whereas the term nanoparticles is used for mixtures of more or less polydisperse large nanoclusters defined by the histogram disclosed by transmission electron microscopy (TEM) measurement [4]. Besides the fundamental aspects of the research around metal nanoparticles, the interest is also governed by the potential applications that they may find. The unique properties of nanosized metal particles make them very attractive materials for various domains of applications including optoelectronics, sensing, biomedicine, energy conversion, storage, and catalysis, as nonexhaustive

Nanoparticles in Catalysis: Advances in Synthesis and Applications, First Edition.
Edited by Karine Philippot and Alain Roucoux.

examples [8–11]. Concerning this last field, several books dedicated to nanocatalysis have been edited in the past 15 years [1, 12]. Indeed, metal nanoparticles are particularly interesting catalytic species because of the high surface-to-volume ratio they display. This ratio is promising when metal nanoparticles are at a size as close as 1 nm, or even below (subnanoparticles), because the number of surface atoms can be >90%, thus providing a great number of potential active sites. Remarkably, the subnanometric particles of late transition metals have been reported as very active, although the optimum catalytic activity is believed to be attained with particles containing between 12 and 20 atoms (i.e. sizes close to 1 nm or slightly below) [13]. The atomic sites exposed in metal nanoparticles may have various reactivities owing to their different coordination numbers according to their situation on the surface (corner, edge, and face atoms). By the way, the development of synthesis tools that enable to produce ultrasmall nanoparticles remains of prime importance in order to promote high surface area combined with an efficient distribution of surface atoms. Indeed, the proportion of edge and corner surface atoms possessing lower coordination numbers increases with the decrease in the particle size. Besides the size, other key morphological parameters need to be controlled. Thus, the crystalline structure is also of great importance because according to the types of crystalline plans that can be exposed at the surface, different catalytic properties could be achieved. Controlling the particles shape also constitutes another route to orientate the crystalline plans exposed [14–16]. The last but not the least key parameter is the composition of the metal nanoparticles that has to be adjusted depending on the targeted catalytic reaction. Apart from the nature of the metallic core that may govern the reactivity (some metals are well known for certain catalytic applications but not for others), the surrounding stabilizer for colloidal catalysis (ionic liquids [ILs], polymers, dendrimers, surfactants, polyols, ligands, etc.) or the support for supported catalysis (metal oxides such as silica, metal organic frameworks [MOFs], carbon derivatives, nanocelluloses, etc.) may also influence or even orientate the catalytic performances. While calcination is typically applied in traditional heterogeneous catalysis in order to suppress any organics and liberate the active sites, such treatment on small nanoparticles can be critical, thus potentially leading to their sintering. Moreover, naked nanospecies are not always optimal catalysts. In modern nanocatalysis, the presence of organic ligands onto the particle surface is not considered as detrimental for catalytic applications but could be a way to improve or even modify the chemoselectivity [17]. Aiming at catalysis by appropriate design, current developments in nanochemistry dedicated to catalytic applications often rely with multifunctionality [18]. This can be achieved by the proper design of nanohybrids, the term hybrid referring to the appropriate association between a metal core and a stabilizing shell or a support. When using typical ligands from coordination chemistry as capping agents, a parallel can be performed with molecular catalysis. Indeed, the interaction of the ligands with the metal atoms on the particle surface can be compared to ligand interactions with the metal centers in homogeneous complexes and is a parameter of paramount importance for stability and catalytic performances (activity and selectivity properties). Thus, ligands can be chosen in order to tune the surface properties of metal nanoparticles through steric and/or electronic effects [19, 20]. The challenge remains to find protective agents that are able to stabilize well-defined nanoparticles while controlling accessibility at the metal surface and reactivity [13, 21]. Strongly bound capping ligands (such as thiols or phosphines) can result in the poisoning of a nanocatalyst at high surface coverage. However, the presence of a limited amount of ligand can be beneficial in terms of catalytic performances (suitable and replicable activities and/or selectivities). The strong coordination of a ligand at a metal surface can also be a way to block selectively some active sites in order to orientate the catalysis evolution. In comparison with the investigation of facet dependency [12, 22], the ligand influence on the catalytic activity has been less intensively studied, but recent results well illustrate the interest to do so [23–27]. Capping agent-stabilized metal nanoparticles can be applied to catalysis as stable colloidal suspensions in various media (water, polyols, and organic solvents) but also in heterogeneous conditions after their deposition on the surface or confinement in the pores of solid supports [28]. Ionic liquids [29] are also very efficient to stabilize metal nanoparticles, and their colloidal suspensions can even be deposited onto inorganic materials [30]. When employing a support, it not only prevents nanoparticle aggregation but may also act in synergy with nanoparticle surface and favor the activation of substrates in a manner comparable to the positive synergy observed between two transition metal atoms in alloys or between a transition metal and a main group element such as nitrogen. Therefore, it was found that N-doped carbon supports were superior to undoped analogs because of such synergy effects [4].

Thus, having in hand synthesis strategies that allow access, in a reproducible manner, to well-defined metal nanoparticles in terms of size, crystalline structure, composition (metal cores and stabilizing agents), chemical order (bimetallic or ternary systems), shape, and dispersion constitutes a beneficial condition to finely investigate the catalytic properties of metal nanoparticles and define structure/properties relationships. Taking advantage of recent developments in nanochemistry that offer efficient tools to reach these objectives, nanocatalysis is now well established as a borderline domain between homogeneous and heterogeneous catalysis. With a molecular approach, nanocatalysts can be considered as assemblies of individual active sites where metal–metal bonds will also have influence [31]. Precisely designed nanohybrids (including the choice of an adequate and noninnocent stabilizer or support) are expected to present benefits from both homogeneous and heterogeneous catalysts, namely, high reactivity and better selectivity [32]. One aim lies in the design of more performant nanocatalysts in order to develop more efficient and eco-compatible chemical production [33]. If huge progress has been performed in the past decade, this topic still remains challenging, in particular with the crucial need to develop multigram-scalable catalytic routes presenting a reduced environmental footprint and economically viable cost for industrial applications. Furthermore, model catalysts are also needed in order to better understand the relationship between the characteristics of metal nanoparticles and their catalytic performances and thus bridge the gap between model surfaces and real catalysts. Each progress that contributes to reduce the gap of knowledge between nanocatalysts and homogeneous catalysts constitutes a step forward the development of more efficient catalytic systems. Efforts have thus to be pursued in order to one day be able to anticipate the design of suitable catalysts for a given reaction.

Various metals are investigated in nanocatalysis toward these principles, with a huge number of studies dedicated to gold, which is highly reputed for CO oxidation and emerges now in hydrogenation catalysis [34, 35] or palladium as a relevant metal for various C–C coupling reactions and also catalytic hydrogenation [36, 37]. Other metals such as rhodium, ruthenium, platinum, iridium, nickel, cobalt, and iron among others are also the subject of numerous studies. Some of these metals, generally called earth abundant, such as iron and cobalt or more recently manganese or copper have attracted increasing attention in the field of nanocatalysis because of their low cost, ready availability, low toxicity, and greater sustainability [38, 39]. Indeed, for many traditional catalytic transformations and challenging reactions, earth-abundant transition metal nanoparticle catalysts have already appeared as attractive alternatives to noble ones.

The purpose of this book is to provide the readers a synthetic overview of the recent advances in research that address the investigation of well-controlled metal nanoparticles in the domain of catalysis in suspension conditions (colloidal catalysis) and supported conditions.

1.2 Metal Nanoparticles: What Is New?

As mentioned above, nanocatalysis, being an exciting research area at the frontier between homogeneous and heterogeneous fields, has known a huge evolution during the past 25 years. This great interest mainly derives from the original catalytic properties that nanoparticles offer but also to the great advances in their synthesis in solution (aqueous, organic, glycerol, polyols, or ionic liquids, as main examples) or onto supports. Also, the combination of complementary analytical techniques from both molecular chemistry and solid chemistry together with the development of computational chemistry tools for the study of their surface state and/or modeling accounts for this still growing interest. Worldwide, numerous scientists are involved in both "nanochemistry" and "nanocatalysis" with a main and common objective, the design of well-controlled nanoparticles in terms of size, dispersion, and surrounding environment to afford high and/or original catalytic performances.

Compared to previous methodologies in which metal nanoparticles were produced almost exclusively via heterogeneous approaches (top-down approach), current strategies offer a better control of the nanoparticle characteristics (bottom-up approach). Nanometer-sized transition metal particles for catalysis, usually called "nanocatalysts," are structure sensitive; their properties in terms of activities and selectivities depend on their core-size, shape, morphologies as well as nature, composition, and their support materials. Nanocatalysis includes the development of metallic and metal oxide nanoscale materials with defined structure and morphology, and their use as active species or as supports of the active phase.

This book will provide the readers a basic knowledge on the current trends in the domain of metal nanoparticle engineering and of derived composite materials, all the research efforts were devoted to reach not only better catalytic performances but also a better understanding of the key parameters in nanocatalysis. A large range of data from the synthesis of metal nanoparticles, either in solution or deposited onto various supports, to their application as nanocatalysts in diverse catalytic reactions have been collected. Actually, each chapter of this book proposes a critical overview on the concerned subject through the discussion of relevant examples from the recent literature.

The 15 chapters of this book cover a wide range of subfields by exploring in detail a great variety of nanocatalysts, catalytic media, catalytic reactions, and potentially their external activation. Following two introductive chapters dedicated to general interests of metal nanoparticles (Chapter 1) and their position between molecular and heterogeneous catalysts (Chapter 2), four thematic ensembles of chapters structure this book (Figure 1.1). The first two

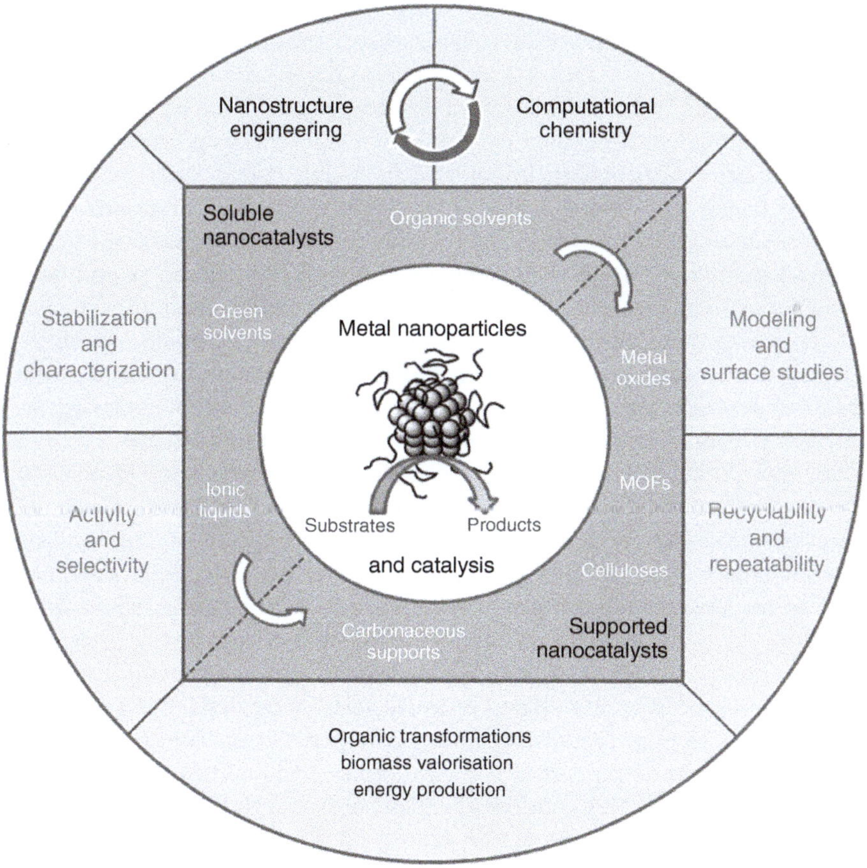

Figure 1.1 Schematic overview of the main but not exhaustive trends currently developed in the area of nanocatalysis and illustrated through the chapters contents of this book.

sets concern nanoparticles as active species in solution (Set I) or deposited onto supports (Set II). Set III is devoted to sustainable applications for energy production or upgrading of renewables. In Set IV, recent advances in physical activation of nanocatalysts for avoiding classical heating are first illustrated. Then, examples on the contribution of computational chemistry dedicated to metal nanostructures are discussed. This subdomain provides theoretical insights and modelings of nanocatalysts for a better understanding of the metal surface environment and of catalytic reactions that occur there.

Chapter 2 constitutes a strategic introduction on the nature of nanoparticle-based catalysts at the interface between molecular complexes and heterogeneous systems. Based on experimental results issued from several analytical techniques such as electron microscopy, mass spectrometry, mercury tests, etc., this chapter reports on the difficulties to reveal the true nature of the active species in dynamic catalysis. According to the catalytic applications, nanoparticles, nanoclusters, mononuclear complexes, and multicenter metal complexes can be simultaneously present, providing cocktail-type systems detailed in a cocktail of metal species and cocktail of catalysts.

The following four chapters are devoted to nanoparticles in solution. In Chapters 3 and 5, the synthesis of nanoparticles in polar and green solvents (water and polyols) is described. In pure water, several applications in biphasic conditions are reviewed including hydrogenation, oxidation, and C–C coupling reactions based on the stabilization of nanoparticles with water-soluble ligands, surfactants, and polymers (Chapter 3). All approaches in the reported chemical transformations are described under the angle of environmentally friendly and economically viable processes. Polyol media such as ethylene glycol or glycerol can act as multitask components (solvents, reductants, and stabilizers) allowing bottom-up strategies called "polyol synthesis" for the synthesis of stable and well-defined nanomaterials. Chapter 5 is mainly centered on metal-based nanoparticles in polyol media synthesized by chemical approaches, but the synthesis of metal nanoparticles by sputtering techniques (top-down methodologies) is also introduced. Chapter 4 reports the synthesis of transition metal nanoparticles in organic phase based on the decomposition of metalloorganic complexes through reduction or ligand displacement from the metal coordination sphere under mild conditions. This synthesis approach allows reaching very well-defined metal nanoparticles that can be used as model systems for a better understanding of the structure/properties relationships. Thus, the influence of the capping ligands onto the surface properties of metal nanoparticles and consequently on their catalytic performance is discussed through a variety of examples related to hydrogenation catalysis. The formation and stabilization of various transition metal nanoparticles in ionic liquids (ILs) and their applications in catalysis under homogeneous and multiphase conditions are described in Chapter 6. Recent investigations of these soluble nanoparticles in hydrogenation of carbon dioxide as well as in the C–C coupling reaction and the hydroformylation of alkenes are highlighted.

The subsequent four chapters are dedicated to metal-based nanoparticles deposited onto a broad variety of supports. In Chapter 7, the interest of cellulose nanocrystals (CNCs) and cellulose nanofibers (CNFs) to support metal NPs is demonstrated, evidencing that this new trend is an active and promising field of research in catalysis. In a sustainable context, nanocellulose derivatives available from biomass prove to be pertinent as renewable supports for the direct deposition and anchoring of metal nanoparticles. The use of unmodified as well as functionalized CNCs is highlighted in catalytic C–C coupling and reduction reactions. Chapter 8 describes the preparation of magnetically recoverable supported metal nanocatalysts. The deposit of various metal nanoparticles is detailed according to the nature of precursors (salts, organometallic complexes, or preformed nanoparticles). The so-produced active species either supported onto simple iron oxide or coated with a protective layer (carbon, silica, or other oxides) exhibit excellent magnetic properties and exciting catalytic activities. Factors influencing the synthesis of metal nanoparticles supported by TEM images as well as their catalytic performances due to the nanoparticle environment (impact of ligands) are discussed in detail. In Chapter 9, recent studies on the synthesis of MOF-supported nanoparticles are reviewed. First, it focuses on the most common methods based on a "bottom-up" approach to generate different types of colloidal metal nanoparticles. Second, the synthetic methods of MOFs are summarized, followed by the integration methods developed to integrate metal nanoparticles into MOFs. This chapter not only provides a good understanding of the concerned topic but also highlights the advantages vs. drawbacks of respective methods. Catalytic applications of several typical metal nanoparticles/MOF composites based on their dimensional properties and structure–property relationships are fully discussed. Focused on Pt nanoparticles supported onto SBA-15, Chapter 10 presents synthetic methodologies to deposit metal nanoparticles over silica. Interestingly, how this support can lead to superior activity as a result of tuning of the pore sizes, their interconnection or surface area for a selection of reactions is understandingly discussed. These heterogeneous nanocatalysts provide high catalytic activity in many reactions, such as oxidation, hydrogenation, and carbon–carbon coupling reactions as reported in detail.

After the review of typical synthesis of nanoparticle suspensions in various solvents in Chapters 3–6 and their deposit or inclusion in different supports (Chapters 7–10), the next three chapters are devoted to nanocatalysis. They focus on recent applications and key reactions for chemical and energy transformations including the production of hydrogen by dehydrogenation of formic acid or water-splitting as well as eco-respectful aspects with the transformation of biomass and CO_2 into fuels or added-value chemicals. Thus, Chapter 11 focuses on supported nanocatalysts for the hydrogenation of CO_2 into fine oxygenated chemicals such as methanol, dimethyl ether, or formic acid. The mechanisms as well as new opportunities and challenges for further advancing in this research field are outlined. Relying with sustainable energy source challenge, Chapter 12 reports on the use of ruthenium-based nanomaterials as catalysts for hydrogen production from water. In comparison with platinum catalysts, ruthenium metal or metal oxide nanoparticles have recently become widespread for the hydrogen evolution reaction (HER). This chapter highlights the key chemical factors that govern the HER performances, and the most promising electrochemical and photochemical systems are critically discussed in order to point to possible future research directions. In the strategic context of the chemical hydrogen storage, Chapter 13 reviews recent advances for hydrogen generation through the catalytic liquid-phase dehydrogenation of formic acid in the presence of nanocatalysts including monometallic or polymetallic nanoparticles. Their advantages vs. disadvantages are discussed according to factors impacting their catalytic performances in terms of activity, selectivity, and durability.

The last two chapters are devoted to recent advances in the light of the new activation methodologies and theoretical modeling developments. Chapter 14 presents the magnetic induction and heating performances of adapted nanoscaled iron alloys to perform catalysis. This recent advance is based on the use of nanoparticles playing simultaneously the role of heating agent through magnetic hyperthermia and the role of the catalyst for a large scope of reactions based on activation of CO, CO_2, and light alkanes. Thanks to a fine control of the magnetic properties of nanomaterials, the perspectives of improvement of these systems are promising and constitute a future active research field. The last chapter (Chapter 15) provides theoretical insights into metal nanocatalyst field as a result of recent progresses in computational catalysis and nanochemistry. Based on the isotropic H/D exchange and the HER activated by ruthenium nanoparticles, this contribution is a concrete case of the positive impact of computational calculations for a better understanding of the effect of the direct environment of metal nanoparticles and the reactivity changes induced by ligands in nanocatalysis.

1.3 Conclusions and Perspectives

Being dedicated to students, young researchers, and confirmed ones, this book gives a comprehensive overview on the recent advances and current trends in the engineering of metal nanoparticles – including their design, synthesis, and characterization by state-of-the-art techniques – together with their various applications in catalysis. A collection of short chapters is proposed to make the reading of the whole book easier and more dynamic, while providing the main key aspects of the use of metal nanoparticles in catalysis today. Coming from a large range of institutions in different countries, the authors have been chosen for the originality of the synthesis method or of understanding approach or/and the interest toward a target catalytic application that they develop.

This book brings together the most recent achievements in the development of metal nanoparticles as catalysts by covering a selection of key aspects in this challenging domain and underlying new trends. Thus, each chapter illustrates one emerging domain of nanocatalysis with concrete applications and understanding of metal nanoparticle behavior through the presentation of the most relevant examples from the recent literature including those of the invited groups.

As it can be seen in most of the chapters of this book, a cleaner chemistry and sustainable developments are at the center of current research efforts. The use of green solvents (for instance, water, polyols, glycerol, and in a certain extent ionic liquids) as synthesis or/and reaction media has already shown to be efficient to provide highly performant nanocatalysts for diverse organic transformations. Also, the study of catalyst lifetimes (in terms of recovering, recycling, durability, and repeatability) is among the objectives of many present works. Concerning the catalytic performances, if highest possible activities are always targeted, high selectivities (both chemo- and stereoselectivity) have become a key point in order to limit as possible product separation steps, which are often costly regarding solvent and energy consumption. Also, the development of nanocatalysts based on earth-abundant and cheap metals is more pertinent nowadays. This allows to face the decrease of noble metal reserves on the Earth and to reduce the catalyst costs. This is particularly relevant with regard to the scale up of catalysts for industrial processes. Beside the objective of catalyst cost reduction, a synergy effect can be expected by associating earth-abundant metals to noble ones. The utilization of biosourced materials (such as celluloses or biochars) to build catalysts, of biosourced chemicals as new reaction media (such as lactate esters, e.g. ethyl lactate, MeTHF) or of abundant and cheap biorenewable substrates (such as terpenes, carbohydrates or agricultural residues, and vegetable oils, among others), as platforms of new synthons by organic transformations has known a huge development in nanocatalysis these past years with promising achievements. This allows not only to take advantage of natural chemicals but also to valorize certain wastes issued from biomass (for instance, woody cellulose, chitin from shellfish shells, etc.). Another important development is the use of raw chemicals issued from chemical industry (like CO_2) or resulting from the increasing human activities. Concerning the improvement of catalytic transformations, which are at the basis of the production of energy vectors such as hydrogen production (either directly from water through the water-splitting process or from dehydrogenation of formic acid that can derive from CO_2), among other present challenges in this domain, nanochemistry already proved to bring powerful solutions. The transformation of CO_2 into value-added chemicals and fuels is a particularly difficult task, where metal nanocatalysts did not really emerge so far, but recent results are very encouraging, and novel nanoscale catalytic systems may meet this challenge in the near future. The application of original heating systems (hot spots obtained by magnetic induction or plasmonic excitation via the use of adequate nanocatalysts) is another strategy to limit energy consumption in catalytic organic transformations. Recent progresses in the domain of computational chemistry methods applied to metal nanoparticles proved these approaches to be very powerful and also to provide a high level of accuracy in nanocatalysis.

All these strategies provide promising perspectives and thus merit to be largely explored in order to find solutions to present challenges that chemistry world is facing and/or to open new research avenues for anticipating the future ones.

References

1 Sabatier, P. (1913). *La Catalyse en Chimie Organique* (ed. C. Béranger).

2 Rampino, L.D. and Nord, F.F. (1941). *J. Am. Chem. Soc.* 63: 2745–2749.

3 Serp, P. and Philippot, K. (eds.) (2013). *Nanomaterials in Catalysis*, 494. Weinheim, Germany: Wiley-VCH.
4 Astruc, D. (2020). *Chem. Rev.* 120: 461–463.
5 Jin, R., Zeng, C., Zhou, M., and Chen, Y. (2016). *Chem. Rev.* 116: 10346–10413.
6 Jin, R., Pei, Y., and Tsukuda, T. (2019). *Acc. Chem. Res.* 52: 1.
7 Kang, X. and Zhu, M. (2019). *Chem. Mater.* 31: 9939–9969.
8 Schmid, G. (ed.) (2010). *Nanoparticles: From Theory to Application*, 2e, 522. Wiley Blackwell.
9 Talapin, D.V. and Shevchenko, E.V. (2016). *Chem. Rev.* 116: 10343–10345.
10 Wu, L., Mendoza-Garcia, A., Li, Q., and Sun, S. (2016). *Chem. Rev.* 116: 10473–10512.
11 Gilroy, K.D., Ruditskiy, A., Peng, H.-C. et al. (2016). *Chem. Rev.* 116: 10414–10472.
12 (a) Zhou, B., Han, S., Raja, R., and Somorjai, G.A. (eds.) (2003). *Nanotechnology in Catalysis*. New York: Springer. (b) Heiz, U. and Landman, U. (eds.) (2007). *Nanocatalysis*. Berlin, Heidelberg: Springer. (c) Roucoux, A. and Philippot, K. (2007). Homogeneous hydrogenation: colloids – hydrogenation with noble metal nanoparticles. In: *The Handbook of Homogeneous Hydrogenation* Chapter 9 (eds. J.G. de Vries and C.J. Elsevier), 217–255. Weinheim, Germany: Wiley-VCH. (d) Astruc, D. (ed.) (2008). *Nanoparticles and Catalysis*. New York: Wiley Interscience; (e) Corain, B., Schmid, G., and Toshima, N. (eds.) (2008). *Metal Nanoclusters in Catalysis and Materials Science: The Issue of Size Control*. Amsterdam, The Netherlands: Elsevier Science. (f) Richards, R., Koodali, R., Klabunde, K., and Erickson, L. (eds.) (2011). *Nanoscale Materials in Chemistry: Environmental Applications*. Washington DC: ACS Publications. (g) Polshettiwar, V. and Asefa, T. (eds.) (2013). *Nanocatalysis: Synthesis and Applications*. Weinheim, Germany: Wiley-VCH. (h) Tao, T. (ed.) (2014). *Metal Nanoparticles for Catalysis: Advances and Applications*. RSC. (i) Prechtl, M.H.G. (ed.) (2016). *Nanocatalysis in Ionic Liquids*. Weinheim, Germany: Wiley-VCH. (j) Van de Voorde, M. and Sels, B. (eds.) (2017). *Nanotechnology in Catalysis: Applications in the Chemical Industry, Energy Development, and Environment Protection*. Weinheim, Germany: Wiley-VCH.
13 Imaoka, T., Kitazawa, H., Chun, W.-J., and Yamamoto, K. (2015). *Angew. Chem. Int. Ed.* 54: 9810–9815.
14 Tao, A.R., Habas, S., and Yang, P. (2008). *Small* 4: 310–325.
15 Xia, Y., Xia, X., and Peng, H.-C. (2015). *J. Am. Chem. Soc.* 137: 7947–7966.
16 Chen, G., Zhang, J., Gupta, A. et al. (2014). *New J. Chem.* 38: 1827–1833.
17 Chen, T. and Rodionov, V.O. (2016). *ACS Catal.* 6: 4025–4033.
18 Boles, M.A., Ling, D., Hyeon, T., and Talapin, D.V. (2016). *Nat. Mater.* 15: 141–153.
19 Zhukhovitskiy, A.V., MacLeod, M.J., and Johnson, J.A. (2015). *Chem. Rev.* 115: 11503–11532.
20 Heuer-Jungemann, A., Feliu, N., Bakaimi, I. et al. (2019). *Chem. Rev.* 119: 4819–4880.
21 Yuan, Y., Yan, N., and Dyson, P.J. (2012). *ACS Catal.* 2: 1057–1069.

22 Zhou, K. and Li, Y. (2012). *Angew. Chem. Int. Ed.* 51: 602–613.
23 Dykeman, R.R., Yan, N., Scopelliti, R., and Dyson, P.J. (2011). *Inorg. Chem.* 50: 717–719.
24 Gonzalez-Galvez, D., Nolis, P., Philippot, K. et al. (2012). *ACS Catal.* 2: 317–321.
25 Stratton, S.A., Luska, K.L., and Moores, A. (2012). *Catal. Today* 183: 96–100.
26 Baker, L.R., Kennedy, G., Krier, J.M. et al. (2012). *Catal. Lett.* 142: 1286–1294.
27 Axet, M.R., Conejero, S., and Gerber, I.C. (2018). *ACS Appl. Nano Mater.* 1: 5885–5894.
28 Costa, N.J.S. and Rossi, L.M. (2012). *Nanoscale* 4: 5826–5834.
29 Janiak, C. (2014). *Metal nanoparticle synthesis in ionic liquids. In: Catalysis in Ionic Liquids: From Catalyst Synthesis to Application*, Chapter 11 (eds. C. Hardacre and V. Parvulescu), 537–577. The Royal Society of Chemistry.
30 Claus, P. and Schwab, F. (2014). Modification of supports and heterogeneous catalysts by ionic liquids: SILP and SCILL systems. In: *Catalysis in Ionic Liquids: From Catalyst Synthesis to Application*, Chapter 6 (eds. C. Hardacre and V. Parvulescu), 391–409. The Royal Society of Chemistry.
31 An, K., Alayoglu, S., Ewers, T., and Somorjai, G.A. (2012). *J. Colloid Interface Sci.* 373: 1–13.
32 Somorjai, G.A. and Li, Y.-M. (2010). *Top. Catal.* 53: 311–325.
33 Freund, H.-J. and Somorjai, G.A. (2015). *Catal. Lett.* 145: 1–2.
34 Bond, G.C., Louis, C., and Thompson, D.T. (2006). *Catalysis by Gold*. World Scientific.
35 Louis, C. and Pluchery, O. (2012). *Gold Nanoparticles for Physics, Chemistry and Biology*. World Scientific.
36 Saldan, I., Semenyuk, Y., Marchuk, I., and Reshetnyak, O. (2015). *J. Mater. Sci.* 50: 2337–2354.
37 Chen, A. and Ostrom, C. (2015). *Chem. Rev.* 115: 11999–12044.
38 Kaushik, M. and Moores, A. (2017). *Curr. Opin. Green Sustainable Chem.* 7: 39–45.
39 Wang, D. and Astruc, D. (2017). *Chem. Soc. Rev.* 46: 816–854.

2

Introduction to Dynamic Catalysis and the Interface Between Molecular and Heterogeneous Catalysts

Alexey S. Galushko[1], Alexey S. Kashin[1], Dmitry B. Eremin[1,2], Mikhail V. Polynski[1,3], Evgeniy O. Pentsak[1], Victor M. Chernyshev[4], and Valentine P. Ananikov[1]

[1]*Russian Academy of Sciences, N.D. Zelinsky Institute of Organic Chemistry, Leninsky Prospect 47, Moscow 119991, Russian Federation*
[2]*University of Southern California, The Bridge@USC, 1002 Childs Way, CA, Los Angeles 90089-3502, USA*
[3]*Saint Petersburg State University, Universitetsky prospect 26, Saint Petersburg 198504, Russian Federation*
[4]*Platov South-Russian State Polytechnic University (NPI), Prosveschenya 132, Novocherkassk 346428, Russian Federation*

2.1 Introduction

Rapid progress in chemistry and nanotechnology made nanoparticle (NP) catalysis a key focus of research and industry. Organic synthesis involving nanoparticles is already well established in many laboratories. Nanoparticles used in the synthesis of drugs, dyes, molecular electronics, etc., are available in a variety of morphologies and types. These can be either zero oxidation-state metal nanoparticles or nanosalts with a positively charged metal bound to heteroatoms [1, 2]. A given type of nanoparticles may exhibit high catalytic activity in a particular family of organic reactions and be inactive in another type of reactions [3, 4]. Investigation of catalytic activity of such nanoparticles and mechanisms of their action are a priority for both organic chemistry and catalytic science [5].

The requirements of green chemistry and improved economic efficiency, as well as the needs of basic research, support the studies of nanoscale catalytic systems. However, studying the behavior of nanoparticles in solution is complicated for several reasons. Just one type of catalyst precursor introduced into a reaction system may produce a variety of particles in solution, i.e. the chemical nature of the introduced catalyst may change under the influence of the reaction medium. Nanoparticles introduced to a solution may be transformed into soluble metal salts, while introduction of metal salts may cause formation of nanoparticles during the reaction.

Classifying the variety of newly formed particles into catalytically active and "dead weight" species is a nontrivial task that requires comprehensive investigations. This chapter provides a brief introduction into dynamic systems (including cocktail-type systems) with a survey of methods suitable for studying the reaction mechanisms.

Nanoparticles in Catalysis: Advances in Synthesis and Applications, First Edition.
Edited by Karine Philippot and Alain Roucoux.

2.2 Dynamic Catalysis

Over the past decades, the scientific community has realized that real catalytic systems are much more complex than described by the classical concept. During the reaction, the catalyst does not retain its initial state but undergoes continuous changes. The catalyst evolution in the reaction medium modulates the reaction pathways and the mechanisms of transformation of the starting reagents into reaction products.

In cross-coupling reactions leading to carbon–carbon bond formation, nanoscale catalysts are widely used in the form of metallic palladium – Pd_n [3, 6, 7]. When Pd_n nanoparticles enter the reaction mixture, they undergo oxidative addition of Ar–X (reactant) to palladium atoms followed by transfer of the resulting organometallic compound into solution [3, 8, 9]. In parallel with this, by interacting with the solvent and ligands, the nanoparticles disintegrate with the formation of nanoclusters and metal complexes (Figure 2.1a). Under the action of Ar–X, the newly formed palladium nanoclusters may further dissolve; otherwise, they may aggregate into Pd_n-type nanoparticles [4, 10]. Disintegration of nanoparticles with the transfer of complexed palladium atoms into solution is called leaching [8, 10, 11]. Mononuclear palladium species are capable of forming molecular complexes with halogen bridges and two or more palladium atoms in a complex [3]. In addition, mononuclear complexes are capable of forming nanoclusters or nanoparticles via atom-by-atom growth or oriented attachment. Multidirectional transitions between the various types of metal-containing entities under the influence of the reaction medium represent the dynamic nature of the catalyst [12].

Reaction systems where nanoparticles, nanoclusters, mononuclear complexes, and multicenter metal complexes are present simultaneously are called cocktail-type systems [3, 12]. Such cocktail-type systems can be divided into two different categories (Figure 2.1b) [3, 4]:

Type 1. A multicomponent system containing various types of metal species with only *one* of them involved in the catalytic conversion of reagents into products is a cocktail of metal species (Figure 2.1b, Eqs. 1–3).

Type 2. A multicomponent system containing various types of metal species with *more than one* of them involved in the catalytic transformation is a cocktail of catalysts (Figure 2.1b, Eqs. 4–7).

For instance, in C–C cross-couplings, palladium halide molecular complexes are inactive and represent a resting state. Replacement of halogen with a different heteroatom gives a new type of nanoscale catalyst, which is effective in another type of transition-metal-mediated reactions. Compounds of the $Pd_n(XAr)_m$ type called nanosalts are nanostructured metal salts, where X stands for heteroatom (O, S, Se, etc.). They exhibit greater efficiency in the carbon–heteroatom bond formation. The main difference of these compounds from palladium nanoparticles is the initially positive oxidation state of the metal.

The use of nanosalts as catalysts efficiently promotes selective Markovnikov-type hydrothiolation of terminal alkynes [13]. The nanosalts are straightforwardly

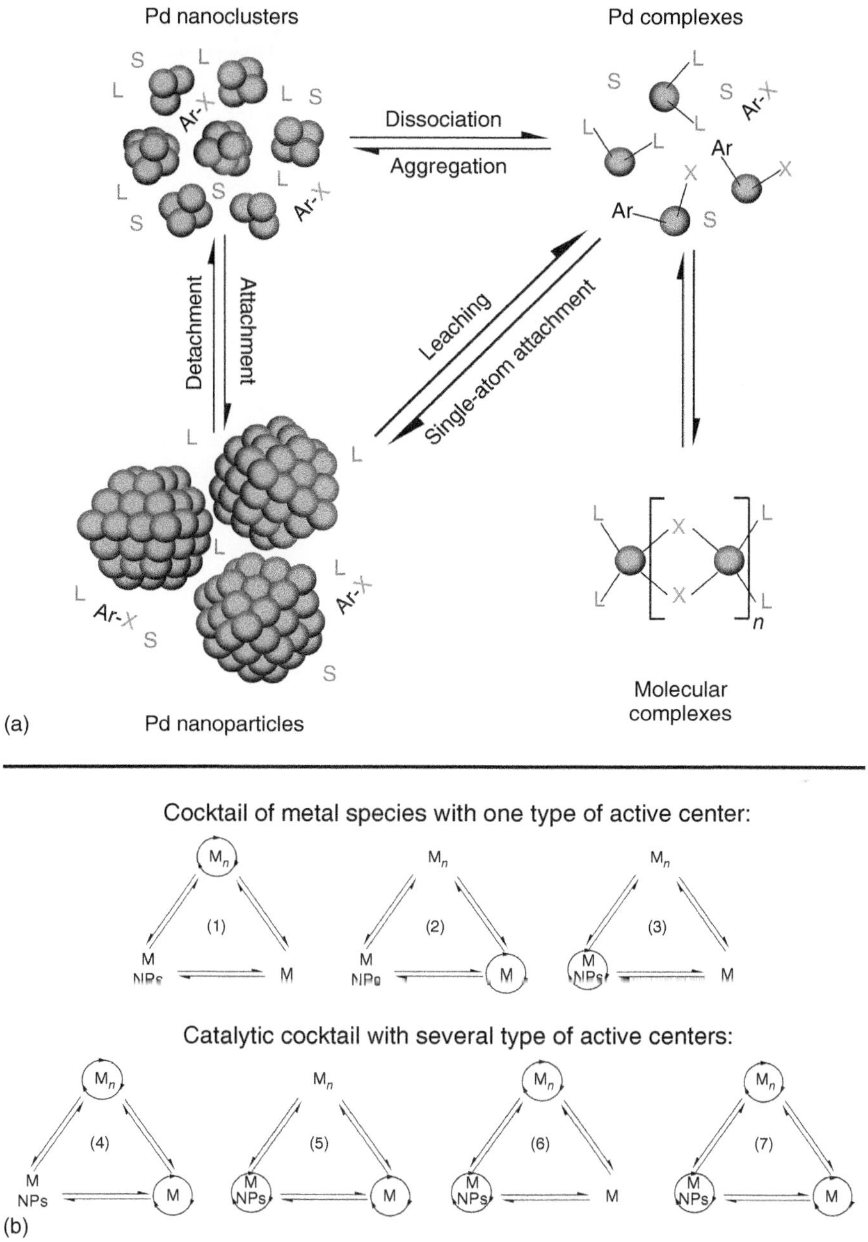

Figure 2.1 (a) Pathways of palladium particle conversion in a catalytic system (L – ligand, S – solvent, Ar – aryl, and X – halogen). (b) Type 1: a cocktail of metal species (1–3) and Type 2: a cocktail of catalysts (4–7). Source: (b) Eremin et al. [3]; Kashin et al. [4].

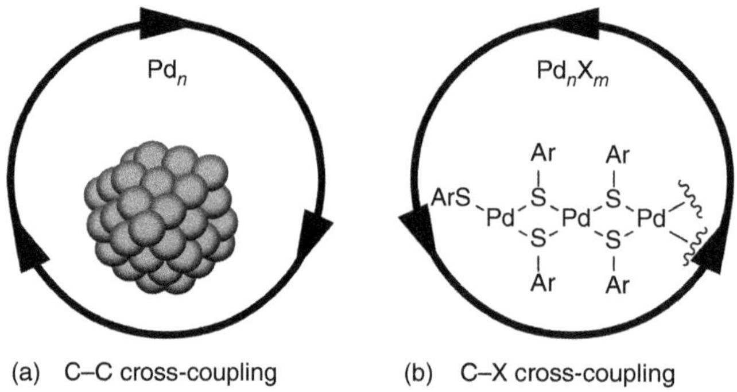

Figure 2.2 Palladium metal nanoparticles catalyze (a) C–C cross-couplings, while palladium nanosalts are effective in (b) C–heteroatom cross-couplings.

obtained by mixing solutions of thiol and palladium diketonate (e.g. palladium acetylacetonate), leading to immediate precipitation of $[Pd(SAr)_2]_n$. $[Pd(SAr)_2]_n$ formation can be carried out *in situ* by introducing palladium acetylacetonate into the reaction system.

The presence of strongly coordinating ligands (e.g. acetylacetone) in the reaction system initiates nanosalt dissolution. In this type of leaching, metal dithiolate is detached from nanosalt to form $[Pd(SAr)_2]_{n-1}$ particles, which undergo further dissolution; 0.1 mol% of the catalyst is enough for the reaction to proceed efficiently.

Besides the catalytic applications, nanosalts can be directly used as sources of reactive organic groups. S-Arylation can be carried out with nanostructured nickel thiolate $[Ni(SAr)_2]_n$ as a starting compound [14]. This nickel salt is insoluble in the reaction medium. However, introduction of a cocatalyst in the form of $Pd(OAc)_2/PPh_3$ promotes gradual dissolution of nickel thiolate accompanied by formation of active palladium-containing species and onset of the reaction. Advanced electron microscopy allows real-time observations of the nanostructured salt dissolution [14, 15].

Thus, palladium nanoparticles with different oxidation states of the metal are effective catalysts for both C—C and C–heteroatom bond formation reactions (Figure 2.2); the catalytic systems based on palladium nanoparticles are dynamic.

In addition to the catalytic systems based on particles and smaller *in situ* generated species, some metallic systems combine particles of different types. An example of such system is $[CuSPh]_n$ copper-nanosalt-catalyzed C–S coupling of thiols with aryl iodides [16]. At the initial stage, nanoparticles of monovalent or divalent copper oxide are mixed with an aromatic thiol in the presence of a base (Cs_2CO_3). The particles of oxide phase mature into copper thiolate nanosalts. Leaching from the surface of nanoscale copper salts in the presence of thiol and base enables them to enter the catalytic cycle; the system is highly active with various substituted aryls and thiols.

Leaching is an important step that allows generation of soluble active species from nanoscale catalyst precursors [3, 8]. Leaching may occur locally, when a smaller derivative (a subnanometric cluster or a single-atom species) emerges at the surface of the nanoparticle but retains a close connection with its close surroundings. In more advanced cases of leaching, the derivative separates and drifts into the solution. Nanoparticles may be further fragmented into nanoclusters and, ultimately, they may completely disintegrate into single-atom species, namely mononuclear complexes. The reverse process of nanoparticle formation from monometallic species may occur by crystallization, atom-by-atom growth, or a mixture of these two modes [3]. Different types of interconversions between the metal species in the solution may coincide; moreover, cocktail-type systems may exhibit different behaviors depending on conditions, being either a cocktail of metal particles (with a single catalytically active species) or a cocktail of catalysts (with multiple active species).

Along with dynamic systems based on palladium and copper, dynamic catalytic systems with ruthenium [17], nickel [18], platinum [19], gold [20], or iron [21] nanoparticles have been reported. The paradigm of dynamic catalysis is currently extended to many catalytic systems that involve nanoparticles.

2.3 Interface Between Molecular and Heterogeneous Catalysts

2.3.1 Direct Observation of Nanoparticle Evolution by Electron Microscopy

The variety of processes occurring at the solid–liquid interface during the nanoparticle-driven catalysis results in the formation of different metal species, from mononuclear molecular complexes to small clusters and bigger nanoparticle fragments. Mechanistic studies for such complicated systems require highly advanced experimental techniques to comprehensively analyze all types of generated metal compounds. From this point of view, electron microscopy is a method of choice for observation of catalytic system structure and dynamics because it allows visualization of the active catalytic phase directly at micro- and nanolevels. For many years, classical solid-phase electron microscopy served as a convenient tool for *ex situ* characterization of heterogeneous catalysts before use and after catalytic transformations. Observed correlations between catalyst structures and their activity contributed to mechanistic insights. Continuous improvement of the electron microscopy equipment eventually made possible the observation of individual metal atoms and small atomic clusters of supported catalysts used in industrial and laboratory settings [22–24]. A principal breakthrough in the use of electron microscopy for the study of catalytic systems occurred owing to the development of environmental transmission electron microscopy (TEM), which, in contrast to the conventional microscopy techniques, allows introducing reactive gases inside the specimen chamber [25–27]. The possibility to visualize solid catalysts under working conditions opened the way for *in situ* dynamic mechanistic studies aimed to reveal the key factors that determine catalyst behaviors in industrially relevant transformations.

Despite the bright achievements in visualization of heterogeneous catalytic processes involving gases, until only recently, liquid reaction mixtures had remained inaccessible for electron microscopy observations because of their incompatibility with harsh conditions inside the microscope. The fast evaporation of liquids and destructive action of electron beam constituted a significant challenge for structural and mechanistic studies. The problem was solved by invention of vacuum-tight specimen holders in the form of enclosed capsules, as well as by the introduction of nonvolatile ionic liquids into the electron microscopy practice [15, 28, 29]. *In situ* liquid-phase electron microscopy quickly gained a prominent position in nanotechnology, catalysis, and electrochemistry. Chemical transformations of metal-containing species occupy a special place among the studied reactions because of their importance in the synthesis of new nanomaterials and design of nanoparticle-based catalytic systems. For example, TEM was successfully used for observations of redox processes involving metal particles, namely, the galvanic replacement [30, 31] and oxidative etching [32, 33] reactions. Mechanistic studies of transition-metal-catalyzed processes are represented, for example, by scanning electron microscopy (SEM) observations of the metal thiolate phase production and consumption during the carbon–sulfur bond formation (Figure 2.3). Along with the use of vacuum-tight capsule approach for the analysis of liquid reaction mixtures (Figure 2.3b–d), the measurements of solidified reaction mixture under the temperature of liquid nitrogen were carried out (Figure 2.3a).

Electron microscopy study of frozen reaction mixture for the copper-oxide-catalyzed C–S cross-coupling reaction (Figure 2.3a) [16] and analysis of catalytic system for the nickel-mediated addition of disulfides to acetylene in a liquid organic environment (Figure 2.3b) [34] demonstrated the key role of polymeric thiolate species $[M(SR^1)_x]_n$ as reservoirs for the soluble catalytically active mononuclear complexes. The developed methodology was further extended to direct observation of the thiolate species leaching under catalytic conditions. For this purpose, nickel thiolates $[Ni(SR^1)_2]_n$ were used as a source of reactive groups for the palladium-catalyzed C–S cross-coupling reaction [14]. The advanced SEM approach allowed not only the detection of structural changes in nickel thiolate during the reaction (Figure 2.3c) but also the recording of electron microscopy video reflecting the formation of reactive points and the development of reaction front within the reactant particles in the course of catalytic transformation (Figure 2.3d).

The unique arsenal of methods for the real-time electron microscopy observations of nanoparticle-mediated reactions in heterogeneous catalytic systems provides an excellent tool to study the evolution of chemical systems at micro- and nanoscales. The relative simplicity of interpretation and huge amount of structural data acquired in a single dynamic experiment make *in situ* electron microscopy a priority method of research in the field of nanocatalysis. Being a relatively new approach, liquid-phase electron microscopy is open to considerable updates, which will expand its resolution to subnanolevel and in every way improves direct observation of dynamic catalysis.

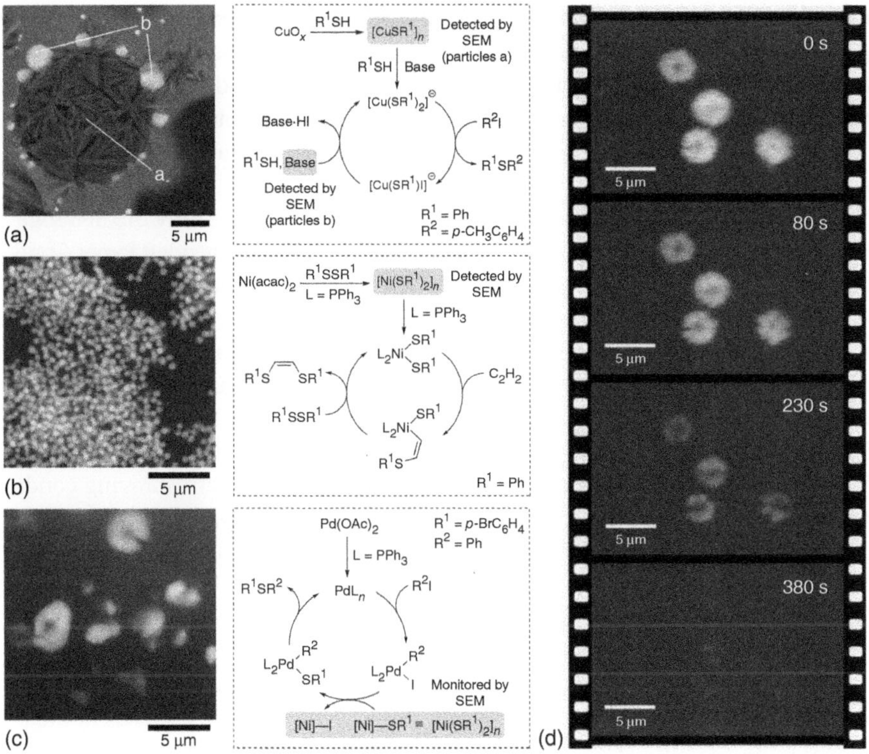

Figure 2.3 (a–c) SEM images of reaction mixtures for various C—S bond formation reactions and corresponding reaction mechanisms. (d) Freeze frames taken from a SEM video of the solid reactant evolution in C–S cross-coupling reaction shown in (c). Components of the reaction mixtures detected by SEM are labeled in each scheme. Source: (a) Adapted with permission from Panova et al. [16]. Copyright ©2016, American Chemical Society. (b) Source: Degtyareva et al. [34]. (c, d) Adapted with permission from Kashin et al. [14]. CC-BY-4.0.

2.3.2 Through the Interface – Detection of Molecular Species by Mass Spectrometry

Thus, the dynamic behavior is an inherent feature of nanosized catalytic systems. Nanoparticles generally evolve in two directions: nano to molecular and molecular to nano, and both ways involve numerous transformations of metal species at the molecular level [3]. Among different methods applied to investigate the dynamic changes in catalysis, mass spectrometry is considered to be a valuable approach for the analysis of complex events at the molecular level because of its high sensitivity and the availability of various ionization methods that allow to study species of different nature [35–38].

Direct detection of nanoparticles is still a challenge for mass spectrometry [39, 40]. The matrix-assisted laser desorption/ionization (MALDI) is one of the most promising ionization methods that can be used to resolve this issue [41, 42]. Nanoparticle size is the major limiting factor. When the size of the nanoparticle increases within a few atom layers, the mass of the expected ion increases exponentially. It makes nanoparticles less polarized and decreases their ability to be vaporized to reach the detector. Novel approaches toward heavy nanoparticles ionization are of high importance to be developed. However, colloidal platinum nanoparticles up to 3.7 nm (the average diameter determined by TEM) were successfully characterized by MALDI mass spectrometry [43].

Meanwhile, mass spectrometry can be successfully applied to study the heterogeneous catalytic systems reactivity by using the Nanoparticle-Embedded Paper-Spray approach developed by Zare and coworkers [44]. This method allowed to study the Pd-catalyzed Suzuki reaction and the reduction of 4-nitrophenol along with the Au-catalyzed aerobic oxidation of D-glucose. The identification of reduction and oxidation intermediates of the reaction was even possible in good agreement with the proposed mechanisms [44].

Ionization sources have been shown to stimulate the dynamic behavior of catalytic systems. Nanoparticle synthesis can be reinforced by electrospraying conditions [45]. Moreover, such nanoparticles retain catalytic activity, as has been shown by Cooks and coworkers for the case of 4-nitrophenol reduction over gold nanoparticles [46].

Palladium(0) complex Pd_2dba_3 is a known precursor for dynamic catalytic systems [47]. However, phosphine ligands can suppress metal aggregation and promote formation of monometallic complexes, as has been reported by McIndoe and coworkers [48]. The absence of strongly binding ligands usually leads to metal speciation. For instance, Pd speciation was recently reported by Koszinowski and coworkers for the Pd/Diene-catalyzed cross-coupling. The interaction of Pd_2dba_3 with isoprene and PhMgX (X = Cl and Br) in THF (tetrahydrofuran) led to clustering of palladium with the formation of Pd_4 clusters after a short period of time (c. 11 minutes) (Figure 2.4) [49]. A similar trend was reported for Co-based catalytic systems for cross-coupling reactions [50].

A key role of the multinuclear palladium–oxygen species formation was demonstrated by Waymouth and coworkers [51, 52]. Monomolecular Pd complexes with neocuproine family ligands tend to aggregate in the course of aerobic oxidation with hydrogen peroxide. Formation of trinuclear species still involved in the catalytic cycle is indicated by a combination of kinetic and mass spectrometry data [51, 52].

Different small-sized metal clusters can be thoroughly characterized by mass spectrometry combined with ion spectroscopy techniques such as infrared multiple photon dissociation spectroscopy [53] or cryogenic ion trap vibrational spectroscopy [54]. Methane activation is one of the areas actively using such approaches in order to address and overcome the challenges of selective C–H cleavage, as has been recently shown by Landman and coworkers [55] and Schwarz [56–59].

Leaching is the opposite process often observed in catalytic systems; its observation indicates the dynamic character of a system. Leached molecular forms can be detected by mass spectrometry, and their detection sheds light on particular complexes present in the solution. In the study of C–C cross-coupling reactions catalyzed by Cu_2O nanoparticles, Andiappan and coworkers have shown that in the

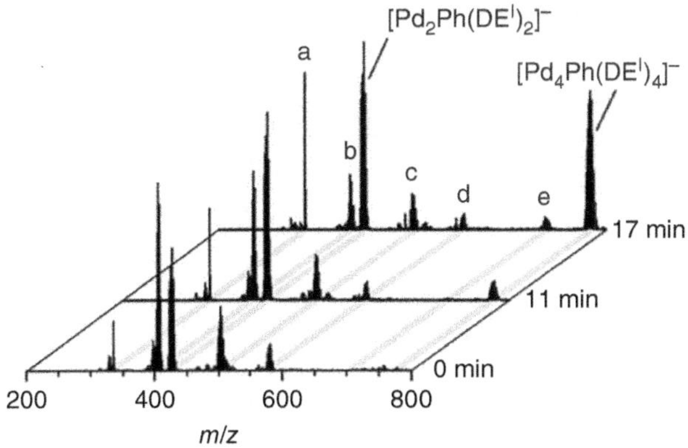

Figure 2.4 Negative ion mode electrospray ionization mass spectra of $[Pd_2dba_3]$ (1.5 mM), isoprene (DE^I, 24 mM), and PhMgCl (12 mM) solution in THF recorded at 0, 11, and 17 minutes after the removal from a 195 K cooling bath, a = $[AlPh_4]^-$, b = $[PdPh_3(DE^I)]^-$, c = $[Pd_2Ph_2(DE^I)_2{-}H]^-$, d = $[Pd_2Ph_3(DE^I)_2]^-$, and e = $[Pd_4Ph(DE^I)_3]^-$. $[AlPh_4]^-$ stems from aluminum impurities in the Grignard reagent. Source: From Kolter and Koszinowski [49]. © 2019. Reprinted with permission of John Wiley & Sons.

presence of a base (K_2CO_3), the reaction can be shifted to the liquid phase, and phenyl acetylene homocoupling intermediate $[CuO(C_8H_5)_2]^-$ can be detected by electrospray ionization mass spectrometry (ESI-MS) [60]. However, no leaching has been observed under base- and ligand-free conditions [60].

Investigation of the C–S cross-coupling reaction on copper oxide has shown that the reaction proceeds independently of the Cu_2O nanoparticle precursor but apparently involves two thiol complexes present in all reaction mixtures: $[Cu(SPh)_2]^-$ and $[CuI(SPh)]^-$. The presence of soluble copper species strongly suggests that this dynamic heterogeneous catalytic system acts homogeneously [16].

The use of nanostructured nickel thiolate complex $[Ni(SAr)_2]_n$ as a source of ArS^- moiety instead of Ar–SH makes the reaction mechanism dependent on the reactant particle morphology and interface properties [14]. Dual behavior of catalytic system has been observed in a reaction with aryl halides promoted by $Pd(OAc)_2/PPh_3$ The reaction efficiently proceeds on the surface of the thiolate polymer and only transmetalation intermediates are leached into the solution and detected by ESI-MS. Molecular species predominate at higher conversions, and all proposed intermediates can be detected in the solution under catalytic conditions, including a bimetallic complex $PdNi(SC_6H_4Br)_2(C_6H_4SO_3{}^-)(Ph_3P)$ at the pretransmetalation step [14].

In the absence of strong ligands, $Pd(OAc)_2$ reduction leads to nanoparticle formation [61]. These highly active nanoparticles can serve as precatalysts at low concentrations, but the actual driving force that predominantly contributes to the product formation is leaching. de Vries and coworkers have reported detection of the leached $[PhPdI_2]^-$ and $[PdI_3]^-$ species in solution [62]. Leaching may be triggered by the catalytic reaction itself [63] and significantly enhanced by the presence of stronger ligands such as N-heterocyclic carbenes (NHCs) [61] or ionic liquids [64]. Imidazolium cations with acidic proton in the second position can act as ligands of carbene type, which was used for palladium chloride. These complexes are stable under air atmosphere and might be more reactive than conventional Pd/NHC complexes because of less bulky substituents on nitrogen atoms [64].

Metallic copper shows exceptional catalytic activity unsupported by any added ligands in the azide–alkyne cycloaddition (CuAAC) reaction [65, 66]. It should be emphasized that in most of the cases, Cu surface does not act as a simple source of the catalytic species in solution [67–69]. The metal bulk also acts as a ligand of Cu(I) species, local leaching effects apparently supporting these Cu(I) species at the metal–liquid interface. If the leaching forces are strong enough, the leached copper may permanently relocate from the surface to solution, as indicated by detection of relatively labile dinuclear copper intermediates similar to those present in the homogeneous-like CuAAC reaction [70].

2.3.3 Pervasiveness of Nanoparticles and the Problem of Catalytic Contamination

Catalyst loading is a very important parameter of catalytic reactions. Transformations of a precatalyst (activation and so forth) result in the active catalyst loadings that may differ from expected values. In theory, the active species concentration may be reliably evaluated by the course of the chemical reaction; however, this approach may fail or be complicated by false positive signals because of the contamination of molecular precatalysts and reactors with nanoparticles. The contamination may occur either because of a partial degradation of the active organometallic species during storage or as a result of unrecognized side processes during the synthesis.

Pd_2dba_3, a well-known precursor of the soluble palladium(0) species, is subject to significant fluctuations in quality [47]. The complex purity may considerably vary, as has been revealed by comparative analysis of the substance from different supplies. The content of nanoparticles may reflect decomposition of up to 40% of the complex. Moreover, the problem is typical not only for complexes of zero-valent metal but for Pd(II) compounds as well. Contamination of $Pd(OAc)_2$ with nanoparticles has been reported [71, 72], and it may represent a significant drawback of the widely used catalyst precursor. Nanoparticle contamination may interfere in Suzuki–Miyaura cross-couplings [73], Negishi cross-couplings [74], cyanation of (hetero)aryl halides [75], asymmetric allylic alkylation of indole-containing allylic carbonates [76], Stille polymerization toward donor–acceptor conjugation [77], catalytic aminations [78], benzyl sulfoxide transformation into diaryl sulfoxides [79], propargyl chlorides cyclization [80], and direct C–H arylation polycondensation of thiophenes [81]. The influence of nanoparticle contamination on enantioselectivity has been demonstrated for the Pd_2dba_3-catalyzed synthesis of (*S*)-(−)-2-allylcyclohexanone, with the role of a supply source emphasized as follows: "The submitters noted that distinctly lower enantioselectivity was obtained when the reagent of other suppliers was used" [82]. In certain cases, contamination of the reaction mixture can be prevented by purification (filtration and recrystallization) of the precursor [83–87].

In contrast with nanoparticle contamination of precatalyst, which frequently varies in a batch-to-batch manner and can be ruled out by analytical check-up and purification before the use, nanoparticle contamination of labware is sometimes harder to detect, and it is most commonly neglected.

Over the past decades, the use of labware with polytetrafluoroethylene (PTFE) coating significantly increased. PTFE is considered an available, chemically stable, and inert material. The PTFE-coated magnetic stir bars are commonplace in chemical laboratories worldwide. Back in 1977, Tölg drew attention to the tendency of PTFE defects generated by exposure to strong acids to adsorb mercury ions [88]. Later in 1991, Naga and Bazsa investigated the effects caused by the contamination of laboratory plastics – tubes, reaction vessels, and stir bars made of or coated with polymers including PTFE, polyethylene, and silicones [89]. Among other things, the authors noted the catalytic effects induced by sorption of metal ions on the damaged PTFE surface. A recent detailed study of metal nanoparticles adsorbed on the PTFE surface of stir bars and their influence during catalytic reactions has been carried out using modern analytical approaches including SEM, energy-dispersive X-ray spectroscopy, and mass spectrometry [90]. This study demonstrates that a randomly selected nondisposable stir bar from a lab shelf is contaminated with a range of metal nanoparticles, including platinum, palladium, gold, cobalt, and iron. Amounts of metal particles adsorbed at the stir bar surface are sufficient to conduct catalytic reactions without the addition of extra catalyst to the reaction mixture. Thus, PTFE labware can become an involuntary carrier of metal from one reaction to another.

Adsorption of nanoparticles by nondisposable laboratory plastics may significantly affect the results and introduce critical biases in the studies of sensitive chemical processes. Finke and coworkers were the first to deliberately draw attention to the need of disposable magnetic stir bars [91–93]. An increasing number of authors raise the need to use virgin labware and plastic parts in accurate catalytic and synthetic studies [94–96]. A smart way to use the adsorption of nanoparticles on labware is to purposefully modify it to create catalytically active surfaces. This approach was used by Janiak and coworkers who attached Rh nanoparticles to a stir bar and used it for hydrogenation of cyclohexene or benzene to cyclohexane [97]. Han and coworkers also used an Au–Pd nanoparticle-coated PTFE surface for the reduction of 4-nitrophenol to 4-aminophenol [98].

In addition to impurities in chemical reagents and on the surfaces of labware, there is another potential source of contamination of reaction mixtures with metal nanoparticles. The environment itself may contain noticeable amounts of catalytically active metal particles [4]. Palladium, platinum, and rhodium nanoparticles from automobile catalytic converters may contaminate soil and water, as well as be contained in the airborne particulate matter; the palladium content of road dust may reach $300\,\mu g\,kg^{-1}$ [99]. Trzeciak and coworkers have demonstrated the catalytic activity of used automobile catalysts and airborne particulate matter in Heck and Suzuki reactions [100]. The catalytic activity of airborne particles containing transition metals can be easily transferred to a regular laboratory environment.

Contamination of reaction mixtures with external nanoparticles may compete with the active catalyst loadings, causing undesired side reactions and resulting in poor efficiency and selectivity. Alternatively, under certain catalytic conditions, the contaminant nanoparticles may provide a major contribution to the observed reactivity of the system. As catalytically active nanoparticles are widely pervasive, careful quality control and analytical testing of initial precatalysts, reagents, and labware are essential.

2.3.4 Computational Modeling of Dynamic Catalytic Systems

2.3.4.1 Equilibrium of Leaching and Recapture

The activity of transition metal nanoparticles in cross-coupling and functionalization reactions is a well-known phenomenon [3, 101–103]. These transformations most frequently involve d^{10} metal catalysts [104], with Ni catalysts being an especially promising and well-established alternative to Pd catalysts. It has been demonstrated that heterogeneous Ni catalysts are active in these transformations, and homogeneous (including ligandless) Ni catalysts can aggregate in situ [105–110]. This leads to a question of whether dynamic interconversions, namely, transformations of soluble Ni species into nanoparticles and vice versa, can occur in the Ni systems. However, theoretical studies considering this phenomenon in Ni catalytic systems seem to be under-represented or even missing in scholarly literature [111]. For this reason, we highlight dynamic processes involving Pd catalysts.

For Pd systems, it has been possible to determine the effect of the surface of preformed nanoparticles on the activity in cross-coupling reactions [112–114]. The activity in the Suzuki–Miyaura reaction is attributed to the propensity to leaching from nanoparticle surface to solution; nanoparticle facets with higher surface energy are prone to leaching into solution [114]. In face-centered cubic Pd nanoparticles, the surfaces cut by (111) and (100) planes (Figure 2.5b) have the lowest surface energy and are the most stable [115]. In line with this, nanoparticles with low-index (100) facets were found to have significantly lower activity (Figure 2.5a).

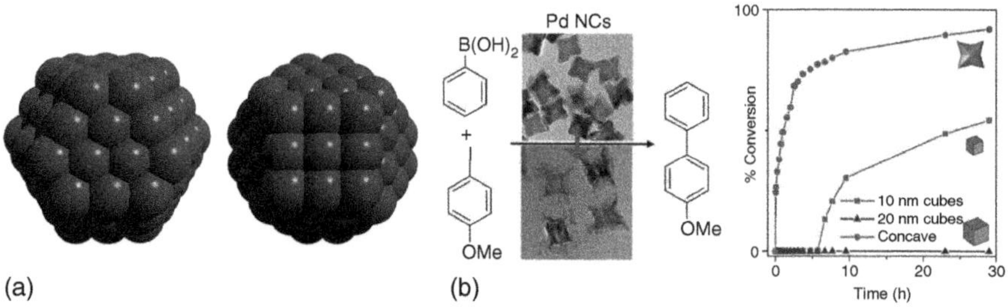

Figure 2.5 Facets of Pd nanoparticles representing (1 1 1) and (1 0 0) surfaces, left and right, respectively (a). Divergent reactivity of concave Pd nanoparticles (NPs) having (7 3 0) facets and cubic Pd nanoparticles having (1 0 0) facets in the Suzuki–Miyaura reaction (b). Source: Reproduced with permission from Collins et al. [114]. Copyright ©2014, American Chemical Society.

On the other hand, Pd species in solution can aggregate under reaction conditions [5, 116]. In an early quantitative structure–property relationship (QSPR) study, extensive analysis of Heck reactions was performed with data science methods and machine learning [117]. The observed importance of Pd loadings for the catalytic activity was attributed to easy Pd aggregation and concomitant precipitation.

These findings show that the balance between the elementary reactions of leaching and recapture determines the catalyst activity. If, for example, a strongly binding phosphine ligand stabilizes metal species in solution, the recapture is prevented by the thermodynamic factor, which makes the leaching irreversible. The relative elementary reaction rates (leaching vs. recapture) are important factors as well. We give a brief account of computational studies on leaching and recapture in cross-coupling and functionalization reactions. By contrast with numerous computational studies in the established field of homogeneous catalysis with Pd [104, 118], these studies are few, which indicates that the field is in its beginning stage.

2.3.4.2 Modeling Leaching, Recapture, and Transformations in Solution

A simplistic model for Pd leaching in solution is leaching in the gas phase driven by a strongly binding gaseous ligand. Rösch and coworkers used the density functional theory (DFT) modeling to study CO-driven leaching of Pd atoms from model slabs representing the surface of a Pd nanoparticle. A detachment of Pd from a clean surface without ligands is highly unfavorable and endothermic by ~270 kJ mol^{-1}. Saturation of the surface with CO drives leaching so that the detachment of Pd sub-carbonyls $Pd(CO)_{2-3}$ becomes thermodynamically favorable. The mechanisms of the detachment were also determined. Surface Pd atoms in model defect sites had lower coordination numbers, which has driven more facile detachment from the defects than from the regular surface (ΔE_{det} depends on the coordination number linearly) [119].

Binding to the Pd centers can drive leaching from Pd clusters in ionic liquids. According to QM/MM (quantum mechanics/molecular mechanics) molecular dynamics (MD) simulations, ionic species PdX^- and $PdAr^+$ can be formed as a result of the addition of aryl halides (ArX) to Pd(0) clusters in [Mmim][BF_4], where [Mmim] is the 1,3-dimethylimidazolium cation [120].

In a series of works, Heinz, Knecht, and coworkers studied leaching from peptide-capped Pd nanoparticles under the Stille reaction conditions with X-ray spectroscopy (EXAFS (extended X-ray absorption fine structure) and SAXS (small-angle X-ray scattering)) and empirical force-field methods [121, 122]. According to the findings, the energy of Pd atoms detachment ΔE_{det} depends on the coordination number; because Pd atoms in different nanoparticle facets have different coordination numbers, a set of possible ΔE_{det} is determined by the nanoparticle surface morphology and size. Heinz and coworkers used ΔE_{det} as an estimate of leaching activation energy and thus obtained the relative rate of atom leaching for each considered Pd nanoparticle. The computed relative rates of abstraction for nanoparticles of different diameters correlated with the observed turnover frequencies (TOFs) in the Stille coupling reaction.

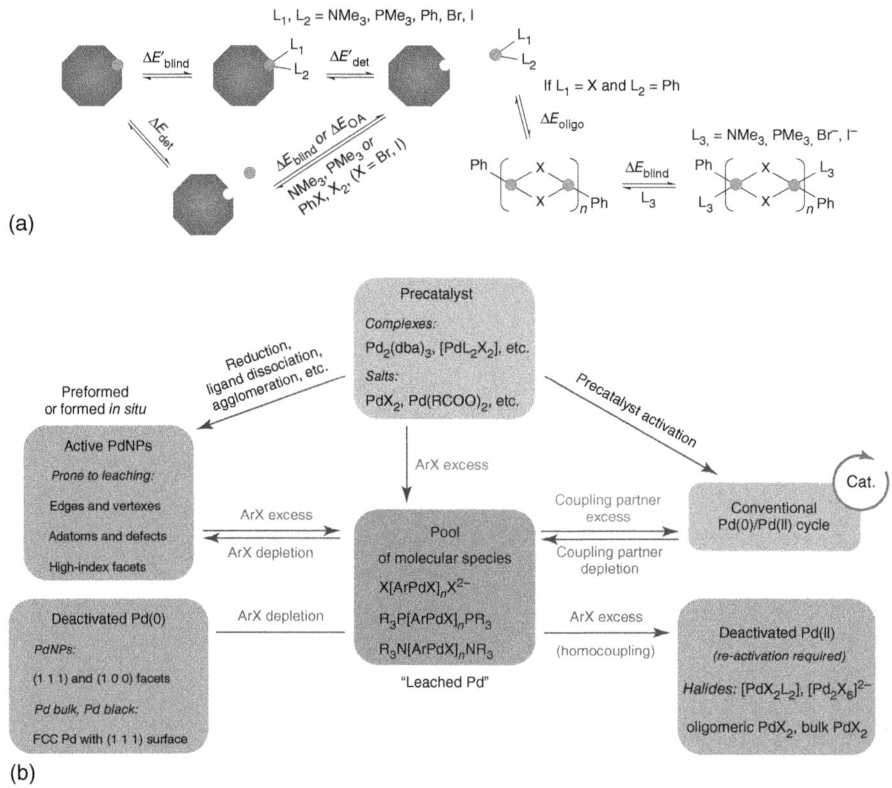

Figure 2.6 Schematic representation of the model pathways of Pd atom detachment and subsequent stabilization (a). The proposed model of the evolution of Pd catalytic systems in cross-coupling and functionalization reactions involving aryl halides (b). Source: Reproduced with permission from Polynski and Ananikov [123]. Copyright 2019, American Chemical Society.

In our study, we used DFT methods to model leaching and recapture as a set of elementary reactions, including Pd atom detachment (ΔE_{det}, endothermic) and subsequent stabilizing transformations in solution (ΔE_{stab}, exothermic) (Figure 2.6a) [123]. The latter includes binding of ligands such as NMe_3, PMe_3, Br^-, I^- (ΔE_{bind}), oxidative addition of PhBr or PhI to Pd species (ΔE_{OA}), and oligomerization of [PhPdX] monomers (ΔE_{oligo}). By comparing the endothermicity of ΔE_{det} with the exothermic effect of ΔE_{stab} (the sum $\Delta E_{OA} + \Delta E_{oligo} + \Delta E_{bind}$) for a chosen stabilization pathway, it is possible to estimate which stabilizing factors overbalance different values of ΔE_{det}. The main stabilizing factor is ΔE_{OA}. Therefore, the proposed model is relevant for cross-couplings and functionalizations involving the oxidative addition of ArX (Figure 2.6b).

The proposed model describes which pools of Pd species may form in coupling reactions, and what stabilizes them. The Pd(0) nanoparticles with high density of surface defects and the $L[PhPdX]_nL$ oligomers (X = Br, I, L = NR_3, PR_3, X, etc.) are two pools that can be observed in active systems; we consider these as reservoirs of catalytically active species, in line with experimental observations [102, 124, 125].

Bulk Pd and Pd nanoparticles with low-energy facets represent inactive forms of Pd. Another pool of deactivated species comprises Pd halides, because these compounds, often used as catalyst precursors, need reduction to Pd(0) to become catalytically active.

A key feature of the model is that it provides a semiquantitative understanding of forces and components that drive the system toward activity or deactivation. For example, ArX can efficiently stabilize the L[PhPdX]$_n$L oligomers. At the same time, ArX, when it is in excess to other system components, is drives the system to the unwanted homocoupling process that results in the formation of PdX_2. Because different *pools* of compounds and species are *interconnected* by elementary reactions (oxidative addition, oligomerization, and ligand binding), Pd catalytic systems in reactions with ArX can be seen as a "cocktail" of species that changes during the reaction.

2.3.5 Nanoparticle Catalysis in Solvent-Free and Solid-State Organic Reactions

Solvents are the main source of waste in organic synthesis. Avoiding the use of solvents makes the chemical industry more environmentally friendly. Therefore, in recent years, the development of "solvent-free" and "solid-state" approaches in organic chemistry has attracted more and more attention of researchers. The application of these approaches could cover wide areas of organic chemistry.

Among the reactions catalyzed by nanoparticles, the use of the solvent-free approach is most intensively studied for relevant transformations as cross-coupling, reduction, and oxidation reactions. Mostly, nanoparticles of transition metals, such as Pd, Pt, Ru, and Au, supported on different substrates such as carbon materials, metal oxides, and clays are used in such reactions.

Several studies have shown that palladium nanoparticles stabilized by pectin [126] or gum arabic [127] can be effectively used as catalysts in Mizoroki–Heck cross-coupling reactions of various aryl halides including chlorides under solvent-free conditions. Nanosized catalysts under solvent-free conditions can exhibit tremendous recycling capacity. In 2015, Mayoral and coworkers showed that palladium nanoparticles supported on synthetic clay retain their activity for more than 50 cycles of the reaction between iodobenzene and butyl acrylate [128]. Another advantage of solvent-free conditions was the ease of isolation of the pure product because the by-product (triethylammonium iodide) stays with the catalyst after the product isolation. Besides, palladium nanoparticles demonstrate good recycling capacity in Suzuki–Miyaura reaction under solvent-free and ligand-free conditions. In one of the cases [129], the nanoparticles are formed *in situ* from $PdCl_2$ in the presence of n-Bu_4NF, which plays the role of a base and is also responsible for Pd species immobilization.

Hydrogenation of organic compounds is one of the most important industrial processes. It is used in manufacturing of dyes, agrochemicals, and pharmaceuticals. The use of large volumes of toxic solvents for this reaction is a significant problem that has initiated the search for alternative catalytic systems excluding the use of solvents. Solid-state catalytic hydrogenations require transition metal catalysis [130]. The possibility of effective application of transition metal nanoparticles for selective hydrogenation of various organic compounds under solvent-free conditions has been demonstrated. For instance, ruthenium nanoparticles can catalyze aromatic ring hydrogenation under mild conditions (4 atm, 75 °C) without a solvent [131]. Palladium nanoparticles supported on the nitrogen-doped ordered mesoporous carbon catalyze hydrogenation of nitroarenes to aromatic amines under solvent-free conditions with high selectivity [132]. Ruthenium nanoparticle-intercalated montmorillonite clay is an excellent catalyst for hydrogenation of alkenes [133]. Iridium nanoparticles on a similar support have been successfully used to hydrogenate an aromatic ring under solvent-free conditions [134]. In all cases, the excellent recycling capacity of the catalyst was combined with straightforward procedure for its isolation.

Oxidation reactions represent another class of important processes eligible for the use of a solvent-free approach to provide more environmentally friendly processes. Stabilized palladium nanoparticles on various supports proved to be excellent catalysts for oxidation of alcohols under solvent-free conditions [135, 136]. In one study, a composite of palladium nanoparticles with magnetic Fe_3O_4 nanoparticles was used to enhance the catalyst recovery. The nanocomposite promotes selective formation of aldehydes from corresponding alcohols under solvent-free conditions. Another interesting approach is the use of bimetallic PdAu particles in oxidation reactions. These nanoparticles were used as a part of composite with graphene oxide and titanium oxide for solvent-free oxidation of benzyl alcohol [137]. Also, bimetallic PdAu nanoparticles supported on various substrates are suitable catalysts for selective solvent-free oxidation of double bonds [138, 139]. Effects of bimetallic gold/copper nanoparticles in a solvent-free styrene oxidation reaction were studied by Blanckenberg et al.; the results show that the acceptable balance between stability and activity of the catalyst can be achieved by adjusting the metal ratio [140].

The use of catalytically active nanoparticles under solvent-free conditions is not limited to the use of transition metal nanoparticles in cross-coupling, hydrogenation, and oxidation reactions. Other interesting solvent-free transformations of organic substances in the presence of nanoparticles are phenol alkylation with zirconium phosphate nanoparticles [141], imidazole synthesis by reacting aldehyde derivatives with 1,2-diaminobenzene in the presence of $Fe_3O_4@SiO_2@(CH_2)_3N^+Me_3I_3^-$ magnetic core–shell nanoparticles [142], and other important reactions [143].

A significant number of organic processes can be carried out as solid-phase syntheses [130]. It is amazing that nanoparticles can be used as catalysts in the reactions between solid reagents. The lack of understandable mass transfer mechanisms in such systems makes the topic extremely interesting. The phenomenon of leaching seems vital to the nanoparticle-driven reactions that proceed in solutions, where catalytically active metal species may pass from nanoparticles into solution and exist in a soluble form. The mode of catalysis by nanoparticles in the absence of liquid reagents seems incomprehensible.

A significant number of the solid-phase metal-catalyzed organic reaction protocols involve processing of the reagents in a ball mill [144]. The procedure facilitates formation of catalytic nanoparticles *in situ* from various initial metal compounds. This phenomenon has been recently described by Ito and coworkers for several cross-coupling reactions such as the solid-state Suzuki–Miyaura reaction [145] and the solid-state C–N coupling reaction [146]. In 2012, Lamaty and coworkers used solid PEG-2000 as a medium for the mechanochemical Mizoroki–Heck reaction [147]. Reduction of $Pd(OAc)_2$ into active zerovalent palladium species, which occurs *in situ* under the mechanically challenging conditions, provides high yields of the product in the reaction of aryl iodides with *tert*-butyl acrylate in the presence of potassium carbonate [147].

Many organic reactions, that have been initially described as solid-phase ones, are of more complex nature; upon closer examination, they involve phase transitions and formation of eutectic mixtures and therefore proceed in a forming liquid phase [148]. In addition, the traceable presence of liquids in a reaction between solid reagents may be associated with the formation of liquid by-products. The solid-phase Suzuki–Miyaura reaction in the presence of Pd/C catalyst [149] is apparently promoted by trace amounts of water generated as a by-product in the course of the reaction. The importance of water formation during arylboronic acid trimerization has been demonstrated for similar reaction in the presence of Pd nanoparticles supported on graphite and carbon nanotubes [150]. The reaction was carried out under the conditions of simple heating, while electron microscopy showed the formation of spherical organic nanoparticles containing the product inside the solid reaction mixture. The advantage of such systems is the possibility of complete elimination of the use of a solvent at all stages including the product isolation by distillation with concomitant recovery of the supported nanoparticles.

Catalysis by nanoparticles in solution is the most common practice; the mechanisms of operation of nanoscale catalysts under these conditions are extensively studied. The solvents are actively involved in leaching of the metal from nanoparticles; this idea is central to the current understanding of the nature of catalysis. In the absence of a solvent, the behavior of metal particles should differ. This makes the studies on the operation of nanoparticles in solvent-free conditions a truly intriguing topic.

The development of methods that avoid the use of toxic solvents is one of the key areas of green chemistry. That is why, solvent-free and solid-state reactions, in which nanoparticles are used as catalysts, become increasingly demanded. At present, the high efficiency of nanoparticles under solvent-free conditions has been observed in a number of important transformations.

The ball mill treatment, which is commonly applied for conducting solid-state reactions, triggers the *in situ* nanoparticle formation; traceable amounts of liquid phase emerge in the course of the reaction; and transitions of the metal to a soluble form cannot be ruled out under these conditions. Thus, the findings indicate the possibility of dynamic processes in the solvent-free and solid-state reactions. The observable catalytic activity of nanoparticles in the absence of solvents is an important issue, and the question of how nanoparticles work in the absence of a solvent remains to be answered.

2.3.6 Applications of the Mercury Test and Other Poisoning Techniques in the Nanoparticle Catalysis Studies

2.3.6.1 Catalyst Poisoning Techniques and Typical Poisons

One of the main problems in the studies of transition-metal-catalyzed reactions is determination of the nature of active centers: are they metal nanoparticles or molecular complexes formed via leaching and dissolution? [5, 11, 151–161].

In practice, the catalyst poisoning methods are very frequently used among various approaches to distinguish between the nanoparticle and truly homogeneous forms of catalysis [5, 11, 152–155, 157, 161–163]. The poisoning methods can be conveniently divided into two main categories: selective catalyst poisoning methods and quantitative poisoning methods ("quantitative kinetic poisoning" and "substoichiometric poisoning" are also frequently used terms) [154, 157, 158].

The concept of selective catalyst poisoning assumes selective inhibition of a specific catalytic species such as nanoparticles. Selective poison should deactivate only a specific catalytic species in the presence of others. Poisoning experiments typically assume the observation of response of a catalytic system to addition of a selective catalyst poison in a large molar excess relative to the total amount of the catalyst. A substantial decrease in the reaction rate (or reagent conversion) after the addition of a poison usually indicates that the catalyst species sensitive to the poison (e.g. nanoparticles) operate as active centers. By contrast, insignificant changes in the reaction rate under the action of a poison indicate that species insensitive to the poison (e.g. molecular complexes) are active centers. Liquid mercury is most frequently used as a selective poison for metal nanoparticles, while polymer-bound phosphines, polyvinylpyridine (PVPy) and [*a*,*e*]-dibenzocyclooctatetrene are used as poisons for molecular metal complexes in the presence of nanoparticles [5, 11, 152, 157, 161].

The quantitative poisoning method has received significant attention owing to the works of Finke and coworkers [158, 164–168]. The method involves nonselective poisons capable of strong binding with active centers of various catalytic species. The distinction between homogeneous and heterogeneous catalysis is based on the relative amount of the poison compared to substrate required for complete suppression of the catalytic activity: $\ll$1 eq. ($\sim$0.1–0.2 eq.) of the poison relative to the metal indicates heterogeneous (nanoparticle) catalysis, while $\geq$1 eq. of the poison indicates molecular catalysis. The explanation is that nanoparticles contain a tiny fraction of active metal atoms located at the surface, while molecular complexes require at least a stoichiometric amount of poison to be completely deactivated [11]. Quantitative poisoning experiments are usually displayed as kinetic calculations and plots of the relative reaction rate against the relative amount of added poison, with the x-intercept of the linear extrapolation corresponding to the amount of poison required to totally inhibit the active catalyst. CS_2, 1,10-phenanthroline, $P(OCH_3)_3$, and tetramethylthiourea are typical poisons used in quantitative experiments with Rh, Ir, Ru, and Pd catalytic systems [152, 158, 164–172]. However, the special thermodynamic parameters of the poisoning of nanoparticles using CS_2 lead to the release of the ligand from the particle at high temperature. This feature distorts the results of the study [11].

2.3.6.2 Mercury Test

Mercury test ("Hg test," "mercury poisoning test," "Hg poisoning test," or "Hg drop test") is the most common selective poisoning method used for distinguishing between homogeneous and nanoparticle catalytic mechanisms [3, 5, 11, 91, 102, 151–161, 173–175]. It has been commonly believed that liquid mercury selectively deactivates metal nanoparticles but is incapable of deactivating molecular metal complexes. Inhibition of a catalytic reaction after addition of a few drops of mercury has been considered as the evidence of a nanoparticle catalytic mechanism given Hg forms amalgams with metal surface (Scheme 2.1) [3, 5, 11, 91, 102, 151–161, 173–175].

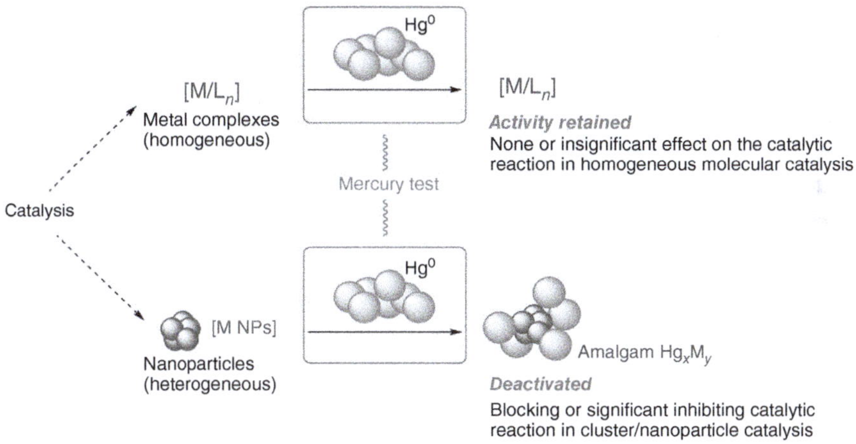

Scheme 2.1 Operating principle of the mercury test in the mechanistic studies of catalytic systems. Source: Reproduced with permission from Chernyshev et al. [176]. Copyright 2019, American Chemical Society.

Wide applications of the mercury tests began after the work of Whitesides and coworkers describing successful use of metallic mercury for suppression of unwanted platinum nanoparticle catalysis in the presence of a molecular platinum catalyst [177, 178]. Since then, the method has been repeatedly criticized [5, 11, 151–161, 165]. Whitesides warned about the limited applicability of the method and the possibility of reactions between mercury and certain organometallic compounds [179, 180]. Several articles reported problems with the use of mercury test for certain organometallic complexes because of their intrinsic reactivity toward mercury [165, 179–181]. Several research groups also reported significant biases associated with the mercury loading and mixing intensity [11, 157, 182].

Nevertheless, despite occasional warning reports, the applicability of the mercury test as a universal method for mechanistic evaluation has not been systematically investigated for a long time [176]. The method was routinely applied in catalytic studies [183], although its accuracy was questionable. Over the past decade, hundreds of articles annually reported the use of mercury test, among which a considerable proportion provided mechanistic considerations using mercury test as the only experimental method [176].

Only recently, the validity of mercury test has been systematically investigated on the example of Pd and Pt catalysis [176]. The inconsistency of the main postulate of the method has been revealed: metallic mercury effectively promotes decomposition of various M^0 and M^{II} complexes of palladium and platinum that are widely used in homogeneous catalysis (Scheme 2.2). The M^0/L complexes are decomposed by mercury to give free ligands and amalgams (Scheme 2.2a), while the M^{II}/L complexes react with mercury as oxidizers to give Hg^{II}/L compounds and reduced amalgamated metals (Scheme 2.2b).

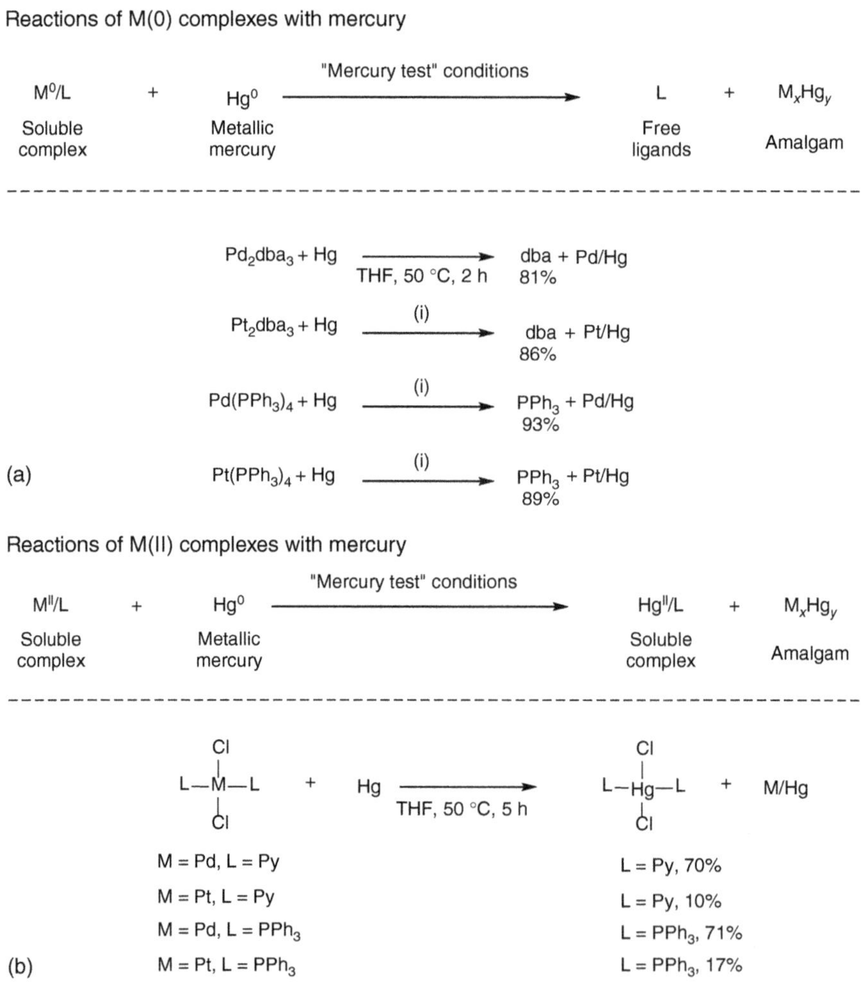

Scheme 2.2 Selected examples of reactions of (a) M^0 and (b) M^{II} complexes with metallic mercury [176].

Even the highly stable complexes of palladium and platinum with NHC ligands show significant reactivity toward mercury under catalytic conditions [176]. Moreover, the strong and poorly predictable dependence of the outcomes on mercury loadings and mixing intensity are confusing; slight shifts in these operational parameters may lead to opposite conclusions for the same catalytic system [176].

To conclude, the mercury test in its simplified variant should be recognized as an unreliable method for mechanistic evaluations. This finding raises questions about more general, fundamental limitations of catalyst poisoning techniques for distinguishing between molecular and nanoparticle catalytic mechanisms in dynamic systems.

2.3.6.3 Fundamental Limitations of the Catalyst Poisoning Techniques for Dynamic Systems

Catalyst poisoning approaches have been criticized in many studies [5, 11, 152–154, 156, 157, 159, 161–163, 184]. One of the main problems is that poison can make a disturbing effect, roughly violating the properties of the whole system [153]. This problem is especially acute in relation to dynamic systems given the rapid interconversions between alternative catalytic species, e.g. metal nanoparticles and molecular metal complexes, via dissolution, leaching, and aggregation [3, 123]. The problem was also analyzed from different points of view [153, 156, 176].

Let us assume that selective nanoparticle poison is added to a catalytic system operating by homogeneous catalysis and comprising nanoparticles (M nanoparticles) capable of mutual interconversions with active molecular complexes ($[M/L_n]$) (Scheme 2.3a). If the rate of equilibrium establishment between the nanoparticles and metal complexes is higher than the rate of selective poisoning, the capture of nanoparticles by the poison can shift the equilibrium and thus decrease the concentration of molecular complex [176]. In this case, even selective poisoning of nanoparticles can deactivate homogeneous catalysis, thus giving a false positive indication of nanoparticle catalysis. In other words, selective poison can be deprived of selectivity in dynamic systems! Similar considerations apply to the selective poisoning of molecular complexes [176]; for the same reasons, incorrect results can be obtained from quantitative poisoning experiments if $r_{eq} \gg r_p$.

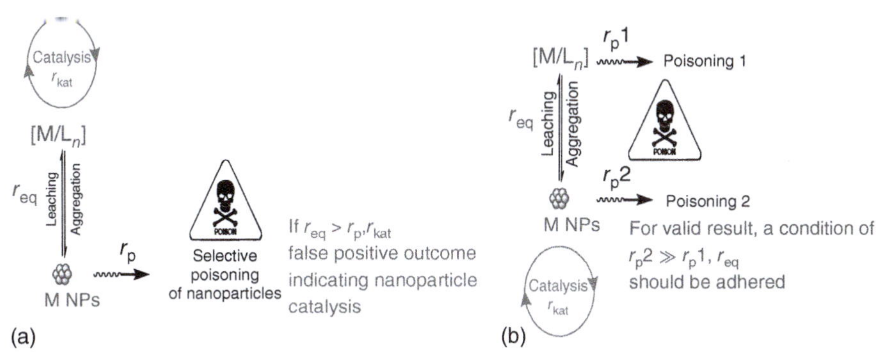

Scheme 2.3 Principal examples of poisoning in dynamic systems. (a) Homogeneous catalysis, selective poisoning, and simplified case and (b) heterogeneous catalysis, poisoning, and general case. Source: Modified from Chernyshev et al. [176].

Returning to the mercury test as a convenient example, we can consider the conditions required to get reliable results (Scheme 2.3b) [176]:

(i) The rate of reaction between the poison (Hg) and metal nanoparticles (r_p2) should be sufficiently higher than the rate of reaction between the poison and metal complexes (r_p1) and the rate of equilibrium establishment (r_{eq}):

$$r_p2 \gg r_p1, r_{eq} \tag{2.1}$$

This is a fundamental limitation for catalytic systems and poisons that, apparently, applies to other known poisoning methods.

(ii) Sufficient contact area and hydrodynamic conditions should be provided to overcome the diffusion limitations.

Without the fulfillment of both requirements, application of poisoning methods may lead to erroneous conclusions.

2.4 Summary and Conclusions

It is now recognized that the boundary between molecular and heterogeneous catalysis cannot be strictly defined. We very often have to deal with unexpected catalytic activities because of the amazing ability of nanoparticles to migrate. Unwanted nanoparticles can get into the reaction in the form of contaminants of reagents, or particles contained in initial precatalysts, together with airborne particles, or from a previous reaction on surface of nondisposable labware. They arise as a jack-in-the-box in most unexpected cases. We have to admit that some classical approaches for determining the homogeneity of a catalytic system, such as the mercury test, are not valid for many systems.

Nanoparticles are enormously pervasive, and the phenomenon of leaching plays a significant role in their pervasiveness. A cocktail of a various metal-containing particles, many of which can exhibit catalytic activity, is formed in the reaction mixture because of leaching. It is interesting that each type of particles that arose during the leaching process can also participate in various interconversions and side processes, interacting not only with reagents and solvents but also with catalyst supports, stir bars, and chemical glassware.

Surprisingly, the activity of nanoparticles was observed even in systems where leaching is not expected – in solvent-free systems and solid-state systems. Apparently, nanoparticles are able to interact with reagents and reaction products even under such conditions. The variety of processes in which metal nanoparticles are involved can be very large, and the transformation paths of metallic species can be incredibly sophisticated and confusing. Indeed, the dynamic nature of catalysis does lie in such a complex and ever-changing kaleidoscope of chemical processes.

Thus, real catalytic systems are multi-component and extremely difficult to characterize. The study of real catalytic systems is a complicated task that cannot be accomplished without “real-time” investigations using mass spectrometry, electron microscopy, and computational modeling.

Electron microscopy has helped to move from a schematic view of reactions to the real-time assessment of catalytic systems evolution at micro- and nanolevels. Special devices and methods allow catalytic reactions to be carried out directly within the specimen chamber of a microscope by reproducing the required conditions (composition, pressure, temperature, and reaction atmosphere) in a limited volume of the vacuum-tight holder.

Real catalytic systems are so complex that it is extremely difficult to draw conclusions on interconversions of the catalytically active centers without direct visualization of the processes. No doubt that the development of new electron microscopy approaches will give rise to new powerful ideas in organic chemistry and catalysis.

Mass spectrometry is a universal tool for studying dynamic catalytic systems at the molecular level. Recent progress in its development provides diverse approaches for studying particular catalytic systems. Different ionization techniques allow detecting and studying reactive intermediates, nano-to-molecular transformations (leaching), and molecular-to-nano transformations (clustering, aggregation). The direct mass spectrometry analysis of nanoparticles is still a major challenge; certain approaches with MALDI ionization allow studying small nanoparticles (<5 nm).

Despite being a method of choice and its versatility for different applications, mass spectrometry should be combined with other microscopy and spectroscopy techniques. A combination of advanced experimental techniques with theoretical modeling will facilitate better understanding of chemical processes at the heterogeneous/homogeneous interfaces in dynamic catalytic systems.

The account of computational approaches given in this chapter demonstrates its efficiency in the modeling of dynamic interconversions. The studies are few and, for now, the computational approach to dynamic catalysis is only to become conventional, as we need to take catalyst and reactant transformations together at an equal level of theory, which can be much more costly than considering the reactant transformations facilitated by a single type of catalytic species alone, either a nanoparticle or a metal complex. The yet first examples considered in this chapter focus on the interconversions of catalytic species and largely avoid comparative analysis of the reactivity by modeling of the corresponding transition states (at the nanoparticle surface, in metal complexes, and with small metal clusters). Extensive modeling of numerous reaction pathways (dozens or more) in one study is currently gaining momentum, but it is still challenging.

Miniaturization of mass spectrometry and NMR equipment, as well as current advances in liquid-phase electron microscopy, allow real-time acquisition of comprehensive data on the catalytic cross-coupling and functionalization reactions. This opens a possibility to automate kinetic and mechanistic studies and, by applying data science and machine learning approaches, to gain new insights into dynamic catalysis. We believe that much awaited applications of machine learning and data science strategies will enormously facilitate future studies in the field.

References

1 Ananikov, V.P. and Beletskaya, I.P. (2011). *Dalton Trans.* 40: 4011–4023.
2 Trzeciak, A.M. and Ziółkowski, J.J. (2007). *Coord. Chem. Rev.* 251: 1281–1293.
3 Eremin, D.B. and Ananikov, V.P. (2017). *Coord. Chem. Rev.* 346: 2–19.
4 Kashin, A.S. and Ananikov, V.P. (2013). *J. Org. Chem.* 78: 11117–11125.
5 Crabtree, R.H. (2011). *Chem. Rev.* 112: 1536–1554.
6 Trzeciak, A.M. and Augustyniak, A.W. (2019). *Coord. Chem. Rev.* 384: 1–20.
7 Leyva-Perez, A. (2017). *Dalton Trans.* 46: 15987–15990.
8 Hii, K.K. and Hellgardt, K. (2015). Catalysis in Flow: Why Leaching Matters, in Organometallic Flow Chemistry. In: *Topics in Organometallic Chemistry*, vol. 57 (ed. T. Noël), 249–262. Cham: Springer.
9 McGlacken, G.P. and Fairlamb, I.J.S. (2009). *Eur. J. Org. Chem.* 2009: 4011–4029.
10 Reay, A.J. and Fairlamb, I.J. (2015). *Chem. Commun.* 51: 16289–16307.
11 Widegren, J.A. and Finke, R.G. (2003). *J. Mol. Catal. A: Chem.* 198: 317–341.
12 Ananikov, V.P. and Beletskaya, I.P. (2012). *Organometallics* 31: 1595–1604.
13 Eremin, D.B., Boiko, D.A., Borkovskaya, E.V. et al. (2018). *Catal. Sci. Technol.* 8: 3073–3080.
14 Kashin, A.S., Degtyareva, E.S., Eremin, D.B., and Ananikov, V.P. (2018). *Nat. Commun.* 9: 2936.
15 Kashin, A.S. and Ananikov, V.P. (2019). *Nat. Rev. Chem.* 3: 624–637.
16 Panova, Y.S., Kashin, A.S., Vorobev, M.G. et al. (2016). *ACS Catal.* 6: 3637–3643.
17 Amenuvor, G., Makhubela, B.C.E., and Darkwa, J. (2016). *ACS Sustainable Chem. Eng.* 4: 6010–6018.
18 Tobisu, M., Nakamura, K., and Chatani, N. (2014). *J. Am. Chem. Soc.* 136: 5587–5590.
19 Moliner, M., Gabay, J.E., Kliewer, C.E. et al. (2016). *J. Am. Chem. Soc.* 138: 15743–15750.
20 Oliver-Meseguer, J., Cabrero-Antonino, J.R., Dominguez, I. et al. (2012). *Science* 338: 1452–1455.
21 Munoz Iii, S.B., Daifuku, S.L., Brennessel, W.W., and Neidig, M.L. (2016). *J. Am. Chem. Soc.* 138: 7492–7495.
22 Corma, A., Concepción, P., Boronat, M. et al. (2013). *Nat. Chem.* 5: 775–781.
23 Liu, W., Zhang, L., Yan, W. et al. (2016). *Chem. Sci.* 7: 5758–5764.
24 Malta, G., Kondrat, S.A., Freakley, S.J. et al. (2017). *Science* 355: 1399–1403.
25 Gai, P.L., Sharma, R., and Ross, F.M. (2008). *MRS Bull.* 33: 107–114.
26 Tao, F. and Crozier, P.A. (2016). *Chem. Rev.* 116: 3487–3539.
27 Jiang, Y., Zhang, Z., Yuan, W. et al. (2018). *Nano Res.* 11: 42–67.
28 Ross, F.M. (2015). *Science* 350: aaa9886.

29 de Jonge, N. and Ross, F.M. (2011). *Nat. Nanotechnol.* 6: 695–704.
30 Sutter, E., Jungjohann, K., Bliznakov, S. et al. (2014). *Nat. Commun.* 5: 4946.
31 Chee, S.W., Tan, S.F., Baraissov, Z. et al. (2017). *Nat. Commun.* 8: 1224.
32 Ye, X., Jones, M.R., Frechette, L.B. et al. (2016). *Science* 354: 874–877.
33 Shan, H., Gao, W., Xiong, Y. et al. (2018). *Nat. Commun.* 9: 1011.
34 Degtyareva, E.S., Erokhin, K.S., Kashin, A.S., and Ananikov, V.P. (2019). *Appl. Catal., A* 571: 170–179.
35 McIndoe, J.S. and Vikse, K.L. (2019). *J. Mass Spectrom.* 54: 466–479.
36 Vikse, K.L. and McIndoe, J.S. (2018). *J. Mass Spectrom.* 53: 1026–1034.
37 Yunker, L.P., Stoddard, R.L., and McIndoe, J.S. (2014). *J. Mass Spectrom.* 49: 1–8.
38 Bailey, G.A. and Fogg, D.E. (2016). *ACS Catal.* 6: 4962–4971.
39 Maya, L., Chen, C.H., Stevenson, K.A. et al. (2002). *J. Nanopart. Res.* 4: 417–422.
40 Liang, C.K., Eller, M.J., Verkhoturov, S.V., and Schweikert, E.A. (2015). *J. Am. Soc. Mass Spectrom.* 26: 1259–1265.
41 Khitrov, G.A. and Strouse, G.F. (2003). *J. Am. Chem. Soc.* 125: 10465–10469.
42 Yang, X., Gan, L., Zhu, C. et al. (2014). *Chem. Commun.* 50: 234–236.
43 Navin, J.K., Grass, M.E., Somorjai, G.A., and Marsh, A.L. (2009). *Anal. Chem.* 81: 6295–6299.
44 Banerjee, S., Basheer, C., and Zare, R.N. (2016). *Angew. Chem. Int. Ed.* 55: 12807–12811.
45 Lee, J.K., Samanta, D., Nam, H.G., and Zare, R.N. (2018). *Nat. Commun.* 9: 1562.
46 Li, A., Luo, Q., Park, S.J., and Cooks, R.G. (2014). *Angew. Chem. Int. Ed.* 53: 3147–3150.
47 Zalesskiy, S.S. and Ananikov, V.P. (2012). *Organometallics* 31: 2302–2309.
48 Janusson, E., Zijlstra, H.S., Nguyen, P.P. et al. (2017). *Chem. Commun.* 53: 854–856.
49 Kolter, M. and Koszinowski, K. (2019). *Chem. Eur. J.* 25: 13376–13384.
50 Kreyenschmidt, F. and Koszinowski, K. (2018). *Chem. Eur. J.* 24: 1168–1177.
51 Ingram, A.J., Walker, K.L., Zare, R.N., and Waymouth, R.M. (2015). *J. Am. Chem. Soc.* 137: 13632–13646.
52 Ingram, A.J., Solis-Ibarra, D., Zare, R.N., and Waymouth, R.M. (2014). *Angew. Chem. Int. Ed.* 53: 5648–5652.
53 Lushchikova, O.V., Huitema, D.M.M., Lopez-Tarifa, P. et al. (2019). *J. Phys. Chem. Lett.* 10: 2151–2155.
54 Schwarz, H. and Asmis, K.R. (2019). *Chem. Eur. J.* 25: 2112–2126.
55 Lang, S.M., Bernhardt, T.M., Chernyy, V. et al. (2017). *Angew. Chem. Int. Ed.* 56: 13406–13410.
56 Debnath, S., Knorke, H., Schollkopf, W. et al. (2018). *Angew. Chem. Int. Ed.* 57: 7448–7452.
57 Sun, X., Zhou, S., Yue, L. et al. (2018). *Angew. Chem. Int. Ed.* 57: 5934–5937.
58 Geng, C., Li, J., Schlangen, M. et al. (2018). *Dalton Trans.* 47: 15271–15277.
59 Schwarz, H. (2017). *Catal. Sci. Technol.* 7: 4302–4314.

60 Addanki Tirumala, R.T., Dadgar, A.P., Mohammadparast, F. et al. (2019). *Green Chem.* 21: 5284–5290.

61 Denisova, E.A., Eremin, D.B., Gordeev, E.G. et al. (2019). *Inorg. Chem.* 58: 12218–12227.

62 de Vries, A.H.M., Parlevliet, F.J., Schmieder-van de Vondervoort, L. et al. (2002). *Adv. Synth. Catal.* 344: 996–1002.

63 Gaikwad, A.V., Holuigue, A., Thathagar, M.B. et al. (2007). *Chem. Eur. J.* 13: 6908–6913.

64 Silarska, E., Trzeciak, A.M., Pernak, J., and Skrzypczak, A. (2013). *Appl. Catal., A* 466: 216–223.

65 Alonso, F., Moglie, Y., and Radivoy, G. (2015). *Acc. Chem. Res.* 48: 2516–2528.

66 Haldon, E., Nicasio, M.C., and Perez, P.J. (2015). *Org. Biomol. Chem.* 13: 9528–9550.

67 Fuchs, M., Goessler, W., Pilger, C., and Kappe, C.O. (2010). *Adv. Synth. Catal.* 352: 323–328.

68 Lipshutz, B.H. and Taft, B.R. (2006). *Angew. Chem. Int. Ed.* 45: 8235–8238.

69 Qiu, R. and Luo, H. (2015). *RSC Adv.* 5: 96213–96221.

70 Worrell, B.T., Malik, J.A., and Fokin, V.V. (2013). *Science* 340: 457–460.

71 Fernandes, A.E., Dirani, A., d'Haese, C. et al. (2012). *Chem. Eur. J.* 18: 16226–16233.

72 Fernandes, A.E., Devillez, S., d'Haese, C. et al. (2012). *Langmuir* 28: 14822–14828.

73 Dufert, M.A., Billingsley, K.L., and Buchwald, S.L. (2013). *J. Am. Chem. Soc.* 135: 12877–12885.

74 Carrillo, J., Gómez, A., Costa, A.M. et al. (2014). *Tetrahedron Lett.* 55: 4623–4627.

75 Senecal, T.D., Shu, W., and Buchwald, S.L. (2013). *Angew. Chem. Int. Ed.* 52: 10035–10039.

76 Pignataro, L., Fiorito, D., Vece, V. et al. (2015). *Eur. J. Org. Chem.* 2015: 6669–6678.

77 Li, W., Yang, L., Tumbleston, J.R. et al. (2014). *Adv. Mater.* 26: 4456–4462.

78 Merritt, J.M., Buser, J.Y., Campbell, A.N. et al. (2013). *Org. Process Res. Dev.* 18: 246–256.

79 Jia, T., Bellomo, A., Montel, S. et al. (2014). *Angew. Chem. Int. Ed.* 53: 260–264.

80 Iwata, A., Inuki, S., Oishi, S. et al. (2015). *Tetrahedron* 71: 6580–6585.

81 Dou, J., Chen, Z., and Ma, C. (2014). *Synth. Met.* 196: 117–124.

82 Braun, M., Meletis, P., and Fidan, M. (2009). *Org. Synth.* 86: 303–314.

83 Imagawa, N., Nagato, Y., Ohmatsu, K., and Ooi, T. (2016). *Bull. Chem. Soc. Jpn.* 89: 649–656.

84 Kendall, A.J., Zakharov, L.N., and Tyler, D.R. (2016). *Inorg. Chem.* 55: 3079–3090.

85 Doran, R. and Guiry, P.J. (2014). *J. Org. Chem.* 79: 9112–9124.

86 Hughes, M., Boultwood, T., Zeppetelli, G., and Bull, J.A. (2013). *J. Org. Chem.* 78: 844–854.

87 Lunn, D.J., Gould, O.E.C., Whittell, G.R. et al. (2016). *Nat. Commun.* 7: 12371.

88 Tölg, G. (1977). *Z. Anal. Chem.* 283: 257–267.

89 Nagy, I.P. and Bazsa, G. (1991). *React. Kinet. Catal. Lett.* 45: 15–25.

90 Pentsak, E.O., Eremin, D.B., Gordeev, E.G., and Ananikov, V.P. (2019). *ACS Catal.* 9: 3070–3081.

91 Lin, Y. and Finke, R.G. (1994). *Inorg. Chem.* 33: 4891–4910.

92 Hornstein, B.J. and Finke, R.G. (2004). *Chem. Mater.* 16: 139–150.

93 Finney, E.E. and Finke, R.G. (2008). *Chem. Mater.* 20: 1956–1970.

94 Daley, R.A., Liu, E.C., and Topczewski, J.J. (2019). *Org. Lett.* 21: 4734–4738.

95 Zhao, G., Roudaut, C., Gandon, V. et al. (2019). *Green Chem.* 21: 4204–4210.

96 Peng, Y., Li, X., Gao, T. et al. (2019). *Green Chem.* 21: 4169–4177.

97 Vollmer, C., Schröder, M., Thomann, Y. et al. (2012). *Appl. Catal., A* 425–426: 178–183.

98 Han, G., Li, X., Li, J. et al. (2017). *ACS Omega* 2: 4938–4945.

99 Boch, K., Schuster, M., Risse, G., and Schwarzer, M. (2002). *Anal. Chim. Acta* 459: 257–265.

100 Mieczyńska, E., Gniewek, A., and Trzeciak, A.M. (2012). *Appl. Catal., A* 421–422: 148–153.

101 Balanta, A., Godard, C., and Claver, C. (2011). *Chem. Soc. Rev.* 40: 4973–4985.

102 de Vries, J.G. (2006). *Dalton Trans.*: 421–429.

103 Collins, G. and Holmes, J.D. (2016). *Adv. Mater.* 28: 5689–5695.

104 Sperger, T., Sanhueza, I.A., Kalvet, I., and Schoenebeck, F. (2015). *Chem. Rev.* 115: 9532–9586.

105 Ma, S., Wang, H., Gao, K., and Zhao, F. (2006). *J. Mol. Catal. A: Chem.* 248: 17–20.

106 Wang, S.-S. and Yang, G.-Y. (2016). *Catal. Sci. Technol.* 6: 2862–2876.

107 Handa, S., Slack, E.D., and Lipshutz, B.H. (2015). *Angew. Chem. Int. Ed.* 54: 11994–11998.

108 Lipshutz, B.H., Frieman, B.A., Lee, C.T. et al. (2006). *Chem. Asian J.* 1: 417–429.

109 Lipshutz, B.H. and Ueda, H. (2000). *Angew. Chem. Int. Ed.* 39: 4492–4494.

110 Tasler, S. and Lipshutz, B.H. (2003). *J. Org. Chem.* 68: 1190–1199.

111 Vogiatzis, K.D., Polynski, M.V., Kirkland, J.K. et al. (2019). *Chem. Rev.* 119: 2453–2523.

112 Wang, F., Li, C., Sun, L.D. et al. (2011). *J. Am. Chem. Soc.* 133: 1106–1111.

113 Collins, G., Schmidt, M., O'Dwyer, C. et al. (2014). *Angew. Chem. Int. Ed.* 53: 4142–4145.

114 Collins, G., Schmidt, M., O'Dwyer, C. et al. (2014). *ACS Catal.* 4: 3105–3111.

115 Popa, C., Zhu, T., Tranca, I. et al. (2015). *Phys. Chem. Chem. Phys.* 17: 2268–2273.

116 Pagliaro, M., Pandarus, V., Ciriminna, R. et al. (2012). *ChemCatChem* 4: 432–445.

117 Burello, E., Farrusseng, D., and Rothenberg, G. (2004). *Adv. Synth. Catal.* 346: 1844–1853.

118 Garcia-Melchor, M., Braga, A.A., Lledos, A. et al. (2013). *Acc. Chem. Res.* 46: 2626–2634.

119 Chang, C.-R., Zhao, Z.-J., Köhler, K. et al. (2012). *Catal. Sci. Technol.* 2: 2238–2248.
120 Zvereva, E.E., Katsyuba, S.A., Dyson, P.J., and Aleksandrov, A.V. (2017). *J. Phys. Chem. Lett.* 8: 3452–3456.
121 Ramezani-Dakhel, H., Mirau, P.A., Naik, R.R. et al. (2013). *Phys. Chem. Chem. Phys.* 15: 5488–5492.
122 Briggs, B.D., Bedford, N.M., Seifert, S. et al. (2015). *Chem. Sci.* 6: 6413–6419.
123 Polynski, M.V. and Ananikov, V.P. (2019). *ACS Catal.* 9: 3991–4005. https://doi.org/10.1021/acscatal.9b00207.
124 Deraedt, C. and Astruc, D. (2014). *Acc. Chem. Res.* 47: 494–503.
125 Sigeev, A.S., Peregudov, A.S., Cheprakov, A.V., and Beletskaya, I.P. (2015). *Adv. Synth. Catal.* 357: 417–429.
126 Khazaei, A., Rahmati, S., Hekmatian, Z., and Saeednia, S. (2013). *J. Mol. Catal. A: Chem.* 372: 160–166.
127 Khazaei, A., Rahmati, S., Ghaderi, A., and Roshani, L. (2013). *J. Iran. Chem. Soc.* 11: 263–269.
128 Martínez, A.V., Leal-Duaso, A., García, J.I., and Mayoral, J.A. (2015). *RSC Adv.* 5: 59983–59990.
129 Tao, L., Xie, Y., Deng, C., and Li, J. (2009). *Chin. J. Chem.* 27: 1365–1373.
130 Kaupp, G. (2016). *Organic solid-state reactions*. In: *Encyclopedia of Physical Organic Chemistry* (eds. Z. Wang, U. Wille and E. Juaristi), 1–79. Weinheim, Germany: Wiley-VCH.
131 Rossi, L.M. and Machado, G. (2009). *J. Mol. Catal. A: Chem.* 298: 69–73.
132 Huang, H., Wang, X., Tan, M. et al. (2016). *ChemCatChem* 8: 1485–1489.
133 Upadhyay, P. and Srivastava, V. (2015). *RSC Adv.* 5: 740–745.
134 Das, P., Sarmah, P.P., Borah, B.J. et al. (2016). *New J. Chem.* 40: 2850–2855.
135 Yan, Y., Chen, Y., Jia, X., and Yang, Y. (2014). *Appl. Catal., B* 156–157: 385–397.
136 Zamani, F. and Hosseini, S.M. (2014). *Catal. Commun.* 43: 164–168.
137 Wang, J., Kondrat, S.A., Wang, Y. et al. (2015). *ACS Catal.* 5: 3575–3587.
138 Dong, H., Xie, R., Yang, L., and Li, F. (2018). *Dalton Trans.* 47: 7776–7786.
139 Alshammari, H. (2016). *React. Kinet. Mech. Catal.* 119: 149–163.
140 Blanckenberg, A., Kotze, G., Swarts, A.J., and Malgas-Enus, R. (2018). *J. Nanopart. Res.* 20: 44.
141 Hajipour, A.R. and Karimi, H. (2014). *Chin. J. Catal.* 35: 1136–1147.
142 Ghodrati, K., Korani, E., and Asnaashari, M. (2018). *Orient. J. Chem.* 34: 1504–1512.
143 Bruss, A.J., Gelesky, M.A., Machado, G., and Dupont, J. (2006). *J. Mol. Catal. A: Chem.* 252: 212–218.
144 Hernández, J.G. and Friščić, T. (2015). *Tetrahedron Lett.* 56: 4253–4265.
145 Seo, T., Ishiyama, T., Kubota, K., and Ito, H. (2019). *Chem. Sci.* 10: 8202–8210.
146 Kubota, K., Seo, T., Koide, K. et al. (2019). *Nat. Commun.* 10: 111.
147 Declerck, V., Colacino, E., Bantreil, X. et al. (2012). *Chem. Commun.* 48: 11778.
148 Rothenberg, G., Downie, A.P., Raston, C.L., and Scott, J.L. (2001). *J. Am. Chem. Soc.* 123: 8701–8708.

149 Monguchi, Y., Fujita, Y., Hashimoto, S. et al. (2011). *Tetrahedron* 67: 8628–8634.
150 Pentsak, E.O. and Ananikov, V.P. (2019). *Eur. J. Org. Chem.* 2019: 4239–4247.
151 Dyson, P.J. (2003). *Dalton Trans.*: 2964–2974.
152 Phan, N.T.S., Sluys, M.V.D., and Jones, C.W. (2006). *Adv. Synth. Catal.* 348: 609–679.
153 Schmidt, A.F. and Kurokhtina, A.A. (2012). *Kinet. Catal.* 53: 714–730.
154 Artero, V. and Fontecave, M. (2013). *Chem. Soc. Rev.* 42: 2338–2356.
155 Yan, N., Yuan, Y., and Dyson, P.J. (2013). *Dalton Trans.* 42: 13294–13304.
156 Schmidt, A.F., Kurokhtina, A.A., and Larina, E.V. (2014). *Catal. Sci. Technol.* 4: 3439–3457.
157 Sonnenberg, J.F. and Morris, R.H. (2014). *Catal. Sci. Technol.* 4: 3426–3438.
158 Stracke, J.J. and Finke, R.G. (2014). *ACS Catal.* 4: 909–933.
159 Cano, R., Schmidt, A.F., and McGlacken, G.P. (2015). *Chem. Sci.* 6: 5338–5346.
160 Meek, S.J., Pitman, C.L., and Miller, A.J.M. (2016). *J. Chem. Educ.* 93: 275–286.
161 Biffis, A., Centomo, P., Del Zotto, A., and Zecca, M. (2018). *Chem. Rev.* 118: 2249–2295.
162 Lin, H. (2018). *Curr. Org. Chem.* 22: 1022–1038.
163 Bourouina, A., Meille, V., and de Bellefon, C. (2019). *Catalysts* 9: 1–27.
164 Hornstein, B.J., Aiken, J.D., and Finke, R.G. (2002). *Inorg. Chem.* 41: 1625–1638.
165 Bayram, E., Linehan, J.C., Fulton, J.L. et al. (2011). *J. Am. Chem. Soc.* 133: 18889–18902.
166 Bayram, E. and Finke, R.G. (2012). *ACS Catal.* 2: 1967–1975.
167 Bayram, E., Lu, J., Aydin, C. et al. (2012). *ACS Catal.* 2: 1947–1957.
168 Bayram, E., Linehan, J.C., Fulton, J.L. et al. (2015). *ACS Catal.* 5: 3876–3886.
169 Zim, D., Nobre, S.M., and Monteiro, A.L. (2008). *J. Mol. Catal. A: Chem.* 287: 16–23.
170 Sonnenberg, J.F., Coombs, N., Dube, P.A., and Morris, R.H. (2012). *J. Am. Chem. Soc.* 134: 5893–5899.
171 Drost, R.M., Rosar, V., Marta, S.D. et al. (2015). *ChemCatChem* 7: 2095–2107.
172 Azua, A., Finn, M., Yi, H. et al. (2017). *ACS Sustainable Chem. Eng.* 5: 3963–3972.
173 Lewis, L.N. (1993). *Chem. Rev.* 93: 2693–2730.
174 de Vries, J.G. (2011). When Does Catalysis with Transition Metal Complexes Turn into Catalysis by Nanoparticles?. In *Selective Nanocatalysts and Nanoscience: Concepts for Heterogeneous and Homogeneous Catalysis* (eds. A. Zecchina, S. Bordiga and E. Groppo), 73–103. Weinheim: Wiley-VCH.
175 Del Zotto, A. and Zuccaccia, D. (2017). *Catal. Sci. Technol.* 7: 3934–3951.
176 Chernyshev, V.M., Astakhov, A.V., Chikunov, I.E. et al. (2019). *ACS Catal.* 9: 2984–2995.
177 Whitesides, G.M., Hackett, M., Brainard, R.L. et al. (1985). *Organometallics* 4: 1819–1830.
178 Foley, P., DiCosimo, R., and Whitesides, G.M. (1980). *J. Am. Chem. Soc.* 102: 6713–6725.

179 Gorunova, O.N., Novitskiy, I.M., Grishin, Y.K. et al. (2018). *Organometallics* 37: 2842–2858.

180 Gorunova, O.N., Livantsov, M.V., Grishin, Y.K. et al. (2013). *J. Organomet. Chem.* 737: 59–63.

181 Stein, J., Lewis, L.N., Gao, Y., and Scott, R.A. (1999). *J. Am. Chem. Soc.* 121: 3693–3703.

182 Weddle, K.S., Aiken, J.D., and Finke, R.G. (1998). *J. Am. Chem. Soc.* 120: 5653–5666.

183 Blanco, V., Leigh, D.A., and Marcos, V. (2015). *Chem. Soc. Rev.* 44: 5341–5370.

184 Schmidt, A.F., Kurokhtina, A.A., and Larina, E.V. (2019). *Kinet. Catal.* 60: 551–572.

Part I

Nanoparticles in Solution

3

Metal Nanoparticles in Water: A Relevant Toolbox for Green Catalysis

Audrey Denicourt-Nowicki, Natalia Mordvinova, and Alain Roucoux

Univ Rennes, Ecole Nationale Supérieure de Chimie de Rennes, CNRS ISCR-UMR 6226, Rennes 35000, France

3.1 Introduction

During the past two decades, nanometer-sized metallic species, finely dispersed in water, have attracted a great interest in the field of catalysis and especially in green catalysis. These catalysts afford relevant, or even original in some cases, catalytic performances owing to their surface reactivity, combined with a potential recycling under adapted conditions such as biphasic catalysis [1, 2]. They have been successfully applied in a wide range of catalytic reactions, including hydrogenation [3], carbon–carbon coupling reactions [4], and more scarcely in the oxidation ones.

Synthetic processes, performed in water, remain of great interest because of the intrinsic properties of this solvent (polarity, dielectric constant, high heat capacity, etc.), being inherently safer as well as available and inexpensive [5]. Moreover, the use of water allows a simple product extraction and an easy recycling of the catalyst. However, water could also have an influence on the catalytic performances, such as an acceleration of the reaction kinetics or selectivity [6–8]. Finally, the use of water for industrial scale-up seems to be a promising alternative to problematic or hazardous solvents [9].

In such a solvent, a bottom-up approach can be followed to prepare aqueous suspensions of metal nanoparticles based on the reduction of commercially available transition metal salts in mild conditions and in the presence of a water-soluble protective agent in order to avoid their agglomeration. The past decade has evidenced an increasing interest for the development of relevant capping agents for the efficient stabilization of metal nanospecies within water, such as ligands possessing water-soluble functions (sulfonated phosphines, carbenes, etc.), surfactants, or again polymers. The recent literature reports many investigations to adapt organosoluble ligands dedicated to homogeneous catalysis into their water-soluble analogs by the introduction of hydrophilic functional polar moieties [10, 11]. The choice of the capping agents and the metal ratio can be crucial not only for the stability of the obtained nanoparticles but also for their catalytic

Nanoparticles in Catalysis: Advances in Synthesis and Applications, First Edition.
Edited by Karine Philippot and Alain Roucoux.

performances because it can influence the reactivity and selectivity achieved in the targeted reaction.

In this nonexhaustive chapter, we will report some pertinent advances in catalytic applications involving aqueous suspensions of metal nanoparticles for hydrogenation, oxidation, and carbon–carbon coupling reactions.

3.2 Protection by Ligands

Ligands, which covalently bind at the particle surface, constitute valuable protective agents for metal nanospecies because they can modulate both their electronic and steric environments, potentially inducing promising surface reactivities. Stabilization can generally be achieved by various ligands such as thiols, amines, carboxylic acids, phosphines, or more recently *N*-heterocyclic carbenes (NHCs). Most of them are used in organic solvents, but in a few cases, the nanoparticle transfer from an organic medium to water was demonstrated to be feasible using a phase transfer agent. Among the water-soluble ligands recently used, phosphorous or nitrogenated ligands, as well as NHCs, were found to be promising candidates for various catalytic applications, including hydrogenation and carbon–carbon coupling reactions (Table 3.1).

3.2.1 Hydrogenation Reactions

3.2.1.1 Phosphorous Ligands

Although the reduction of metal salts is a common approach to prepare nanospecies directly in water, the team of Philippot used organometallic complexes to prepare ruthenium and platinum nanoparticles (NPs) with 1,3,5-triaza-7-phosphaadamantane (PTA) as a stabilizing agent, which allowed their easy dispersion in water (Figure 3.1) [12]. The obtained NPs display low size dispersity and very small mean diameters of 1.4 and 1.1 nm, respectively, for ruthenium and platinum species. The authors evidenced the strong coordination of the ligand at the particle surface with dodecanethiol exchange experiments, followed by NMR spectroscopy. The aqueous colloidal suspensions proved active for hydrogenation of simple olefins (octene and dodecene) and arenes (toluene and *m*-methylanisole) under mild conditions (20 °C, 1–10 bar H_2).

The same team also reported the organometallic synthesis of ruthenium nanoparticles stabilized by water-soluble alkyl sulfonated diphosphines as ligands (Figure 3.2) [13]. Sulfonated diphosphines with different numbers of carbon atoms (n) between the two phosphorous atoms (1,4-bis[(di-*m*-sulfonatophenyl)phosphino]-butane (or dppb-TS) for $n = 4$, –propane (dppp-TS) for $n = 3$, and –ethane (dppe-TS) for $n = 2$) were used. The obtained ruthenium NPs with small diameters in the range of 1.2–1.5 nm depending on the [L]/[Ru] ratio and stable for some months were investigated in the reduction of unsaturated substrates (tetradecene, styrene, and acetophenone) in biphasic liquid–liquid conditions. In a first set of experiments, a comparison of the three catalysts obtained according to the ligand

Table 3.1 Overview of recent catalytic applications with ligand-stabilized metal nanoparticles in neat water.

Type of ligands	Examples of ligands	Metal	Reaction	References
Phosphorous	Phosphaadamantane, alkyl sulfonated phosphines	Ru, Pt	Hydrogenation of unsaturated substrates (olefins and arenes)	[12, 13]
	Tricyclohexylphosphine	Pd	Suzuki–Miyaura reaction	[14]
Nitrogenated	NTPA, CyDTA, EDTA, etc.	Pd	Hydrogenation of polyunsaturated methyl esters	[15]
	Ammonium-functionalized bidentate nitrogen-containing ligands	Pd	Suzuki–Miyaura reaction	[16]
Carbonaceous	*N*-heterocyclic carbenes Sulfonated carbenes	Au, Pd	Hydrogenation of olefins	[17, 18]
	Imidazole-containing ligands	Pd	Suzuki–Miyaura reaction	[14]

used was made for styrene hydrogenation, showing a decrease in the following order: dppb-TS > dppp-TS > dppe-TS. This reactivity was explained by a higher flexibility, thus providing a suitable dynamic coordination mode at the particle surface in solution. The best nanocatalyst, Ru-dppb-TS NPs, was also extended to the reduction of more lipophilic tetradecene and acetophenone. Compared to styrene, longer reaction times were needed for tetradecene under atmospheric hydrogen pressure, owing to its low water solubility, which limits the accessibility of the reducible C=C double bond to the active sites at the metal surface. In the case of acetophenone, a complete conversion was achieved in 20 hours with interesting selectivities into phenylethan-1-ol (84%) and cyclohexylethan-1-ol (15%).

More sophistically, ruthenium nanoparticles stabilized by a mixture of a sulfonated phosphine and a randomly methylated cyclodextrin proved to be active in the reduction of some unsaturated substrates such as styrene, acetophenone, or *m*-methylanisole [19]. These systems showed an increased performance as the result of a shuttle effect of cyclodextrin, as known in homogeneous catalysis.

3.2.1.2 Nitrogenated Ligands

In 2014, Papadogianakis and coworkers investigated the use of nitrogen-containing ligands as protective agents, considering their great advantages such as low prices and high hydrophilicity and stability compared to phosphorous analogs

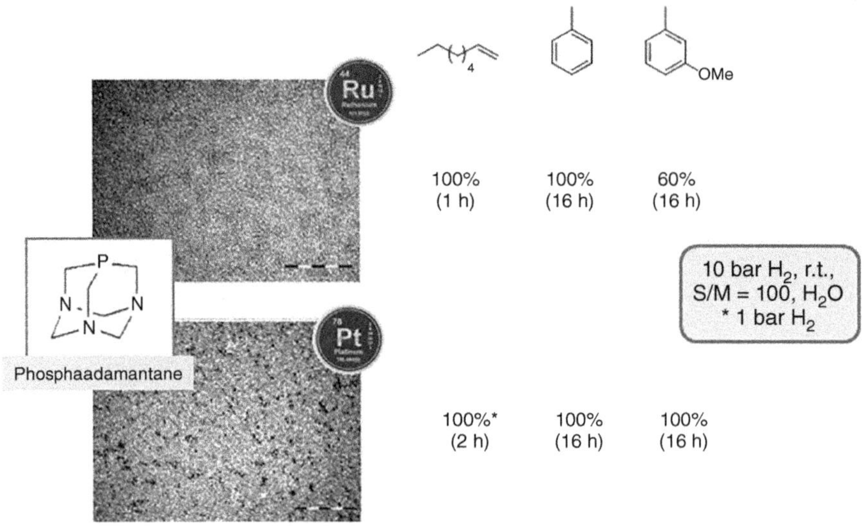

Figure 3.1 1,3,5-Triaza-7-phosphaadamantane-protected ruthenium and platinum NPs for hydrogenation of olefins and arenes. Source: Adapted from Debouttière *et al.* [12].

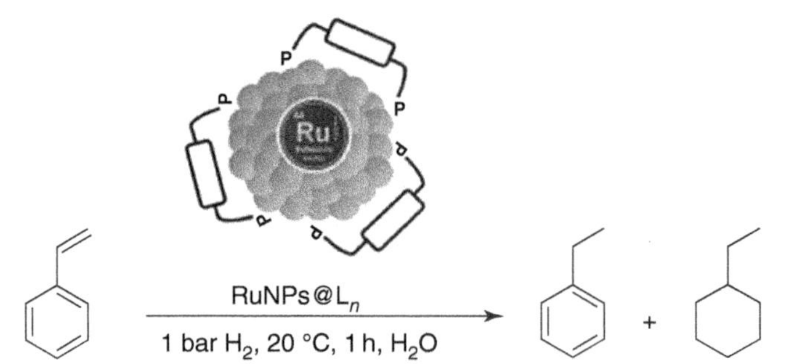

Ligand	ST (%)	EB (%)	EC (%)
L_1	25	75	0
L_2	59	40	1
L_3	75	24	1

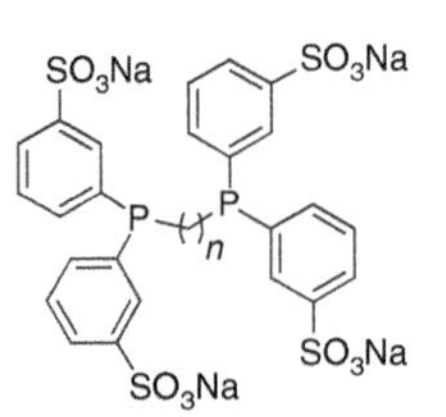

n = 2, 3, 4 (**L_3**, **L_2**, **L_1**) **ST** = Styrene, **EB** = Ethylbenzene, **EC** = Ethylcyclohexane

Figure 3.2 Diphosphine-coated Ru NPs. Comparison in styrene hydrogenation. Source: Adapted from Guerrero *et al.* [13].

[15]. Thus, water-soluble palladium NPs protected by nitrogenated ligands were synthesized in situ during the aqueous/organic selective hydrogenation of renewable methyl esters of soybean oil into their monounsaturated analogs and compared to Pd/TPPTS reference system (Figure 3.3). Outstanding high catalytic activities (TOF = 59 000–68 000 h^{-1}) and high selectivities into C18:1 esters up

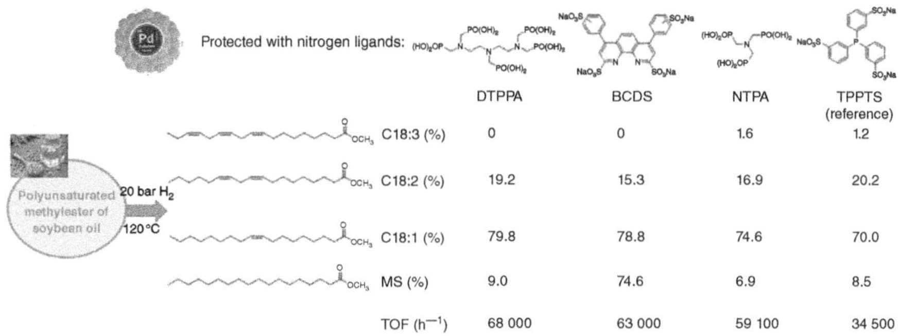

Figure 3.3 Hydrogenation of polyunsaturated methyl esters of soybean oil with Pd NPs. Nitrogen ligands *vs.* phosphorous one as protective agents. Source: Adapted from Bouriazos *et al.* [15]. DTPPA: diethylenetriaminepentakis(methylphosphonic) acid, BCDS: bathocuproine disulfonic acid disodium salt, NPTA: nitrolis(methylphosphonic)acid, TPPTS: trisodium 3,3′,3-phosphinetriyltribenzenesulfonate.

to 80% were achieved when using palladium dichloride precursor ($PdCl_2$) with nitrogenated ligands such as diethylenetriaminepentakis(methylphosphonic) acid (DTPPA), bathocuproine disulfonic acid disodium salt (BCDS), and nitrolis (methylphosphonic) acid (NTPA). A recycling experiment at 120 °C showed that the palladium(0) NPs are stable and active, even at this temperature, thus making this system promising to obtain upgraded biodiesel fuel.

3.2.1.3 Carbon Ligands

NHCs have recently found great interest as protective agents of metal nanospecies owing to their strong σ-donor effect and their easy modification with various functional groups, thus giving rise to tunable metal surfaces [20].

Glorius and coworkers developed the synthesis of palladium and gold NPs in water through a ligand exchange method. For that purpose, they used a library of negatively charged NHCs possessing various functional groups, such as sulfonates or carboxylates (Figure 3.4) [17]. This approach consisted in displacing a weakly binding thioether (i.e. didodecyl sulfide) with the charged NHC in a biphasic hexane/DMF media and then redispersing the particles in water. The ligand exchange was followed by NMR experiments and the negative zeta potential values. The so-obtained NHC-decorated palladium showed relevant catalytic activities in the hydrogenation of various olefins (styrene, 1-decene, 3-methyl-2-cyclohexenone, and citronellol). Among the reaction media, water seems the most promising, leading to reduction of all compounds with good yields (TON up to 2500 and TOF up to 2000 h^{-1}) under mild reaction conditions and low metal loadings. These catalysts were also used for the hydrogenation of more complex substrates. The γ,δ-olefin bond of sorbinaldehyde was reduced with a high degree of selectivity (>98%). Pulegone and (+)-carvone were also hydrogenated with a complete conversion and a 1 : 1 diastereoisomeric ratio.

In a similar approach, Chaudret and coworkers used sulfonated NHC ligands for the synthesis of stable water-soluble palladium nanospecies through three different approaches (thermal decomposition, reduction under ^{13}CO atmosphere,

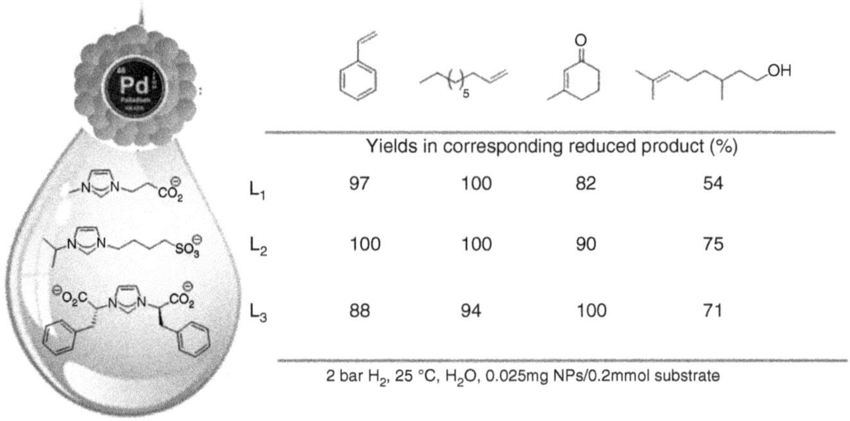

Figure 3.4 Water-soluble Pd NPs stabilized with sulfonated NHC ligands as relevant catalysts for hydrogenation of olefins. Source: Adapted from Ferry *et al.* [17].

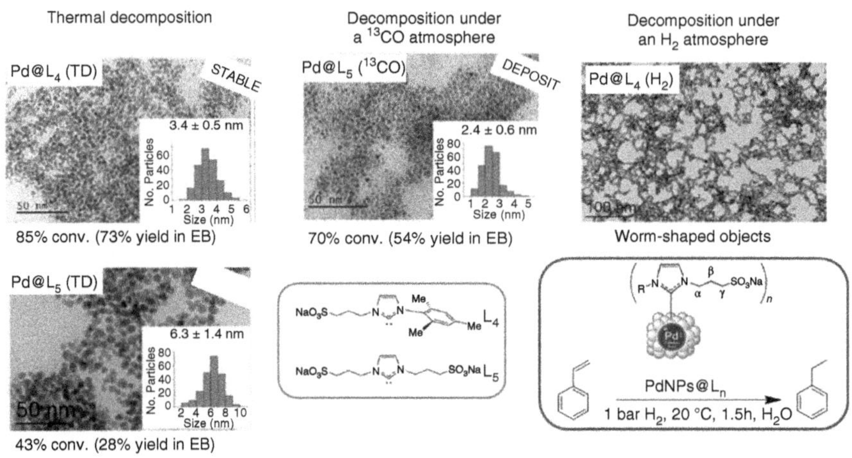

Figure 3.5 Comparison of sulfonated NHC-stabilized Pd NPs prepared by various methods. Investigation in styrene hydrogenation. Source: Adapted from Asensio *et al.* [18].

and reduction with H_2) [18]. According to the method, particles with various shapes and sizes, ranging from 1.5 to 7 nm, were obtained and compared in the chemoselective hydrogenation of styrene (Figure 3.5). Best catalytic results were achieved with Pd@L_4 catalyst prepared by thermal decomposition of dimethyl palladium complexes, affording a 97% conversion after three hours without precipitation of metal Pd. This catalyst could be recycled up to 10 times without significant loss of the catalytic activity.

3.2.2 Suzuki–Miyaura Coupling Reactions

3.2.2.1 Nitrogenated Ligands

Wang and coworkers reported the easy synthesis of an ammonium-functionalized bidentate imidazole-based ligand (1-[*N*,*N*′,*N*″-trimethyl-(4-butyl)ammonium]-2-

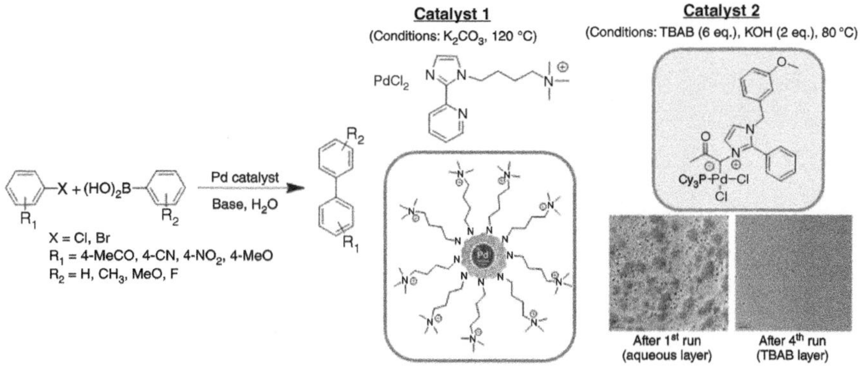

Figure 3.6 Water-soluble Pd NPs for Suzuki–Miyaura coupling reaction. Source: Adapted from Lee *et al.* [14].

(2-pyridyl) imidazole chloride) and its corresponding palladium complex as a catalyst precursor or a mere reservoir of Pd(0) for carbon–carbon coupling reactions (Figure 3.6, Catalyst 1) [16]. After the catalytic reaction between bromoacetophenone and phenylboronic acid at 120 °C, palladium nanospecies with size diameters of about 3 nm were observed. The obtained catalysts were highly efficient for the Suzuki–Miyaura reaction of substituted aryl chlorides and bromides with phenylboronic acid derivatives at 120 °C in water, with yields ranging from 60% to 100% according to the substituents on the substrates.

3.2.2.2 Carbonaceous and Phosphorous Ligands

Lee *et al.* prepared a robust and highly electron-rich zwitterionic complex based on a dichloropalladium(II) complex, possessing a tricyclohexylphosphine and an imidazolium-based carbon donor as ligands [14]. The in situ formed nanoparticles were investigated for the Suzuki–Miyaura coupling reaction in water with a large scope of substrates, such as aryl-bromides or chlorides and sterically hindered arylboronic acid derivatives (Figure 3.6, Catalyst 2). The use of *n*-tetrabutylammonium bromide (TBAB) was required as a coprotective agent of particles in water, as well as a phase transfer agent for substrate solubility. Moreover, this ammonium improves the catalyst's durability with a possible reuse over several runs, with similar catalytic performances (yields up to 100%). The authors observed a transfer of the particles in the liquid TBAB layer after the fourth run. However, one drawback of this system remains the high metal loading (2 mol%).

3.3 Stabilization by Surfactants

Surfactants are well-suited protective agents to stabilize metal nanospecies in water, affording rather strong electrosteric interactions with the particle surface that avoid their aggregation. Among them, ammonium-type surfactants, which possess a highly hydrophilic hydroxylated polar head combined with a sufficiently lipophilic

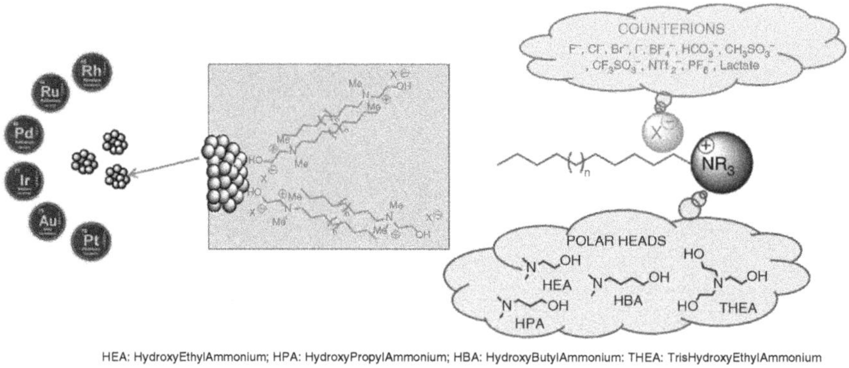

Figure 3.7 *N*-(Hydroxyalkyl)-*N*-alkylammonium salts as efficient bilayer protective agents of various metal nanoparticles in water. Source: Adapted from Denicourt-Nowicki and coworkers [21, 22].

chain, proved to be relevant particle stabilizers because of the formation of an electrosteric double layer around the metal cores (Figure 3.7) [21, 23].

For that purpose, the group of Roucoux has developed a library of easily accessible *N*,*N*-dimethyl-*N*-alkyl-*N*-(2-hydroxyethyl)ammonium salts (HEA salts) with various alkyl chain lengths (up to 18 carbon atoms), possessing mono- or poly-hydroxylated polar heads (HEA, HPA, HBA, and THEA) and different counteranions X^- with X = Br, Cl, I, F, CH_3SO_3, BF_4, HCO_3, and CF_3SO_3 (Figure 3.7) [21, 22]. These surfactants were efficiently used as protective agents for nanoparticles of diverse metals (rhodium, palladium, ruthenium, iridium, gold, platinum, etc.). These aqueous suspensions of metal NPs were usually synthesized by chemical reduction of the corresponding chloride metal salts with sodium borohydride in dilute aqueous suspensions of quaternary hydroxylated ammonium salts and were investigated in various polyphasic catalytic reactions, including hydrogenation and oxidation reactions.

3.3.1 Hydrogenation Reactions

A systematic study on the influence of the surfactant features, including the length of the lipophilic chain [24], the nature of the counterion [23, 25], and the polar head [26, 27], on the stability and activity of the obtained nanocatalysts was performed in arene hydrogenation in neat water under mild conditions. Some of the relevant catalytic performances are shown in Figure 3.8. First, among the noble metals investigated, rhodium nanospecies proved to be the most active [22], while ruthenium and iridium ones required higher pressure conditions [28, 29]. Concerning the lipophilic chain of the surfactant, only 16- or 18 carbon-containing chains afforded efficient stabilization of the particles within the aqueous phase, compared to C12 and C14 ones, with higher catalytic performances of C16 for anisole hydrogenation [24]. The influence of the counterion of the HEA16 surfactants on the rhodium NPs features (surface tension, morphology, and size of the nanospecies) was also studied [23]. These compounds were easily synthesized by a one-step ion exchange reaction,

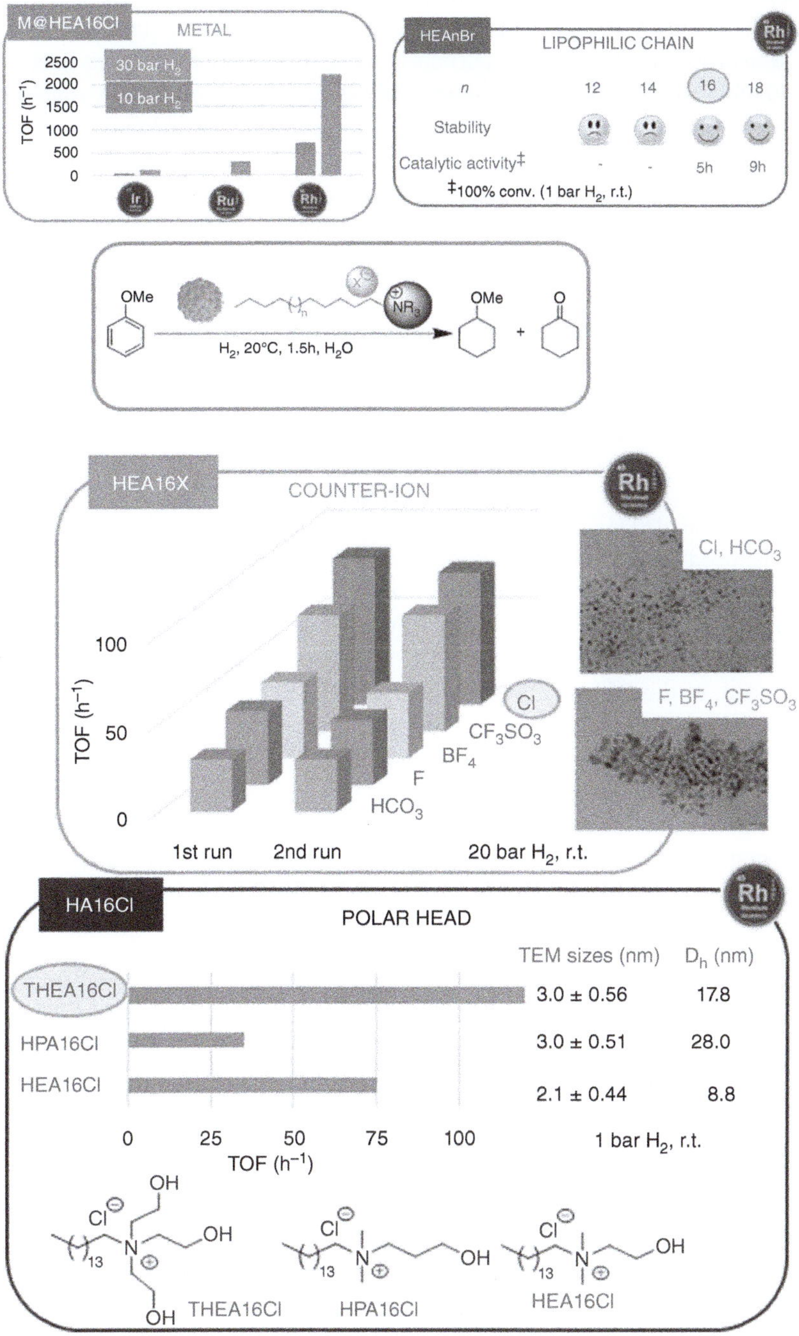

Figure 3.8 Optimization of the surfactant features in the reduction of anisole with noble NPs. HEA: HydroxyEthylAmmonium, THEA: TrisHydroxyEthylAmmonium, and HPA: HydroxyPropylAmmonium. Source: Adapted from Denicourt-Nowicki and coworkers [22, 23].

starting from HEA16Cl salt. According to the weak stabilization at the metal surface, fluorinated ions (such as fluoride F^-, tetrafluoroborate BF_4^-, or triflate $CF_3SO_3^-$) gave rise to worm-like rhodium(0) particles, whereas spherical colloids were observed with chloride (Cl^-) and hydrogenocarbonate (HCO_3^-) ions being stronger protective agents. The spherical particles arising from HCO_3^- anion provided higher catalytic activity compared to the $CF_3SO_3^-$ one. This can be explained by the different nanoparticle sizes and morphologies (spherical or dendrites, respectively) exposing different active sites. Rhodium NPs coated with HEA16CF_3SO_3 and HEA16BF_4 afforded lowest activities, considering their extended dendrite morphology where higher energy facets possess lower density. Best results in terms of catalytic performances were achieved with rhodium particles stabilized with HEA16Cl, which also allowed an efficient catalyst recycling. The influence of the polar hydroxylated head of the surfactant on physicochemical and catalytic properties was also investigated [26, 27]. Based on a cetylammonium chloride salt as common skeleton, two novel surfactants bearing either a longer polar chain (HPA16Cl for *N,N*-dimethyl-*N*-cetyl-*N*-(3-hydroxypropyl)ammonium chloride) or a trishydroxylated head (THEA16Cl for *N*-cetyl-*N*-tris-(2-hydroxyethyl)ammonium chloride) were used to prepare rhodium(0) particles and compared to the usual HEA16Cl stabilizer in terms of geometry, steric hindrance, surface area per molecule, and dynamic sizes. In the anisole hydrogenation at room temperature (r.t.) and 1 bar of H_2, the catalytic activity was found to be influenced by the nature of the polar head, in the following order: THEA16Cl > HEA16Cl > HPA16Cl. This tendency was attributed to the larger hydrodynamic diameter of rhodium(0) species stabilized by the more longer hydroxypropyl head group (HPA), which can decrease the substrate's accessibility onto the metallic surface.

Based on this optimization work, the Rh@HEA16Cl catalyst was evaluated in the hydrogenation of various arenes, including alkylated, functionalized, and disubstituted ones [26], oxygen or nitrogen-containing heteroaromatics [30], and halogenoarenes [31] under mild biphasic conditions, with outstanding turnover catalytic activities (TOF h^{-1} values). An overview of the results obtained is presented in Figure 3.9.

Concerning disubstituted arenes such as xylene isomers, the selectivity is in favor of the thermodynamically less stable *cis*-product, as generally observed in heterogeneous catalysis [32, 33], owing to a "continuous" coordination of the substrate at the metal surface and consequently the addition of the dihydrogen *via* only one face of the substrate [34, 35]. In the case of halogenoarenes, such as chlorobenzene or chloroanisole, a concomitant dehalogenation occurred within the reduction of the aromatic ring with rhodium(0) nanoparticles [31]. This rhodium-catalyzed tandem process is particularly relevant for the removal of toxic and persistent halogeno-anisole or phenol derivatives present in the industrial effluents or polluted groundwater. In many cases, the catalyst could be reused by a simple extraction of the organic phase with an appropriate solvent (alkanes or ether) over several runs, with similar catalytic performances. In the case of catalyst aggregation, the colloidal suspensions could be easily heterogenized onto inorganic supports such as silica or titanium dioxide [36, 37], as well as in mesoporous materials

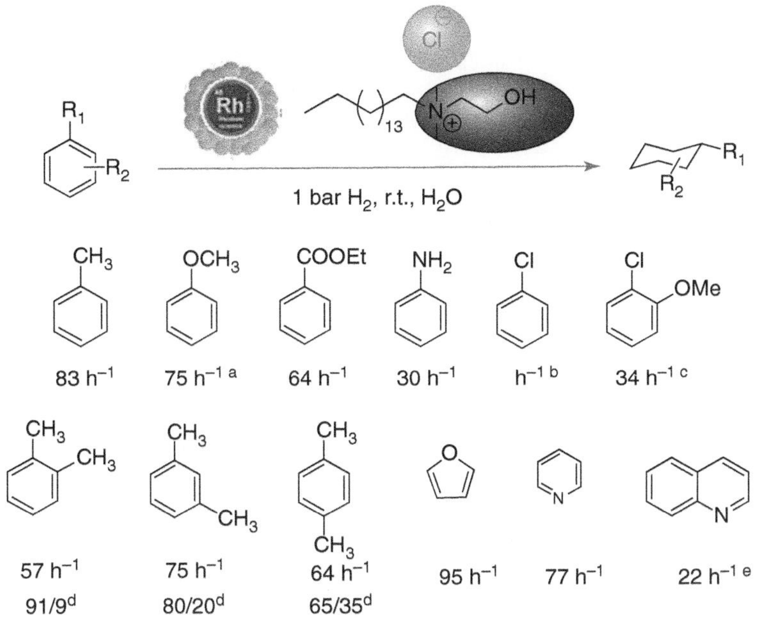

[a] With 30% cyclohexanone formed, [b] Cyclohexane obtained, [c] 53% methylcyclohexane and 47% cyclohexane; [d] Cis major isomer, [e]1,2,3,4-tetrahydroquinoline

Figure 3.9 Turnover catalytic activities (TOF h^{-1} values) in the reduction of various arenes catalyzed by Rh@HEA16Cl.

[38] or nonfunctionalized magnetic supports (γ-Fe_2O_3) [39, 40]. Finally, it can be underlined that host–guest inclusion complexes between hydroxylated ammonium surfactants and cyclodextrins were successfully used to protect ruthenium or rhodium NPs [41], providing interesting catalysts for arene hydrogenations.

The use of effective catalytically active nanospecies for asymmetric reactions is a challenging research area [42]. Based on their methodology, namely, the use of hydroxylated ammonium salts as nanoparticle stabilizers, the group of Roucoux has followed two main approaches for the asymmetric hydrogenation of ethylpyruvate in neat water. On the one hand, they synthesized a library of water-soluble ammonium salts based on an optically active skeleton derived from *N*-methylephedrine or *N*-methylprolinol, associated with various counterions (Br^-, HCO_3^-, and (*S*)- or (*R*)-lactate) (Figure 3.10) [44, 45]. Aqueous suspensions of rhodium NPs, prepared by the usual approach, with a size diameter between 1.75 and 3 nm, were evaluated in the reduction of ethylpyruvate at r.t. under 40 bar of H_2. A promising asymmetric induction up to 13% was achieved, using optically active NMeEph12X salts with a dodecyl lipophilic chain, in combination with bromide or (S)-(−)-lactate ion. These nanocatalysts also proved efficient for the reduction of *m*-methylanisole, a challenging transformation to access optically enriched cyclohexyl derivatives. High cis selectivity was achieved, but with no asymmetric induction. On the other hand, colloidal suspensions of platinum were prepared by chemical reduction of H_2PtCl_6 with

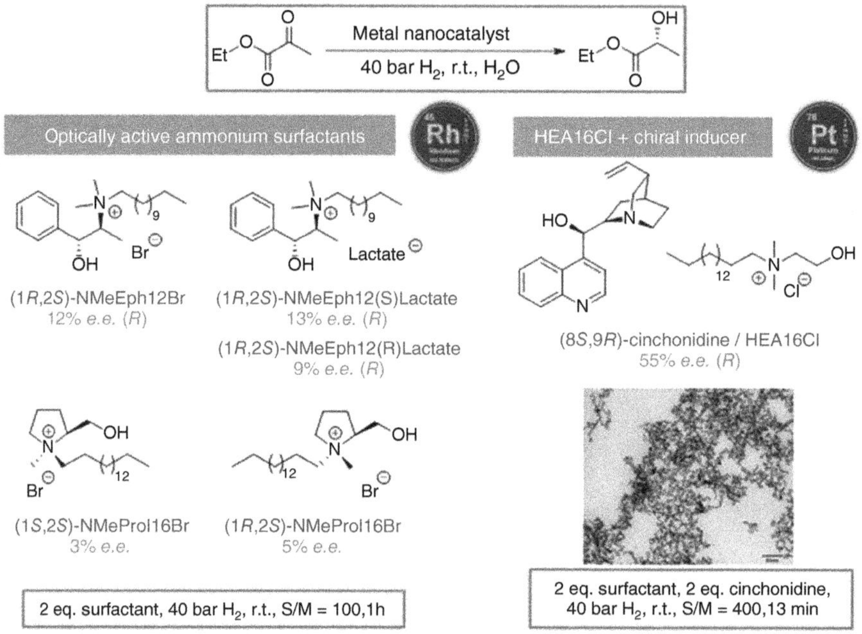

Figure 3.10 Optically active aqueous suspensions of metal nanoparticles – Toward asymmetric hydrogenation of ethylpyruvate. Source; Adapted from Guyonnet Bilé *et al* [45] and from Mévellec *et al.* [43].

dihydrogen in the presence of the usual HEA16Cl surfactant, followed by the adsorption of the optically pure cinchonidine onto the metal surface. The so-obtained catalyst, stabilized with the ammonium surfactant and modified with the *cinchona* salt, proved to efficiently transform ethylpyruvate in (*R*)-ethyllactate with a complete conversion and up to 55% enantiomeric excess under 40 bar of H_2 in only 13 minutes [43]. Moreover, the catalyst was easily recycled and reused with no loss of catalytic performances, leading upon time to a high-order superstructure. Unfortunately, the use of randomly methylated cyclodextrins grafted with chiral amino acid moieties, alone or in combination with optically active ammonium surfactants, as stabilizers for ruthenium nanoparticles did not allow to achieve asymmetric induction [46, 47].

3.3.2 Oxidation Reactions

More recently, these ammonium surfactants bearing a hydroxylated polar head were used as protective agents of ruthenium nanospecies for oxidation of terpenic olefins in water [48]. The aqueous suspensions of Ru(0) species were prepared starting from ruthenium chloride ($RuCl_3{\cdot}3H_2O$). A systematic study was carried out using hydroxyethylammonium (HEA) derivatives with various counterions (Cl, Br, BF_4), a trishydroxyethyl ammonium (THEA) surfactant, two *N*-oxide (NO12 and NO16) compounds, and the commercially available CTABr surfactant and PVP polymer for comparison purpose (Figure 3.11). The obtained ruthenium colloids (*ca.* 2 nm) proved to be active and selective in the oxidation of α-pinene in the

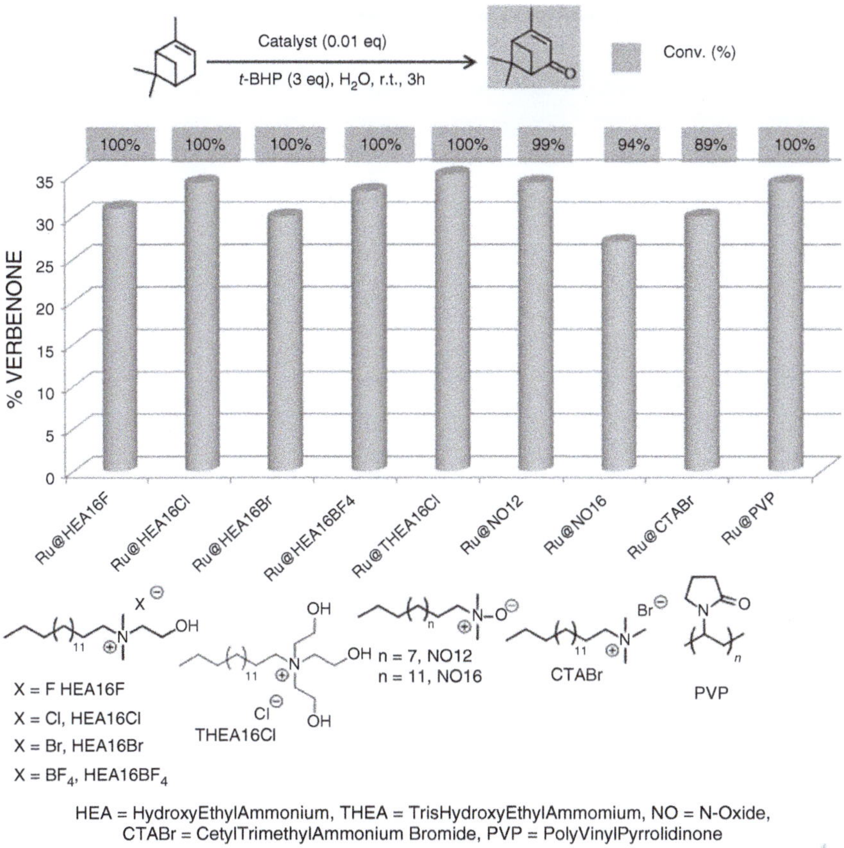

Figure 3.11 Ammonium surfactant-stabilized Ru NPs for oxidation of α-pinene **1** in neat water. Source: Adapted from Rauchdi *et al.* [48].

presence of *tert*-butylhydroperoxide in neat water and at r.t. As a product of great interest for fine chemistry, verbenone was obtained as a major product with a yield up to 41%, in the presence of Ru@HEA16Cl catalyst. This catalytic system was extended to valencene, affording the corresponding nootkatone with a 60% isolated yield in water at room temperature. This result is quite similar to the one achieved with a Cu–Al–Ox catalyst in acetonitrile at 80 °C [49]. This oxyfunctionalized molecule constitutes an important flavoring constituent because of its citrusy, woody, peely, and grapefruit-like aroma profile [50] and a natural insect repellent [51].

3.3.3 Other Reactions

In an original way, TBAB-coated palladium nanospecies were proved active for the chemoselective cyclization reaction of a large scope of α-allenol derivatives in water, leading to dihydrofurans and carbazoles as promising synthons for fine chemistry. The catalytic performances obtained with pregenerated particles

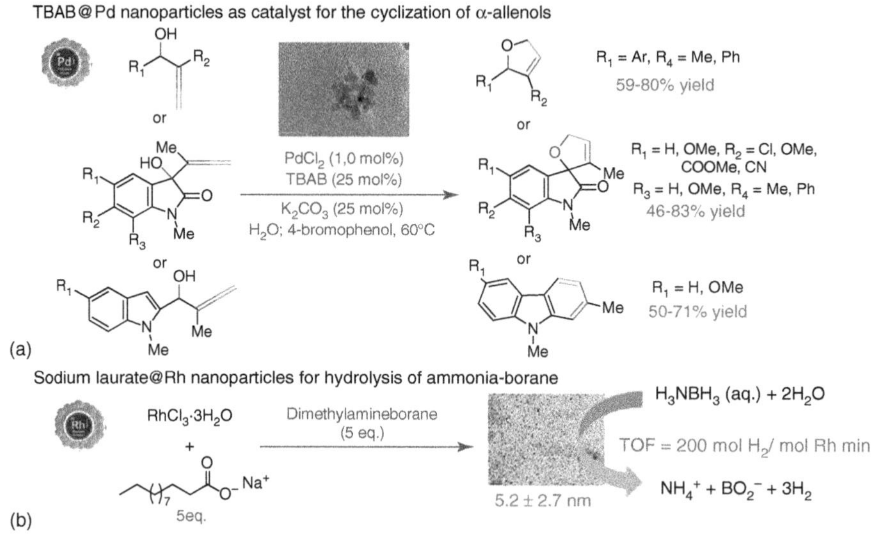

Figure 3.12 Surfactant-stabilized metal nanoparticles in other catalytic applications. (a) TBAB@Pd nanoparticles as a catalyst for the cyclization of α-allenols. (b) Sodium laurate@Rh nanoparticles for hydrolysis of ammonia-borane. Source: (a) Adapted from Alcaide *et al.* [52]. (b) Adapted from Durap *et al.* [53].

were higher than those achieved with palladium precursor ($PdCl_2$) or in situ generated nanoparticles, combined with an efficient durability over several runs (Figure 3.12a) [52]. However, the addition of bromophenol as a costabilizing agent seems unavoidable.

Sodium laurate-protected rhodium NPs, possessing a mean size of 5.2 nm, were synthesized in situ from the dimethylamine-borane reduction [54]. These catalysts were found highly effective and durable in the hydrolysis of ammonia-borane at r.t. and even at low catalyst concentration, with a total turnover number (TTO) of 80 000 mol H_2/mol Rh and a turnover frequency (TOF) of 200 mol H_2/mol Rh min (Figure 3.12b).

3.4 Stabilization by Polymers

Water-soluble polymers also constitute a suitable library of stabilizing agents for metal nanoparticles, providing a steric protection in water. The nanocatalysts obtained can be applied in various catalytic reactions including hydrogenation, carbon–carbon coupling, and oxidation reactions (Table 3.2).

3.4.1 Hydrogenation Reactions

Polyvinylpyrrolidone (PVP)-stabilized metallic nanoparticles are among the most used nanocatalysts, demonstrating excellent activities and stabilities particularly in hydrogenation reactions, as shown in Figure 3.13. In fact, well-defined

Table 3.2 Overview of recent catalytic applications with polymer-stabilized metal nanoparticles in neat water.

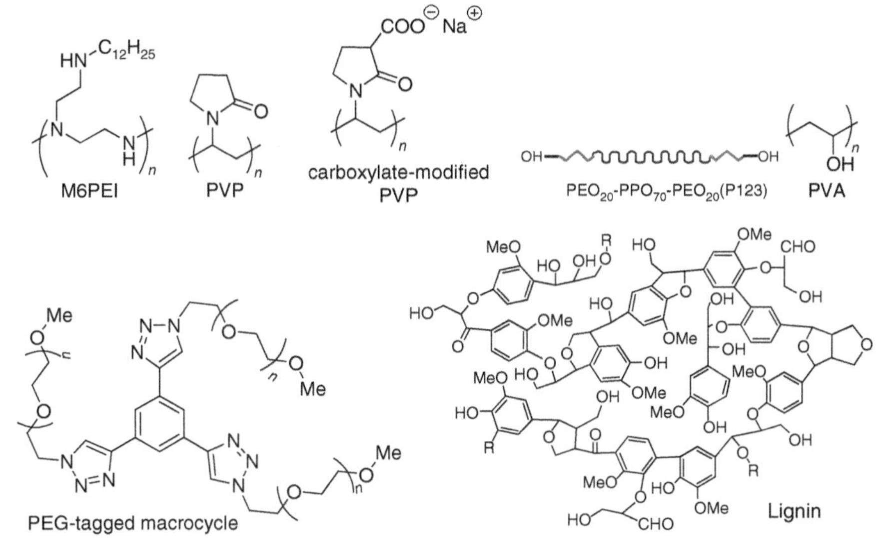

Reactions	Polymers	Metal	References
Hydrogenation of olefins	Polyethylenimine (M6PEI), pluronic block P123 copolymer (P123), dendrimers, cucurbit[6]uril	Pd, Ru, Pt	[55–58]
Hydrogenation of arenes	PVP, carboxylate-modified PVP, cucurbit[6]uril	Rh	[53, 59, 60]
Hydrogenation of nitroarenes	PVA, cucurbit[6]uril	Pd	[58, 61]
Hydrogenation of epoxides	Cucurbit[6]uril	Pd	[58]
Hydrogenation of CO	PVP, cucurbit[6]uril	Rh, Pd	[58, 59]
C–C coupling reactions	PEG, cucurbit[6]	Pd	[62]
Stille reaction	Dendrimers	Ni	[63]
Tsuji–Trost allylation	PVP	Pd	[64]
Oxidation	PVP	Au–Pt–Ag	[65, 66]

PVP-protected rhodium(0) nanospecies, with a mean size diameter of 2.2 nm, were synthesized by the decomposition under hydrogen of the organometallic $[Rh(\eta^3\text{-}C_3H_5)_3]$ complex in THF in the presence of PVP [59]. This nanocatalyst afforded excellent activity and selectivity in the hydrogenation of aromatics possessing electron-donating or -withdrawing substituents, in water under mild conditions (r.t., 1 bar of H_2), with conversions from 74% to 100%. Selectivities superior to 97% were achieved for substrates presenting another reducible function (*i.e.* oxo group), except for acetophenone with the major formation of

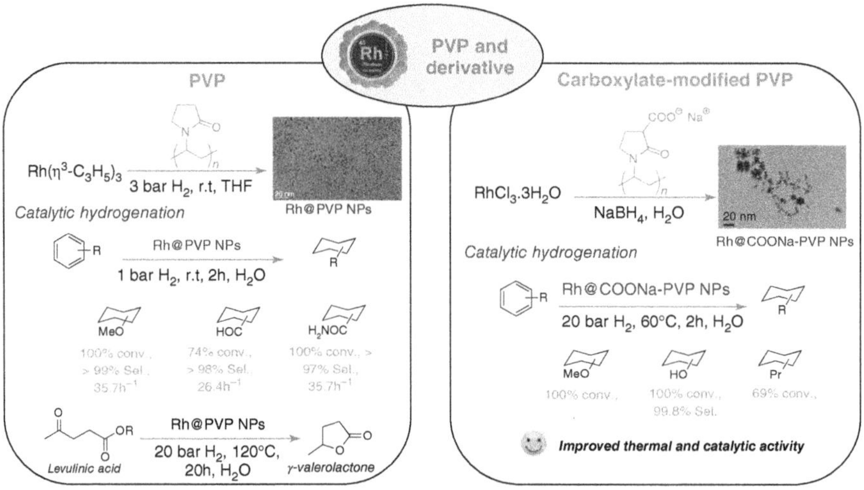

Figure 3.13 PVP and carboxylate-modified PVP-coated Rh NPs for arene hydrogenation. Source: (a) Adapted from Ibrahim *et al.* [59]. (b) Adapted from Yan *et al.* [53].

1-phenylethanol. The Rh@PVP catalyst was also active for the reduction of levulinic acid and its corresponding methyl ester, leading to the quantitative production of γ-valerolactone, a value-added platform for chemical, flavor, and biofuel industries, under moderate reaction conditions (120 °C, 20 bar of H_2).

In a similar approach, Dyson and coworkers developed a carboxylate-modified PVP, which allows an increased stability of the rhodium colloids compared to their PVP-coated analogs, because of the weak coordination of the carboxylate group at the metal surface. This additional electrostatic stabilization with the formation of a protective electrical double layer was assessed by an increase in the zeta potential value from −20 mV for PVP to −64 mV for the modified one [53]. This catalyst was active in the hydrogenation of various functionalized arenes under 20 bar of H_2 and at 60 °C, with an improved thermal and catalytic stability upon recycling, compared to PVP system for which a loss of activity was observed after the fifth run.

This team also studied the influence of water-soluble phosphines on the chemoselectivity in the hydrogenation of phenylacetone catalyzed by PVP-coated rhodium nanospecies (Figure 3.14) [60]. After optimization, a relevant chemoselectivity toward the reduction of the aromatic ring was increased from 70% (with no ligand) to 92% with $\mathbf{P_c}$, probably owing to site blocking effects as well as to the hydrophobicity of this ligand, which avoids the approach of the carbonyl group.

Very recently, palladium nanoparticles confined in polyvinylalcohol (PVA) were synthesized by reduction of palladium acetate under dihydrogen, with a mean size of *ca.* 4.8 nm [61]. They proved active for the selective reduction of nitrobenzene into aniline with a 99.3% conversion in three hours, under 20 bar of H_2 and at 45 °C, with a high substrate/metal ratio of 500 (Figure 3.15). Moreover, this catalytic system could be easily recycled without any reactivation treatments, thereby with a decrease in the catalytic activity.

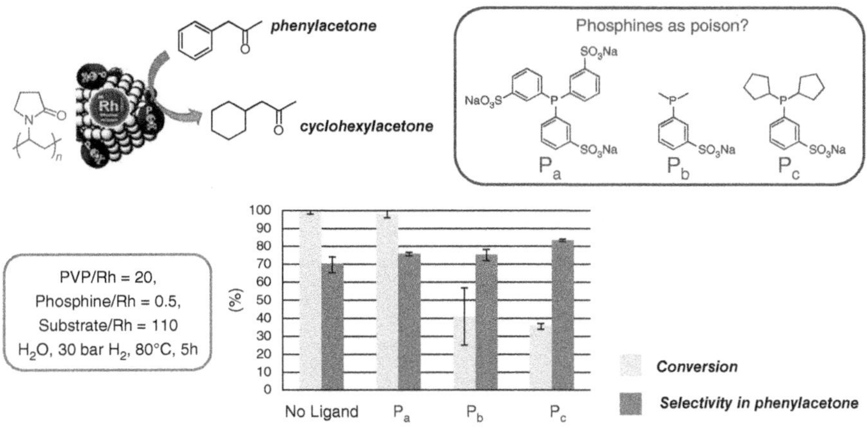

Figure 3.14 Influence of various phosphines on PVP-coated Rh NPs on the chemoselectivity in phenylacetone hydrogenation. Source: Adapted from Snelders *et al.* [60].

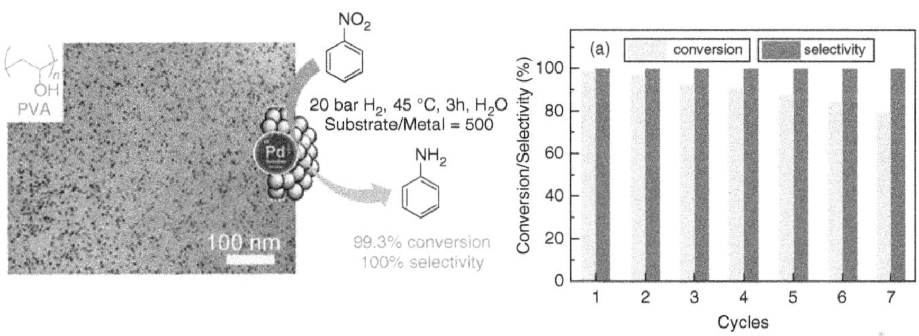

Figure 3.15 PVA-protected palladium colloids for the selective hydrogenation of nitrobenzene. Source: Adapted from Wang *et al.* [61].

A cheap and nontoxic polyoxoethylene–polyoxopropylene–polyoxoethylene triblock copolymer (P123) was used as a protective agent of ruthenium NPs in water [55]. Optimization studies showed that P123 with a molecular weight of 5800 gave the best compromise between efficient stability within aqueous phase and relevant catalytic performances in the hydrogenation of α-pinene into *cis*-pinane with a complete conversion (99.9%) and a 98.9% selectivity (Figure 3.16). This catalyst could be efficiently recycled by a biphasic approach over eight runs, before a drop of the conversion.

In 2016, supramolecular self-assemblies of polyethylenimine (M6PEI) amphiphilic polymer with the previously described hydroxyethylammonium (HEA16Cl) surfactant were used as stabilizers for aqueous palladium NPs [56]. The size of the metal cores and the hydrodynamic diameter of the nanospecies were determined by transmission electron microscopy (TEM) and small-angle X-ray scattering (SAXS) and dynamic light scattering (DLS), as shown in Figure 3.17. The catalytic performances of these Pd NPs were investigated in the biphasic hydrogenation of alkenes (i.e. cyclooctene) and α,β-unsaturated ketones (i.e.

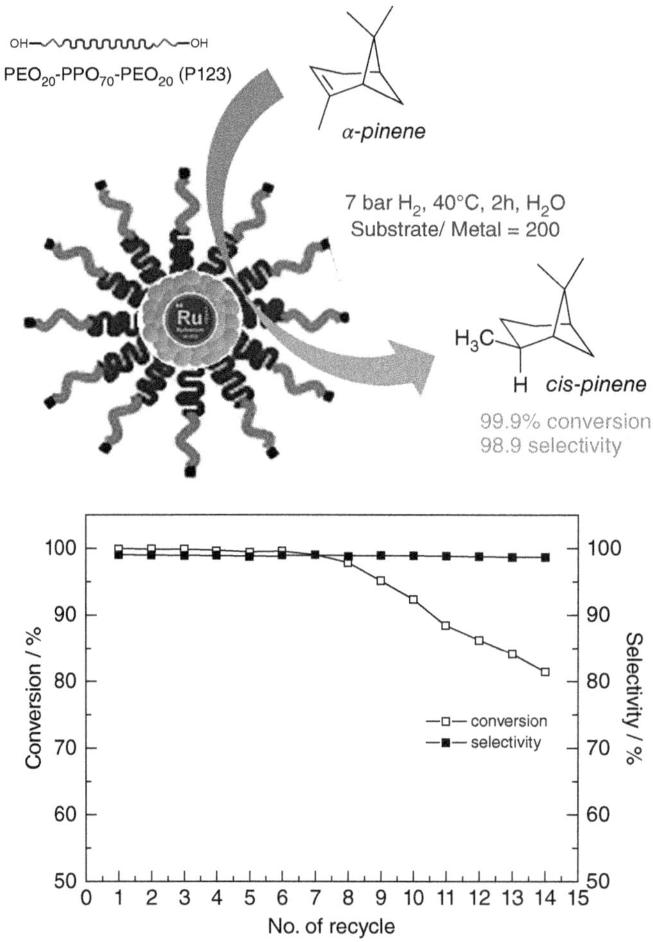

Figure 3.16 P123-protected ruthenium(0) NPs for the reduction of α-pinene into cis-rich pinane. Source: Adapted from Hou *et al.* [55].

5-methylhex-3-en-2-one) under mild conditions (r.t., 1 bar H_2). A comparison study was performed in order to determine the influence of various mixtures of HEACl-M6PEI with those of M6PEI or HEA salts alone as references. This study showed that the specific activity (SA), determined on the basis of an optimized time for complete conversion, increases in the following order: M6PEI < Mix 3 < Mix 2 < Mix 1 < HEA16Cl (Figure 3.17). Although best catalytic performances were obtained with Pd@HEA16Cl catalyst owing to a better access of the substrates to the particle surface according to a dynamic behavior because of surfactant fluxional exchanges, these self-assemblies of an ammonium salt within a polyethylenimine provided high stability and relevant catalytic activities.

Platinum nanospecies, coated by a glycol-dendrimer bearing modified xylose branches, were also active for the hydrogenation of the olefinic bond of α,β-unsaturated ketones, such as isophorone and (*R*)-(+)-pulegone in water, under

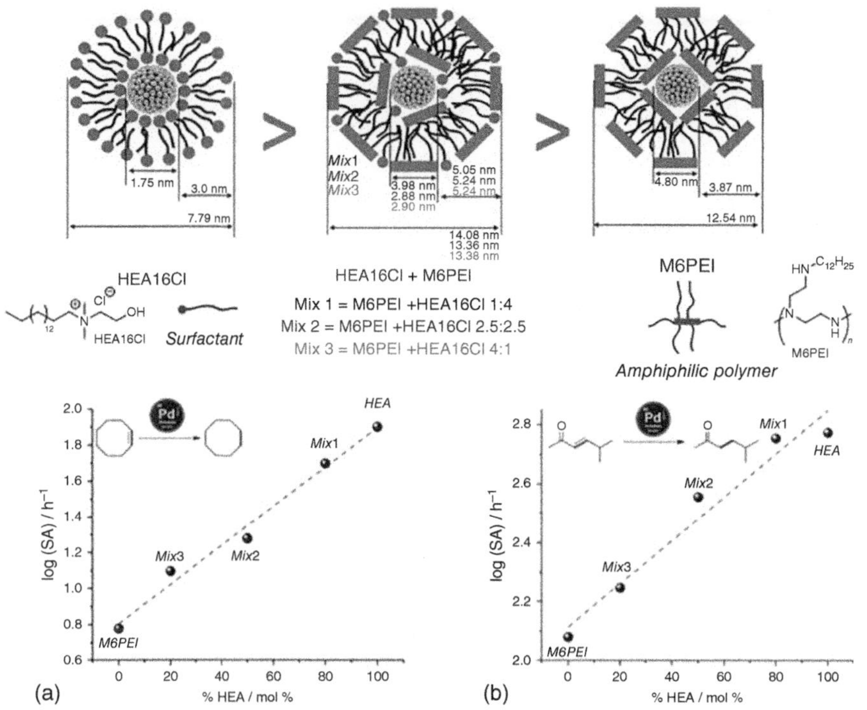

Figure 3.17 M6PEI/HEA16Cl self-assembly-stabilized Pd(0) colloids. Correlation of specific activity (SA) for the reduction of (a) cyclooctene and (b) 5-methyl-hex-3-en-2-one according to HEA16Cl amount. Source: Adapted from Albuquerque *et al.* [56].

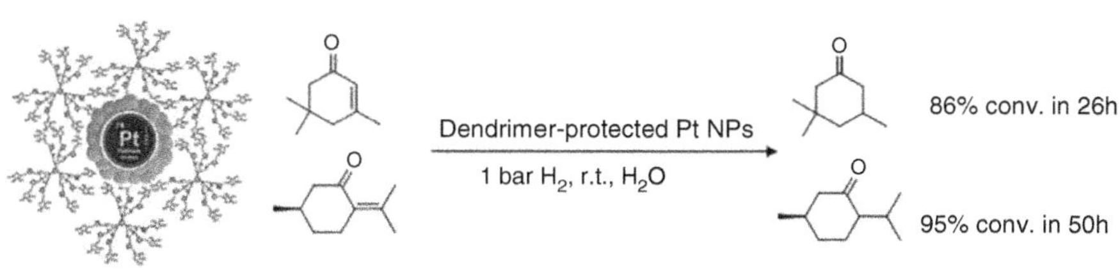

Figure 3.18 Glycodendrimer-stabilized Pt nanoparticles as a catalyst for hydrogenation of olefins. Source: Adapted from Gatard *et al.* [57].

1 bar of H_2, at r.t. under long reaction times (Figure 3.18) [57]. For comparison, the Pd@HEA16Cl nanocatalyst, previously described, afforded the corresponding saturated ketones with a 90% conversion in less than one hour.

Cucurbit[6]uril (CB[6]), a macrocycle in which six glycoluril units are joined with twelve methylene bridges, was reported to stabilize metal NPs, either by forming a passive layer or acting as a protecting agent [58]. A CB[6] material loaded with 3 nm-sized palladium nanospecies (Pd content: 2.5 wt%) was obtained by an impregnation method of the $PdCl_2$ precursor on the support, followed by a reduction step. This catalyst showed remarkable conversion, selectivity (>99%), and high TOF (up to 205 882 h^{-1}) for the chemoselective hydrogenation of various

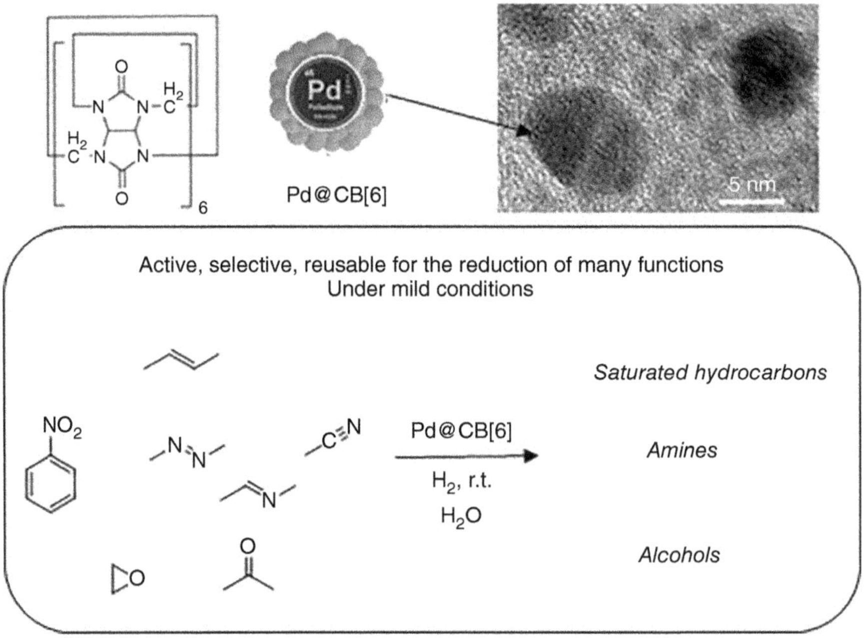

Figure 3.19 Cucurbit[6]uril-coated Pd NPs for hydrogenations. Source: Adapted from Nandi *et al.* [58].

reducible groups, including epoxides, olefins, carbonyls, nitriles, azo, imines, and nitroarenes (Figure 3.19). The durability of the catalyst was successfully studied on several runs (>10 times) with any significant loss in conversion nor selectivity.

3.4.2 Carbon–Carbon Coupling Reactions

Carbon–carbon coupling reactions using palladium nanospecies as catalysts constitute key reactions of organic chemistry.

PEG-coated palladium NPs, easily prepared from a Fischer carbene complex, proved effective for a wide range of carbon–carbon coupling reactions such as Suzuki, Heck, and copper-free Sonogashira and Stille reactions with good catalytic activities (Figure 3.20) [62].

Nearly at the same time, the team of Pleixats reported the use of a PEG-tagged compound, possessing three long polyoxyethylenated chains and potentially coordinating triazole moieties as a protective agent for palladium nanospecies [67]. The so-obtained nanoparticles showed pertinent activities in the Suzuki coupling reactions of aryl iodides, as well as the more challenging bromides, with a large scope of aryl and heteroaryl substrates (Figure 3.21).

Polyvinylpyrrolidone (PVP)-stabilized palladium NPs, obtained by a polyol reduction method under microwave, were successfully investigated in the Tsuji–Trost allylation of a series of ethyl acetoacetates with allyl methyl carbonates in water, with complete conversion in a few hours at a very low metal loading (Figure 3.22)

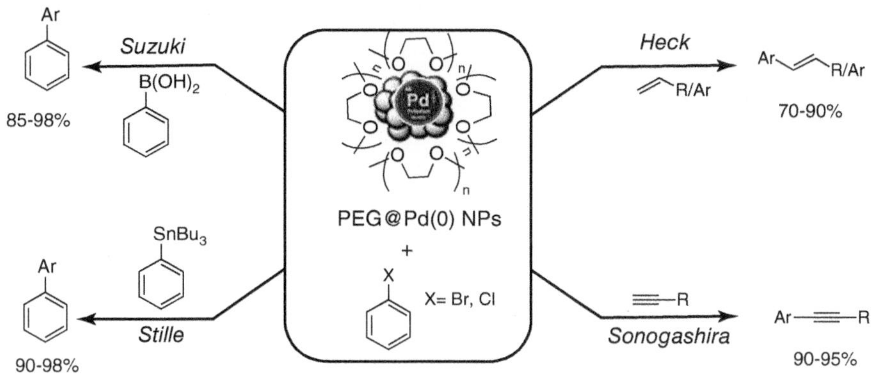

Figure 3.20 PEG@Pd NPs, efficient catalyst for carbon–carbon couplings. Source: Adapted from Sawoo *et al.* [62].

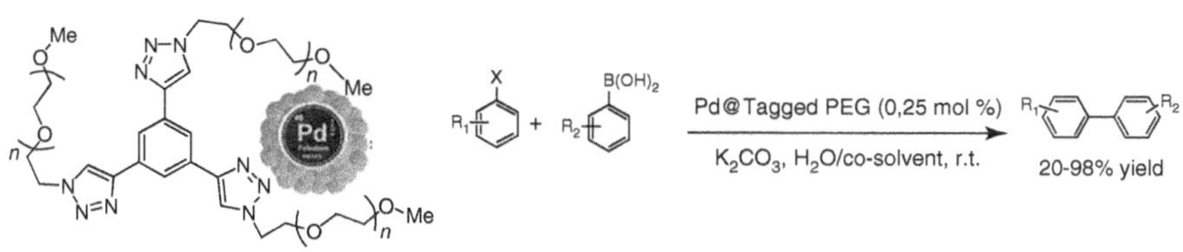

Figure 3.21 PEG-tagged capping agent of Pd NPs for Suzuki couplings. Source: Adapted from Mejías et al. [67].

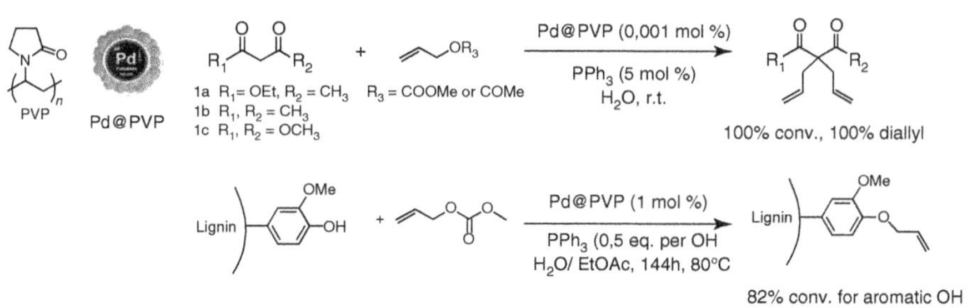

Figure 3.22 PVP-protected Pd NPs for Tsuji–Trost allylations. Source: Adapted from Llevot *et al.* [64].

[64]. The addition of triphenylphosphine was crucial to help the transfer of the nanospecies from aqueous media to the aqueous/organic interphase. Furthermore, this catalyst was also active in the allylation of lignin with a higher selectivity toward the aromatic OH groups (with conversion up to 82%).

Dendrimer (G_3DenP)-coated nickel colloids, prepared by chemical reduction of the nickel dichloride with sodium borohydride, proved active and recyclable for the Stille coupling reaction in neat water, without cocatalyst [68]. This catalytic system was applied to a large scope of trichlorophenylstannanes with a series of bromo or chloroarenes, leading to the formation of the corresponding biaryls with good to excellent yields (Figure 3.23).

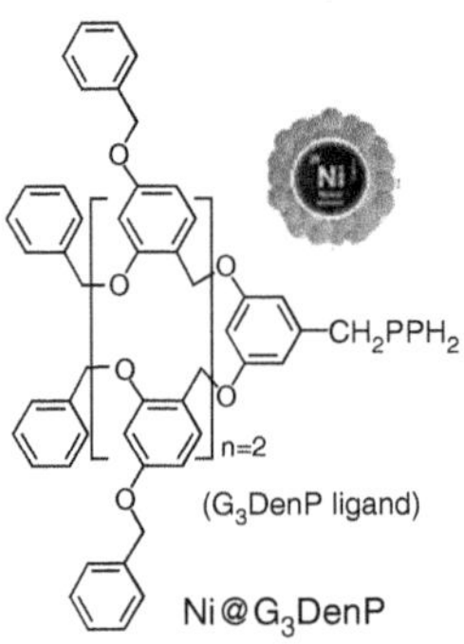

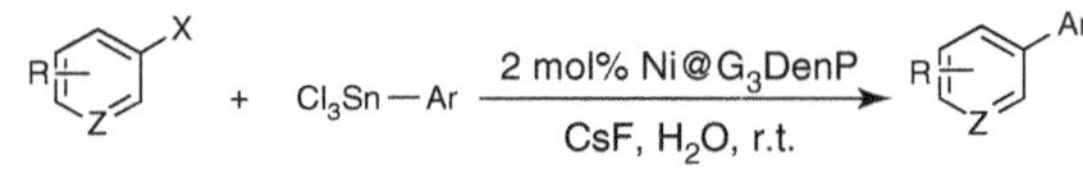

R	X/Z	Ar	t (h)	Yield (%)
p-COOMe	Br/C	Ph	3	99
p-CH_3	Br/C	Ph	4	97
p-NMe_2	Br/C	Ph	4	98
p-COOMe	Cl/C	Ph	5	98
H	2-Br/N	Ph	6	89
o-CH_3	Cl/C	Thiophene	6	90

Figure 3.23 G_3DenP-capped Ni NPs for Stille coupling reactions. Source: Adapted from Wu *et al.* [68].

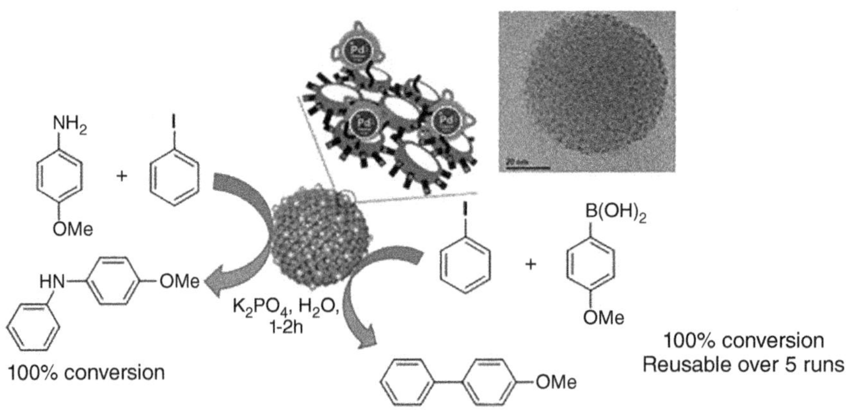

Figure 3.24 Pd@CB-PN-catalyzed C–C and C–N bond forming reactions. Source: Adapted from Yun et al. [63].

Stable aqueous dispersions of palladium NPs on CB[6] nanocapsules (PN), possessing "disulfide loops," were synthesized through a two-step process and proved active with complete conversion in one to two hours and reusable in carbon–carbon or carbon–nitrogen bond forming reactions in aqueous media (Figure 3.24) [63].

3.4.3 Oxidation Reactions

Bimetallic and trimetallic nanoparticles with various compositions and structures are particularly relevant in glucose oxidation compared to their monometallic counterparts. Among the various bi- or trimetallic alloys (Au–Ag, Au–Cu, Au–Pt, Au–Pd, Au–Rh, and Au–Pt–Ag) prepared by rapid injection at 0 °C of $NaBH_4$ into the corresponding metallic ion solutions in the presence of PVP as a protective agent, the Au_{70}–Pt_{80}–Ag_{10} trimetallic NPs (*ca.* 1.5 nm) showed a high and durable catalytic activity for the aerobic glucose oxidation (Figure 3.25) [65, 66], which

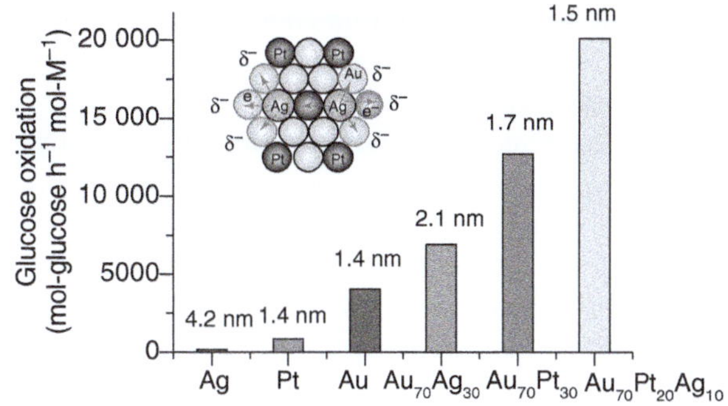

Figure 3.25 Comparison of various bi- or trimetallic alloys on the glucose oxidation. Source: Adapted from Zhang *et al.* [65, 66].

was 5 times higher than that of AuNPs and about 1.5–3 times higher than that of bimetallic alloys. The quite different catalytic performances observed for the trimetallic alloy were attributed to different factors, namely, the small particle size, long-term stability (due to thermodynamically stable alloyed structure), and the formation of negatively charged Au atoms (due to electron donation of Ag neighboring atoms) acting as catalytically active sites for the desired reaction.

3.5 Conclusions and Perspectives

In the current context of the development of eco-responsible chemical processes, the synthesis of metallic nanoparticles, stabilized in neat water, presents a great interest. The development of water-soluble capping agents, such as ligands (as used in homogeneous catalysis), ammonium surfactants, as well as polymers, has already led to promising results. Mainly noble metal nanospecies were synthesized through a bottom approach under quite mild conditions in the presence of these stabilizers. As reported in this chapter, many nanocatalysts showed relevant catalytic performances (both in terms of activity and selectivity) for diverse applications, including hydrogenation of various unsaturated substrates (arenes, olefins, and α,β-unsaturated ketones), oxidation processes, as well as carbon–carbon coupling reactions (such as Suzuki, Sonogashira, Heck, Stille reactions, etc.). In most cases, the studied catalytic systems could be easily recycled through a biphasic approach.

This chapter evaluates different wet chemical methods for the production of metal nanoparticles. This classical approach is relatively simple, modular, and scalable and leads to a set of interesting properties for target catalytic applications. Thus, during the past two decades, the scientific community has focused with success on the control over the nanostructures with well-defined composition, shapes, and sizes as well as selective surface or controlled crystallographic phases.

Based on the diversity of synthetic methodologies and the induced kinetic properties, catalysis and nanomaterials have been inextricably linked to each other for a long time. Although the synthesis of novel, highly efficient, poisoning-resistant, or low-cost catalysts to improve productivity and selectivity has evidenced as an economic priority, the environmental impact is a more recent concern. Consequently, the development of an ideally greener nanocatalyst, able to provide a large-scale production of fine chemical intermediates, as well as its scalable and adaptable synthesis in order to always satisfy the growing industrial demand, still remains a great challenge.

Along with reducing the environmental impact of organic synthesis by atom economy, energy efficiency, as well as waste prevention, the use of greener media remains a crucial parameter. Alternatives to usual organic solvents, such as 2-methyltetrahydrofuran, 1,3-dioxolane or polyethylene glycol, or the use of renewable solvents such as 1,3-propanediol or glycerol, have been investigated. Nevertheless, using water as a reaction medium seems to be an ideal solution because of its unique properties in terms of economy and environment. Water thus appears as a promising alternative to problematic or hazardous solvents by simplifying operations, allowing mild reaction conditions and enabling easy extraction and purification work-up of products from the reaction medium.

If the use of water as the reaction medium for industrial scale-up is still underway, the growing interest of the industry for cleaner processes provides good opportunities in order to combine nanoscience and catalysis with an emphasis on green chemistry concerns. Finally but not the least, the synthesis of nanometer-sized catalysts based on cheap, abundant first-row transition metals is certainly of great interest in the drive toward low environmental footprint processes.

References

1 Yan, N., Xiao, C., and Kou, Y. (2010). *Coord. Chem. Rev.* 254 (9): 1179–1218.
2 Denicourt-Nowicki, A. and Roucoux, A. (2013). *Nanomaterials in Catalysis* (eds. P. Serp and K. Philippot), 55–98. Weinheim, Germany: Wiley-VCH Verlag GmbH & Co. KGaA.
3 Bulut, S., Fei, Z., Siankevich, S. et al. (2015). *Catal. Today* 247: 96–103.
4 Ohtaka, A. (2019). *Curr. Org. Chem.* 23 (6): 689–703.
5 Blackmond, D.G., Armstrong, A., Coombe, V., and Wells, A. (2007). *Angew. Chem. Int. Ed.* 46 (21): 3798–3800.
6 Genin, E. and Michelet, V. (2015). *Green Process Engineering. From Concept to Industrial Applications* (eds. M. Poux, P. Cognet and C. Gourdon), 292–324. New-York: CRC Press, Tailor & Francis Group.
7 Simon, M.-O. and Li, C.-J. (2012). *Chem. Soc. Rev.* 41 (4): 1415–1427.
8 Zhou, F. and Li, C.-J. (2019). *Chem. Sci.* 10 (1): 34–46.
9 Prat, D., Wells, A., Hayler, J. et al. (2016). *Green Chem.* 18 (1): 288–296.
10 Velazquez, H.D. and Verpoort, F. (2012). *Chem. Soc. Rev.* 41 (21): 7032–7060.
11 Tyler, D.R. (2019). *Inorg. Chim. Acta* 485: 33–41.
12 Debouttière, P.-J., Coppel, Y., Denicourt-Nowicki, A. et al. (2012). *Eur. J. Inorg. Chem.* 2012 (8): 1229–1236.
13 Guerrero, M., Roucoux, A., Denicourt-Nowicki, A. et al. (2012). *Catal. Today* 183 (1): 34–41.
14 Lee, J.-Y., Tzeng, R.-J., Wang, M.-C., and Lee, H.M. (2017). *Inorg. Chim. Acta* 464: 74–80.

15 Bouriazos, A., Sotiriou, S., Stathis, P., and Papadogianakis, G. (2014). *Appl. Catal., B* 150–151: 345–353.

16 Zhou, C., Wang, J., Li, L. et al. (2011). *Green Chem.* 13 (8): 2100–2106.

17 Ferry, A., Schaepe, K., Tegeder, P. et al. (2015). *ACS Catal.* 5 (9): 5414–5420.

18 Asensio, J.M., Tricard, S., Coppel, Y. et al. (2017). *Chem. Eur. J.* 23 (54): 13435–13444.

19 Guerrero, M., Coppel, Y., Chau, N.T.T. et al. (2013). *ChemCatChem* 5 (12): 3802–3811.

20 Hopkinson, M.N., Richter, C., Schedler, M., and Glorius, F. (2014). *Nature* 510: 485.

21 Denicourt-Nowicki, A. and Roucoux, A. (2015). *Catal. Today* 247: 90–95.

22 Denicourt-Nowicki, A. and Roucoux, A. (2016). *Chem. Rec.* 16 (4): 2127–2141.

23 Guyonnet Bilé, E., Sassine, R., Denicourt-Nowicki, A. et al. (2011). *Dalton Trans.* 40 (24): 6524–6531.

24 Schulz, J., Roucoux, A., and Patin, H. (2000). *Chem. Eur. J.* 6 (4): 618–624.

25 Roucoux, A., Schulz, J., and Patin, H. (2003). *Adv. Synth. Catal.* 345 (1–2): 222–229.

26 Pélisson, C.-H., Hubert, C., Denicourt-Nowicki, A., and Roucoux, A. (2013). *Top. Catal.* 56 (13): 1220–1227.

27 Hubert, C., Denicourt-Nowicki, A., Guégan, J.-P., and Roucoux, A. (2009). *Dalton Trans.* 36: 7356–7358.

28 Nowicki, A., Le Boulaire, V., and Roucoux, A. (2007). *Adv. Synth. Catal.* 349 (14–15): 2326–2330.

29 Mévellec, V., Roucoux, A., Ramirez, E. et al. (2004). *Adv. Synth. Catal.* 346 (1): 72–76.

30 Mévellec, V. and Roucoux, A. (2004). *Inorg. Chim. Acta* 357 (10): 3099–3103.

31 Hubert, C., Guyonnet Bilé, E., Denicourt-Nowicki, A., and Roucoux, A. (2011). *Appl. Catal., A* 394 (1): 215–219.

32 Jason Bonilla, R., James, B.R., and Jessop, P.G. (2000). *Chem. Commun.* 11: 941–942.

33 Fonseca, G.S., Umpierre, A.P., Fichtner, P.F.P. et al. (2003). *Chem. Eur. J.* 9 (14): 3263–3269.

34 Hu, T.Q., Lee, C.-L., James, B.R., and Rettig, S.J. (1997). *Can. J. Chem.* 75 (9): 1234–1239.

35 Hagen, C.M., Vieille-Petit, L., Laurenczy, G. et al. (2005). *Organometallics* 24 (8): 1819–1831.

36 Hubert, C., Denicourt-Nowicki, A., Beaunier, P., and Roucoux, A. (2010). *Green Chem.* 12 (7): 1167–1170.

37 Hubert, C., Guyonnet Bilé, E., Denicourt-Nowicki, A., and Roucoux, A. (2011). *Green Chem.* 13 (7): 1766–1771.

38 Boutros, M., Denicourt-Nowicki, A., Roucoux, A. et al. (2008). *Chem. Commun.* (25): 2920–2922.

39 Pélisson, C.-H., Denicourt-Nowicki, A., Meriadec, C. et al. (2015). *ChemCatChem* 7 (2): 309–315.

40 Pélisson, C.-H., Denicourt-Nowicki, A., and Roucoux, A. (2016). *ACS Sustainable Chem. Eng.* 4 (3): 1834–1839.
41 Hubert, C., Denicourt-Nowicki, A., Roucoux, A. et al. (2009). *Chem. Commun.* 10: 1228–1230.
42 Roy, S. and Pericàs, M.A. (2009). *Org. Biomol. Chem.* 7 (13): 2669–2677.
43 Mévellec, V., Mattioda, C., Schulz, J. et al. (2004). *J. Catal.* 225 (1): 1–6.
44 Guyonnet Bilé, E., Denicourt-Nowicki, A., Sassine, R. et al. (2010). *ChemSusChem* 3 (11): 1276–1279.
45 Guyonnet Bilé, E., Cortelazzo-Polisini, E., Denicourt-Nowicki, A. et al. (2012). *ChemSusChem* 5 (1): 91–101.
46 Chau, N.T.T., Guégan, J.-P., Menuel, S. et al. (2013). *Appl. Catal., A* 467: 497–503.
47 Chau, N.T.T., Menuel, S., Colombel-Rouen, S. et al. (2016). *RSC Adv.* 6 (109): 108125–108131.
48 Rauchdi, M., Ait Ali, M., Roucoux, A., and Denicourt-Nowicki, A. (2018). *Appl. Catal., A* 550: 266–273.
49 Islam, M., Hossain, D., Mondal, P. et al. (2010). *Bull. Korean Chem. Soc.* 31: 3765–3770.
50 Mac Leod, W.D. Jr., and Buigues, N.M. (1964). *J. Food Sci.* 29 (5): 565–568.
51 Zhu, B.C.R., Henderson, G., Sauer, A.M. et al. (2003). *J. Chem. Ecol.* 29 (12): 2695–2701.
52 Alcaide, B., Almendros, P., González, A.M. et al. (2016). *Adv. Synth. Catal.* 358 (12): 2000–2006.
53 Yan, N., Yuan, Y., and Dyson, P.J. (2011). *Chem. Commun.* 47 (9): 2529–2531.
54 Durap, F., Zahmakıran, M., and Özkar, S. (2009). *Appl. Catal., A* 369 (1): 53–59.
55 Hou, S., Xie, C., Zhong, H., and Yu, S. (2015). *RSC Adv.* 5 (109): 89552–89558.
56 Albuquerque, B.L., Denicourt-Nowicki, A., Mériadec, C. et al. (2016). *J. Catal.* 340: 144–153.
57 Gatard, S., Liang, L., Salmon, L. et al. (2011). *Tetrahedron Lett.* 52 (16): 1842–1846.
58 Nandi, S., Patel, P., Jakhar, A. et al. (2017). *Chemistry Select* 2 (31): 9911–9919.
59 Ibrahim, M., Poreddy, R., Philippot, K. et al. (2016). *Dalton Trans.* 45 (48): 19368–19373.
60 Snelders, D.J.M., Yan, N., Gan, W. et al. (2012). *ACS Catal.* 2 (2): 201–207.
61 Wang, X., Huang, C., Li, X. et al. (2019). *Chem. Asian J.* 14 (13): 2266–2272.
62 Sawoo, S., Srimani, D., Dutta, P. et al. (2009). *Tetrahedron* 65 (22): 4367–4374.
63 Yun, G., Hassan, Z., Lee, J. et al. (2014). *Angew. Chem. Int. Ed.* 53 (25): 6414–6418.
64 Llevot, A., Monney, B., Sehlinger, A. et al. (2017). *Chem. Commun.* 53 (37): 5175–5178.
65 Zhang, H. and Toshima, N. (2013). *Catal. Sci. Technol.* 3 (2): 268–278.

66 Zhang, H., Okuni, J., and Toshima, N. (2011). *J. Colloid Interface Sci.* 354 (1): 131–138.

67 Mejías, N., Pleixats, R., Shafir, A. et al. (2010). *Eur. J. Org. Chem.* 2010 (26): 5090–5099.

68 Wu, L., Zhang, X., and Tao, Z. (2012). *Catal. Sci. Technol.* 2 (4): 707–710.

4

Organometallic Metal Nanoparticles for Catalysis

M. Rosa Axet and Karine Philippot

Université de Toulouse, UPS, INPT, CNRS, LCC (Laboratoire de Chimie de Coordination), UPR8241, 205 route de Narbonne, F-31077, Toulouse Cedex 4, France

4.1 Introduction

Assimilated to heterogeneous catalysts, metal nanoparticles (MNPs) have been known for a long time, but they have been received a renewed interest for the past three decades because of the necessity to design better defined nanocatalysts [1]. Common issues of heterogeneous catalysts are the size dispersity (e.g. 5% in even highly monodispersed samples), the uncertain surface state, the sometimes unknown core/ligand interfaces, the defects and elusive structures in 2D materials, and the still missing information on alloy patterns in bi- and multimetallic nanoparticles (NPs). Such disadvantages preclude deep understandings of many fundamental aspects, among which is the knowledge on atomic-level mechanisms that can happen at their surface, a key point if one wants to develop more performant nanocatalysts [2]. Looking for synthesis strategies that permit to access, in a reproducible manner, to well-defined MNPs in terms of size, crystalline structure, composition (metal cores and stabilizing agents), chemical order, shape, and dispersion is thus a prerequisite in order to finely study their properties and conclude on the relationships between structural features and catalytic performances. In this context, the bottom-up liquid-phase techniques are helpful because of their high versatility and ease of use, and also, the more straightforward equipments required than for the physic routes. Recent advances in solution nanochemistry allow having at disposal efficient synthesis tools to reach these objectives. Such evolution in the preparation of MNPs makes nanocatalysis a key domain, at the borderline between homogeneous and heterogeneous catalyses [1]. Interestingly, ligand-stabilized MNPs offer a metal surface with both an interface close to that of molecular complexes, as isolated surface atoms can be seen as metal centers, with their coordination sphere and neighboring metal atoms as in pure metal surfaces. In these NPs, both coordinated ligands and neighboring metal atoms can affect the surface properties and consequently their catalytic performances. Ligand-capped MNPs are thus very attractive systems to perform surface studies as done with

Nanoparticles in Catalysis: Advances in Synthesis and Applications, First Edition.
Edited by Karine Philippot and Alain Roucoux.

homogeneous catalysts while keeping the benefit of metal atom assemblies. In parallel of nanochemistry progress, computational chemistry developments offer powerful tools to access to MNP structural information, including information on the capping ligands such as their coordination modes and their influence on MNP surface state, as well as on their reactivity through mechanistic studies. Recent advances bring computational chemistry applied to small MNPs to the same level of accuracy and relevance as in molecular chemistry [3]. Thus, the alliance between nanochemistry and computational chemistry enables to have a precise view of the surface properties of MNPs [4].

On the basis of the above, this chapter aims at first illustrating the interests of the organometallic approach for the synthesis of MNPs with a precise control of their characteristics. Then, a selection of recent results obtained by our group, investigating nanocatalysts engineered following these concepts, will provide a view of the possibilities offered in terms of catalytic properties in hydrogenation reactions.

4.2 Interests of the Organometallic Approach to Study Stabilizer Effect on Metal Surface Properties

The synthesis of MNPs in organic solvents proved to be a versatile route to have a control on sizes, shapes, compositions, and structures for catalytic studies. Being part of current nanochemistry, the organometallic approach (Figure 4.1) aims at using the molecular chemistry concepts, particularly those of coordination and organometallic chemistries, in order to develop efficient tools for a reproducible synthesis of well-defined soluble MNPs and explore their surface features. Being now applied by several academic research groups in the world, this approach is a well-established method to access model MNPs for investigations in catalysis. This strategy is based on the use of organometallic or metal–organic complexes as the precursors of metal atoms together with adequate stabilizers such as ionic liquids, polymers, ligands, and also inorganic and carbonaceous supports. It permits building diverse nano-objects with modulable sizes including ultrasmall size (c. 1–10 nm) and a metallic surface free of contaminants, such as halides or other ions, and functionalizable as per the requirement. A major advantage with organometallic or metal–organic complexes is their facile decomposition in mild conditions (1–3 bar of H_2; $T \leq 150\,°C$) by reduction or ligand displacement from the metal coordination sphere in the presence of a suitable amount of a stabilizer, usually in an organic medium, but also described in aqueous media [5]. Olefinic complexes are generally preferred because when exposed to H_2 pressure, they lead to clean metal surfaces, as the formed alkanes are inert toward the growing MNP surface and can be easily eliminated by simple evaporation and/or by precipitation of MNPs and further filtration. Through this method, monodisperse assemblies of MNPs with an effective control of size, shape, and surface state can be synthesized. The so-obtained MNPs can be finely characterized before their application in catalysis.

Choosing the stabilizer is also essential as it intervenes at different levels, namely, the growth, stability, solubility, and surface properties and consequently on catalytic

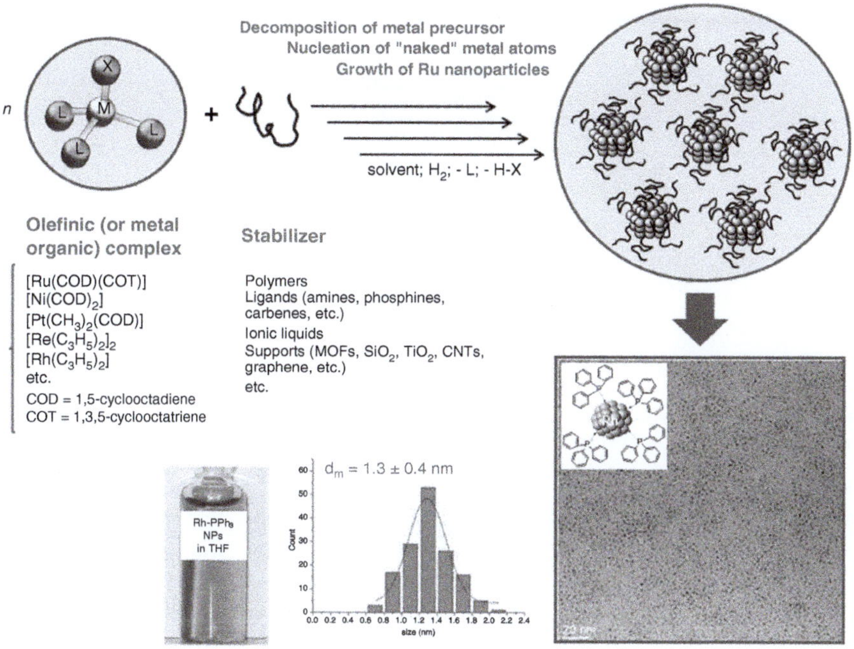

Figure 4.1 Organometallic approach for the synthesis of metal nanoparticles. Source: Adapted from Ibrahim et al. [18b].

performance of the MNPs. Besides organic polymers (such as polyvinylpyrrolidone, PVP) or ionic liquids [6] that provide steric and electrosteric stabilization, respectively, and weak interaction with the metal surface, a large variety of organic ligands coordinating via N, O, S, Si, P, or C atoms (as amines, carboxylic acids, thiols, silanes, phosphines, carbenes, fullerenes, but also more sophisticated ligands) to the metal surface have been described [7]. By using water-soluble polymers (such as PVP [8]) and ligands (such as 1,3,5-triaza-7-phosphaadamantane [9], or sulfonated phosphines [10], and also cyclodextrins [11]), the MNPs become transferable into water, thus leading to aqueous colloidal suspensions that offer other opportunities in catalysis as reported in Chapter 3 [12]. The organometallic approach is also very effective for accessing to controlled bimetallic NPs. Indeed, the adequate choice of the reaction conditions, such as metal precursors and reducing agents, allows governing the kinetics of metal atom generation to reach the target bimetallic chemical order: either alloyed, core–shell, or surface-decorated systems [13]. Immobilization of MNPs into alumina, silica, or carbon materials [14] allowed improving stability and recovery of the nanomaterials and also taking benefit of the support properties during catalysis. This can be performed either by impregnation of a given support with a colloidal solution of preformed MNPs or by their direct synthesis in the presence of the support. Improvement of the NP grafting and stability was observed when using functionalized supports decorated at their surface with chemical groups similar to those involved in common stabilizing ligands (such as amines for instance).

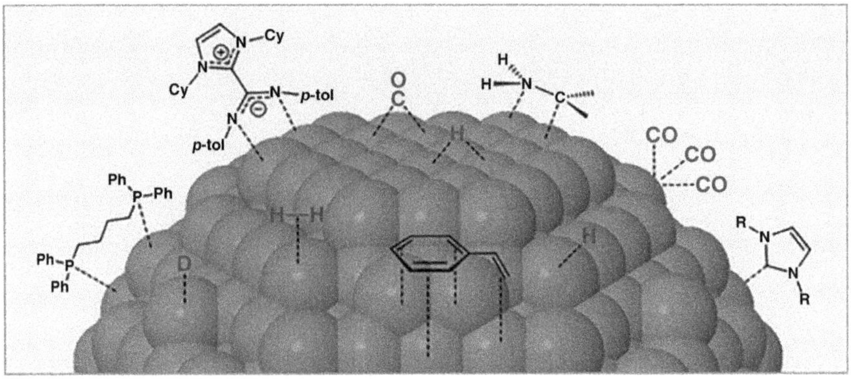

Figure 4.2 Schematic view of some surface studies performed on ruthenium nanoparticles. Source: Reproduced with permission from [15a]. Copyright 2018 American Chemical Society.

Fine characterization of the NPs is carried out by using combined techniques from both solid and molecular chemistry such as transmission electron microscopy (TEM), high-resolution electron microscopy (HREM), scanning transmission electron microscopy (STEM) coupled with energy-dispersive X-ray (EDX), scanning transmission electron microscopy at high-angle annular dark field (STEM-HAADF), powder X-ray diffraction (XRD), wide-angle X-ray scattering (WAXS), liquid- and solid-state nuclear magnetic resonance (NMR), Fourier transform infrared (FTIR), thermogravimetric analysis (TGA), inductively coupled plasma (ICP), elemental analysis, magnetic measurements, and X-ray photoelectron spectroscopy (XPS), among others. The combination of these tools allows obtaining detailed information of the NP systems. Even if the surface properties of Ru NPs have been extensively studied and described by means of those techniques (Figure 4.2) [4a, 15], the same methodology has been successfully applied to characterize other MNPs, such as Pt [16], Pd [17], Rh [18], Ir [19], or Ni [20].

Using H_2 as a reducing agent to synthesize MNPs gives rise to the adsorption of hydrogen atoms at the metal surface, a clear advantage for reduction catalysis. It has been shown that the surface hydride content can vary depending on the other species present on the surface of Ru NPs and it is generally higher than 1 (1–1.3 H per surface Ru atom) [21] even into water [10]. However, on Ru NPs capped by carboxylate ligands, the number of hydrides per surface Ru atom was found to be significantly lower (c. 0.4 H/surface Ru atom) by both experimental and theoretical techniques, as a result of the coordination mode of the carboxylate groups [4a]. These surface hydrides can also be displaced by coordination of CO molecules at the NP surface. Solid-state ^{2}H NMR evidenced H-D exchange between the surface of Ru NPs and ligand sites, for instance, the incorporation of ^{2}H atoms in the alkyl chain of hexadecylamine (HDA), as a result of a C–H activation [22]. Such reactivity has been exploited catalytically further for deuteration of different substrates, such as nitrogen [8, 12a, 23], phosphorus [24], and sulfur [25], containing compounds using Ru NP-based catalysts.

^{13}CO being easily detected by IR and magic angle spinning (MAS) NMR spectroscopy techniques, it is an ideal probe to provide indirect information on location and mobility of ligands at metal surface [26]. The strong coordination of phosphine ligands at Ru NP surface was evidenced to block CO mobility contrarily to the few, weak bonds involved when using a polymer as a stabilizer [27]. The location of carbene [28] or betaine adduct of N-heterocyclic carbene (NHC) and carbodiimide [29] ligands at Ru NP surface was determined by similar strategies. ^{13}CO adsorption was also applied in order to investigate the surface of Pt NPs [16a].

Besides the interest of catalysis per se, it can be used as a complementary tool in order to determine the structure and composition of MNPs. Benchmark catalytic reactions can be chosen for that purpose. Surface-state investigations of MNPs based on CO adsorption together with the comparison of the catalytic performances in CO oxidation allowed having a better insight into the effect of ligands on the surface of MNPs. In the case of PVP-stabilized or phosphine-capped Ru NPs, CO oxidation occurred at room temperature (r.t.), but the former Ru system presented a rapid deactivation because of the formation of RuO_2, while the latter were effectively protected against bulk oxidation [27]. CO adsorption and oxidation reactions also provided interesting information of the chemical order in bimetallic RuRe NPs [30]. The PVP-stabilized RuRe NPs of alloy type showed better resistance to oxidation than the PVP-stabilized RuRe NPs displaying a Re-enriched surface and were more active toward CO dissociation than monometallic Re/PVP NPs [30a], probably as a result of the synergic effect between Ru and Re. Interestingly, the dissociation of CO was not observed with PVP-stabilized RuRe NPs displaying a Re-enriched surface [30b]. Selective hydrogenation reactions are very attractive in order to correlate the structural parameters of MNPs to their catalytic performances and vice versa. The numerous bonds able to be reduced by MNPs, such as C=C, C=O, nitro, and azo groups, among others, give large possibilities to study the MNP surfaces accurately. For instance, the hydrogenation of styrene, a compound that contains alkene and arene moieties, provides a suitable system in this sense. Thus, the influence of ancillary ligands or CO on the surface of the MNPs, or the impact of the addition of a second metal in Ru NP-based catalysts, was studied by performing selective hydrogenation catalytic tests. For instance, using PVP-stabilized or dppb-stabilized Ru NPs (dppb = bisdiphenylphosphinobutane) as a catalyst, styrene was slowly hydrogenated at r.t., first into ethylbenzene and then into ethylcyclohexane [27]. Although the selective poisoning of the NPs with bridging CO groups led to catalysts only able to reduce the vinyl group of styrene as a result of the coordination of CO onto NP faces, a full poisoning of the surface with both terminal and bridging CO groups led to inactive catalysts. These results revealed that bridging CO groups and arenes compete for the same sites on the NP surface, presumably on faces. Similarly, bimetallic NP could be studied by this means [31]. Tin-decorated Ru NPs were prepared by reacting preformed Ru NPs, stabilized by a polymer, Ru/PVP, or a phosphine ligand, Ru/dppb, with [$(n\text{-}C_4H_9)_3SnH$] in order to foment a tin deposit at their surface [31b]. Styrene hydrogenation investigations allowed rationalizing the influence of the surface tin on the surface chemistry and catalytic activity of the Ru NPs. RuPt NPs stabilized with the dppb ligand were found to be more active (turnover frequency [TOF] = $2.1\,h^{-1}$) than the monometallic Ru/dppb ones (TOF = $1.5\,h^{-1}$) in styrene hydrogenation reaction [31a]. Interestingly, these RuPt/dppb NPs were more active toward arene hydrogenation even after CO poisoning, thus pointing to the interest of preparing bimetallic species. In the same line, more information about the composition of RuPt catalysts could be ascertained by analyzing the results obtained in the selective hydrogenation of *trans*-cinnamaldehyde, as detailed below [32].

4.3 Application of Organometallic Nanoparticles as Catalysts for Hydrogenation Reactions

Ligand effects play a key role in the development and application of nanocatalysts. Electronic or steric effects or the combination of both are important parameters to design efficient catalysts. The understanding of these effects is crucial, and combining characterization analytical techniques together with computational methods has proved to be efficient to reach this aim. The versatility of the organometallic approach allowed producing a plethora of MNPs bearing a diversity of ligands suitable for catalysis. Next, a description of the use of phosphorus-, amine-, carboxylic acid-, and NHC group-containing compounds, among others, as ligands to stabilize MNPs for their application as catalysts is analyzed. A few examples will also describe the effect of the presence of a second metal or of a support on catalysis.

4.3.1 Metal Nanoparticles Stabilized with Phosphorus Ligands

The surface properties of Rh NPs prepared by hydrogenation of [$Rh(\eta^3\text{-}C_3H_5)_3$] complex in the presence of PVP (2.2 nm), PPh_3 (1.3 nm), or dppb (1.7 nm) were compared in hydrogenation of cyclohexene as a model catalytic reaction, both in colloidal and supported conditions, after immobilization onto amino-functionalized silica-coated magnetite support [18b]. PVP-stabilized Rh NPs were the most active catalyst regardless of the catalytic conditions, when compared to phosphine-stabilized ones; this fact attributed to the blockage of some Rh surface sites by the strong coordination of the phosphine ligands, thus limiting the reactivity of the metallic surface. Even after several recycles of the supported nanocatalysts, Rh NPs containing phosphorus ligands displayed lower activity, underlining the strength of the ligand coordination. The PVP-stabilized Rh NPs were also evaluated in the hydrogenation of various arenes as well as levulinic acid and methyl levulinate [33]. Excellent activity and selectivity toward aromatic ring hydrogenation in mild reaction conditions (r.t.; 1 bar of H_2) were achieved. They also showed high performance in the hydrogenation of levulinic acid and methyl levulinate in water, quantitatively leading to the fuel additive γ-valerolactone under moderate reaction conditions. Interestingly, a series of very small Rh NPs (1.1–1.7 nm) were obtained using ferrocenyl phosphines as stabilizing ligands [18d]. Evaluation of these Rh NPs in styrene hydrogenation showed good performance with selectivities up to 99% toward ethylbenzene, and activities differed depending on the substituents of the phosphine ligand. These results evidenced that the metal surface is not blocked despite the steric bulk of the ferrocenyl ligands, reasonably indicating a degree of flexibility of these ligands at the metal surface.

Roof-shaped phosphine ligands were used to stabilize Ru NPs in order to evaluate how ancillary phosphine ligands may influence their reactivity when applied as catalysts for the hydrogenation of arenes [34]. These NPs (c. 1.1–2.1 nm) were synthesized by hydrogen-assisted decomposition of $[Ru(\eta^4\text{-}C_8H_{12})(\eta^6\text{-}C_8H_{10})]$ in THF in the presence of a series of mono- and diphosphines at different ligand/Ru ratios. ^{31}P-HRMAS NMR evidenced the coordination of the phosphines at the metal surface. High influence of the phosphine nature on the NP catalytic performances in the hydrogenation of *o*-methylanisole was observed. Triarylphosphine-capped Ru NPs were not or very poorly active, while those with dialkylarylphosphines led to the full hydrogenation (TOF = 212 h^{-1} for *o*-methylanisole hydrogenation and for benzene reduction: TOF > 5000 h^{-1} under 3 bar of H_2 and 295 K and TOF > 60 000 h^{-1} under 40 bar of H_2 and 393 K). This behavior was attributed to the electronically richer NPs produced with dialkylarylphosphines and to the higher steric hindrance of triarylphosphines. Ru NPs stabilized with a secondary phosphine oxide (SPO) were applied in the hydrogenation of several aromatic model substrates in acidic or basic conditions, as such a parameter is key in SPO-based homogeneous catalysis [35]. The absence of such an effect for SPO-stabilized Ru NPs evidenced the noninvolvement of the oxygen atom. High activity was observed for aromatic derivatives, the highest TOFs being achieved in neat conditions (TOFs up to 2700 h^{-1}). Chiral SPO ligand (4,5-dihydro-3*H*-dinaphtho[2,1-*c*:1′,2′-*e*] phosphepine-4-oxide) was successfully used to stabilize Ir NPs (synthesized by H_2 reduction of $[Ir(OMe)(C_8H_{12})]_2$; c. 1.4 nm), which proved to be active for the enantioselective hydrogenation of prochiral ketones (Table 4.1) [19]. Extensive characterization of these Ir NPs provided interesting information on the surface chemistry of the particles and on the role of the SPO ligand. Specifically, MAS-NMR experiments through coordination of ^{13}CO highlighted the proximity of the asymmetric ligands to free active sites, located in edges and apexes, thus enabling the enantioselective formation of products. SPO is strongly bound to the NP surface, and the resulting rigidity may orientate the approach of the substrate because of the hydrogen bonding, thus inducing the enantioselectivity.

The deoxygenation of fatty acids was investigated with supported Pd NPs capped with PPh_3 in comparison with other supported Pd NPs stabilized with PVP or without any stabilizer on their surface (direct synthesis on the carbon support) [36]. Pd@carbon nanomaterials were found active and stable catalysts for the deoxygenation of oleic acid with different results according to the reaction process (batch or flow). Under batch conditions (20 bar of H_2; 573 K), the reaction efficiency depended on the Pd loading, the treatment of the carbon support (heating or acidic conditions), and the stabilization mode of the Pd NPs (no stabilizer; PVP or PPh_3). Aromatic compounds, which can occur through highly saturated compounds by hydrogen transfer reactions or Diels–Alder reaction of C_6–C_8 olefins, were not detected with the Pd@carbon catalysts prepared without a stabilizer, with octadecanol and octadecane being observed in large extents, thus indicating a deoxygenation mechanism in which the hydrocarbons are produced via both decarbonylation/decarboxylation and dehydration routes. For the Pd/PPh_3@carbon and Pd/PVP@carbon nanocatalysts, the decarbonylation/decarboxylation was the only deoxygenation route. Under

Table 4.1 Enantioselective hydrogenation of prochiral ketones catalyzed by Ir NPs[a)] and chiral secondary phosphine oxide (**L**) employed in this study (*S* enantiomer is depicted).

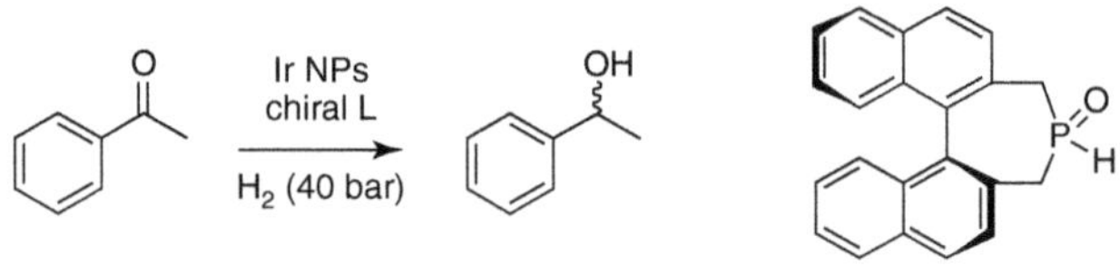

Substrate	SPO ligand	Conversion (%)[b)]	ee (%)[b)]	Product configuration
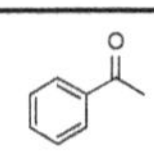	R	88	55	S
	S	47	52	R
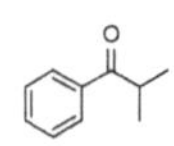	R	38	30	S
	S	11	29	R
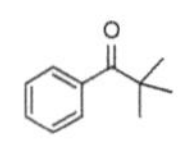	R	90	22	S
	S	38	20	R
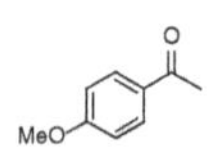	R	63	56	S
	S	14	39	R
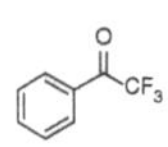	R	100	10	S
	S	100	8	R
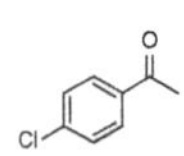	R	98	50	S
	S	51	34	R
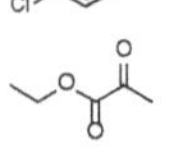	R	100	26	S
	S	100	30	R

a) Reagents and conditions: Ir NP (0.0025 mmol of Ir assuming % of Ir from elemental analysis), substrate (0.25 mmol), THF (0.75 ml), 18 hours, r.t., and 40 bar H_2.
b) Conversion and %ee were determined by GC (average of two runs).
Source: Adapted with permission from Ref. [19]. Copyright 2016, Royal Society of Chemistry.

flow conditions (4 bar of H_2; 573 K), except for the Pd@carbon catalyst, the conversion of stearic acid (SA) was superior to that observed in batch conditions. The product mixture contained over 20% heptadecane, and no octadecanol, octadecane, or aromatic compounds were observed. ICP-OES measurements indicated no leaching of palladium and simple washing of catalysts with mesitylene allowing their recycling without any change in conversion or product distribution.

4.3.2 Metal Nanoparticles Stabilized with N-Heterocyclic Carbenes

The ability of NHCs to stabilize metal NPs was first investigated on Ru NPs using classical NHCs [28]. Then, the synthesis of Ru NPs was enlarged to NHC ligands displaying a variety of backbones (Figure 4.3) [37] and further extended to the stabilization of Pt NPs [16b, 38]. The decomposition of [Ru(η^4-C_8H_{12})(η^6-C_8H_{10})] in pentane (3 bar of H_2; r.t.) in the presence of a NHC, i.e. 1,3-bis(2,6-diisopropylphenyl)imidazol-2-ylidene (IPr) (1.5 nm/0.5 eq.;

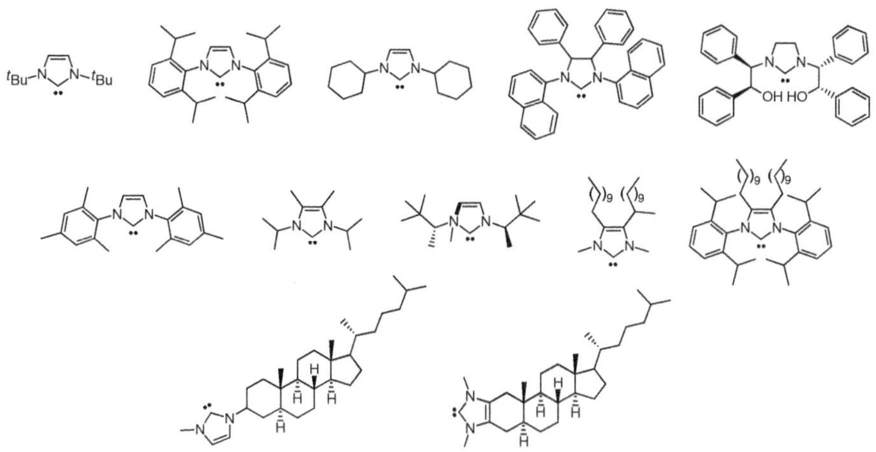

Figure 4.3 N-heterocyclic carbenes used as ligands to stabilize metal nanoparticles.

1.7 nm/0.2 eq. IPr) or *N*,*N*-di(*tert*-butyl)imidazol-2-ylidene (I^{t}Bu) (1.7 nm/0.5 eq. I^{t}Bu) [28] led to the synthesis of small NHC-stabilized Ru NPs. Full characterization, in particular by IR and NMR, proved the coordination of the ligand by the carbenic carbon and a large metal surface coverage. These carbene-capped Ru NPs were active in styrene hydrogenation with a rate significantly slower than that with PVP-stabilized Ru NPs, which is attributed to high coverage of the metal surface. NHC-stabilized Ru NPs also catalyzed hydrogenation of other substrates (benzene derivatives, methylanisole, and acetophenone) [37a], where an interesting ligand effect was observed; Ru/IPr NPs were more active than Ru/I^{t}Bu NPs, which was attributed to steric effects.

Several NHCs with different substituents and/or backbones in their structure, including chirality, showed high ability to stabilize small Ru NPs [37c–e]. Interesting ligand effects were evidenced, but enantioselectivity was elusive when using these Ru-based nanocatalysts. NHC-stabilized Pt NPs (1.5–1.8 nm) synthesized by hydrogenation of [Pt(dba)$_2$] (dba = dibenzylideneacetone) (3 bar of H_2; r.t.) in the presence of two different carbenes (IPr or 1,3-diisopropyl-4,5-dimethylimidazol-2-ylidene (I^{i}Pr$_2$Me$_2$)) were very active for the hydrogenation of nitroaromatics in mild conditions (30 °C; 1 bar of H_2) [16b]. IPr-capped Pt NPs achieved chemoselective reduction of various functionalized nitroarenes including groups such as hydroxyl, benzyloxy, carbonyl, and olefinic moieties as well as halogens. Also, long-chain NHCs efficiently stabilized Pt NPs [38]. The bulkiness of the ligand substituents influenced the catalytic properties of the Pt NPs. For example, the hydroboration of several acetylenes was effectively catalyzed by Pt NPs bearing a bulky carbene, while Pt NPs stabilized by less steric bulk demanding long-chain NHC displayed low catalytic performances. Nevertheless, the good selectivity for the *trans*-hydroborated isomer displayed with the more bulky NHC might be attributed to homogeneous species.

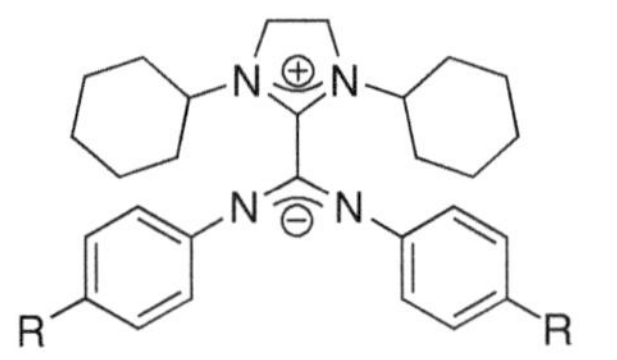

R = H, Me, OMe, Cl

Figure 4.4 Zwitterionic imidazolium-amidinate (betaine) ligands.

4.3.3 Metal Nanoparticles Stabilized with Zwitterionic Ligands

As mentioned above, the selectivity in styrene hydrogenation can depend on the availability of faces at NP surface, a size reduction being expected to increase the selectivity toward the reduction of the vinyl group. This has been evidenced with betaine-stabilized Ru NPs (Figure 4.4) [29]. The larger NPs (c. 1.3 nm) totally transformed styrene into ethylcyclohexane, whereas the smaller NPs (c. 1 nm) provided a partial selectivity to ethyl benzene. Such behavior is expected for molecular complexes, indicating that such c. 1 nm Ru NPs display reactivity at the frontier between the molecular and the solid states.

The same zwitterionic imidazolium-amidinate ligands also stabilized Pt NPs [16c]. The strong coordination of the ligands to platinum surface was unequivocally determined by a combination of spectroscopic techniques, which were endorsed by theoretical calculations. The small Pt NPs (1.9–2.3 nm) were revealed efficient in the hydrogenation of several substrates containing various functional groups such as olefinic bonds, carbonyl groups, and aromatic rings, showing that electron-rich Pt NPs yield faster catalysis.

4.3.4 Metal Nanoparticles Stabilized with Fullerenes

Ru nanocatalysts were prepared in a straightforward manner by decomposition of $[Ru(\eta^4\text{-}C_8H_{12})(\eta^6\text{-}C_8H_{10})]$ under H_2 in the presence of C_{60} [39]. Fullerene C_{60} behaves like an electron-deficient alkene giving electron-deficient properties to MNPs. Raman, XPS, EXAFS, and CO adsorption experiments on Ru/C_{60} NPs synthesized using several Ru/C_{60} ratios pointed out that Ru on these samples is electron poor. These electron-deficient Ru NPs supported on Ru-fulleride nanospheres allowed the successive and chemoselective hydrogenation of nitrobenzene to aniline and then to cyclohexylamine. The reaction was studied at 30 bar of H_2 at 80 °C in alcoholic solvents and appeared to be solvent sensitive, proceeding faster in methanol than in other alcohols. The scope of the reaction was investigated on several substituted nitroarenes, and in all cases, the reaction proceeded very selectively as observed for nitrobenzene, first the hydrogenation of the nitro group, followed by the hydrogenation of the arene ring. Density functional theory calculations suggest that the observed chemoselectivity is mainly governed by the presence of surface hydrides on the electron-deficient Ru NPs. The adsorption energy through the nitro group or the arene moiety of nitrobenzene was calculated at several coverages of surface hydrides (Figure 4.5). It was found that at a threshold value of 1.5 H per Ru surface atom, the formation of aniline is favored because of the

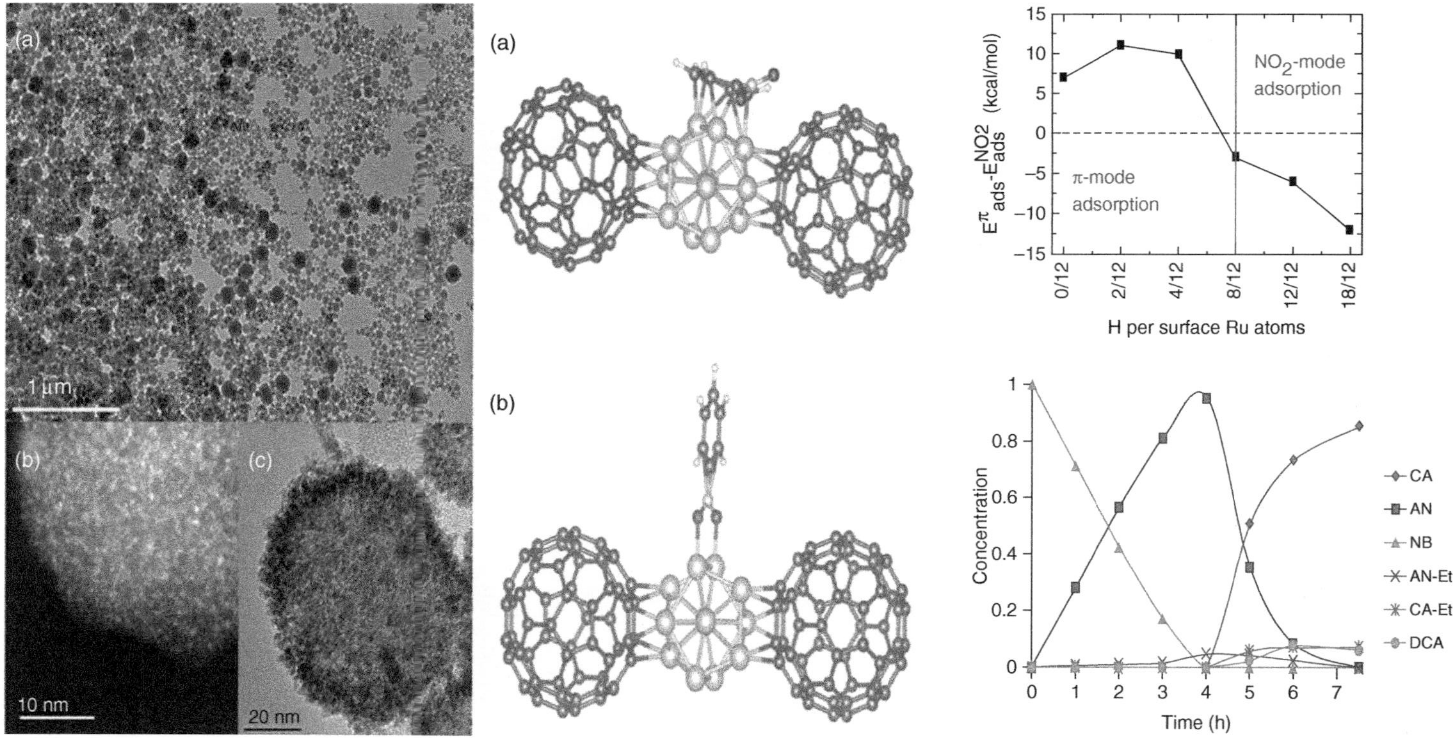

Figure 4.5 (a) TEM micrograph of Ru/C_{60} 1/1. (b) STEM of Ru/C_{60} 1/1 (scale bar 10 nm). (c) HREM of Ru/C_{60} 20/1 (scale bar 20 nm). (d) Side view of the π-mode coordination of a nitrobenzene molecule on a facet of a naked $2C_{60}$-Ru_{13} molecular complex. (e) Side view of the NO_2-mode coordination of a nitrobenzene molecule on the edge of a naked $2C_{60}$-Ru_{13} molecular complex. (f) Evolution of the energy difference between the two adsorption modes with respect to the ratio of H per Ru surface atoms present on the metallic cluster. (g) Time–concentration curve for nitrobenzene hydrogenation with Ru/C_{60} (Ru/C_{60} = 10/1), nitrobenzene (NB), aniline (AN), cyclohexylamine (CA), N-ethylaniline (AN-Et), dicyclohexylamine (DCA), and N-ethylcyclohexylamine (CA-Et). Source: Reproduced with permission from [39b]. Copyright 2016, American Chemical Society.

net preference of the NO_2 coordination. However, a ligand effect was ascertained in a further study (see Section 4.3.6) [40], in which it was demonstrated that the unique electronic properties of the fullerene C_{60}-based Ru NPs affect the catalytic properties generating a very efficient catalysis for this catalytic reduction. These species also catalyzed the hydrogenation of *trans*-cinnamaldehyde [41]. The main product obtained using methanol and a base was cinnamyl alcohol, while an aprotic and apolar solvent decreased the activity and produced hydrocinnamaldehyde.

Similarly, Rh/C_{60} NPs were prepared from $[Rh(\eta^3\text{-}C_3H_5)_3]$, giving rise to similar species, i.e. Rh-fulleride nanospheres decorated with Rh NPs, in which Rh was found to be electron deficient by Raman and XPS analyses [18a]. This property was exploited in the selective reduction of quinoline, in which Rh/C_{60} NPs exhibited excellent activity and produced selectively the partially hydrogenated product, 1,2,3,4-tetrahydroquinoline. Thermal treatment of the samples led to a more robust catalyst, as determined by recycling tests and characterization of the spent catalysts, and also to an increase of the catalytic activity (TOF = 488 h^{-1}). Theoretical calculations indicated again the crucial role of the hydride coverage of the metal NPs in the adsorption modes of the substrate and intermediates in the quinoline hydrogenation and also that these adsorption modes are modulated by the presence of fullerene C_{60}, thus affecting the activity and selectivity in the catalysis.

4.3.5 Metal Nanoparticles Stabilized with Carboxylic Acids

It has been shown that carboxylic acid groups coordinate strongly to Ru NP surfaces [42], and their coordination mode has been investigated by means of spectroscopic techniques together with density functional theory calculations [4]. It has been demonstrated that –COOH species easily lose protons onto the Ru NP surface, thus providing carboxylate species and surface hydrides, as a result of the small activation barrier of O—H bond dissociation [4a]. For example, ethanoic acid coordinates to the surface in an ethanoate form [4a]. Similarly, 1,3-adamantanedicarboxylic acid coordinates as the carboxylate form [4b], but it also acts as a bridging ligand in which both O atoms coordinate to two surface Ru atoms of the NPs. The robustness of the coordination of carboxylates has been exploited in order to achieve stable catalysts. In particular, the use of multicarboxylate ligands led to the straightforward production of 3D networks of Ru NPs, providing an interesting way to create confined spaces for catalysis. The fullerene-based hexa-adduct ligand $C_{66}(COOH)_{12}$ provided 3D networks of Ru NPs (c. 1.6 nm) with an interparticle distance of c. 2.8 nm [43]. These Ru NP networks were obtained in one-step synthesis by reduction of $[Ru(\eta^4\text{-}C_8H_{12})(\eta^6\text{-}C_8H_{10})]$ under H_2 in the presence of $C_{66}(COOH)_{12}$ hexa-adduct. Small-angle X-ray scattering (SAXS) and electron tomography analyses showed that carboxylate-bridged 3D networks of Ru NPs display a short range order (Figure 4.6). These Ru assemblies catalyzed the hydrogenation of nitrobenzene. A similar methodology allowed preparing other networks of Ru NPs by using two types of functionalized polymantane compounds, bearing carboxylic acid or amine functions [4b]. Ru NP networks built up from adamantanes with carboxylic acid groups revealed to be more robust catalysts than those prepared with amine-based

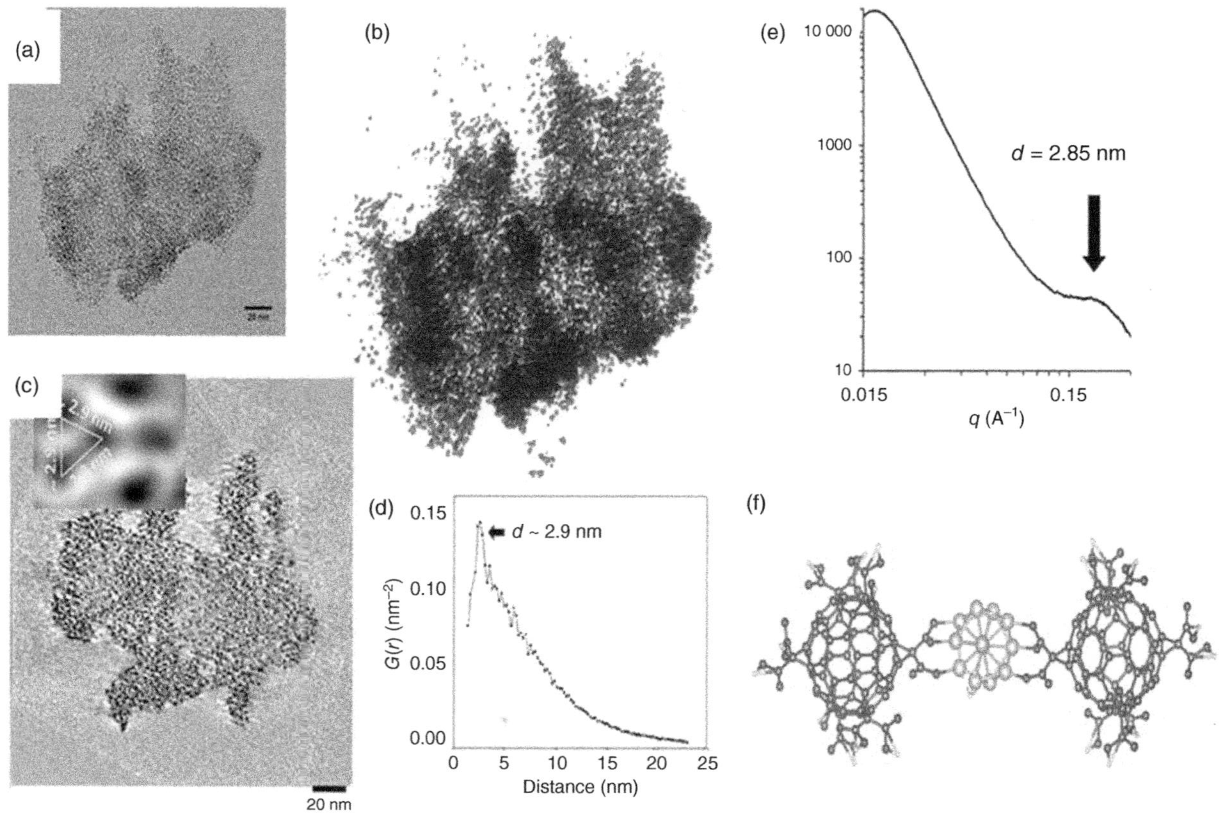

Figure 4.6 (a–d) Electron tomography analysis of a representative aggregate from the Ru/$C_{66}(COOH)_{12}$ 12 : 1 sample. (a) TEM image from the tilt series at 0° tilt. (b) 3D model of the reconstructed volume, showing the spatial distribution of all nanoparticles forming the aggregate. (c) Typical longitudinal slice extracted from the reconstruction volume. The inset image shows the distribution of a few NPs around a reference one. The repetitive distance is around 2.9 nm. (d) Pair distribution function of the distances between NPs calculated from their 3D coordinates extracted from electron tomography data. Primary peak shows a short-range order around 2.9 nm. (e) SAXS spectrum of Ru/$C_{66}(COOH)_{12}$ 12 : 1. (f) Optimized structure of the $C_{66}(COOH)_{12}$–Ru_{13}–$C_{66}(COOH)_{12}$ species. Source: Reproduced with permission from [43]. Copyright 2017 Wiley.

ligands. In addition, interesting ligand effects on catalysis were observed (see Section 4.3.6), but especially, confinement effects were shown to increase the catalytic activity in the hydrogenation of phenyl acetylene. Indeed, not only the electronic effects due to the coordination of ligands of different nature at the metal surface affected the activity and the selectivity but also the Ru interparticle distance governed the catalyst activity. Short distances between NPs were correlated with a higher catalytic activity.

4.3.6 Metal Nanoparticles Stabilized with Miscellaneous Ligands

The ligands coordinated at the surface of MNPs have a large impact on catalysis. Nevertheless, few works are still reported, in which a large diversity of the structure or the nature of the coordination groups of the capping agents are compared at once. In fact, other crucial parameters for catalysis than the surface properties such as size, chemical order, crystal structure, and so on can be tuned when using different stabilizing agents, making the comparison arduous. Some efforts have been done by using Ni [20] and Ru [4b, 40] NP-based catalysts. A series of Ni NPs (5–6 nm) prepared by H_2 decomposition of $[Ni(C_8H_{12})_2]$ complex in the presence of different stabilizers (HDA, PVP, PVP/PPh_3 mixture, octanoic acid (OA, and stearic acid (SA) (Figure 4.7) selectively catalyzed the hydrogenation of α,β-unsaturated carbonyl compounds under mild reaction conditions (THF; 3 bar of H_2; 60 °C; and low catalyst loading) [20]. All nanocatalysts reduced only the C=C bond of the α,β-unsaturated carbonyl compounds, this chemoselectivity being attributed to the high metallic character of the Ni NP surface, as reflected by the high magnetic properties measured. The value of M_s (emu g^{-1}; 2 K) and of the magnetic moments (given in brackets; μ_B) determined for each system are 63.7 (0.67), 58.7 (0.61), 49.3 (0.52), 45.7 (0.48), and 34.9 (0.37) for Ni-OA@SiO_2, Ni-SA, Ni-PVP, Ni-HDA, and Ni-PVP/PPh_3 systems, respectively. Moreover, the hydrogenation reaction rate appeared to be sensitive to ligand nature with the carboxylic acid-stabilized systems providing the best performances (Figure 4.7). A full kinetic investigation on the *t*-chalcone chemoselective reduction of the C=C bond, with the best catalyst (OA-stabilized Ni NP acid), revealed that the rate-determining step is the hydrogenation of the adsorbed substrate on the NP surface, following a Horiuti–Polanyi mechanism type. Regarding sustainable chemistry concerns, deposition of OA-capped Ni NPs onto silica (OA@SiO_2) led to an easily recoverable catalyst by magnetic separation that could be reused up to 10 times without significant loss of activity (Figure 4.7).

A series of Ru NPs capped with different ligands were used as catalysts in the selective hydrogenation of nitrobenzene to cyclohexylamine [40]. Ru/C_{60}, Ru/PVP, Ru/HDA, and Ru/IPr were chosen owing to the different electronic properties of their surface, with HDA and IPr acting as donor ligands, PVP as a polymer with small interaction with the Ru NP surface, and fullerene C_{60} as an electron-attractor ligand. This was confirmed by the frequency of the CO band observed by IR after CO adsorption onto all the NP surfaces, while their size and structure remained similar. In all cases, the nitrobenzene reduction proceeded in a stepwise manner,

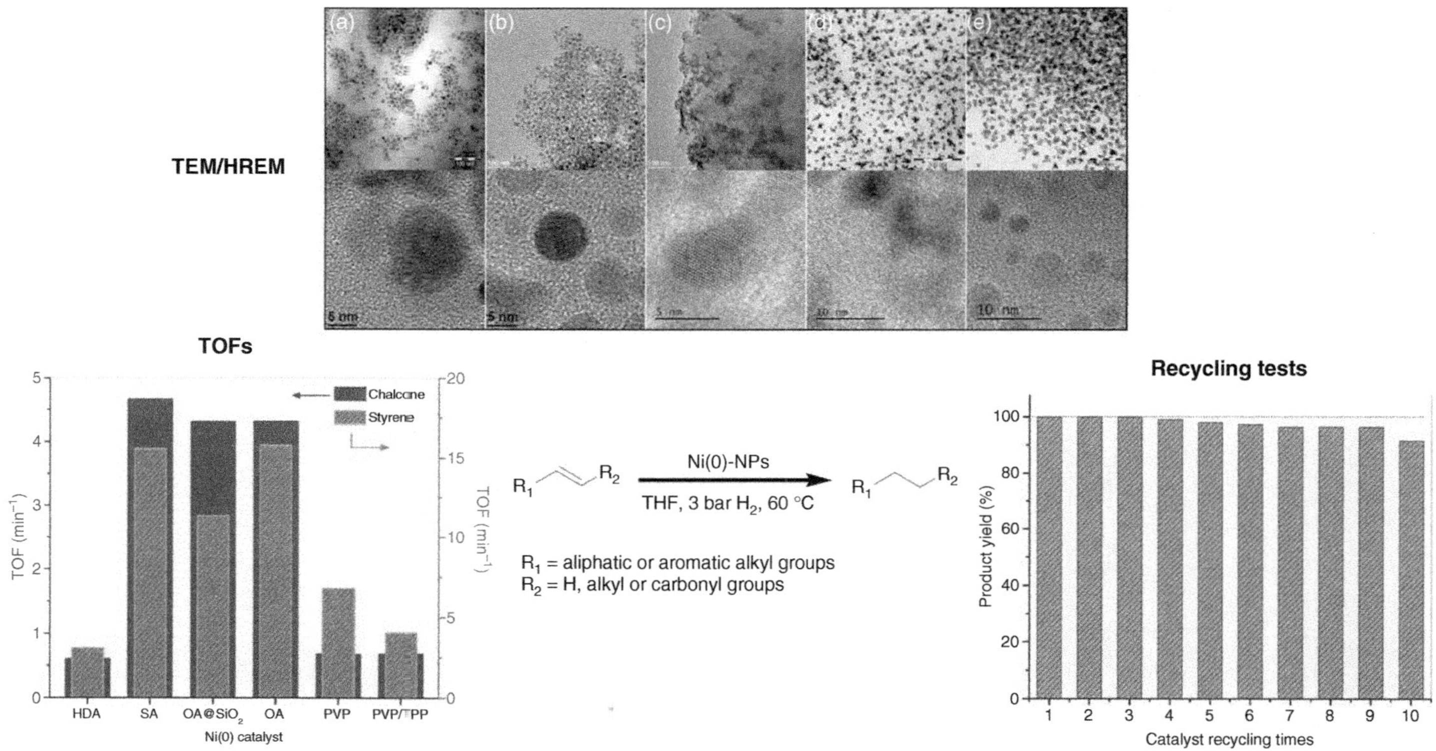

Figure 4.7 TEM (top) and HRTEM (bottom) images of Ni NP: (a) Ni-HDA, (b) Ni-SA, (c) Ni-OA@SiO$_2$, (d) Ni-PVP, and (e) Ni-PVP/PPh$_3$; TOF as a function of Ni(0)-NP sample for *t*-chalcone (dark grey bars) and styrene (light grey bars) (TOF = mol product converted/(mol of surface Ni × time) calculated from the slope of plots of TON vs. time at low substrate conversions (up to 20%), reaction conditions: [substrate] = 0.416 mol l^{-1}, 12.5 μmol Ni, 3 ml THF, 3 bar H$_2$, and 60 °C); recycling tests: product conversion (GC) as a function of recycling times for the hydrogenation of *t*-chalcone by Ni-OA@SiO$_2$ catalyst (reaction conditions: [*t*-chalcone] = 0.416 mol l^{-1}, 62.5 μmol Ni, 3 ml of THF, 3 bar H$_2$, and 60 °C). Source: Reproduced with permission from [20]. Copyright 2017 Royal Society of Chemistry.

producing aniline first and then cyclohexylamine, which agrees with the fact that the reaction selectivity is mainly governed by surface hydrides present onto the Ru NP surface, as pointed out by theoretical calculations. Nevertheless, the surface ligand strongly affected the activity and the selectivity of the catalysis. Less donor ligands promoted the hydrogenation of the N-phenylhydroxylamine (PHA) intermediate, rate-determining step of the reaction, while the Ru NP systems bearing more donating ligands were less active and less selective.

Ru NPs capped with carboxylic acid or amine functional groups, displaying the same polymantane backbone, i.e. adamantane, bis-adamantane, and diamantane, and structurally similar in terms of size, were used as catalysts in the selective hydrogenation of phenyl acetylene [4b]. All catalysts reached good selectivity toward styrene, and both selectivity and activity were significantly affected by Ru NP interparticle distance, as mentioned above, and electronic ligand effects. Ru NPs capped with σ-donor amine ligands, which provide a higher electronic density on the Ru NPs than carboxylic acid ligands, led to higher activity but with a lower selectivity toward styrene. Carboxylic acid ligands also provided more robust catalysts than amine ligands, taking into account the analyses of the spent catalysts.

4.3.7 Bimetallic Nanoparticles

The organometallic approach is also an effective way for the engineering of bimetallic NPs. Adequate selection of the reaction conditions (precursor, stabilizer, reactant, and temperature) allows accessing bimetallic NPs of controlled chemical order, i.e. alloy, core–shell, or even MNPs decorated with another metal (as tin or platinum for instance). Compared to their monometallic counterparts, bimetallic NPs can offer synergetic effect and other selective catalytic performances because of the different surface properties induced by the metal order. Such an effect was achieved when controlling the synthesis of RuPt NPs by an adequate choice of the metal precursors based on their kinetics of decomposition. At r.t. and in the presence of PVP as a stabilizer, the codecomposition of $[Ru(\eta^4\text{-}C_8H_{12})(\eta^6\text{-}C_8H_{10})]$ and $[Pt(dba)_2]$ led to a RuPt alloy of fcc structure [44], while using $[Pt(CH_3)_2(C_8H_{12})]$ instead of $[Pt(dba)_2]$ gave rise to core–shell RuPt NPs [45]. The chemical segregation toward core–shell RuPt is governed by kinetic and thermodynamic (preferred location of each metal in the particle) parameters, but the steric properties of the polymer which has little or no chemical interaction with the growing nanoparticles is also favorable. When keeping the same metal precursors (i.e. $[Ru(\eta^4\text{-}C_8H_{12})(\eta^6\text{-}C_8H_{10})]$ and $[Pt(CH_3)_2(\eta^4\text{-}C_8H_{12})]$) and reaction conditions and replacing the steric stabilizer by a strongly coordinating phosphorus ligand (dppb), RuPt NPs with a ruthenium-rich core and a disordered shell with the two metals were formed [31a]. The structure of the RuPt NPs raised from the high chemical affinity of the diphosphine ligand for both metals.

Moreover, the preformed Ru NPs and Pt NPs were used as seeds for the preparation of core–shell RuPt and PtRu NPs by a two-step procedure. Several metal Ru/Pt ratios were used, and 4-(3-phenylpropyl)pyridine (PPP) was introduced as a capping ligand [32]. Besides interesting results, the implementation of these bimetallic RuPt

Table 4.2 Cinnamaldehyde hydrogenation with bimetallic NPs Ru_1Pt_1, Ru_1Pt_2, and Ru_1Pt_4.

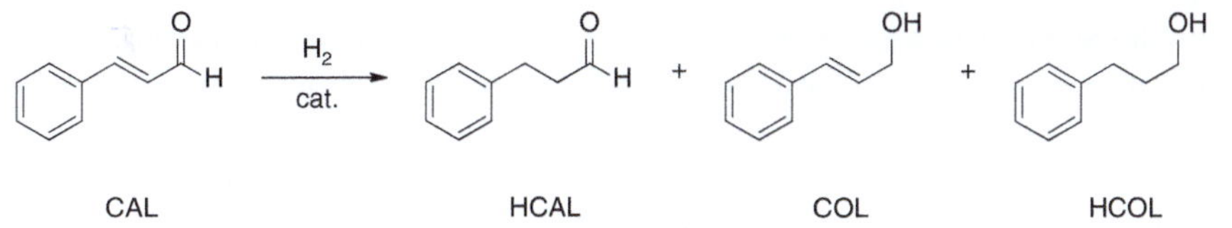

Catalyst	Time (h)	Conv. (%)	TON	TOF (h^{-1})	HCAL (%)	HCOL (%)	COL (%)	Acetal (%)
Ru_1Pt_1	4	27	190	47	49	29	22	n.d.
	6	42	298.5	50	45	25	30	n.d.
	22	64	449	20	27	28	41	4
Ru_1Pt_2	4	18	172	43	68	25	7	n.d.
	6	24	226	38	58	25	15	2
	22	53	501	23	38	16	37	9
Ru_1Pt_4	4	34	216	54	74	11	15	n.d.
	6	44	282	47	68	13	18	1
	22	87	556	25	51	26	22	1
Pt_1Ru_1	4	29	271	68	65	14	19	2
	6	34	314	52	57	15	25	3
	22	82	763	35	30	23	33	14
Pt_1Ru_2	4	43	283	71	71	12	16	1
	6	45	300	50	66	13	19	2
	22	80	527	24	38	24	30	8
Pt_1Ru_4	4	14	94	23	59	19	22	n.d.
	6	22	143	24	59	18	23	n.d.
	22	53	348	16	38	21	40	1

Reaction conditions: 2.5 mg of bimetallic NP, cinnamaldehyde (7.5 mmol), isopropanol (50 ml), $P_{H_2} = 20$ bar, and $T = 70\,°C$. Yields were determined by GC analysis using nonane (3.7 mmol) as an internal standard. n.d.: not detected. TOF = mol substrate converted/mol of metal. TOF = mol substrate converted/(mol of metal × time).
Source: Adapted with permission from Ref. [32]. Copyright 2014 Royal Society of Chemistry.

NPs as catalysts in the hydrogenation of *trans*-cinnamaldehyde greatly contributed to finely determine the composition and metal distribution in these NPs and complementarily to usual characterization techniques (FTIR, NMR, TEM, and WAXS). First, a higher activity was observed with higher content of platinum onto the surface (best activity for Ru_1Pt_4 NPs in RuPt NPs and for Pt_1Ru_1 in PtRu NPs) (Table 4.2). The composition and structure of the particles also influenced the selectivity. A higher selectivity was reached when both ruthenium and platinum were present at the surface of core–shell RuPt nanoparticles (partially covered core–shell RuPt NP; Ru_1Pt_1 and Ru_1Pt_2), illustrating a synergistic effect on the selectivity between both metals.

Tin-decorated Ru NPs were prepared by reacting preformed Ru NPs, stabilized by a polymer, Ru/PVP, or a diphosphine ligand, Ru/dppb, with $[(n\text{-}C_4H_9)_3SnH]$ in order to foment a tin deposit at their surface [31b]. Different Sn/Ru molar ratios were applied. The formation of μ^3-bridging "SnR" groups on the ruthenium surface was evidenced by different techniques (HREM, WAXS, IR, NMR, and Mössbauer data) and corroborated by theoretical calculations. Styrene hydrogenation investigations allowed rationalizing the influence of the surface tin on surface chemistry and catalytic activity of the Ru nanoparticles. For dppb-stabilized Ru NPs, the reaction with the tin precursor was limited by the surface coverage induced by the coordination of the bulky diphosphine ligand. On the contrary, the amount of tin deposited on the Ru surface could be adapted using PVP instead of dppb as a stabilizer. The resulting modification of the ruthenium surface led to a tuning of their surface properties, as observed through the coordination of CO as well as the catalytic hydrogenation of styrene. Ru/PVP/Sn cannot accommodate CO after the deposit of even a very low amount of tin while Ru/dppb/Sn can still accommodate terminal CO groups. The Ru/PVP/Sn NP showed a gradual variation of selectivity with increasing tin loading while only a small quantity of tin was sufficient to impede the arene hydrogenation ability of Ru/dppb nanoparticles. These results evidenced that the presence of tin adatoms on the metallic surface blocks some reactive sites.

4.3.8 Supported Nanoparticles

The application of supported MNPs as catalysts in organic synthesis is largely explored because of the synergistic effects that can arise from the association between a metal and a support. Also, immobilizing MNPs onto solid supports remains an efficient alternative for their recyclability and recovery from the reaction medium and may favor widespread implementation of NPs in catalysis. In this context, the organometallic approach for the synthesis of MNPs provides efficient tools for the preparation of composite nanomaterials, using alumina membranes, mesoporous silica, and carbon materials as nonexhaustive examples of supports. The inclusion of MNPs can be performed either by impregnation of the support using a colloidal solution of preformed NPs or by direct synthesis of the NPs in the presence of the support, with or without a ligand. There are also synthetic tools that allow the direct synthesis of MNPs in the presence of a support in a controlled manner, among which the functionalization of the support inspired by the ligands used to stabilize nanoparticles in solution reinforces the nanoparticle anchorage [46].

Decomposition of $[Pd_2(dba)_3]$ directly over amino-modified silica-coated magnetite particles (see Chapter 8) as a support led to a nanomaterial, a palladium magnetic nanocatalyst containing well-dispersed Pd NPs (3.5 ± 0.8 nm) exclusively on the support surface. This Fe_3O_4@silica-Pd nanocatalyst was applied in different hydrogenation reactions. It is proved to be highly active in the reduction of alkenes and highly selective in the semihydrogenation of alkynes to alkenes without the addition of metal promoters, as reported for other Pd-based catalysts [47]. Interestingly, it was also chemo- and stereoselective with more challenging

Table 4.3 Catalytic hydrogenation of propargylamines with the magnetic Pd nanocatalyst.

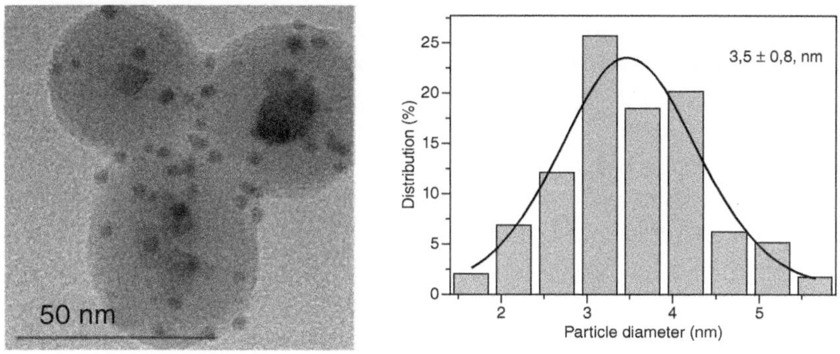

Substrate	Conversion (selectivity)[a]	Isolated yield (%)	Product
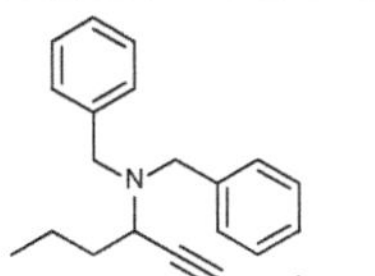 4a	100% (88%)	68	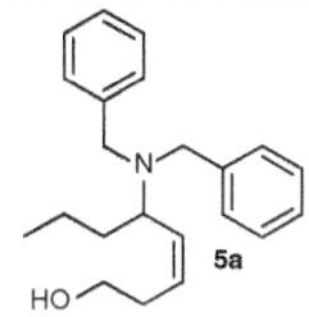 5a
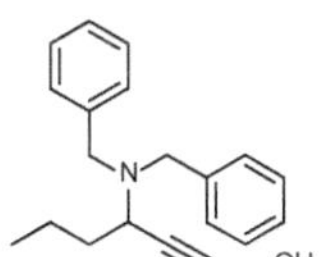 4b	91% (97%)	53	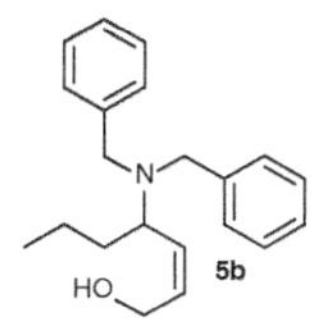 5b
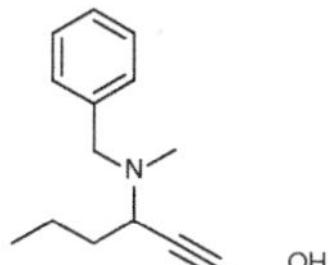 4c	n.d.[b]	61	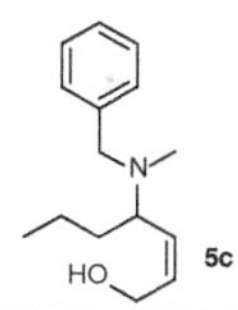 5c

Top: TEM image of the magnetic Pd nanocatalyst used with size histogram.
Reaction conditions: 1 mmol of alkyne, 2 mg of catalyst (0.04 mol%), 5 ml of cyclohexane, 27 °C, two hours, and 3 bar H_2.
a) Determined by GC.
b) Not determined.
Source: Adapted with permission from Ref. [47]. Copyright 2015, Royal Society of Chemistry.

benzylated propargylamine substrates, as only alkene product was obtained and side reactions such as isomerization or debenzylations were not detected, showing good overall yields under 3 bar of H_2 at 27 °C after two hours (Table 4.3). Comparative studies with commercial Pd/C evidenced relevant activity and selectivity, easy recoverability from the reaction media by simple application of an external magnet, negligible metal leaching, and reusability up to seven cycles, maintaining its activity and selectivity. Finally, Fe_3O_4@silica-Pd nanocatalyst allowed optimizing the last step of a multistep protocol (combining biocatalysis, heterogeneous metal catalysis, and magnetic nanoparticles) in order to access the bitter-taste dipeptide Ala-Phe, a potential substitute for caffeine as a food additive [48]. The Pd nanocatalyst allowed removing the protective Z group of the Z-Ala-Phe-OH peptide (Z = benzyloxycarbonyl group; to block of amino groups) under hydrogen in mild conditions (27 °C; 1 bar of H_2). The esterified amino acid derivative was converted to the unprotected dipeptide Ala-Phe in nearly quantitative yield (99%; two hours), releasing Phe-OMe and CO_2 with almost no catalyst pollution, thanks to an efficient magnetic separation of the catalyst.

Size-controlled bimetallic NiPd NPs (codecomposition of $[Ni(C_8H_{12})_2]$ and $[Pd_2(dba)_3]$; toluene; 10 eq. HDA; 3 bar of H_2; and 110 °C) of different Ni/Pd ratios were used to prepare supported NiPd catalysts by sol-immobilization onto an amino-modified magnetic silica [49]. These Fe_3O_4@silica-NiPd nanocatalysts were applied in the hydrogenation of cyclohexene and compared to Ni and Pd monometallic systems in the same reaction conditions. A higher activity than that provided by the Pd monometallic counterpart was obtained with bimetallic nanocatalysts, the 1 : 9 Ni/Pd composition achieving the highest initial TOF > 50 000 h^{-1}. In addition to the synergetic effect, which allows saving of noble metal, magnetic separation greatly facilitated catalyst separation from the liquid products without metal leaching, leading to an efficient recycling. Direct decomposition of $[Pd_2(dba)_3]$ over Fe_3O_4@silica functionalized with terpyridine pendant groups provided well-dispersed Pd NPs of c. 2.5 nm on the magnetic silica support, with interesting catalytic properties in cyclohexene and β-myrcene hydrogenation [46]. High activity in mild conditions was observed for the hydrogenation of cyclohexene (TOFs up to 129 000 h^{-1}) as well as high selectivity for hydrogenation of β-myrcene, in which the formation of monohydrogenated compounds by hydrogenation of less hindered double bonds was observed. Activity and selectivity were largely increased in comparison with those obtained with a similar nanomaterial containing Pd NP supported onto an amino-modified magnetic silica, thus evidencing the influence of the functional groups used for the anchoring of the NPs onto the support, similarly to ligand effects in colloidal catalysis. Carbon materials are also largely used for the immobilization of MNPs because they offer multiple advantages including easy availability, relatively low cost, high mechanical strength, and chemical stability. As for silica supports, their functionalization favors the dispersion and the anchoring of the NPs onto the surface. NP@carbon nanocatalysts were achieved by impregnation of different mesoporous carbons with colloidal solutions of Ru or Pd NPs (decomposition of $[Ru(\eta^4\text{-}C_8H_{12})(\eta^6\text{-}C_8H_{10})]$ or $[Pd_2(dba)_3]$ in THF; 3 bar of H_2; r.t.) stabilized by PPP or triphenyl- and trioctylphosphine, respectively [50]. Well-dispersed Ru and Pd NPs on the carbon materials (mean sizes of c. 1.2–1.3 and 1.9–2.2 nm, respectively) were obtained following this procedure. Investigation on the oxidation of benzyl alcohol (in water; 80 °C) led to excellent conversion and selectivity (>99%) toward the aldehyde for both metals. Recyclability studies performed with Pd catalysts evidenced no significant loss in activity up to eight successive catalytic runs. With oxidized carbon support, hydrophilicity was shown to affect the catalytic activity, with significance depending on the hydrophobic character of the ligand used to stabilize the NPs. These carbon-supported Pd NPs proved to be efficient dual catalysts for the hydrogenation of furfural and

Table 4.4 Hydrogenation of furfural catalyzed by carbon-supported Pd NP under microwave conditions.

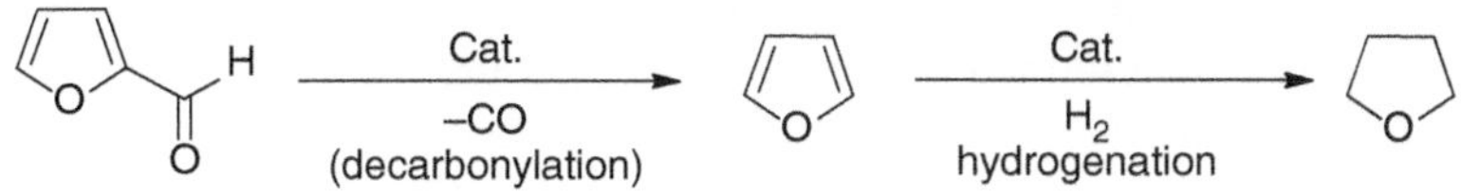

Nanocatalyst	Reaction time (min)	Conversion (mol%)	Selectivity in furan (mol%)	Selectivity in THF (mol%)
No catalyst	60	—	—	—
MB-H_2O_2	60	—	—	—
MB-1500	60	—	—	—
Pd(TOP)/MB-H_2O_2	30	>90	15	80
Pd(TOP)/MB-1500	30	75	20	72
Pd(PPh_3)/MB-H_2O_2	30	69	10	85
Pd(PPh_3)/MB-1500	30	65	<10	80
Pd/MB-H_2O_2	45	70	30	70
Pd/MB-1500	45	45	35	65

Reaction conditions: 2 mmol of furfural, 1.5 ml of formic acid, 1.5 ml of water, 0.1 g of catalyst, 100 °C, microwave power = 100–150 W, and P = 150–200 psi (developed in the systems). MB, mesoporous carbon beads; TOP, trioctylphosphine.
Source: Adapted with permission from Ref. [51]. Copyright 2012, Royal Society of Chemistry.

the selective oxidation of benzyl alcohol in water under microwave irradiation. Nanocatalysts based on trioctylphosphine and triphenylphosphine-stabilized Pd NPs based on oxidized carbon supports appeared the most active ones [51]. The presence of oxygen groups on the surface of the carbon support, particularly those of acidic character, improved the immobilization of the Pd NPs as well as the water affinity, providing high catalytic performances (Table 4.4).

The hydrogenation of anthracene with carbon-supported Pd NPs was also investigated, and nearly complete conversions as well as high selectivity toward the hydrogenated products without cracking or open-ring-derived products (naphtenics) were achieved [52]. Influence of the surface and textural properties of the carbon supports owing to their activation treatment (heating or acidic conditions) was studied. No cracking was observed for nonacidic carbon supports. Product distribution was observed as a function of reaction temperature (250–300 °C), with 9,10-dihydroanthracene being the main product at lower temperatures and deeper hydrogenation being achieved at intermediate temperatures (300 °C). Pd NPs supported onto a carbon material heated at 2000 °C, although less active than the other catalysts at short reaction times, appeared the most effective catalyst at 300 °C regarding the selective production of perhydroanthracene, namely, the final hydrogenated product from anthracene.

Using surface-modified carbon nanotubes (CNTs) as support, the influence of the confinement of RuPt nanoparticles inside CNT cavities on their catalytic properties in the hydrogenation of cinnamaldehyde was studied. The synthesis of RuPt/L/CNTs was performed in different reaction conditions from $[Ru(\eta^4\text{-}C_8H_{12})(\eta^6\text{-}C_8H_{10})]$ and $[Pt(CH_3)_2(C_8H_{12})]$ in the presence of PPP as a ligand [53]. Colloidal suspension of RuPt NPs was prepared in the presence of the ligand and then used to impregnate CNTs. A second route consisted in the codecomposition of the two complexes in the presence of both the ligand and the CNTs. The best results were obtained by the impregnation method using amide-functionalized CNTs (c. 2–2.5 nm NPs located inside the CNTs; 80% of NPs inside the CNTs for a 23 wt% of metal), where the confinement of the NPs inside the cavities was more important owing to limited access at the CNT surface. Catalytic results in the hydrogenation of cinnamaldehyde were compared to those achieved with the nonsupported RuPt/PPP NPs, showing higher catalytic activity and selectivity for CNT-confined RuPt/PPP NPs. A remarkable selectivity toward the formation of cinnamyl alcohol (95%) was achieved with the catalyst presenting the highest location of NPs inside the CNTs, attributed to a confinement effect.

4.4 Conclusions

The advantages of the organometallic approach for the design and synthesis of precisely defined mono- and bimetallic nanoparticles destined to catalysis have been illustrated through a selection of significant examples. If only examples of organic colloidal catalysis are presented here, the adapted synthesis tools such as the use of amphiphilic ligands can also provide aqueous colloidal suspensions in water, which proved to be of interest for catalysis, as reported in Chapter 3. Moreover, our synthetic approach also allows obtaining well-defined supported NPs either by impregnation or by direct synthesis in the presence of a given support. This allows taking advantage of the support either for a synergy effect and/or for easier recovering/recycling of the nanocatalysts. If the ability of stabilizing ligands to increase the nanocatalyst performance has been long underestimated, recent advances evidence that tailor-made systems can improve their activity, selectivity, stability, solubility, and recyclability. Varying the nature of the capping ligands enables to tune the nanoparticle surface state at both electronic and steric hindrance levels. Beside the state-of-the art techniques in solid chemistry, using usual characterization tools of molecular chemistry allows to dress a mapping of the nanoparticle surface at the molecular level. Computational chemistry dedicated to MNPs contributed greatly to the understanding of the influence of coordination chemistry at nanoparticle surface (see Chapter 15). The surface chemistry of MNPs becomes more understood. The recent advances offer a strong basis to continue to apprehend catalytic processes on the metal surface and explore more how these can be affected by the stabilizing molecules and/or supports. Our own experience makes us believe that continuing to take inspiration from the concepts of the different fields of catalysis (homogeneous, heterogeneous, and enzymatic) should contribute to the development of more efficient and selective nanocatalysts.

References

1 Corain, B., Schmid, G., and Toshima, N. (eds.) (2008). *Metal Nanoclusters in Catalysis and Materials Science: The Issue of Size Control*, 458. Amsterdam, The Netherlands: Elsevier Science.

2 Serp, P. and Philippot, K. (eds.) (2013). *Nanomaterials in Catalysis*, 494. Wiley-VCH Verlag GmbH & Co. KGaA.

3 Cusinato, L., Martinez-Prieto, L.M., Chaudret, B. et al. (2016). *Nanoscale* 8: 10974–10992.

4 (a) Gonzalez-Gomez, R., Cusinato, L., Bijani, C. et al. (2019). *Nanoscale* 11: 9392–9409. (b) Min, Y., Nasrallah, H., Poinsot, D. et al. (2020). *Chem. Mater.* 32 (6): 2365–2378.

5 Cushing, B.L., Kolesnichenko, V.L., and O'Connor, C.J. (2004). *Chem. Rev.* 104: 3893–3946.

6 (a) Salas, G., Campbell, P.S., Santini, C.C. et al. (2012). *Dalton Trans.* 41: 13919–13926. (b) Chacon, G. and Dupont, J. (2019). *ChemCatChem* 11: 333–341.

7 Lara, P., Philippot, K., and Chaudret, B. (2013). *ChemCatChem* 5: 28–45.

8 Taglang, C., Martinez-Prieto, L.M., del Rosal, I. et al. (2015). *Angew. Chem. Int. Ed.* 54: 10474–10477.

9 Debouttiere, P.-J., Martinez, V., Philippot, K., and Chaudret, B. (2009). *Dalton Trans.*: 10172–10174.

10 Guerrero, M., Roucoux, A., Denicourt-Nowicki, A. et al. (2012). *Catal. Today* 183: 34–41.

11 Guerrero, M., Coppel, Y., Chau, N.T.T. et al. (2013). *ChemCatChem* 5: 3802–3811.

12 (a) Martinez-Prieto, L.M., Baquero, E.A., Pieters, G. et al. (2017). *Chem. Commun.* 53: 5850–5853. (b) Debouttiere, P.-J., Coppel, Y., Denicourt-Nowicki, A. et al. (2012). *Eur. J. Inorg. Chem.* 2012: 1229–1236.

13 Amiens, C., Ciuculescu-Pradines, D., and Philippot, K. (2016). *Coord. Chem. Rev.* 308: 409–432.

14 Kinayyigit, S. and Philippot, K. (2014). Chapter 4: Organometallic approach for the synthesis of noble metal nanoparticles: towards application in colloidal and supported nanocatalysis. In: *Metal Nanoparticles for Catalysis: Advances and Applications* (ed. F. Tao), 47–82. The Royal Society of Chemistry.

15 (a) Martinez-Prieto, L.M. and Chaudret, B. (2018). *Acc. Chem. Res.* 51: 376–384. (b) Axet, M.R. and Philippot, K. (2020). *Chem. Rev.* 120: 1085–1145. (c) Creus, J., Drouet, S., Surinach, S. et al. (2018). *ACS Catal.* 8: 11094–11102.

16 (a) Kinayyigit, S., Lara, P., Lecante, P. et al. (2014). *Nanoscale* 6: 539–546. (b) Lara, P., Suarez, A., Colliere, V. et al. (2014). *ChemCatChem* 6: 87–90. (c) Martinez-Prieto, L.M., Cano, I., Marquez, A. et al. (2017). *Chem. Sci.* 8: 2931–2941. (d) Baquero, E.A., Tricard, S., Flores, J.C. et al. (2014). *Angew. Chem. Int. Ed.* 53: 13220–13224. (e) Lara, P., Philippot, K., and Suarez, A. (2019). *ChemCatChem* 11: 766–771.

17 Asensio, J.M., Tricard, S., Coppel, Y. et al. (2017). *Chem. Eur. J.* 23: 13435–13444.

18 (a) Luo, Z., Min, Y., Nechiyil, D. et al. (2019). *Catal. Sci. Technol.* 9: 6884–6898. (b) Ibrahim, M., Garcia, M.A.S., Vono, L.L.R. et al. (2016). *Dalton Trans.* 45: 17782–17791. (c) Garcia, M.A.S., Ibrahim, M., Costa, J.C.S. et al. (2017). *Appl. Catal., A* 548: 136–142. (d) Ibrahim, M., Wei, M.M., Deydier, E. et al. (2019). *Dalton Trans.* 48: 6777–6786.

19 Cano, I., Tschan, M.J.L., Martinez-Prieto, L.M. et al. (2016). *Catal. Sci. Technol.* 6: 3758–3766.

20 Zaramello, L., Albuquerque, B.L., Domingos, J.B., and Philippot, K. (2017). *Dalton Trans.* 46: 5082–5090.

21 Garcia-Anton, J., Axet, M.R., Jansat, S. et al. (2008). *Angew. Chem. Int. Ed.* 47: 2074–2078.

22 (a) Pery, T., Pelzer, K., Buntkowsky, G. et al. (2005). *ChemPhysChem* 6: 605–607. (b) Rothermel, N., Roether, T., Ayvali, T. et al. (2019). *ChemCatChem* 11: 1465–1471.

23 (a) Palazzolo, A., Feuillastre, S., Pfeifer, V. et al. (2019). *Angew. Chem. Int. Ed.* 58: 4891–4895. (b) Pieters, G., Taglang, C., Bonnefille, E. et al. (2014). *Angew. Chem. Int. Ed.* 53: 230–234.

24 Rothermel, N., Bouzouita, D., Roether, T. et al. (2018). *ChemCatChem* 10: 4243–4247.

25 Gao, L., Perato, S., Garcia-Argote, S. et al. (2018). *Chem. Commun.* 54: 2986–2989.

26 Gutmann, T., Bonnefille, E., Breitzke, H. et al. (2013). *Phys. Chem. Chem. Phys.* 15: 17383–17394.

27 Novio, F., Monahan, D., Coppel, Y. et al. (2014). *Chem. Eur. J.* 20: 1287–1297.

28 Lara, P., Rivada-Wheelaghan, O., Conejero, S. et al. (2011). *Angew. Chem. Int. Ed.* 50: 12080–12084.

29 Martinez-Prieto, L.M., Urbaneja, C., Palma, P. et al. (2015). *Chem. Commun.* 51: 4647–4650.

30 (a) Ayvali, T., Lecante, P., Fazzini, P.-F. et al. (2014). *Chem. Commun.* 50: 10809–10811. (b) Ayvali, T., Fazzini, P.-F., Lecante, P. et al. (2017). *Dalton Trans.* 46: 15070–15079.

31 (a) Lara, P., Ayvali, T., Casanove, M.-J. et al. (2013). *Dalton Trans.* 42: 372–382. (b) Bonnefille, E., Novio, F., Gutmann, T. et al. (2014). *Nanoscale* 6: 9806–9816.

32 Qi, X., Axet, M.R., Philippot, K. et al. (2014). *Dalton Trans.* 43: 9283–9295.

33 Ibrahim, M., Poreddy, R., Philippot, K. et al. (2016). *Dalton Trans.* 45: 19368–19373.

34 Gonzalez-Galvez, D., Nolis, P., Philippot, K. et al. (2012). *ACS Catal.* 2: 317–321.

35 Rafter, E., Gutmann, T., Loew, F. et al. (2013). *Catal. Sci. Technol.* 3: 595–599.

36 Dragu, A., Kinayyigit, S., Garcia-Suarez, E.J. et al. (2015). *Appl. Catal., A* 504: 81–91.

37 (a) Gonzalez-Galvez, D., Lara, P., Rivada-Wheelaghan, O. et al. (2013). *Catal. Sci. Technol.* 3: 99–105. (b) Martinez-Prieto, L.M., Ferry, A., Lara, P. et al. (2015). *Chem. Eur. J.* 21: 17495–17502. (c) Lara, P., Martinez-Prieto, L.M., Rosello-Merino, M. et al. (2016). *Nano-Struct. Nano-Objects* 6: 39–45. (d) Martinez-Prieto, L.M., Ferry, A., Rakers, L. et al. (2016). *Chem. Commun.* 52:

4768–4771. (e) Martinez-Prieto, L.M., Ferry, A., Rakers, L. et al. (2018). *Chem. Commun.* 54: 7070–7073.

38 Martinez-Prieto, L.M., Rakers, L., Lopez-Vinasco, A.M. et al. (2017). *Chem. Eur. J.* 23: 12779–12786.

39 (a) Leng, F., Gerber, I.C., Lecante, P. et al. (2016). *RSC Adv.* 6: 69135–69148. (b) Leng, F., Gerber, I.C., Lecante, P. et al. (2016). *ACS Catal.* 6: 6018–6024.

40 Axet, M.R., Conejero, S., and Gerber, I.C. (2018). *ACS Appl. Nano Mater.* 1: 5885–5894.

41 Leng, F., Gerber, I.C., Axet, M.R., and Serp, P. (2018). *C.R. Chim.* 21: 346–353.

42 Machado, B.F., Oubenali, M., Axet, M.R. et al. (2014). *J. Catal.* 309: 185–198.

43 Leng, F., Gerber, I.C., Lecante, P. et al. (2017). *Chem. Eur. J.* 23: 13379–13386.

44 (a) Pan, C., Dassenoy, F., Casanove, M.-J. et al. (1999). *J. Phys. Chem. B* 103: 10098–10101. (b) Dassenoy, F., Casanove, M.J., Lecante, P. et al. (2001). *Phys. Rev. B: Condens. Matter* 63: 235407, 1–7.

45 Lara, P., Casanove, M.-J., Lecante, P. et al. (2012). *J. Mater. Chem.* 22: 3578–3584.

46 Guerrero, M., Costa, N.J.S., Vono, L.L.R. et al. (2013). *J. Mater. Chem. A* 1: 1441–1449.

47 Uberman, P.M., Costa, N.J.S., Philippot, K. et al. (2014). *Green Chem.* 16: 4566–4574.

48 Ungaro, V.A., Liria, C.W., Romagna, C.D. et al. (2015). *RSC Adv.* 5: 36449–36455.

49 Costa, N.J.S., Guerrero, M., Colliere, V. et al. (2014). *ACS Catal.* 4: 1735–1742.

50 Garcia-Suarez, E.J., Tristany, M., Garcia, A.B. et al. (2012). *Microporous Mesoporous Mater.* 153: 155–162.

51 Garcia-Suarez, E.J., Balu, A.M., Tristany, M. et al. (2012). *Green Chem.* 14: 1434–1439.

52 Pinilla, J.L., Garcia, A.B., Philippot, K. et al. (2014). *Fuel* 116: 729–735.

53 Castillejos, E., Debouttiere, P.-J., Roiban, L. et al. (2009). *Angew. Chem. Int. Ed.* 48: 2529–2533.

5

Metal Nanoparticles in Polyols: Bottom-up and Top-down Syntheses and Catalytic Applications

Trung Dang-Bao[1], Isabelle Favier[2], and Montserrat Gómez[2]

[1] Ho Chi Minh City University of Technology, VNU-HCM, Faculty of Chemical Engineering, 268 Ly Thuong Kiet, District 10, Ho Chi Minh City, Vietnam
[2] Université Toulouse 3–Paul Sabatier, CNRS UMR 5069, Laboratoire Hétérochimie Fondamentale et Appliquée,, 118 Route de Narbonne, 31062 Toulouse Cedex 9, France

5.1 Introduction

Well-defined metal nanoparticles (MNPs), with dimensions in the range of 1–100 nm, have undergone a colossal development over the past four decades because of their distinctive physical and chemical properties in comparison to materials exhibiting higher (such as coordination polymers or bulk metals) or lower (such as molecular complexes) sizes. Actually, this unique behavior is mainly due to the electronic and structural effects of nanomaterials, in particular enabling a wide range of catalytic transformations [1]. Even if MNPs have been extensively used in heterogeneous catalysis, with important contributions mainly associated to oil sector [2], nanocatalysis has appeared at the end of the twentieth century, combining colloidal nanoparticles in various solvents and engineering, allowing a control of morphology and surface state of the as-prepared nano-objects. The advances in deep characterization of nano-objects using cutting-edge techniques provide a rational design of catalysts at the nanometric scale [3]. Certainly, reproducible synthetic methodologies leading to MNPs with a close-fitting control of structure, morphology, and surface state are decisive for their applications.

The methodologies to prepare MNPs are commonly cataloged in chemical (bottom-up) and physical (top-down) approaches [4]. Chemical methods can be classified into chemical vapor deposition [5] and colloidal synthesis [4b, 6], including different activation modes (conventional heating, ultrasounds, microwaves, photolysis, electrochemistry, etc.). On the other hand, physical procedures are based on the transformation of bulk materials into nano-objects by different means (photons, heat, ions, mechanical milling, etc.) [7], producing highly pure nanomaterials. Both approaches are complementary and can be combined in order to afford MNPs with control of morphology, structure, and purity. In this frame, sputtering deposition of metal atoms or small clusters on a liquid phase seems to be an appropriate methodology to prepare distinct nanomaterials [4a, 8]. For

Nanoparticles in Catalysis: Advances in Synthesis and Applications, First Edition.
Edited by Karine Philippot and Alain Roucoux.

that purpose, liquids showing negligible vapor pressure should be involved, such as silicone oils [9], vegetable oils [10], poly(ethylene glycol) (PEG) [11], and ionic liquids [12], among the substrates the most studied.

This chapter focuses on the preparation of nonsupported MNPs in polyols with a potential for their applications in catalysis, covering both bottom-up and top-down synthetic methodologies.

5.2 Bottom-up Approach: Colloidal Synthesis in Polyols

With the aim of controlling the size, shape, and composition of MNPs that induce their unique physicochemical properties, bottom-up methodologies to synthesize MNPs in liquid phase have been successfully developed [4b, 6, 13]. In this context, chemical reduction and/or decomposition of metallic salts and organometallic complexes have been frequently applied, using different types of reducing agents, such as hydrogen, carbon monoxide, hydride salts ($NaBH_4$, $LiAlH_4$, etc.), hydrazine, alcohols, citrate salts, etc. Many solvents have been involved in these approaches, from conventional (organic solvents, generally coming from fossil sources) to alternative ones (showing a lower environmental impact), such as water [14], ionic liquids [15], $scCO_2$ [16], or polyols [17]. Some of these sustainable solvents, in contrast to common volatile organic ones, often perform diverse tasks, acting as stabilizers, reductants, and/or liquid phase able to confine the nanoclusters due to their supramolecular structures, as in the case of ionic liquids [18] and glycerol [19]. Thirty years ago, the polyol process was for the first time applied by Fiévet's group. Thus, metal salts in liquid polyols (mainly using ethylene glycol) lead to the formation of well-defined metal-based nanoparticles [20]. The polyol acts as a polyfunctional component, i.e. solvent of the metal precursors (relative high dielectric constant), reducing agent (relative low reduction potentials), and in some cases stabilizer (mostly for higher polyols, such as polysaccharides, polyphenols, etc. [1d, 21]), avoiding the agglomeration and the oxidation of the as-prepared MNPs [17a, 22]. In addition, the high viscosity of polyols can limit the diffusion of reactants, influencing the particle growth and in consequence controlling the size, shape, and structure of MNPs [17a, 23]. Moreover, polyols are appropriate solvents in microwave-assisted synthesis of MNPs because of their polarity and high boiling points [24]. This section aims at describing the synthesis of MNPs for catalytic purposes, involving as solvents ethylene glycol, PEG, glycerol, and also polysaccharides, which exhibit a high stability [25].

5.2.1 Ethylene Glycol and Poly(ethylene glycol)

Ethylene glycol is the most frequently used polyol to prepare MNPs. In the pioneering work of Fiévet, Lagier, and Figlarz reported in 1989 [20a, b], MNPs of abundant metals, such as Co, Ni, and Cu, were prepared by mixing the corresponding metal salt with ethylene glycol in the acid medium.

A polyol process consists in reducing a metal precursor (commonly a salt or metal oxide) by a polyol even if an additional reductant is sometimes necessary because of the low reduction potential of the corresponding alcohol [17a]. The identification of the oxidation products coming from the polyol is a key aspect for elucidating the redox mechanism involved in the polyol synthesis. The oxidation products of ethylene glycol are known (Scheme 5.1): glycol-aldehyde is generated by oxidation of glycol (the first two-electron oxidation step) and then oxidized to CO_2 through different two-electron oxidation steps [26]. Moreover, the MNPs can in turn cause the polyol oxidation through an autocatalytic process [27]. Actually, polyols play an important role in both nucleation and growth of MNPs. The control of size, shape, composition, and crystallinity of obtained MNPs could be modified by tuning the reaction parameters, such as nature of the polyol, stabilizer (if other than polyol), reducing agent, and nature and concentration of metal precursors [28, 29].

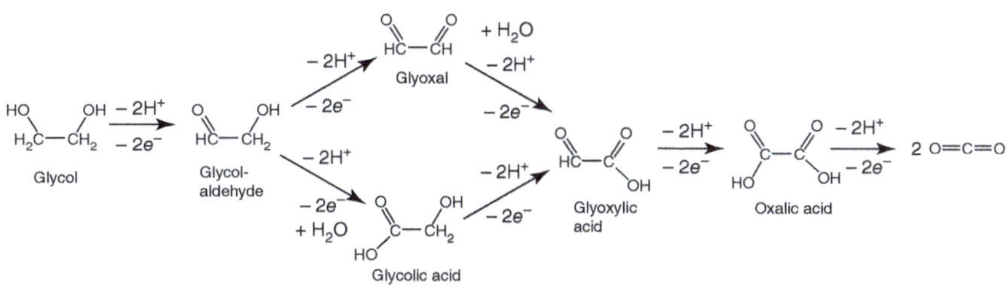

Scheme 5.1 Stepwise oxidation of ethylene glycol to CO_2. Source: Reprinted from Fiévet et al. [17a] with permission of Royal Society of Chemistry via Copyright Clearance Center (license number: 4595920111837) 2018.

The group of Thünemann reported the synthesis of AgNPs in ethylene glycol, stabilized by different macromolecules and characterized by (HR)TEM, SAXS, DLS (dynamic light scattering), TGA (thermal gravimetric analysis), IR, and UV–vis. PAA (poly(acrylic acid)), BSA (bovine serum albumin), and GSH (glutathione) were also used as stabilizers in order to study their influence on the NP surface state and consequently on catalysis (Figure 5.1) [30]. The as-prepared AgNPs were applied in the catalytic reduction of 4-nitrophenol toward 4-aminophenol, using $NaBH_4$ as a reducing agent. AgNPs stabilized by PAA were the most active, in relation to those stabilized by BSA and GSH (activity parameter: $436 \pm 24\,l\,g^{-1}\,s^{-1}$ [PAA], $77.6 \pm 0.9\,l\,g^{-1}\,s^{-1}$ [GSH], and $3.47 \pm 0.50\,l\,g^{-1}\,s^{-1}$ [BSA]), evidencing the passivation of the AgNP surface caused by GSH and BSA, not correlated with the shell thickness of the stabilizer.

Jiu et al. reported the synthesis of triangular hexagonal-shaped single-crystal Au nanosheets by reduction of $H[AuCl_4]$ in ethylene glycol at 150 °C in the presence of poly(vinylpyrrolidone) (PVP) as a capping agent and additional salts (such as $FeCl_3$, NaCl, and $CuCl_2$), which influence the growth of NPs [31]. Both the stabilizer and salt additives played a crucial influence on the formation of Au sheet-shaped nanostructures, together with the concentration of $H[AuCl_4]$ and PVP, as evidenced by scanning electron microscopy (SEM) analyses. Ethylene glycol derivatives, such as di-, tri-, and tetra-substituted derivatives, including PEG, have also been widely investigated in the synthesis of MNPs. Among them, PEG is commonly used because its physicochemical properties can be tuned by changing its molecular weight [32]. The pioneering work applying PEG as both a reducing agent and a

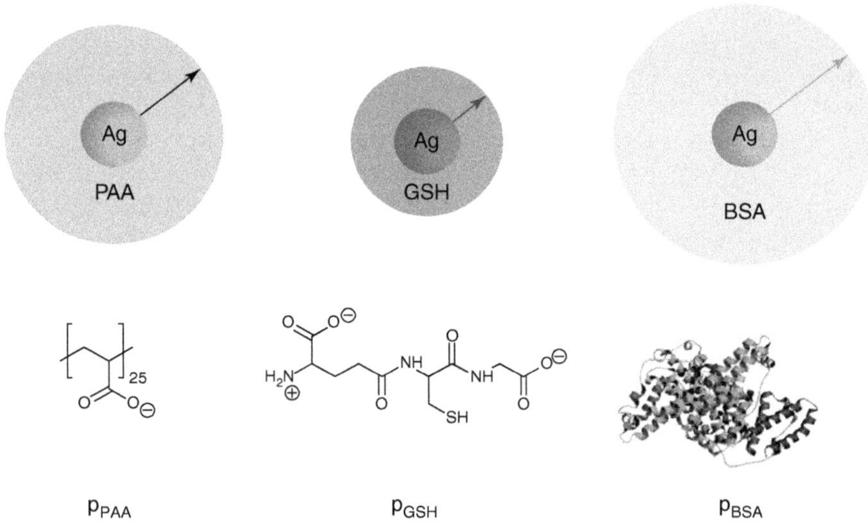

Figure 5.1 Schematic illustration of AgNP core (macro)molecule shell. The mean shell thicknesses are as follows: for PPA, 7.0 nm; for GSH, 3.8 nm; and for BSA, 8.9 nm. PAA means poly(acrylic acid); GSH, glutathione; and BSA, bovine serum albumin. Source: Reprinted with permission from Kästner et al. [30]. Copyright 2016, American Chemical Society.

solvent was reported by Pillai et al. [33]. In this study, PdNPs dispersed in PEG400 were synthesized by reduction of $Pd(OAc)_2$ under hydrogen at room temperature (r.t.), using phenanthroline as a stabilizer. Transmission electron microscopy (TEM) micrographs showed the formation of well-defined PdNPs of 2–6 nm mean diameter; however, bigger PdNPs of 500–1000 nm were observed in the absence of phenanthroline. These PdNPs were efficiently applied in the hydrogenation of a wide scope of olefins (conversion and selectivity up to 100%) under mild conditions (H_2 balloon, 30–50 °C).

Following a similar procedure, Zhang and Wang performed the synthesis of PdNPs dispersed in PEG2000, by reducing $Pd(OAc)_2$ at 80 °C [34]. In this case, PEG acts as both a reducing agent and a stabilizer. The obtained PdNPs were characterized by PXRD (powder X-ray diffraction), TEM, and UV–vis. Interestingly, the convenient reduction of Pd^{2+} to Pd^0 was correlated with the PEG chain length. The as-prepared PdNPs were used as efficient catalysts for Heck cross-coupling reactions. The catalytic phase was recycled up to six times without significant loss of reactivity (Scheme 5.2).

Zhao and coworkers reported the in situ reduction of $H_2[PtCl_6]$ in PEG2000 without addition of any additive at 100 °C, affording spherical PtNPs (4–12 nm mean diameter) [35]. As stated above, PEG acts as a reducing agent, stabilizer, and solvent. They studied the effect of the molecular weight of PEGs on particle size and observed that the larger the PEG is, the bigger the particle is, as proven by TEM. The as-prepared PtNPs-PEG2000 were applied in the selective hydrogenation of *ortho*-chloronitrobenzene to *ortho*-chloroaniline (full conversion and 99.7% selectivity), exhibiting better catalytic behavior than Pt/C (83.4% selectivity). Moreover, the PtNP-PEG2000 catalyst was recycled up to five times.

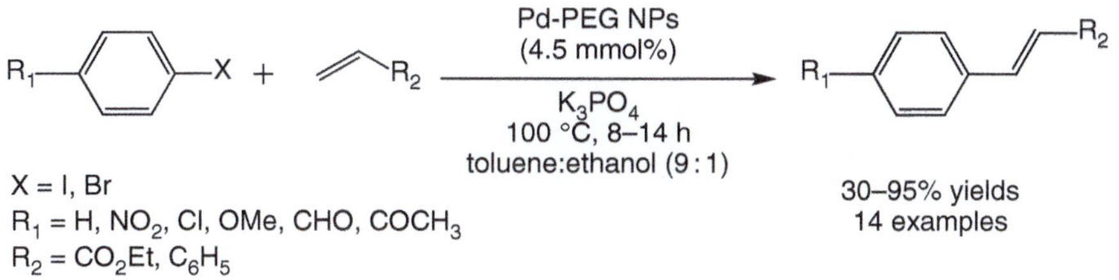

Scheme 5.2 Heck cross-coupling reactions of aryl halides and alkenes catalyzed by PdNPs. Source: Adapted from Luo et al. [34].

Somorjai and coworkers described the synthesis of PtNPs in ethylene glycol by reduction of $H_2[PtCl_6]{\cdot}6H_2O$ in NaOH at 160 °C for three hours. The as-prepared PtNPs (1.7 nm) were incorporated into mesoporous silica and used as catalysts for the hydrogenation of ethylene and ethane hydrogenolysis [36]. The same group also reported the synthesis of PtNPs starting from a mixture of Pt(IV) and Pt(II) precursors, i.e. $(NH_4)_2[PtCl_6]$ or $H_2[PtCl_6]{\cdot}6H_2O$ and $(NH_4)_2[PtCl_4]$, in the presence of tetramethylammonium bromide and PVP, heating at 180 °C for 20 minutes [37]. PtNPs exhibiting different sizes and shapes (nanocubes and nanopolyhedra) were formed by controlling the reducing rate of platinum ions. These NPs were immobilized on mesoporous silica and successfully applied in ethylene hydrogenation.

Without any additional reducing agent, the reduction of first-row metal transition ions can occur in polyols but at high temperatures (>230 °C). However, under these conditions, the decomposition of polyols is frequently observed, together with the reoxidation of as-prepared MNPs during the cool down treatment [22]. In order to lower the temperature conditions, a reducing agent is often added. Therefore, Kawasaki et al. described the synthesis of monodispersed spherical CuNPs (c. 3.5 nm) in ethylene glycol from $Cu(OAc)_2$ (high metal concentration, up to 0.6 M of copper salt) at r.t., using hydrazine as a reducing agent and 1-amino-2-propanol as a stabilizer [38]. Luo's group also developed the synthesis of CuNPs by chemical reduction of $CuCl_2$ in ethylene glycol at 100 °C, using L-ascorbic acid as a reducing agent in the absence of any other capping agent [39]. In this work, highly dispersed CuNPs (<6 nm) were obtained and characterized by (HR)TEM, UV–vis, and EDX (energy dispersive X-ray analysis). The rate of the reagent by addition of CuNPs was remarkable, providing smaller CuNPs by slow addition of L-ascorbic acid into $CuCl_2$ solution than mixing both reagents together. Similarly, the synthesis of zero-valent CuNPs was carried out by reduction of $CuSO_4{\cdot}5H_2O$ at r.t. in ethylene glycol in the presence of PVP or sodium dodecyl sulfate (SDS) as a stabilizer [40]; vanadium(II) sulfate was used as a reductant and also as a oxide scavenger in the acidic medium. The as-prepared CuNPs exhibited a mean size of c. 50 nm in the absence of oxidized copper (no oxygen detected by EDX analysis).

Jeyadevan and coworkers prepared CoNPs by reduction of $Co(OAc)_2$, $CoCl_2$, or $Co(OH)_2$ in ethylene glycol in the presence of sodium metal or sodium hydroxide at high temperature (>190 °C) [41]. Recently, Balachandran et al. reported the mechanism of the formation of CoNPs by polyol process based on Co(II)-ethylene glycol

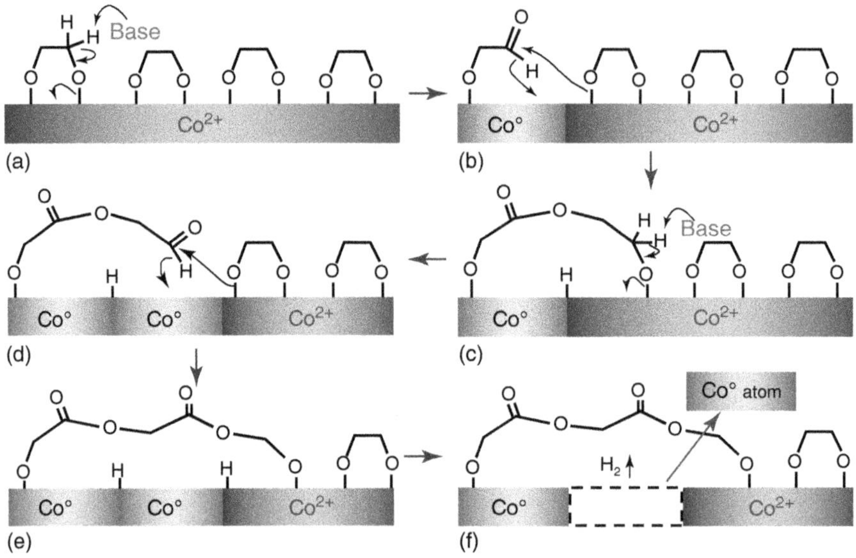

Figure 5.2 Schematic representation of the overall redox mechanism for the synthesis of CoNPs in ethylene glycol. (a) Base reacts with the proton of Co(II) glycolate. (b) An ester bond is formed through bond exchange with partial reduction of Co(II) ions to Co(0) atoms. (c) Base reacts with the proton of a neighboring Co(II) glycolate. (d) An ester bond is formed through bond exchange. (e) The hydrogen atom of the aldehyde group shifts to the Co(0) surface as a hydride ion. (f) H_2 evolution from the surface with dissolution of Co(0) atoms into the solution. Source: Reprinted from Takahashi et al. [42] with permission of Royal Society of Chemistry via Copyright Clearance Center (license number: 4665971417281) 2016.

system [42]. CoNPs were prepared in ethylene glycol at 190 °C using Co(II)-glycolate as a cobalt precursor in the presence of different additives (NaOH, PVP, PPh_3, or KI). SEM analyses exhibited the formation of CoNPs in the size range of 340–970 nm; this wide range of particle sizes could be due to the superficial adsorption of additives on the metal surface. Intermediate products (including Co-based products and the oxidation products of ethylene glycol) formed during the redox reaction of Co(II)-ethylene glycol system were identified by different techniques (PXRD, NMR (nuclear magnetic resonance), FT-IR, and ESI-TOF-MS), permitting to elucidate the overall redox mechanism (Figure 5.2). The presence of basic ions such as hydroxyl or acetate ions is crucial to generate monoanionic ethylene glycol species and then facilitates the formation of Co(II)-glycolate compounds before reducing Co(II) ions to Co(0) metal particles.

Other than transition metals, BiNPs have also been prepared in ethylene glycol in the presence of PVP, leading to the formation of nanoparticles of different morphologies (nanocubes, nanospheres, and triangles) depending on the concentration of the stabilizer, base, and reducing agent (ethylene glycol or sodium hypophosphite) [43]. PXRD analyses showed rhombohedral phase of Bi(0). These as-prepared BiNPs appeared interesting for electrochemical stripping analyses of heavy metals.

5.2.2 Glycerol

Glycerol (propane-1,2,3-triol) is produced in high amounts as a coproduct in the production of biodiesel [44]. Since the beginning of the twenty-first century, glycerol has been considered as a sustainable solvent for organic transformations and in particular for the synthesis of MNPs [45]. Glycerol exhibits several advantages, such as low cost, low toxicity, nonflammability, biodegradability, high boiling point, negligible vapor pressure, high solubility for both organic and inorganic compounds, high polarity, and low miscibility with other organic solvents [17b, c, 46]. Especially, the hydroxyl groups lead to a complex supramolecular network [19, 47], in which MNPs are trapped and dispersed, thus avoiding their agglomeration. Until now, mainly second- and third-row transition MNPs have been synthesized in glycerol. For instance, Harada and coworkers reported the synthesis of PdNPs, RhNPs, RuNPs, and PtNPs in ethylene glycol and glycerol in the presence of PVP, applying the microwave-assisted polyol process, under batch and continuous-flow conditions [48].

In our group, we have developed the synthesis of PdNPs in neat glycerol, using different palladium precursors (Pd(II) and Pd(0) ones) and different stabilizers. PdNPs stabilized by TPPTS (Trisodium 3,3′,3″-phosphinetriyltribenzenesulfonate) [49], PTA (1,3,5-Triaza-7-phosphaadamantane) ammonium derivatives (N-substituted 1,3,5-triaza-7-phosphaadamantane ionic ligands) [50], and cinchona alkaloids [49c, 51] were successfully prepared in glycerol, leading to stable colloidal solutions constituted by small nanoparticles (from c. 1.3 to 3.5 nm, depending on the nature of the stabilizer) (Figure 5.3). The synthetic parameters were optimized, such as metal concentration, nature of the stabilizer, ratio of metal/stabilizer, temperature, and H_2 pressure. The as-prepared PdNPs were characterized by different techniques, including (HR)TEM, EDX, PXRD, XPS (X-ray photoelectron spectroscopy), IR, and ICP (inductively coupled plasma). These PdNPs were efficiently applied in a variety of Pd-catalyzed transformations, such as C–C cross-coupling (Suzuki–Miyaura, Heck–Mizoroki, Sonogashira, and Hiyama reactions), C-heteroatom bond formation (Michael conjugate additions using amines, phosphines, and thiols), and hydrogenation (nitro, nitrile, formyl derivatives, and C-C multiple bonds) processes [49–51]. Multistep processes in order to synthesize heterocyclic compounds were also developed, affording the target products without isolating intermediates [49b]. More significantly, the immobilization of PdNPs in glycerol permitted the recycling of the catalytic phase up to 10 times without metal leaching as proven by ICP analyses.

Regarding first-row transition metals, to the best of our knowledge, very few works report on Cu, Ni, and Mn, leading to mixtures of zero-valent metal and metal oxide NPs. Nisaratanaporn and coworkers described the preparation of ultrafine Cu powders from $Cu(NO_3)_2{\cdot}3H_2O$ in a NaOH/glycerol solution (at molar ratios of NaOH:$Cu(NO_3)_2{\cdot}3H_2O$ in the range of 0 : 1 to 5 : 1) at 120–160 °C. The size, shape, and composition of the obtained NPs were tuned by the concentration of NaOH and the reaction time. PXRD and SEM analyses evidenced the formation of Cu(0) and Cu_2O particles with mean particle sizes in the range of 60–400 nm [52]. Recently, Khan's group prepared CuNPs in glycerol by reduction of $CuCl_2$ with hydrazine as a reducing agent [53]. TEM analysis revealed c. 5 nm sized NPs and

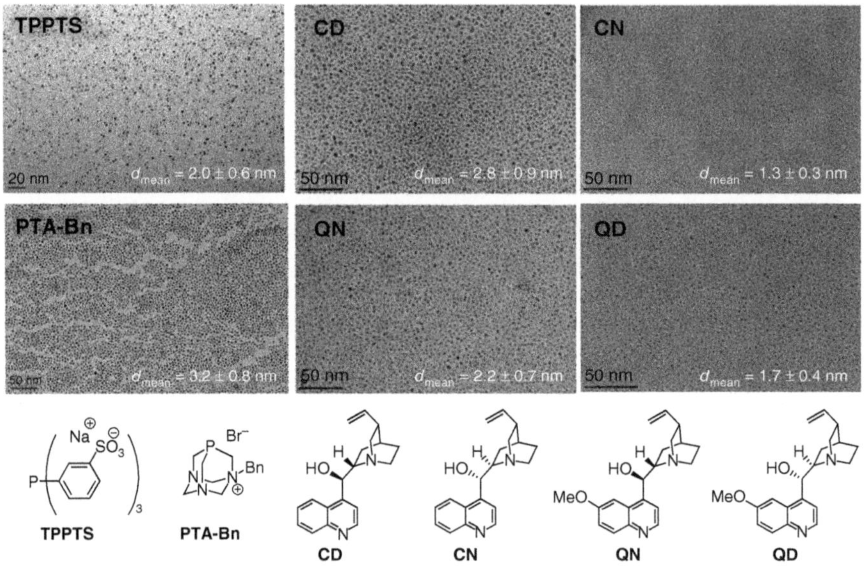

Figure 5.3 TEM images of PdNPs in neat glycerol stabilized by different ligands: phosphines such as TPPTS and PTA-Bn, and cinchona-based derivatives, such as cinchonidine (CD), cinchonine (CN), quinine (QN), and quinidine (QD). Source: Chahdoura et al. [49a, b, 50]; Reina et al. [49c, 51].

XANES studies (X-ray absorption near-edge spectroscopy) evidenced the presence of mixtures of Cu(0) and CuO NPs. Bartůněk and coworkers reported the synthesis of CuNPs stabilized in a polyvinyl alcohol-glycerol matrix. PXRD analysis showed the presence of crystalline Cu(0) (fcc structure) of c. 37 nm, but the presence of amorphous copper oxides in the samples could not be excluded [54].

In our group, we reported the selective preparation of Cu_2O NPs (d_{mean} = 4.7 ± 1.5 nm) immobilized in glycerol by reduction of $Cu(OAc)_2$ with H_2 (3 bar) at 100 °C, using PVP as a stabilizer [55]. A full characterization ((HR)TEM, XRD (X-ray diffraction), FT-IR, and EDX) evidenced the exclusive formation of Cu_2O phase. These Cu_2O NPs were applied in C-heteroatom couplings and azide–alkyne cycloadditions (CuAAC), leading to the expected products in high isolated yields (86–97%). Furthermore, the catalytic phase was recycled up to 10 times showing a good catalytic lifetime. Based on the above strategy, small and spherical zero-valent CuNPs were immobilized in glycerol in the presence of PVP as a stabilizer and hydrogen as a reducing agent and fully characterized by (HR)TEM, EDX, UV–vis, IR, PXRD, XPS, and cyclic voltammetry [56] (Figure 5.4). CuNPs dispersed in glycerol proved to be a robust and versatile catalyst for a diversity of C=N bond formation reactions, synthesis of di- (via cross-dehydrogenative coupling), tri- (via aldehyde–amine–alkyne A^3 coupling), and tetra-substituted propargylic amines (via ketone–amine–alkyne KA^2 coupling). Moreover, various heterocyclic compounds, such as indolizines, benzofurans, and quinolines, were prepared by tandem A^3 coupling-cycloisomerization processes using ortho-functionalized benzaldehydes as substrates. Interestingly, the

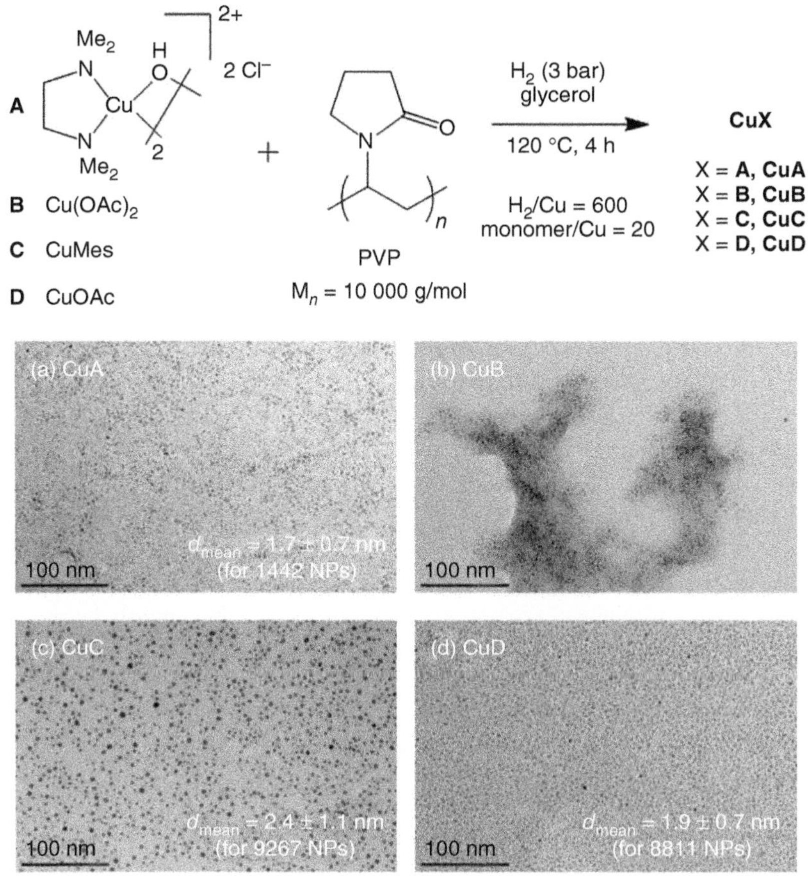

Figure 5.4 (A) Synthesis of Cu(0) nanoparticles **CuX** (X denotes the copper precursor, X = **A–D**) in glycerol in the presence of PVP under hydrogen atmosphere (top). (B) TEM analyses (recorded in the liquid phase) of zero-valent CuNPs dispersed in glycerol, synthesized at 120 °C from Cu(II) (**CuA** and **CuB**) (a,b) and from Cu(I) metallic precursors (**CuC** and **CuD**) (c,d). Source: Reproduced from Dang-Bao et al. [56] with permission of John Wiley and Sons and Copyright Clearance Center (license number: 4595311324325) 2017.

catalytic glycerol phase was recycled more than five times, preserving its activity [56].

Concerning nickel nanoparticles, the group of Shen reported the synthesis of Ni and Ni/NiO core–shell NPs (c. 12–30 nm) by the polyol approach, from $Ni(OAc)_2{\cdot}4H_2O$ or $Ni(NO_3)_2{\cdot}6H_2O$. After calcination at 400 °C under nitrogen atmosphere, Ni(0)NPs were obtained. However, at higher temperature (400–600 °C) and under air, Ni/NiO core–shell NPs exhibiting different structures were formed [57].

We succeeded to synthesize colloidal Ni(0)NPs in glycerol under smoother conditions, starting from [Ni(1,5-cyclooctadiene)$_2$]. A full characterization ((HR)TEM, EDX, PXRD, XPS, IR, and magnetization) evidenced that Ni(0)NPs (c. 1 nm) were exclusively formed [58]. This catalytic nanomaterial was efficiently applied in the synthesis of *Z*-alkenes by semihydrogenation processes of alkynes, as well as in the hydrogenation of nitro, nitrile, and formyl groups to the corresponding anilines, benzylamines, and benzylalcohols (77–95% yields) and was recycled up to 10 times.

The synthesis of $Mn(OH)_2$ (c. 20 nm) and $MnCO_3$ (c. 60 nm) nanocrystals was also developed by Li's group by hydrothermal reduction of $KMnO_4$, using glycerol and glycol as solvents, reductants, and stabilizing agents [59].

Remarkably, we recently reported the synthesis of well-dispersed Pd/Cu bimetallic NPs (PdCu NPs) in glycerol [60], representing an innovative way to obtain stable zero-valent PdCu NPs in a liquid phase. These PdCu NPs were prepared by coreduction of $Pd(OAc)_2$ $[Cu(TMEDA)(\mu\text{-}OH)]_2Cl_2$ (TMEDA = tetramethylethylendiamine) under hydrogen pressure (3 bar) and fully characterized (PXRD, FTIR, XPS, cyclic voltammetry, HR(TEM), HAADF-STEM, and EDX (scanning lines on isolated NPs)). Depending on the Pd/Cu ratios employed, PdNPs coated by a nonuniform Cu shell leading to a cluster-in-cluster structure (Pd/Cu = 1/1, **Pd1Cu1**), disordered alloy (Pd/Cu = 1/2, **Pd1Cu2**), or mainly a mixture of monometallic PdNPs and CuNPs (Pd/Cu = 2/1, **Pd2Cu1**) were obtained (Figure 5.5). The structure–reactivity correlation was evidenced by the reactivity observed in the selective formation of alkenes by hydrogenation of alkynes. Thus, only **Pd1Cu1** led to the formation of alkenes, proving the influence of copper on palladium in the NPs, mainly because of electronic effects. Interestingly, **Pd1Cu1** acted as a multitask catalytic system, involving Cu-catalyzed alkyne azide cycloaddition (AAC) and Pd-catalyzed C–C cross-couplings (Suzuki–Miyaura or Sonogashira). Because of the different rates exhibited by the AAC reaction (faster) and C–C couplings (slower), these tandem processes permitted to obtain the desired products in high yields [60b].

5.2.3 Carbohydrates

The polyol processes often require high reaction temperatures and addition of a hazardous reducing agent. The use of biomolecules as reducing agents is an efficient approach to induce the formation of MNPs, often called green synthesis or biosynthesis of MNPs. Among them, carbohydrates are the most frequently used in the reduction of noble metal ions (Ag, Au, Pd, and Pt).

Alditol bearing polysaccharides (chitosan and saccharide derivatives) allowed the synthesis of AgNPs from $AgNO_3$ at r.t. in the absence of any reducing agent [61]. The influence of $AgNO_3$ concentration and alditol nature on the formation of AgNPs was examined. TEM analysis revealed the formation of AgNPs of c. 10 nm along with few bigger particles (c. 30 nm). The formation of lactones coming from the

Figure 5.5 (A) PdCu NPs (**Pd1Cu1**) prepared by coreduction of metal precursors in glycerol with the corresponding TEM image and the size distribution histogram (below). (B) HAADF-STEM images of single nanoparticles with the corresponding EDX scan following the gray line for **Pd1Cu1** (a) and **Pd1Cu2** (b). Source: Reprinted with permission from Dang-Bao et al. [60b]. Copyright 2019 American Chemical Society.

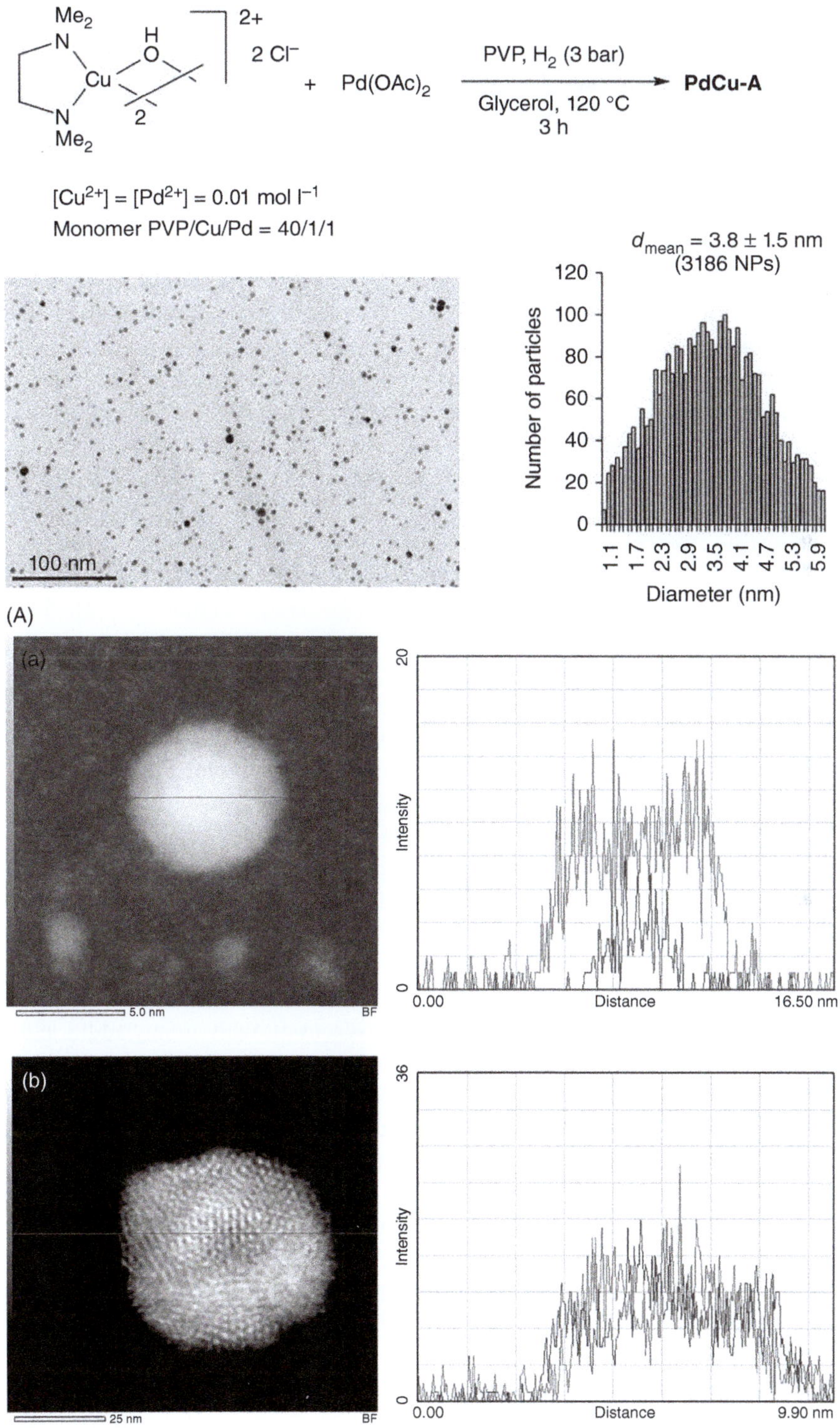

2+
2 Cl⁻
+ Pd(OAc)2
PVP, H2 (3 bar)
Glycerol, 120 °C
3 h
PdCu-A
[Cu2+] = [Pd2+] = 0.01 mol l−1
Monomer PVP/Cu/Pd = 40/1/1
dmean = 3.8 ± 1.5 nm
(3186 NPs)
Number of particles
Diameter (nm)
100 nm
(A)
(a)
5.0 nm
Intensity
Distance
0.00
16.50 nm
(b)
25 nm
9.90 nm
(B)

polysaccharide oxidation was observed by ^{13}C NMR without the cleavage of C—C bonds as observed in most of the polyols used in the synthesis of MNPs.

The group of Lee and Park reported the synthesis of AuNPs from $H[AuCl_4]$ (0.5 mM) using various glycosides as reducing agents in aqueous NaOH solution (7 mM) at r.t. [62]. Eight sugar-containing reductants (glycoside, glucose, and glucuronic acid) were evaluated in the formation of AuNPs. The selective oxidation of C6–OH of glycosides to carboxylic acid ($C5–CO_2H$) was crucial for the reduction of Au(III) precursor. Thus, the use of arbutin, phenyl β-D-glucoside, and methyl β-D-glucoside as reducing agents permitted to obtain nanosized particles (c. 8–40 nm). However, the oxidation of C6–OH of salicin was not possible because of the steric effect of $C2'–CH_2OH$, affording aggregates. The presence of a carboxylic acid group in C5 avoided the formation of AuNPs using phenyl β-D-glucuronide and methyl β-D-glucuronide as polysaccharides. Interestingly, D-glucose and D-glucuronic acid could be oxidized to the corresponding aldehyde via anomerization (because of their weak glycosidic bonds), reducing $H[AuCl_4]$ toward AuNPs.

Following this protocol, both Au(III) ions and *N*-methylmorpholine *N*-oxide were reduced, and cellulose-thiosemicarbazides provided self-assembly immobilization to the surfaces of gold nano-objects via S—Au bond formation (Figure 5.6) [63]. This procedure afforded AuNPs of c. 10–20 nm and AgNPs of c. 5–30 nm. Other sugars such as chitosan, cellohexose, cellobiose, maltose, and lactose also appeared efficient for the synthesis of carbohydrate–AuNPs conjugates [64].

Polyphenols possessing hydroxyl groups can act as both reducing and protective agents for the synthesis of MNPs. Sivasubramanian and Muthuraman reported the synthesis of AgNPs from $AgNO_3$ at r.t. in aqueous extract of *Clerodendrum serratum* leaves thanks to the rich composition of polyphenol glycosides (e.g. baicalein and quercetin 3-O-β-D-glucoside). The as-prepared AgNPs were characterized by UV–vis (absorption band at 400 nm), SEM, TEM (spherical NPs of c. 5–30 nm), and PXRD techniques (*fcc* structure of Ag) [65].

Epigallocatechin-3-gallate (EGCG, a typical plant polyphenol) grafted on the surface of collagen fiber (CF) was used as a reducing-cum-stabilizing agent, permitting to obtain size-controlled AgNPs from $AgNO_3$ at r.t. without any additional reductant [66]. The particle size of AgNPs was easily controlled by varying the ratio of EGCG/CF, affording AgNPs in a size range of 5–22 nm. Recently, Zhao and Huo developed the straightforward synthesis of colloidal PtNPs using black wattle tannin (BWT, a typical plant polyphenol) as an amphiphilic stabilizer [67]. Its hydrophobic and rigid backbone can preclude the aggregation of PtNPs, leading to well-dispersed NPs, when using at least 15 mg of BWT (Figure 5.7). The as-prepared PtNPs were characterized by UV–vis, FT-IR, XPS, XRD, and TEM analyses, evidencing the formation of spherical and small NPs (c. 1.8 nm). PtNPs-BWT were applied in the biphasic aerobic oxidation of alcohols in aqueous media, giving from moderate to high yields (12–59% for aliphatic alcohols and >80% for aromatic alcohols). The hydrophilic property of BWT provides good stabilization and water solubility to the so-obtained PtNPs, thus promoting the catalytic reaction. The catalyst was recycled up to seven times, preserving the catalytic reactivity, in contrast to those observed

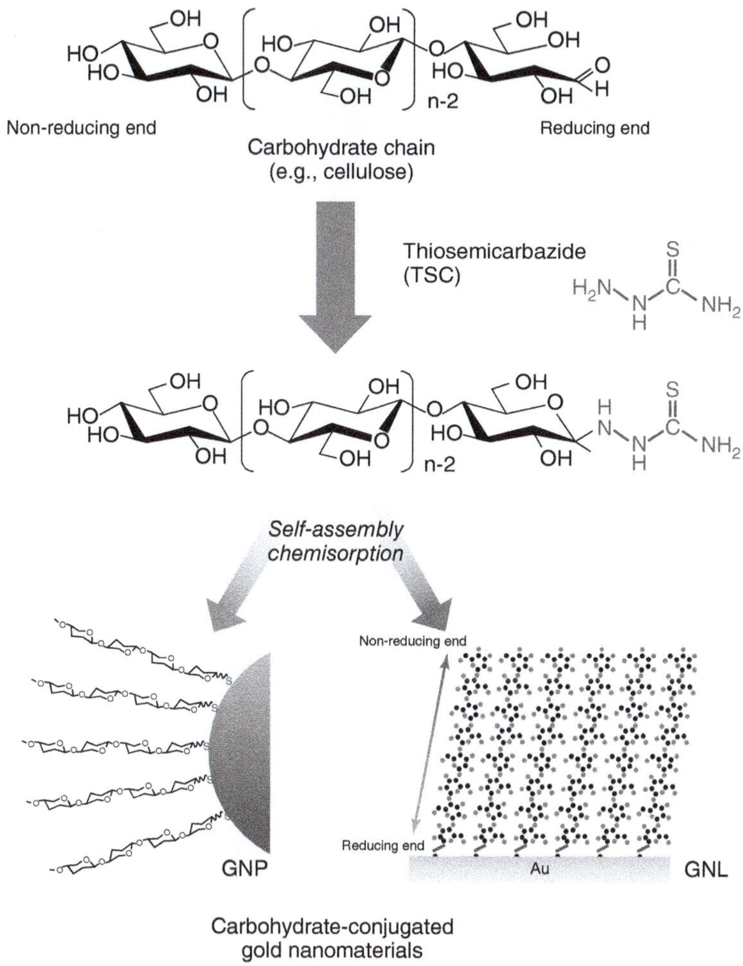

Figure 5.6 Schematic representation of the synthesis of carbohydrate-thiosemicarbazide gold nanoparticles (GNPs) and gold nanolayers (GNLs). Source: Reproduced from Kitaoka et al. [63] with permission of Elsevier and Copyright Clearance Center (license number: 4595350105994) 2011.

with PtNPs stabilized by PVP and PEG, which showed a lower efficiency together with a lack of recyclability because of the NP aggregation and metal leaching.

Veisi and coworkers reported the biosynthesis of PdNPs from $PdCl_2$ in the aqueous solution of oak gum at 100 °C for two hours [68]. Polyphenols contained in the gum of oak fruit can reduce metal ions toward zero-valent metal and stabilize MNPs. The reduction of Pd(II) ions was monitored by UV–vis spectroscopy. The solid material was centrifuged and then characterized by TEM, SEM, PXRD, and EDX, evidencing the formation of spherical PdNPs with a mean diameter of c. 5–7 nm. Suzuki–Miyaura coupling reactions of aryl halides and phenylboronic acid were chosen to evaluate the catalytic reactivity of PdNPs/Oak gum, revealing the efficiency of this catalyst, using iodo-arenes (92–96% isolated yields, in 1–4 hours), bromo-arenes

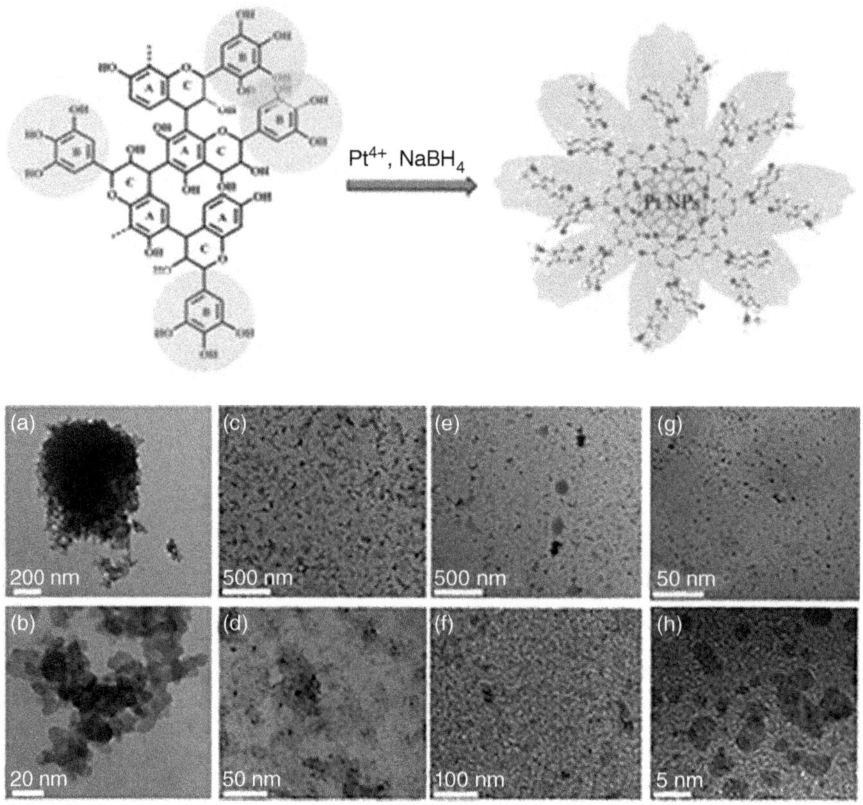

Figure 5.7 (A) Synthesis of PtNPs using the polyphenol BWT as amphiphilic stabilizer. (B) TEM micrographs of PtNPs stabilized by different BWT amounts: (a,b) 2 mg, (c,d) 5 mg, (e,f) 10 mg, and (g,h) 15 mg. Source: Reprinted from Mao et al. [67] with permission of Royal Society of Chemistry via Copyright Clearance Center (license number: 4595360416853) 2016.

(90–96% isolated yields, in 2–7 hours), and chloro-arenes (65–75% isolated yields, in 15–18 hours); the catalytic phase was recycled up to five cycles without any significant loss of reactivity. These as-prepared PdNPs/Oak gum were also efficient in the reduction of nitroarenes (90–96% isolated yields, in 1–2.5 hours).

Similarly, polyols present in aqueous bark extracts of *Ulmus davidiana* and aqueous leaf extracts of *Artemisia annua* were applied as reducing and capping agents for the synthesis of MNPs [69]. Using the aqueous bark extracts of *U. davidiana*, spherical FeNPs (c. 50 nm in diameter) and PdNPs (c. 5 nm in diameter) were synthesized starting from Fe_2O_3 and $PdCl_2$, respectively, at 60 °C for two hours. The obtained NPs were characterized by UV–vis, PXRD, TGA, FT-IR, and (HR)TEM analyses. More interestingly, Fe@Pd NPs were also prepared by a sequential process (the reduction of Fe_2O_3 followed by that of $PdCl_2$) at 60 °C, using the bark extract solution. Their characterization showed a Pd shell deposited on the outer surface of a Fe core structure (Figure 5.8). The as-prepared Fe@Pd NPs were applied as a

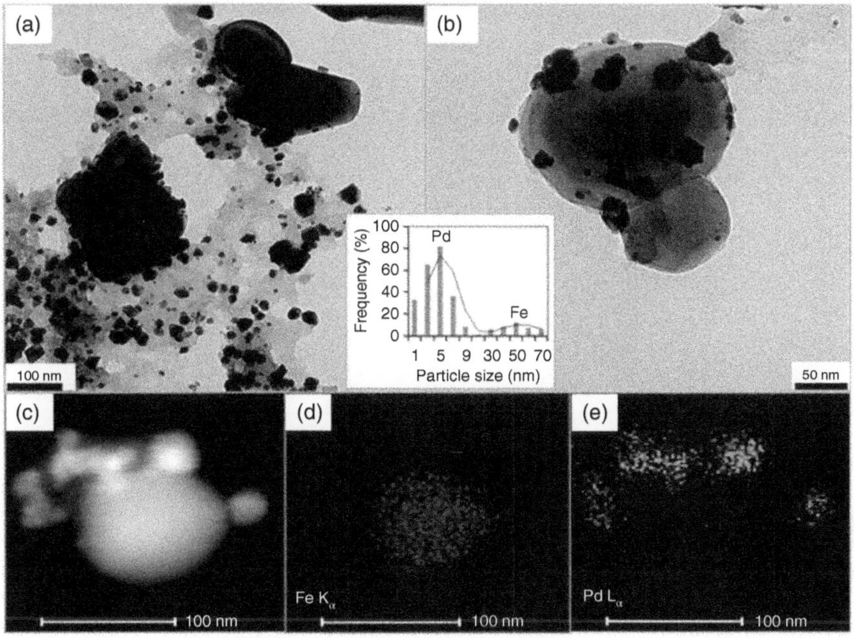

Figure 5.8 TEM micrographs of FePd NPs at different magnifications (a) 100 nm and (b) 50 nm with the corresponding particle size distribution (inset); (c–e) STEM-EDS elemental maps of Fe–Pd bimetallic nanoparticles. Source: Reproduced from Mishra et al. [69a] with permission of Royal Society of Chemistry via Copyright Clearance Center (license number: 4595361481278) 2015.

nanocatalyst in the [3+2] cycloaddition of 1,4-naphthoquinones with β-ketoamides (Scheme 5.3a), showing better reactivity compared to the respective monometallic NPs; the Fe@Pd nanocatalyst was recycled up to five times without loss of catalytic performance because of intrinsic magnetic properties (easy recovery by applying an external magnet) [69a]. Moreover, Fe_3O_4 NPs synthesized from $FeCl_3$ and aqueous leaf extracts of *A. annua* were applied in the synthesis of benzoxazinones and benzothioxazinone derivatives (80–94% isolated yields) (Scheme 5.3b) [69b].

5.3 Top-down Approach: Sputtering in Polyols

Top-down methodologies consist in the transformation of a bulk metal into nanometric species via interaction with photons (such as laser ablation [7a, 70] or gas ions [71]), thermal activation [7b] or mechanical milling [72]. This section aims at describing the synthesis of MNPs by sputtering techniques, which represent the major approaches concerning top-down approaches in liquid phase, in particular involving polyols [73].

Vacuum sputtering is a well-known physical technique to produce NPs immobilized on a solid substrate by bombardment of a target with energetic gas ions

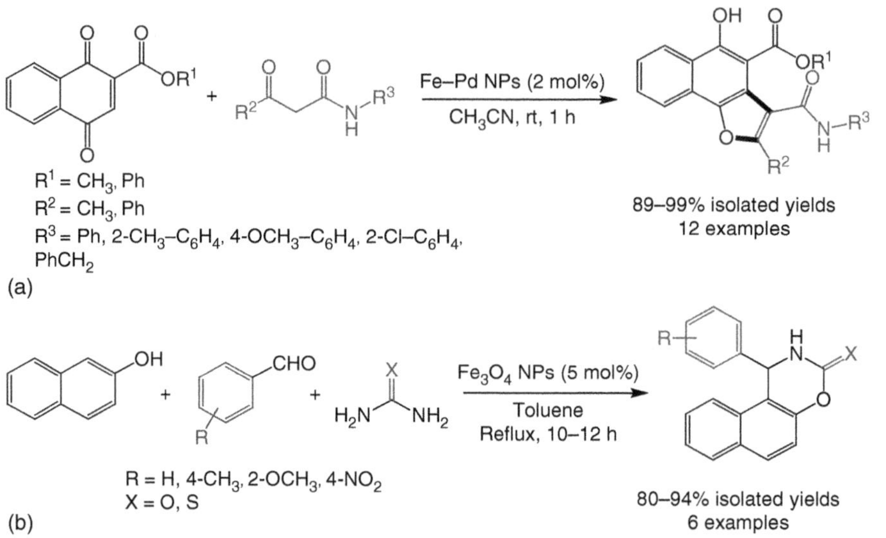

Scheme 5.3 (a) [3+2] Cycloadditions of 1,4-naphthoquinones with β-ketoamides catalyzed by FePd NPs and (b) synthesis of benzoxazinones and benzothioxazinone derivatives catalyzed by Fe_3O_4 nanoparticles. Source: Adapted from Mishra et al. [69].

(using Ar, O_2, N_2, etc.) [71]. These gas ions can be generated by direct current, radio frequency, or magnetron sputtering [74]. Ye et al. reported the first sputtering process in a liquid as a substrate, using silicone oil [75]. A low vapor pressure liquid was used (ionic liquids, ionic surfactants, poly(ethylene) glycol, vegetable oils, etc.) to trap the atoms and clusters produced by the gas bombardment on the corresponding target [76]. The sputtering process in a liquid is particularly attractive in order to form MNPs because it combines a physical technique leading to highly pure angstromic-sized species (metal seeds) and chemical steps by confinement of these seeds in a liquid phase, permitting to control their growth in a better way than under classical sputtering on solids [8c, 76, 77]. Furthermore, the liquids used as substrates generally show high viscosity at ambient conditions, allowing a good dispersion of MNPs and in consequence avoiding their aggregation.

With regard to polyols acting as a liquid substrate, the published reports mainly concern noble metals (Ag and Au) as described below.

AuNPs and AgNPs were prepared by sputtering using anhydrous glycerol as a liquid substrate (conditions: 300 second deposition time, 30 mA current, 7 Pa Ar pressure, and 5 cm target–solution distance) [78]. TEM analyses revealed homogeneous dispersions of MNPs exhibiting small sizes (c. 3.5 nm for both AuNPs and AgNPs). The colloidal solutions of AuNPs were stable, but UV–vis spectroscopy evidenced some aggregates after three months. In contrast, AgNPs were stable even at longer times.

Jang and coworkers prepared AuNPs by combining the use of an ion coater under vacuum (1.10^{-5} atm) and glycerol [79]. Before the AuNPs synthesis, a gold film, with a thickness of c. 130 nm, was prepared on the silicon substrate in the ion coater. Afterward, this gold film was used as a target and glycerol as a liquid substrate, leading to the formation of small AuNPs, depending on the discharge current with diameter from 3.5 to 6.4 nm. At low sputter intensity (3 mA), the detached atoms from the gold film cannot penetrate into glycerol because of their low kinetic energy; thus, nucleation/growth processes occur at the surface of the liquid. Higher discharge currents led to larger AuNPs because of the fast diffusion into glycerol (higher kinetic energy), favoring their growth.

Nishikawa and coworkers studied the temperature effect on size and shape in the synthesis of AuNPs by sputter deposition in liquid PEG, PEG600 [80]. The mean size of AuNPs increased from c. 2 to 8 nm when the temperature increased (from 20 to 60 °C). The temperature also showed an important effect on shape: at low temperature, AuNPs were spherical, while at higher temperature, the anisotropy increased, as stated by TEM. This anisotropy was also confirmed by UV–vis spectra showing SPR (surface plasmon resonance) peaks shifted to higher wavelengths and by SAXS (small-angle X-ray scattering) patterns.

Staszek et al. also studied the temperature influence (from −5 to 40 °C) on the size of AgNPs prepared by sputtering using PVP as a stabilizer and glycerol as a liquid support (280–305 V voltage, 30 mA current, 300 s deposition time, and 0.06 Pa Ar pressure) [81, 82]. In the temperature range 5–20 °C, AgNPs showed an average diameter between c. 13 (at 5 °C) to 21 nm (at 20 °C). At higher temperature (25–40 °C), the size decreased, reaching for the smallest ones, c. 8 nm; this trend could arise from the significant vaporization of glycerol at high temperature, which leads to lower concentration of the incident metal atoms because of the increase of vapor pressure of substrate. This group also synthesized various noble MNPs (AuNPs, AgNPs, PtNPs, and PdNPs) under the same conditions, obtaining spherical and small NPs (c. 2–6 nm of mean diameter) with different degrees of aggregation.

One advantage in using liquid substrates is the possibility to add stabilizers in this phase. Actually, Yonezawa and coworkers synthesized AgNPs by sputter deposition on PEG600 (30 mA current, 2.0 Pa of Ar pressure, 20 minutes) in the presence of 11-mercaptoundecanoic acid as a capping agent, which exhibits a low vapor pressure [83]. UV–vis spectroscopy (SPR absorption bands) and STEM-EDX analyses evidenced the formation of well-dispersed AgNPs with the mercapto-based acid adsorbed at the metal surface. A series of small NP sizes were observed (c. 2–4 nm of mean diameter), evidencing the influence of the stabilizer concentration. Thus, sizes of 3.0 ± 0.6 nm and 2.2 ± 0.5 nm at 5.2×10^{-3} M and 5.2×10^{-1} M of the stabilizer were respectively obtained. The same group applied the unchanged conditions to synthesize AuNPs in PEG600, except the addition of α-thioglycerol as a volatile stabilizer in order to control the growth of AuNPs [84]. The stabilizer was evaporated inside the chamber, providing its coordination to gold atoms and small clusters, generated from the substrate, in the gas phase (nucleation stage). The mean diameter of AuNPs capped by α-thioglycerol was c. 2–3 nm and c. 5 nm in the absence of a stabilizer. The small AuNPs ($d_{mean} < 3$ nm) led to a relative high fluorescence quantum yield (16%). The authors justified the fluorescence behavior of the as-prepared AuNPs by their structure, which certainly

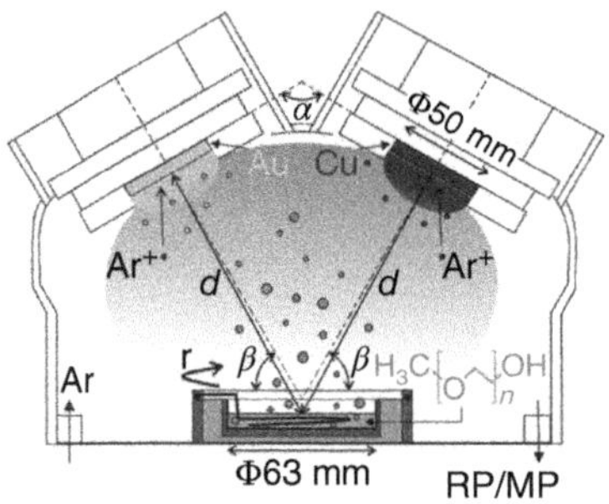

Figure 5.9 Scheme of the sputtering vacuum chamber used for the synthesis of AuCu nanoalloys produced in the gas phase by Ar plasma bombardment of the double target (Cu and Au), using PEG as a liquid substrate; rotary, and molecular pumps were employed. Source: Reprinted with permission from Nguyen et al. [86b]. Copyright 2017 American Chemical Society.

corresponds to multinuclear gold components, formed by aggregation of very small Au clusters.

The same authors applied a matrix sputtering technique to prepare inorganic/organic optical materials that led to AuNPs/thiourethane and AuNPs/urethane hybrid resins with high transparency [85]. The resins were prepared by sputtering of gold target into pentaerythritol tetrakis(3-mercaptopropionate (PEMP) or pentaerythritol ethoxylate (PEEL) (conditions: 40 mA under 10 Pa of air pressure at r.t. in a sputter coater during 5–15 minutes), followed by polymerization of AuNPs/PEEL and AuNPs/PEMP in the presence of *m*-xylylene diisocyanate. AuNPs/urethane resins did not exhibit surface plasmon absorption band; however, they showed luminescence at 690 nm. On the other hand, AuNPs/urethane resins showed surface plasmon absorption bands, depending on the size of AuNPs/PEEL. Nguyen and Yonezawa also applied sputtering methods to synthesize multimetallic nanostructures involving liquids as substrates, both by single- and double (or multiple)-target sputtering [76]. Bimetallic NPs were obtained by using a double target on a liquid polymer, such as PEG, both in the absence [86] and in the presence of additional stabilizers in the liquid phase (Figure 5.9) [87]. In neat PEG, Ag/Au [86a] and Au/Cu [86b] random alloys were obtained, as proven by STEM-HAADF and EDX mapping analyses on individual NPs. The composition of Au/Cu was observed to be dependent on the applied sputter current, and ICP analyses indicated that the sputter rate of gold was roughly three times higher than that of copper. XPS and Auger data showed the presence of some copper oxide. In the presence of 11-mercaptoundecyl-*N,N,N*-trimethylammonium bromide (MUTAB) in PEG, Ag/Au nanoalloys capped by MUTAB were produced, providing photoluminiscent properties that can be tuned for potential applications in biomedicine, from blue to near infrared depending on the NP composition [87]. Moreover, mono (Pt) and bimetallic (NiPt) NPs supported on carbon were obtained by metal sputtering on a substrate constituted of carbon dispersed in PEG [88]. These materials were fully characterized, revealing c. 2.0 nm of mean diameter, for both mono- and bimetallic NPs and a random alloy structure for the NiPt system (Figure 5.10). These NPs were interestingly used as electrocatalysts for the oxygen reduction reaction, being

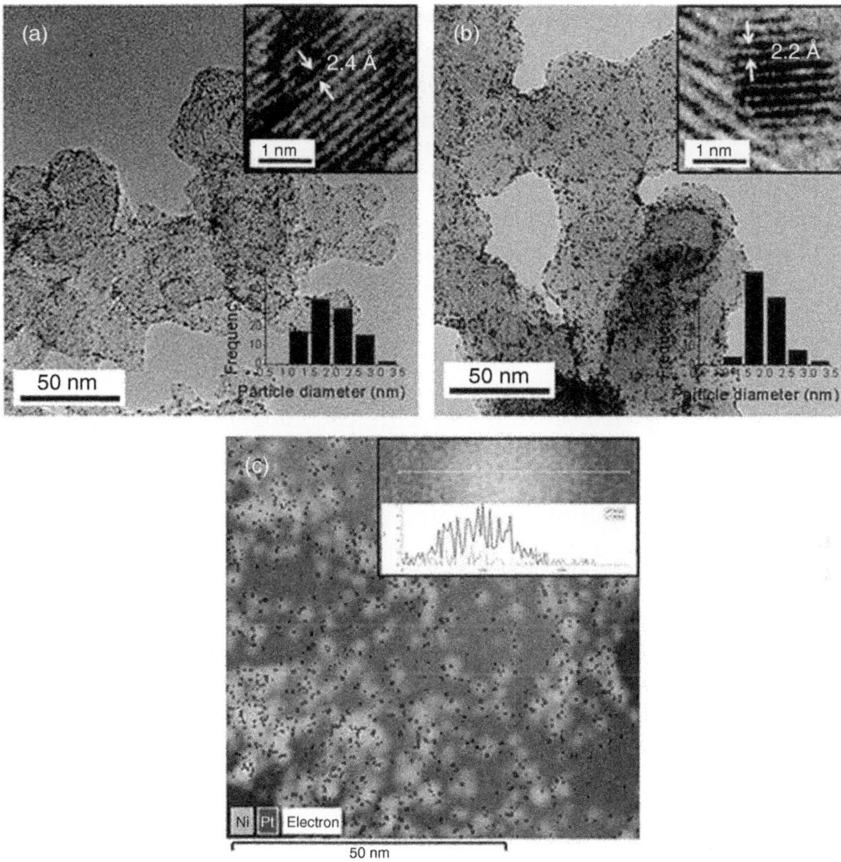

Figure 5.10 TEM images of PtNPs (a) and PtNi bimetallic nanoparticles (b) on carbon, using PEG as a liquid substrate in the presence of carbon; insets: magnified HR-TEM images (upper) and size distribution histograms (lower). STEM and EDS mapping image of PtNi/C (c), with an inset showing the EDS line scan of one isolated nanoparticle. Source: Reprinted from Cha et al. [88] with permission of Royal Society of Chemistry via Copyright Clearance Center (license number: 4595911167560) 2014.

NiPt/C materials the most efficient; this behavior was in agreement with the electron transfer process from Ni to Pt evidenced by XANES analysis.

5.4 Summary and Conclusions

The literature survey presented above clearly highlights the crucial significance of polyols in the synthesis of metal-based NPs. One of the distinctive advantages of polyols is their versatility, behaving as multitask constituents (reducing agent of metallic precursors, solvent, and sometimes as a stabilizer). Additionally, polyols favor the dispersion of the prepared NPs mostly due to their supramolecular assemblies. Polyols have been classically used for the synthesis of "noble" MNPs (metals exhibiting relative high electronegativity, such as Ag, Au, Pd, Pt, etc.). Nevertheless, polyols are also convenient solvents for first-row transition metals in the presence of strong reductants leading to zero-valent NPs; their stability probably results from the low solubility of molecular oxygen in polyol media. More recently, polyols from plant extracts have been applied in the synthesis of well-defined MNPs; the main concern is related to the surface state control of these NPs, particularly dependent on the composition of biopolyols. Polyols together with stabilizers (often polymers, ligands, or surfactants) can efficiently control the size and shape at nanometric scale. As reported in this contribution, metal-based NPs in polyol medium have mostly been synthesized by bottom-up approaches, i.e. decomposition or reduction of metallic precursors to form in situ "naked" metal atoms that quickly generate nuclei, which produce nano-sized structures under well-defined conditions (temperature, pressure, concentration, stabilizer, etc.). However, top-down approaches have appeared more recently in the literature. In particular, sputtering seems to be the most versatile approach, leading to successful synthesis of both mono- and bimetallic NPs, with high purity and remarkable size and composition control. The main limitation is the vapor pressure of polyols because sputtering techniques require high vacuum; this limitation is overcome using high molecular weight polyols, such as PEG600.

Combining the two methodologies, single- or multitarget sputtering deposition on a polyol substrate with chemical reactivity in the liquid phase (by addition of stabilizers, dispersion of solid supports, metallic compounds, etc.) opens new opportunities to prepare well-defined NPs with accurate control at the different stages of their formation, i.e. nucleation and growth, and thus the tuning of their physical and chemical properties, which are particularly interesting for catalytic applications.

Acknowledgments

Financial support from the Centre National de la Recherche Scientifique (CNRS) and the Université Toulouse 3 – Paul Sabatier are gratefully acknowledged.

References

1 (a) Schmid, G. (ed.) (2011). *Nanoparticles: From Theory to Application*. Weinheim, Germany: Wiley-VCH Verlag. (b) Serp, P. and Philippot, K. (eds.) (2013). *Nanomaterials in Catalysis*. Weinheim, Germany: Wiley-VCH Verlag GmbH. (c) Martínez-Prieto, L.M. and Chaudret, B. (2018). *Acc. Chem. Res.* 51: 376–384. (d) Adil, S.F., Assal, M.E., Khan, M. et al. (2015). *Dalton Trans.* 44: 9709–9717. (e) Chaudhuri, R.G. and Paria, S. (2012). *Chem. Rev.* 112: 2373–2433.

2 Zhou, B., Hermans, S., and Somorjai, G.A. (eds.) (2004). *Nanotechnology in Catalysis*. New York, United States: Kluwer Academics/Plenum Press.

3 (a) An, K., Alayoglu, S., Ewers, T., and Somorjai, G.A. (2012). *J. Colloid Interface Sci.* 373: 1–13. (b) Corain, B., Schmid, G., and Toshima, N. (eds.) (2008).

Metal Nanoclusters in Catalysis and Materials Science: The Issue of Size Control. Amsterdam, The Netherlands: Elsevier Science. (c) Astruc, D. (ed.) (2008). *Nanoparticles and Catalysis*. Weinheim, Germany: Wiley-VCH Verlag GmbH; (d) Heiz, U. and Landman, U. (eds.) (2007). *Nanocatalysis*. Berlin, Germany: Springer.

4 (a) Torimoto, T., Kameyana, T., and Kuwabata, S. (2017). *Nanocatalysis in Ionic LiquidsWiley-VCH* (ed. M.H.G. Prechtl), 171–205. Weinheim, Germany. (b) Roucoux, A., Schulz, J., and Patin, H. (2002). *Chem. Rev.* 102: 3757–3778.

5 (a) Hao, Y.F., Meng, G.W., Ye, C.H., and Zhang, L.D. (2005). *Cryst. Growth Des.* 5: 1617–1621. (b) Okumura, M., Tsubota, S., Iwamoto, M., and Haruta, M. (1998). *Chem. Lett.*: 315–316. (c) Reina, A., Jia, X.T., Ho, J. et al. (2009). *Nano Lett.* 9: 30–35. (d) Sivula, K., Le Formal, F., and Gratzel, M. (2009). *Chem. Mater.* 21: 2862–2867.

6 (a) Burda, C., Chen, X.B., Narayanan, R., and El-Sayed, M.A. (2005). *Chem. Rev.* 105: 1025–1102. (b) Amiens, C., Chaudret, B., Ciuculescu-Pradines, D. et al. (2013). *New J. Chem.* 37: 3374–3401.

7 (a) Gelesky, M.A., Umpierre, A.P., Machado, G. et al. (2005). *J. Am. Chem. Soc.* 127: 4588–4589. (b) Dai, Z.R., Pan, Z.W., and Wang, Z.L. (2003). *Adv. Funct. Mater.* 13: 9–24. (c) Wasa, K., Kitabatake, M., and Adachi, H. (2004). *Thin Film Materials Technology: Sputtering of Compound Materials*. Berlin, Germany: Springer-Verlag.

8 (a) Wender, H., de Oliveira, L.F., Migowski, P. et al. (2010). *J. Phys. Chem. C* 114: 11764–11768. (b) Torimoto, T., Okazaki, K., Kiyama, T. et al. (2006). *Appl. Phys. Lett.* 89: 1, 243117–3. (c) Wender, H., Migowski, P., Feil, F.A. et al. (2013). *Coord. Chem. Rev.* 257: 2468–2483.

9 (a) Ye, G.-X., Michely, T., Weidenhof, V. et al. (1998). *Phys. Rev. Lett.* 81: 622–625. (b) Anantha, P., Cheng, T., and Wong, C.C. (2014). *Philos. Mag.* 94: 1967–1981.

10 (a) Mensah, M.B., Awudza, J.A.M., and O'Brien, P. (2018). *R. Soc. Open Sci.* 5: 180824. (b) Wender, H., de Oliveira, L.F., Feil, A.F. et al. (2010). *Chem. Commun.* 46: 7019–7021.

11 (a) Aschenbrenner, O., Supasitmongkol, S., Taylor, M., and Styring, P. (2009). *Green Chem.* 11: 1217–1221. (b) Krieger, U.K., Siegrist, F., Marcolli, C. et al. (2018). *Atmos. Meas. Tech.* 11: 49–63.

12 (a) Janiak, C. (2014). *Catalysis in Ionic Liquids: From Catalyst Synthesis to Application*, RSC Catalysis Series No. 15 (eds. C. Hardacre and V. Parvulescu) Chapter 11, 537–577. (b) Richter, K., Campbell, P.S., Baecker, T. et al. (2013). *Phys. Status Solidi B* 250: 1152–1164.

13 Cushing, B.L., Kolesnichenko, V.L., and O'Connor, C.J. (2004). *Chem. Rev.* 104: 3893–3946.

14 (a) Bulut, S., Fei, Z., Siankevich, S. et al. (2015). *Catal. Today* 247: 96–103. (b) San, K. and Shon, Y.-S. (2018). *Nanomaterials*: 8, 346-1–346-21.

15 Prechtl, M.H.G. (ed.) (2016). *Nanocatalysis in Ionic Liquids*. Weinheim, Germany: Wiley-VCH.

16 (a) Dahl, J.A., Maddux, B.L.S., and Hutchison, J.E. (2007). *Chem. Rev.* 107: 2228–2269. (b) Patete, J.M., Peng, X., Koenigsmann, C. et al. (2011). *Green Chem.* 13: 482–519. (c) Zhang, Y. and Erkey, C. (2006). *J. Supercrit. Fluids* 38: 252–267.

17 (a) Fiévet, F., Ammar-Merah, S., Brayner, R. et al. (2018). *Chem. Soc. Rev.* 47: 5187–5233. (b) Favier, I., Pla, D., and Gómez, M. (2018). *Catal. Today* 310: 98–106. (c) Chahdoura, F., Favier, I., and Gómez, M. (2014). *Chem. Eur. J.* 20: 10884–10893.

18 (a) Favier, I., Madec, D., and Gómez, M. (2013). *Nanomaterials in Catalysis* (eds. K. Philippot and P. Serp), 203–249. Weinheim, Germany: Wiley-VCH. (b) Favier, I., Teuma, E., and Gómez, M. (2016). *Nanocatalysis in Ionic Liquids* (ed. M.H.G. Prechtl), 125–146. Weinheim, Germany: Wiley-VCH. (c) Consorti, C.S., Suarez, P.A.Z., De Souza, R.F. et al. (2005). *J. Phys. Chem. B* 109: 4341–4349. (d) Dupont, J., Suarez, P.A.Z., De Souza, R.F. et al. (2000). *Chem. Eur. J.* 6: 2377–2381. (e) Canongia Lopes, J.N.A., Costa Gomes, M.F., and Padua, A.A.H. (2006). *J. Phys. Chem. B* 110: 16816–16818.

19 Kusukawa, T., Niwa, G., Sasaki, T. et al. (2013). *Bull. Chem. Soc. Jpn.* 86: 351–353.

20 (a) Fievet, F., Lagier, J.P., Blin, B. et al. (1989). *Solid State Ionics* 32-33: 198–205. (b) Fievet, F., Lagier, J.P., and Figlarz, M. (1989). *MRS Bull.* 14: 29–34.

21 (a) Golinska, P., Wypij, M., Ingle, A.P. et al. (2014). *Appl. Microbiol. Biotechnol.* 98: 8083–8097. (b) Alam, M.N., Roy, N., Mandal, D., and Begum, N.A. (2013). *RSC Adv.* 3: 11935–11956.

22 Dong, H., Chen, Y.C., and Feldmann, C. (2015). *Green Chem.* 17: 4107–4132.

23 Nguyen, T.-D. (2013). *Nanoscale* 5: 9455–9482.

24 (a) Gabriel, C., Gabriel, S., Grant, E.H. et al. (1998). *Chem. Soc. Rev.* 27: 213–224. (b) Zhu, Y.-J. and Chen, F. (2014). *Chem. Rev.* 114: 6462–6555. (c) Baghbanzadeh, M., Carbone, L., Cozzoli, P.D., and Kappe, C.O. (2011). *Angew. Chem. Int. Ed.* 50: 11312–11359.

25 (a) Molnár, Á. and Papp, A. (2014). *Catal. Sci. Technol.* 4: 295–310. (b) Garrett, C.E. and Prasad, K. (2004). *Adv. Synth. Catal.* 346: 889–900.

26 Livshits, V., Philosoph, M., and Peled, E. (2008). *J. Power Sources* 178: 687–691.

27 (a) Skrabalak, S.E., Wiley, B.J., Kim, M. et al. (2008). *Nano Lett.* 8: 2077–2081. (b) Aït Atmane, K., Michel, C., Piquemal, J.-Y. et al. (2014). *Nanoscale* 6: 2682–2692.

28 Eid, K., Wang, H., and Wang, L. (2017). *Supra-materials Nanoarchitectonics* (eds. K. Ariga and M. Aono), 135–171. Amsterdam, The Netherlands: Elsevier Science.

29 Hu, Y., Ge, J., Lim, D. et al. (2008). *J. Solid State Chem.* 181: 1524–1529.

30 Kästner, C. and Thünemann, A.F. (2016). *Langmuir* 32: 7383–7391.

31 Jiu, J., Suganuma, K., and Nogi, M. (2011). *J. Mater. Sci.* 46: 4964–4970.

32 Yan, N., Xiao, C., and Kou, Y. (2010). *Coord. Chem. Rev.* 254: 1179–1218.

33 Pillai, U.R. and Sahle-Demessie, E. (2004). *J. Mol. Catal. A: Chem.* 222: 153–158.

34 Luo, C., Zhang, Y., and Wang, Y. (2005). *J. Mol. Catal. A: Chem.* 229: 7–12.

35 Cheng, H., Xi, C., Meng, X. et al. (2009). *J. Colloid Interface Sci.* 336: 675–678.

36 Song, H., Rioux, R.M., Hoefelmeyer, J.D. et al. (2006). *J. Am. Chem. Soc.* 128: 3027–3037.

37 Tsung, C.-K., Kuhn, J.N., Huang, W.Y. et al. (2009). *J. Am. Chem. Soc.* 131: 5816–5822.

38 Hokita, Y., Kanzaki, M., Sugiyama, T. et al. (2015). *ACS Appl. Mater. Interfaces* 7: 19382–19389.

39 Li, M., Xiang, K., Luo, G. et al. (2013). *Chin. J. Chem.* 31: 1285–1289.

40 Reverberi, A.P., Salerno, M., Lauciello, S., and Fabiano, B. (2016). *Materials*: 9, 809-1–809-11.

41 Matsumoto, T., Takahashi, K., Kitagishi, K. et al. (2015). *New J. Chem.* 39: 5008–5018.

42 Takahashi, K., Yokoyama, S., Matsumoto, T. et al. (2016). *New J. Chem.* 40: 8632–8642.

43 Cadevall, M., Ros, J., and Merkoçl, A. (2015). *Electrophoresis* 36: 1872–1879.

44 Knothe, G., Gerpen, J.V., and Krahl, J. (eds.) (2010). *The Biodiesel Handbook*. Urbana, United States: AOCS Press.

45 Wolfson, A., Dugly, C., Tavor, D. et al. (2006). *Tetrahedron: Asymmetry* 17: 2043–2045.

46 (a) Díaz-Álvarez, A.E., Francos, J., Lastra-Barreira, B. et al. (2011). *Chem. Commun.* 47: 6208–6227. (b) Gu, Y. and Jérôme, F. (2010). *Green Chem.* 12: 1127–1138.

47 (a) Geukens, I. and De Vos, D.E. (2003). *Langmuir* 29: 3170–3178. (b) Tagliapietra, S., Orio, L., Palmisano, G. et al. (2015). *Chem. Pap. - Chem. Zvesti* 69: 1519–1531. (c) Shi, S., Peng, X., Liu, T. et al. (2017). *Polymer* 111: 168–176.

48 Harada, M. and Cong, C. (2016). *Ind. Eng. Chem. Res.* 55: 5634–5643.

49 (a) Chahdoura, F., Pradel, C., and Gómez, M. (2013). *Adv. Synth. Catal.* 355: 3648–3660. (b) Chahdoura, F., Mallet-Ladeira, S., and Gómez, M. (2015). *Org. Chem. Front.* 2: 312–318. (c) Reina, A., Serrano-Maldonado, A., Teuma, E. et al. (2018). *Catal. Commun.* 104: 22–27.

50 Chahdoura, F., Favier, I., Pradel, C. et al. (2015). *Catal. Commun.* 63: 47–51.

51 Reina, A., Pradel, C., Martin, E. et al. (2016). *RSC Adv.* 6: 93205–93216.

52 Chokratanasombat, P. and Nisaratanaporn, E. (2012). *Eng. J.* 16: 39–46.

53 Ong, H.R., Khan, Md.M.R., Ramli, R. et al. (2015). *RSC Adv.* 5: 24544–24549.

54 Dobrovolný, K., Ulbrich, P., Švecová, M. et al. (2017). *J. Alloys Compd.* 697: 147–155.

55 Chahdoura, F., Pradel, C., and Gómez, M. (2014). *ChemCatChem* 6: 2929–2936.

56 Dang-Bao, T., Pradel, C., Favier, I., and Gómez, M. (2017). *Adv. Synth. Catal.* 359: 2832–2846.

57 Li, Y., Cai, M., Rogers, J. et al. (2006). *Mater. Lett.* 60: 750–753.

58 Reina, A., Favier, I., Pradel, C., and Gómez, M. (2018). *Adv. Synth. Catal.* 360: 3544–3552.

59 Wang, X. and Li, Y. (2003). *Mater. Chem. Phys.* 82: 419–422.

60 (a) Dang-Bao, T., Pla, D., Favier, I., and Gómez, M. (2017). *Catalysts* 7 (207): 1–33. (b) Dang-Bao, T., Pradel, C., Favier, I., and Gómez, M. (2019). *ACS Appl. Nano Mater.* 2: 1033–1044.

61 Donati, I., Travan, A., Pelillo, C. et al. (2009). *Biomacromolecules* 10: 210–213.
62 Jung, J., Park, S., Hong, S. et al. (2014). *Carbohydr. Res.* 386: 57–61.
63 Kitaoka, T., Yokota, S., Opietnik, M., and Rosenau, T. (2011). *Mater. Sci. Eng., C* 31: 1221–1229.
64 Yokota, S., Kitaoka, T., Opietnik, M. et al. (2008). *Angew. Chem. Int. Ed.* 47: 9866–9869.
65 Raman, R.P., Parthiban, S., Srinithya, B. et al. (2015). *Mater. Lett.* 160: 400–403.
66 Wu, H., He, L., Gao, M. et al. (2011). *New J. Chem.* 35: 2902–2909.
67 Mao, H., Liao, Y., Ma, J. et al. (2016). *Nanoscale* 8: 1049–1054.
68 Veisi, H., Nasrabadi, N.H., and Mohammadi, P. (2016). *Appl. Organomet. Chem.* 30: 890–896.
69 (a) Mishra, K., Basavegowda, N., and Lee, Y.R. (2015). *Catal. Sci. Technol.* 5: 2612–2621. (b) Basavegowda, N., Magar, K.B.S., Mishra, K., and Lee, Y.R. (2014). *New J. Chem.* 38: 5415–5420.
70 Park, H.-Y., Jang, I., Jung, N. et al. (2015). *Sci. Rep.* 5: 1, 14245–7.
71 (a) Kelly, P.J. and Arnell, R.D. (2000). *Vacuum* 56: 159–172. (b) Bräuer, G., Szyszka, B., Vergöhl, M., and Bandorf, R. (2010). *Vacuum* 84: 1354–1359.
72 Yadav, T.P., Yadav, R.M., and Singh, D.P. (2012). *Nanosci.Nanotechnol.* 2: 22–48.
73 Ealias, A.M. and Saravanakumar, M.P. (2017). *IOP Conf. Ser.: Mater.Sci. Eng.* 263: 1, 032019–15.
74 Alexeeva, O.K. and Fateev, V.N. (2016). *Int. J. Hydrogen Energy* 41: 3373–3386.
75 Ye, G., Zhang, Q., Feng, C. et al. (1996). *Phys. Rev. B: Condens. Matter* 54: 14754–14757.
76 Nguyen, M.T. and Yonezawa, T. (2018). *Sci. Technol. Adv. Mater.* 19: 883–898.
77 Wender, H., Gonçalves, R.V., Feil, A.F. et al. (2011). *J. Phys. Chem. C* 115: 16362–16367.
78 Siegel, J., Kvítek, O., Ulbrich, P. et al. (2012). *Mater. Lett.* 89: 47–50.
79 Lee, S.H., Jung, H.K., Kim, T.C. et al. (2018). *Appl. Surf. Sci.* 434: 1001–1006.
80 Hatakeyama, Y., Morita, T., Takahashi, S. et al. (2011). *J. Phys. Chem. C* 115: 3279–3285.
81 Staszek, M., Siegel, J., Polívková, M., and Švorčík, V. (2017). *Mater. Lett.* 186: 341–344.
82 Staszek, M., Siegel, J., Rimpelová, S. et al. (2015). *Mater. Lett.* 158: 351–354.
83 Ishida, Y., Nakabayashi, R., Matsubara, M., and Yonezawa, T. (2015). *New J. Chem.* 39: 4227–4230.
84 Sumi, T., Motono, S., Ishida, Y. et al. (2015). *Langmuir* 31: 4323–4329.
85 Shishino, Y., Yonezawa, T., Udagawa, S. et al. (2011). *Angew. Chem. Int. Ed.* 50: 703–705.
86 (a) Nguyen, M.T., Yonezawa, T., Wang, Y., and Tokunaga, T. (2016). *Mater. Lett.* 171: 75–78. (b) Nguyen, M.T., Zhang, H., Deng, L. et al. (2017). *Langmuir* 33: 12389–12397.
87 Corpuz, R.D., Ishida, Y., Nguyen, M.T., and Yonezawa, T. (2017). *Langmuir* 33: 9144–9150.
88 Cha, I.Y., Ahn, M., Yoo, S.J., and Sung, Y.-E. (2014). *RSC Adv.* 4: 38575–38580.

6

Catalytic Properties of Metal Nanoparticles Confined in Ionic Liquids

Muhammad I. Qadir, Nathália M. Simon, and Jairton Dupont

Institute of Chemistry-Universidade Federal do Rio Grande do Sul-UFRGS, Av. Bento Gonçalves, 9500, Porto Alegre, RS 91501-970, Brazil

6.1 Introduction

A green environment is the desire of this modern world that forces the chemical society to find evidence-based solutions. The classical solutions/solvents have great drawbacks related to health and safety issues. In this regard, ionic liquids (ILs) have emerged as green solvents that typically consist of organic cations with organic or inorganic anions [1]. ILs constitute an innovative class of solvents with huge potential to have an impact across many areas of scientific and engineering research. These fluids demonstrate negligible vapor pressure, high chemical and thermal stability, a wide electrochemical window, nonflammability, high ionic conductivity, acceptable biocompatibility, good capability of dissolving various organic/inorganic materials, and high boiling point and stability [2]. They often have a melting point below 100 °C, which is why they are also called room temperature ILs, RTILs. In the beginning, ILs were designated as molten salts [1a]. The very first molten salt, ethylammonium nitrate, was prepared in 1914 with a low melting point of 12 °C. It took almost 40 years to report alkylimidazolium aluminate IL that was prepared by mixing imidazolium chloride with aluminum chloride [3]. However, it was very sensitive to moisture being easily degraded to generate other side products including HCl and Al_2O_3. In the 1990s, air-stable ILs involving tetrafluoroborate (BF_4^-) and hexafluorophosphate (PF_6^-) were synthesized [4]. As these ILs are easily stored and handled outside an air environment, the interest in them followed an exponential growth.

Nonfunctionalized ILs mainly consist of imidazolium, pyridinium, pyrrolidinium, and phosphonium cations with the desired anions. Among them, imidazolium-based ILs are the most popular and studied ones. Although, in various cases, the physical chemical properties and/or the outcome of the processes in these liquids differ significantly from those performed in “classical” dipolar organic solvents, they are still regarded as merely homogeneous solvents. There is a concept that pure 1,3-dialkylimidazolium ILs are better described as hydrogen-bonded polymeric supramolecules of the type $\{[(DAI)_x(X)_{x-n})]^{n+}[(DAI)_{x-n}(X)_x)]^{n-}\}_n$, where

Nanoparticles in Catalysis: Advances in Synthesis and Applications, First Edition.
Edited by Karine Philippot and Alain Roucoux.

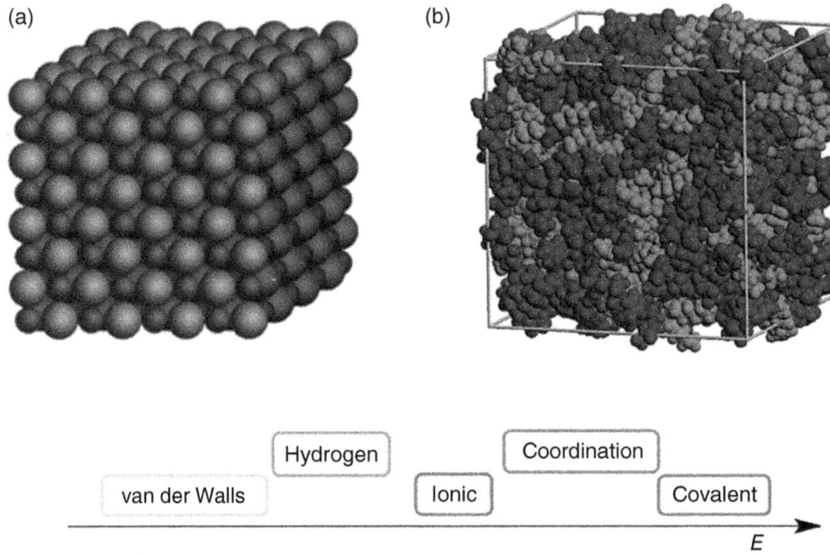

Figure 6.1 3-D structural arrangement representations of an ionic crystal displaying a classical charge ordering structure (a) and a "modern" 1-alkyl-3-methylimidazolium IL (b) that presents polar and nonpolar nanodomains. Source: Reproduced with permission from Canongia Lopes and Pádua [7b]. Copyright 2006, American Chemical Society.

DAI is the 1,3-dialkylimidazolium cation and X is the anion [5]. This structural pattern is a general trend for the solid phase and is maintained largely in the liquid phase, and even into the gas phase. It is believed that when imidazolium-based ILs are much diluted in other solvents, they still remain in their contact ion pairs through the hydrogen bonding of their cations and anions. Other prominent characteristics of these ILs are the Coulombic interactions akin to the classical salt systems (NaCl). One of the main differences between RTILs and simple molten salts is the molecular asymmetry built into at least one of the ions. This asymmetry opposes the strong charge ordering because of ionic interactions, which normally, in the case of classical molten salts, would cause the system to crystallize. Another difference is the presence of a cooperative network of hydrogen bonds between the cations and anions that induce structural directionality ("entropic effect") [6]. Therefore, modern ILs possess preorganized structures, mainly through hydrogen bonds, that induce structural directionality. In contrast, classical salt aggregates are only formed through ionic bonds (i.e. charge-ordering structures) (Figure 6.1). Therefore, imidazolium-based ILs consist of multiple intermolecular interactions such as van der Waals, hydrogen, ionic, coordination, and covalent [8].

6.2 Stabilization of Metal Nanoparticles in ILs

Small-sized (1–10 nm) transition metal nanoparticles (NPs) have gained a lot of attention because of the presence of large surface-to-volume ratios and quantum size effects [9]. As the NPs are only kinetically stable, they are unstable in classical organic solvents and undergo agglomeration, leading to the formation of bulk metal. Aggregation can be prevented by the use of different kinds of protective agents, such as water-soluble polymers, quaternary ammonium salts, surfactants, and polyoxoanions, which provide electrostatic and/or steric protection [10]. However, these stabilizing agents may have drawbacks because of their binding at the surface of NPs and hence may decrease their catalytic efficiency (not always). In this context, ILs not only provide stability to the NPs through the weak interactions (van der Waals, electrostatic, structural, solvophobic, steric, and hydrogen bonding) but also provide easily accessible coordination sites at the surface of NPs for catalysis because of the easy diffusion of the reactants.

Based on experimental studies, different theories have been proposed about the stabilization of NPs in ILs. It has been postulated that the ILs stabilize the NPs by an electric double layer (the Derjaguin–Landau–Verwey–Overbeek model) in which a first solvation shell of anions surrounds the metal cluster, followed by a less ordered layer of cations, and so on. However, other studies have evidenced close interactions of the NPs with the cations through deuterium exchange on positively charged imidazolium rings, studied through surface-enhanced Raman spectroscopy on gold NPs in imidazolium liquids [11]. Other studies suggest that NPs are solvated in nonpolar regions formed by aggregation of the hydrophobic alkyl side chains of the ions, as there is a relationship between the length scale of the structural heterogeneities of the IL and the size of nanoparticles synthesized therein [12]. Also, transmission electron microscopy (TEM), X-ray photoemission spectroscopy (XPS), and small-angle X-ray scattering (SAXS) analyses clearly showed the interactions of the IL with the metal surface, demonstrating the formation of a semiorganized IL-protective layer surrounding platinum NPs [13].

6.3 Synthesis of Soluble Metal Nanoparticles in ILs

There are two well-known methods to prepare metal NPs in ILs; (i) *in situ* chemical synthesis and (ii) physical synthesis (magnetron sputtering). In the chemical method, NPs (1–10 nm) can be prepared by dissolving a transition metal complex into ILs, followed by the decomposition/reduction in the presence of hydrogen (Scheme 6.1). The ideal is to use zero-valent metal complexes possessing olefinic ligands that, by decomposition under hydrogen, generate alkanes as the only by-products without coordinating properties to the metal surface and which can be easily removed from the reaction mixture [9, 13a, 14]. It is believed that this synthesis method of NPs follows the autocatalytic mechanism that involves two steps: nucleation and surface growth [15]. The physicochemical nature of the ILs plays an important role in controlling the shape and size of the NPs. The narrow size distribution of NPs *in situ* prepared in ILs, and the link between their size and characteristic lengths of the IL (represented by either the anion volume or the size of structural heterogeneities), means that the length scales of the IL do affect the nucleation and growth of the NPs and also the stability of the resulting colloid [15].

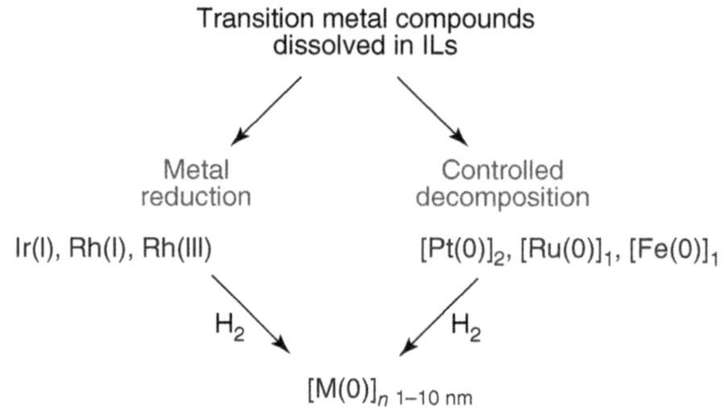

Scheme 6.1 Schematic representation of the preparation of metal NPs in ILs by chemical methods using transition metal precursors.

Simple thermal decomposition of zero-valent metal carbonyl $M_x(CO)_y$ complexes in IL can yield NPs without the use of $NaBH_4$ and H_2 atmosphere [16]. Moreover, NPs can also be prepared in ILs by means of photochemical reduction, or electroreduction/electrodeposition, of metal salts where the metal atom is in a formally positive oxidation state [17].

A well-known physical method, magnetron-sputtering deposition, has also emerged as a simple and easy technique to produce well-defined (size, size distribution, and shape) surface-clean (naked) NPs in ILs [18]. A large variety of Pt, Au, Ag, Ru, and Pd NPs prepared from bulk metals (sputtering targets) and the study of the effect of ILs on the size and shape of NPs have been reported [19]. This versatile method can generate highly pure supported metal NPs in both solid and liquid supports, as opposed to classical chemical and electrochemical methods that usually require further purification steps. ILs act as a template with cavities that help to control the size and location of the NPs depending on their chemical structures. For example, different kinds of nonfunctionalized and functionalized, imidazolium-based (containing –CN, –SH, and $-CO_2H$ moieties) ILs were used to study the size, shape, and location of Au, Ag, and Pd NPs [20]. It was observed that the NP's location can, in principle, be modulated by the presence of coordination groups attached to the imidazolium cation. The nucleation step appears to occur preferentially at the IL surface, regardless of the coordination group attached to the imidazolium cation. In ILs possessing weakly coordinating groups, i.e. less anisotropic fluids, the NP growth occurs mainly within the IL, whereas in those containing high σ donor groups (SH, for instance), both nucleation and growth occur preferentially at the IL surface. In the case of ILs containing σ donor and π-acceptor functions (–CN), the NP growth may take place both in the bulk IL and at the IL surface [20b].

6.4 Catalytic Application of NPs in ILs

It is well known that the choice of the IL cations and anions impacts the geometric and electronic properties of metal NPs supported in ILs and that has strong influence

R_1= Et; R_2= H; X= Br; C_2MIm.Br
R_1= ^{n}Prop; R_2= CH_3; X= NTf_2; C_3CNMMIm.NTf_2
R_1= ^{n}Prop; R_2= H; X= NTf_2; C_3CNMIm.NTf_2
R_1= ^{n}Bu; R_2= CH_3; X= TPPM[a]; BMMIm.TPPM
R_1= ^{n}Bu; R_2= CH_3; X= TPPT[a]; $(BMMIm)_3$.TPPT
R_1= ^{n}Bu; R_2= H; X= BF_4; BMIm.BF_4
R_1= ^{n}Bu; R_2= CH_3; X= BF_4; BMMIm.BF_4
R_1= ^{n}Bu; R_2= H; X= PF_6; BMIm.PF_6
R_1= ^{n}Bu; R_2= CH_3; X= PF_6; BMMIm.PF_6
R_1= ^{n}Bu; R_2= H; X= Br; BMIm.Br
R_1= $CH_2OCH_2CH_2CH_3$; R_2= CH_3; X= Cl; MPMIm.Cl
R_1= ^{n}Oc; R_2= H; X= $CF_3CF_2CF_2CF_2SO_3$; C_8MIm.$CF_3(CF_2)_3SO_3$

R_1= $CH_3(CH_2)_7(OCH_2CH_2)_2$; R_2= $CH_3(OCH_2CH_2)_2$; R_3= CH_3; R_4= CH_3; X= CH_3SO_3; NMPEG.CH_3SO_3
R_1= $CH_3(OCH_2CH_2)_{16}$; R_2= CH_3CH_2; R_3= CH_3CH_2 R_4= CH_3CH_2; X= CH_3SO_3; NEtPEG.CH_3SO_3
R_1= R_2= R_3= R_4= ^{n}Bu; X= Br; Bu_4N.Br
R_1= R_2= R_3= R_4= ^{n}Bu; X= OAc; Bu_4N.OAc

Figure 6.2 Structure and abbreviations of ILs discussed in this chapter.

in catalysis, for instance, on the residence time/diffusion of the reactants, intermediates, and products in the generated nanoenvironment [14c, 21]. In the following parts of this chapter, a short preview of the catalytic application of the NPs in ILs (Figure 6.2) will be given. The use of these systems in the hydrogenation of simple double bonds, such as C=C and C=O, will not be covered because there are several recently published extensive reviews available on this subject [9, 22]. We will focus on the use of IL-soluble NPs according to their behavior during the catalytic process: (i) either the reaction occurs mainly at the NP surface and the selectivity is strongly influenced by either the nature of the IL NPs surrounding layers or by the addition of external ligands as in the case of the hydrogenation of arenes (mainly benzene) or (ii) the IL-soluble NPs are a simple reservoir of monometallic or low-ligated metal clusters, catalytically active species such as observed in the C–C couplings [14a], and hydroformylation of alkenes [23].

6.4.1 Catalytic Hydrogenation of Aromatic Compounds

The catalytic hydrogenation of unsaturated hydrocarbons, especially the benzene-related ones, is of current interest in the industrial sector because of more strict environmental concerns. The obtained hydrogenated products (cyclohexene and cyclohexane) can be used as intermediate materials for producing adipic acid and ε-caprolactam exploited in the manufacture of nylon-6 or nylon-6,6, respectively, and other fine chemicals. Thermodynamic and kinetic barriers strongly control the partial hydrogenation of benzene, which is the most investigated and challenging reaction. Thermodynamically, the semihydrogenation of benzene to cyclohexene is not favorable because the cyclohexane is at least 75 kJ mol^{-1} more stable than cyclohexene [24]. The kinetics of the liquid-phase hydrogenation of benzene is still under debate as the mechanism deals with the chemical state of the catalytically active hydrogen and the role of the reaction intermediates. Two hypotheses exist regarding active hydrogen species for this hydrogenation either in a dissociated form on the catalyst surface or as adsorbed molecular hydrogen (Figure 6.3) [25]. Cyclohexene was detected as a by-product during the

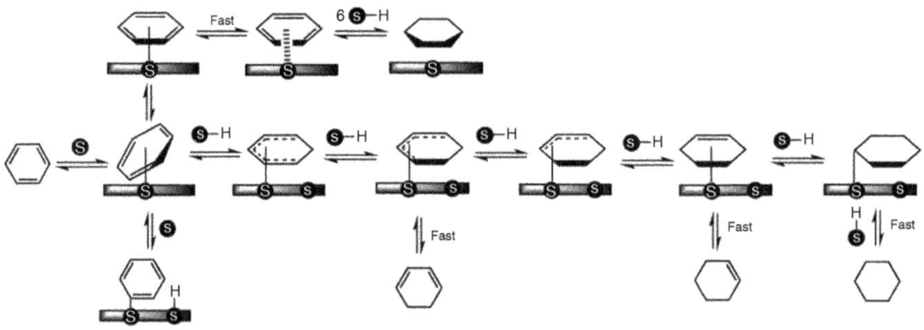

Figure 6.3 Mechanistic representation of catalytic benzene hydrogenation. S and s represent benzene and hydrogen coordinating active sites, respectively, and the wide dashed line represents van der Waals interactions. Source: Reproduced with permission from Foppa and Dupont [25]. Copyright 2015, Royal Society of Chemistry.

hydrogenation of benzene using Ru and Ni catalysts, which arises the question of whether cyclohexadiene and cyclohexene are the real reaction intermediates or not. Although the hydrogenation kinetics can be described with these intermediates, a doubt remains on the thermodynamic consistency of the rate models based on cyclohexene and cyclohexadiene because their hydrogenation rates are known to be very high compared to that of benzene.

It is assumed that the hydrogenation of benzene initiates with its coordination to the metal surface, followed by its reduction to 1,3-cyclohexadiene, then to cyclohexene, and finally to the thermodynamically more stable cyclohexane. Theoretical studies (DFT) of the mechanism of benzene hydrogenation over Pt(111) suggest that 1,3-cyclohexadiene and cyclohexene are in low amounts along with the most abundant cyclohexane because their generation does not follow the dominant reaction path [26]. However, 1,3-cyclohexadiene and cyclohexene have been obtained during hydrogenation of benzene promoted by lanthanide NPs in ammonia, although in virtually stoichiometric conditions. Interestingly, 1,3-cyclohexadiene is also usually observed as an intermediate during the dehydrogenation of cyclohexene or cyclohexane by Pt-based catalysts [27].

Solubility of the aromatic compounds is essential for their hydrogenation into more valuable semihydrogenated and/or completely hydrogenated ones. For this purpose, 1,3-dialkylimidazolium ILs can be modulated by simple changes in the *N*-alkyl imidazolium substituents and/or in the anion in order to improve the substrate solubility. For instance, benzene is highly soluble in 1-*n*-butyl-3-methylimidazolium hexafluorophosphate ($BMIm.PF_6$) IL at room temperature, whereas alkenes and alkanes are sparingly soluble in this medium. A high selectivity about 39% in cyclohexene was thus obtained at a very low benzene conversion (1%) by using Ru NPs in $BMIm.PF_6$, but it decreased with the increase of benzene conversion. $BMIm.PF_6$ is believed to generate a cage around the Ru NPs that not only stabilizes the NPs but could also repel the formed cyclohexene and thus decrease its readsorption and further hydrogenation to cyclohexane [28]. The conversion of benzene reached 73% with a 100% selectivity to cyclohexane (Table 6.1, entry 2) [28]. Ultrafine Pt NPs (2–3 nm) were also invested under the

Table 6.1 Examples of catalytic hydrogenation of aromatic compounds in ILs using various metal NPs without additives.

Entry	Substrate	NPs	IL	Major product	Conversion/ selectivity (%)	References
1	Benzene	Ru/Pt	$BMIm.PF_6$	1,3-Cyclohexadiene	5/21	[29]
2	Benzene	Ru	$BMIm.PF_6$	Cyclohexane	73/100	[28]
3	Benzene	Ru	$BMIm.BF_4$	Cyclohexane	30/100	[28]
4	Benzene	RuO_2	$BMIm.PF_6$	Cyclohexane	97/100	[30]
5	Benzene	RuO_2	$BMIm.BF_4$	Cyclohexane	3/100	[30]
6	Benzene	Ru	$BMIm.NTf_2$	Cyclohexane	100/100	[14b]
7	Benzene	Rh	$BMIm.PF_6$	Cyclohexane	100/100	[28]
8	Benzene	Ir	$BMIm.PF_6$	Cyclohexane	100/100	[28]
9	Benzene	Ni	$BMIm.NTf_2$	Cyclohexane	15/100	[31]
10	Benzene	Ni	$BPy.NTf_2$	Cyclohexane	18/100	[31]
11	Benzene	Pt	$BMIm.PF_6$	Cyclohexane	46/100	[13]
12	Toluene	Ru	$BMIm.NTf_2$	Methylcyclohexane	85/100	[14b]
13	*o*-Xylene	Ru	$BMIm.NTf_2$	1,2- Dimethylcy-clohexane	32/100	[14b]
14	Toluene	Rh	$BMIm.NTf_2$/ $BIHB.NTf_2$	Methylcyclohexane	100/100	[32]
15	Toluene	Rh	$BMIm.NTf_2$/ bipy	Methylcyclohexane	68/100	[32]
16	Toluene	Rh	$BMIm.NTf_2$/ $BIMB.(NTf_2)_2$	Methylcyclohexane	18/100	[32]
17	Ethylbenzene	Rh	$BMI.BF_4$/poly [NVP-co-VBIM.Cl]	Ethylcyclohexane	100/>99	[33]
18	Styrene	Pd	$BMMIm.PF_6$	Ethylcyclohexane	100/100	[34]
19	Quinoline	Ru	BMMIm.TPPM/ $BMMIm.BF_4$	1,2,3,4- Tetrahy-droquinoline	95/98	[35]

same reaction and IL ($BMIm.PF_6$), which provided 46% conversion of benzene to cyclohexane (selectivity of 100%) without the detection of cyclohexene (Table 6.1, entry 11) [13a].

The selective and quantitative generation of 1,3-cyclohexadiene by the partial hydrogenation of benzene remains a challenge. Interestingly, bimetallic Ru/Pt NPs in $BMIm.PF_6$ produced 1,3-cyclohexadiene with an unprecedented 21% selectivity at 5% benzene conversion (Table 6.1, entry 1) [29]. Moreover Pt or Ru NPs, and isolated bimetallic Ru/Pt NPs (without IL), showed very low selectivity for partially hydrogenated products under the same reaction conditions (biphasic system with *n*-heptane as a cosolvent, 60 °C, and 6 bar of H_2). The low selectivity (5%) toward 1,3-cyclohexadiene occurred when the reaction was performed by adding cosolvents such as dichloromethane, which is miscible with $BMIm.PF_6$.

Although noble metal catalysts have more activity and stability for the hydrogenation of benzene, the development of nickel-based catalysts is attractive because of the availability and low cost of this metal. Ni NPs were reported with relatively low activity for the hydrogenation of benzene to cyclohexane with 15% and 18% conversion using BMIm.NTf_2 and BPy.NTf_2 ILs, respectively [31], because of the easy oxidation of Ni NPs. Different additives such as $AlCl_3$, bipyridines, HF, and polyvinylpyrrolidone have been applied to hydrogenate benzene with desired selectivity. Studies and further details have been largely described in recent reviews [25, 36].

Metal NPs in ILs have also been successfully applied for the complete hydrogenation of benzene derivatives such as styrene, ethylbenzene, and quinolines (Table 6.1, entries 13–19). Thus, Ru NPs (2.1–3.5 nm) in ILs were used for liquid–liquid biphasic hydrogenation of arenes under mild reaction conditions (50–90 °C and 4 bar H_2). The apparent activation energy (EA) was estimated to be 42.0 kJ mol^{-1} for the hydrogenation of toluene in the biphasic liquid–liquid system with RuNPs/BMI.NTf_2 [14b]. The same catalyst also presented efficient activity for the hydrogenation of *o*-xylene with a 32% conversion and 100% selectivity to 1,2-dimethylcyclohexane (Table 6.1, entry 13). Dyson and coworkers reported on the use of Rh NPs stabilized in the imidazolium-functionalized bipyridine compounds, 4,4′-bis-[7-(2,3-dimethylimidazolium)heptyl]-2,2′-bipyridine^{2+} ($[BIHB]^{2+}$), and 4,4′-bis[(1,2-dimethylimidazolium)methyl]-2,2′-bipyridine^{2+} ($[BIMB]^{2+}$), as nanocatalysts in the biphasic hydrogenation of various arene substrates [32]. The catalytic activity was strongly influenced by the stabilizer employed with a trend $[BIHB]^{2+}$ > bipy > $[BIMB]^{2+}$ (Table 6.1, entries-14–16). NPs stabilized by BIHB.NTf_2, with a C7 alkyl chain separating the imidazolium functionality from the pyridine backbone, were considerably more active than the bipy-stabilized system (with an increase in conversion up to 68%). However, the NPs protected by the BIHB.NTf_2 stabilizer, with one CH_2 group between the imidazolium and the pyridine, resulted in the lowest activity.

N-Heterocyclic compounds and their hydrogenated derivatives, for example, 1,2,3,4-tetrahydroquinoline (THQ), indoline, or piperidine, are structural units present in many basic bioactive natural products [35]. Zheng and coworkers developed a low-cost system using a Ru nanocatalyst stabilized by EGMMIm.NTf_2. The best performance was achieved in the hydrogenation of quinoline to 1,2,3,4-THQ with up to 99% selectivity [35]. FT-IR measurements allowed to confirm that this transformation is induced by the formation of hydrogen bonds between the Lewis basic substrate and the hydroxyl groups of the IL. These results clearly evidence that the catalytic performance of IL-soluble NPs can be modulated by the introduction of ligands into IL structure akin to the classical monometallic coordination catalysis.

6.4.2 Coupling Reactions in ILs

The products resulted by the C–C coupling are intermediate structures present in agrochemicals, pharmaceuticals, natural products, and polymers. Because of the importance of these industrial sectors, the search for efficient catalytic systems to obtain such coupling products is one of the main objectives of modern organic

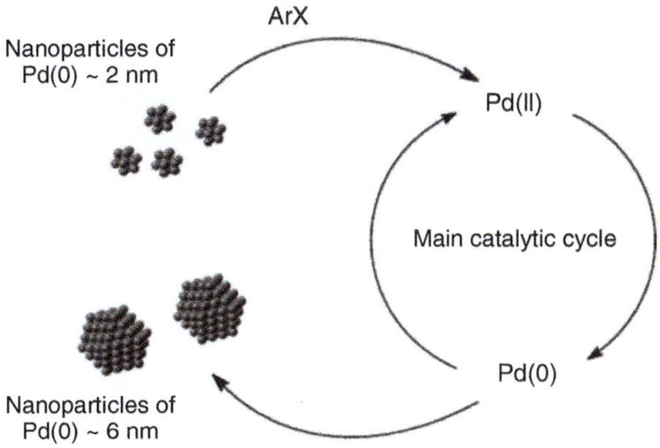

Scheme 6.2 Possible pathways involved in the Heck reaction promoted by Pd NPs dispersed in imidazolium ILs. Source: Reproduced with permission from Cassol et al. [14a]. Copyright 2005, American Chemical Society.

synthesis [37]. In this sense, methodologies involving metal NPs and ILs have been explored. Among the recent literature reports, few address the use of unsupported nanocatalysts, which are the focus of this section. In addition, all reported works are based on the use of Pd NPs, indicating that palladium is the most popular metal used in cross-coupling reactions [38]. In 1996, ILs were used for the first time in the Heck coupling reaction of iodoarenes and alkene/alkyne in a mixture of 1,3-di-*n*-butylimidazolium bromide (BBIm.Br) and 1,3-di-*n*-butylimidazolium tetrafluoroborate (BBIm.BF_4), driven by Pd NPs.

Ligand-free soluble Pd NPs immobilized in BMIm.PF_6 IL were investigated for the coupling reaction of aryl halides with *n*-butylacrylate [14a]. However, it was observed that Pd NPs in the IL act in fact as a reservoir of molecular catalytically active Pd species. It is assumed that the reaction starts with the oxidative addition of iodoarene on the NP surface, thus forming Pd(II) species that are released from the NP surface and can start a catalytic cycle. After β-hydride and reductive elimination steps, the formed zero-valent palladium specie can either continue in the catalytic cycle or return to the NP reservoir (Scheme 6.2).

Nacci and coworkers developed a catalytic system in which Pd NPs were *in situ* generated from $Pd(OAc)_2$ in a molten mixture of Bu_4N.Br as the solvent and Bu_4N.OAc as the base [39]. These Pd NPs catalyzed one-pot sequential Heck and Suzuki coupling of aryl dihalides to afford unsymmetrically substituted arenes (4-bromochlorobenzene, butylacrylate, and phenylboronic acid) with yields up to 92%. A similar methodology was previously applied in the coupling of aryl chlorides with deactivated olefins yielding 72–96% of β,β-diaryl acrylates [40]. Gaikwad et al. studied the ability of *in situ* generated Pd NPs to perform Heck and Suzuki coupling reactions. The multifunctionalized 4-amino-1-(2,3-dihydroxypropyl) pyridinium hydroxide (ADPPy.OH) ILs were designed to perform multiple roles. The hydroxyl group acted as a reducing agent for the formation of Pd NPs as well as the base required for the Suzuki coupling reaction. The amino group served as a stabilizing agent for the produced nanoparticles [41]. The same group previously designed a hydrophobic fluorous IL (1-octyl-3-methylimidazolium nonafluorobutanesulfonate). It was the reaction medium for the Heck reaction catalyzed by Pd NPs [42]. In both cases, the reactions occurred under mild and ligand-free conditions with excellent yields up to 98%. Aryl bromides, chlorides, and iodides were tested that yielded 58–92% coupled products, showing the versatility of the methodologies. Moreover, the catalysts could be reused for five cycles with superior activity.

The catalytic systems developed by Trzeciak and coworkers differ from those presented above because the use of ILs did not aim at synthesizing Pd NPs but rather Pd molecular complexes [43]. Even so, the presence of Pd NPs in the reaction's medium was confirmed by different analysis methods and their role in the catalytic process was investigated. Dimeric complexes of the type $[Pd(\mu\text{-}X)X(NHC)]_2$ with N-heterocyclic carbene (NHC) ligand, synthesized in the ILs (BMIm.Br, C_2MIm.Br, and HIMes.Cl), were employed in Suzuki–Miyaura cross-coupling. According to the authors, the reaction proceeded through the formation of a composite made of Pd NPs surrounded by a layer of a Pd(II) species, such as anions $[PdBr_4]^{2-}$ or $[PdBr_3(NHC)]^-$, which protect them against agglomeration (confirmed by XPS and TEM analyses). The formation of $[PdBr_3(NHC)]^-$ reinforces the importance of NHC ligands and, consequently, of the ILs that originated them. In an earlier study, $[RIm]_2[PdX_4]$-type complexes were obtained in a reaction between $PdCl_2$ with diverse imidazolium chloride ILs. After optimization, $[MPMIm]_2[PdCl_4]$ and $[HIPr]_2[PdCl_4]$ were applied at Suzuki–Miyaura cross-coupling for the purpose of mechanistic studies. Results indicated that Pd NPs were present during the reaction and acted mainly as a source of catalytically active soluble palladium species (confirmed by XPS and TEM analyses). Once again, the presence of IL in the system is remarkable because imidazolium cations and NHCs were considered as key intermediates [44].

6.4.3 Hydroformylation in ILs

Hydroformylation is one of the most important industrial processes for the manufacture of aldehydes from alkenes; usually, propene (about 10 million tons yr^{-1}) is effectively driven by homogeneous catalysts in the presence of carbon monoxide and hydrogen. Nevertheless, this reaction can suffer of several drawbacks such as the regioselectivity control (ratio normal/iso of aldehyde isomers), operating pressures, recycling and recovery of expensive catalysts, and complex multiphase reaction engineering. Thus, investigations involving metal NPs confined in ILs can contribute to overcome such problems. Wang and coworkers investigated Rh NP-driven hydroformylation in a thermoregulated PEG-functionalized IL/organic biphasic system that enables the separation of the catalysts from the products. In optimized conditions, the system NEtPEG.CH_3SO_3/toluene tested with 1-octene, cyclohexene, and styrene selectively produced aldehydes with yields above 85% in 99% of conversion [23]. In the same manner, the system *N*,*N*-dimethyl-*N*-(2-(2-methoxyethoxy)ethyl)-*N*-(2-(2-octyloxyethoxy)ethyl) ammonium methanesulfonate/cyclohexane, applied for hydroformylation of

1-octene, 1-decene, 1-dodecene, and cyclohexene, provided aldehydes with yields above 98% in 99% of conversion [45].

6.4.4 Fischer–Tropsch Synthesis in ILs

Because of the depletion of fossil fuels, Fischer–Tropsch synthesis (FTS) has emerged as an alternative method to generate energy fuels by the conversion of syngas (CO/H_2). Supported and nonsupported Co-, Ru-, and Fe-based nanocatalysts have been successfully applied for FTS [46]. Among them, catalysts made of nanostructured cobalt are considered to be the best candidates owing to their high activity and wide availability of cobalt. Co NPs (c. 7.7 nm) dispersed in $BMIm.NTf_2$ were evaluated under 20 atm of syngas (CO/H_2, 1 : 2) at 210 °C. Mainly higher alkanes (7–30 carbons) were produced in liquid phase [47]. The reaction followed the Anderson–Shulz–Flory (ASF) type of distribution, where a 0.90 ASF factor was estimated for the products (C_8–C_{30}) [47].

6.4.5 Catalytic Carbon Dioxide Hydrogenation in ILs

Catalytic hydrogenation of carbon dioxide (CO_2) into hydrocarbons (formic acid, methanol, methane, or long-chain hydrocarbons) is one of the promising approaches to combat CO_2-induced climate change. CO_2 can be converted to higher hydrocarbons, following two consecutive processes: (i) CO generation via reverse water gas shift (RWGS) reaction [48], and its further reduction into methane [49], or (ii) generation of heavier HC/oxygenates through the Fischer–Tropsch process [50]. This is equivalent to declaring that the carbon source in this FT process will be carbon dioxide instead of the usual gasified coal, methane, or biomass. The use of FT catalytic systems using CO_2, mainly based on iron catalysts, has been successfully reported [51]. In fact, Fe-based catalysts are considered to be ideal candidates for CO_2 hydrogenation because of intrinsic WGS and RWGS activity and also high availability of this metal [52]. Methane and HCs can be selectively produced from CO_2 hydrogenation, either via the CO or the higher HC pathways, depending on the thermodynamic/kinetic balance, which is a delicate parameter, and also the fine-tuning of the electronic and geometric properties of the catalysts. In this vein, ILs have emerged as ideal media through which not only the conversion of CO_2 can be increased but also the selectivity of the hydrocarbons can be controlled.

Ru NPs dispersed in 1-octyl-3-methylimidazolium bis trifluoromethanesulfonylimide, ($OMIm.NTf_2$) IL have been reported for the CO_2 hydrogenation into methane at high temperature (150 °C). A turnover number (TON) of 72% and 69% CH_4 yield was obtained [53]. Recently, we have developed an efficient method to hydrogenate CO_2 into various hydrocarbons in the range C_2–C_{21} using RuFe alloy bimetallic NPs through the judicious choice of IL [14c]. The hydrophobic or hydrophilic behavior of a series of ILs was studied on $BMIm.BF_4$, BMIm.OAc, BMIm.FAP, and $BMIm.NTf_2$, and this parameter appeared to be a key point in the selectivity of the reaction. Thus, RuFe NPs displayed outstanding abilities in the

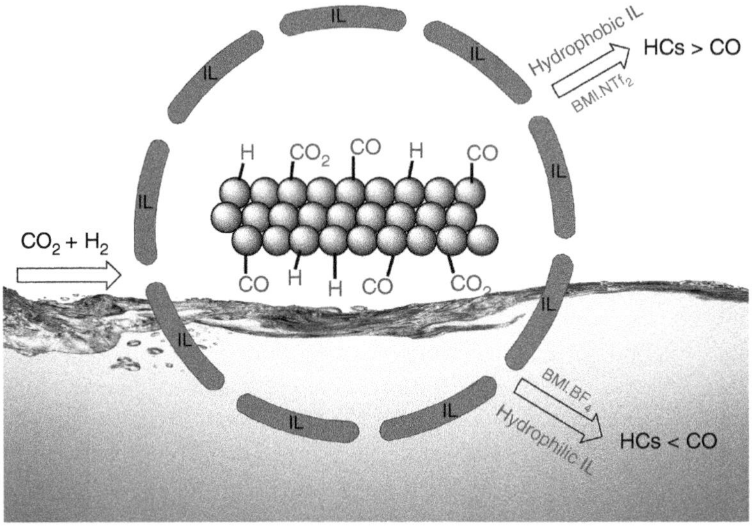

Scheme 6.3 CO_2 hydrogenation to the higher hydrocarbons driven by Ru/Ni NPs in ILs. Source: From Qadir et al. [54]. Reprinted with permission of Elsevier.

formation of long-chain HCs (57% C_2–C_4, 31% C_5–C_6, and 10% C_7–C_{21}) with efficient catalytic activity (12% conversion) in the BMIm.NTf$_2$ hydrophobic IL, while in the hydrophilic ILs (BMIm.BF$_4$ and BMIm.OAc), they led to high production of CO (>39%). The hydrophobic IL repels the formed water from catalytic active phase to increase the FT synthesis. We assume that IL hydrophilic nature not only reduces the rate of CO_2 hydrogenation but may also reduce the FT catalytically active surface species with the dominant CO path formation, thus favoring the formation of high hydrocarbons. Interestingly, when the IL with basic anion (acetate) was applied, formic acid was generated with a molar ratio of 2.03 M with a TON of 400.

In 2019, we obtained a high CO_2 conversion (30%) into C_1–C_6 hydrocarbons (79% alkanes and 16% olefins) with only 5% of CH_4 when investigating RuNi bimetallic core–shell NPs in BMIm.NTf$_2$ hydrophobic IL (Scheme 6.3) [54]. The segregation/migration of the Ni atoms from core to the ruthenium shell was observed, which resulted in the increase in Ni concentration at the surface of NPs after catalysis, as indicated by synchrotron XPS analysis at 1840 eV. It is assumed that the segregated Ni atoms play a key role not only in enhancing the RWGS pathways but also in controlling the selectivity of hydrocarbons. According to the increasing ratio of Ni onto the surface of the catalysts, an increase of CH_4 production was observed because of the more positive impact of Ni in methanation than Ru species [55].

6.5 Conclusions

Soluble, small, stable, and ligand-free metal NPs in ILs can be easily prepared by either chemical or physical methods. These colloids can be used as catalysts and/or precursors for various reactions such as hydrogenation, carbonylation, and C–C coupling reactions in both homogeneous and multiphase conditions. It appears that under reductive conditions, the reactions occur mainly at the NP surface, and the catalytic performance can be controlled by either the nature of the IL, the addition of ligands, or the diffusion of the substrates and products to the catalytic sites, such as in the hydrogenation of arenes and carbon dioxide. Under oxidative conditions, the metal NPs are simple reservoirs of catalytically active species composed of "ligandless" monometallic or low-ligated metal clusters, as in the case of C–C coupling and hydroformylation reactions. The association of these colloids with other supports, i.e. hybrids, will certainly open new opportunities for the development of more active and selective catalysts or "catalyst delivery systems."

Acknowledgments

The authors would like to thank CAPES (158804/2017-01 and 001), FAPERGS (16/2552-0000 and 18/2551-0000561-4), and CNPq (406260/2018-4, 406750/2016-5, and 465454/2014-3) for their financial support.

References

1 (a) Dupont, J., de Souza, R.F., and Suarez, P.A.Z. (2002). *Chem. Rev.* 102: 3667–3692. (b) Holbrey, J.D. and Seddon, K.R. (1999). *Clean Prod. Processes* 1: 223–236.

2 (a) Earle, M.J., Esperança, J.M.S.S., Gilea, M.A. et al. (2006). *Nature* 439: 831–834. (b) Chambreau, S.D., Schneider, S., Rosander, M. et al. (2008). *J. Phys. Chem. A* 112: 7816–7824. (c) Dupont, J. and Spencer, J. (2004). *Angew. Chem. Int. Ed.* 43: 5296–5297. (d) Shiddiky, M.J.A. and Torriero, A.A.J. (2011). *Biosens. Bioelectron.* 26: 1775–1787. (e) Gomes, J.M., Silva, S.S., and Reis, R.L. (2019). *Chem. Soc. Rev.* 48: 4317–4335.

3 Hurley, F.H. and WIer, T.P. (1951). *J. Electrochem. Soc.* 98: 203–206.

4 (a) Chauvin, Y., Mussmann, L., and Olivier, H. (1996). *Angew. Chem. Int. Ed.* 34: 2698–2700. (b) Suarez, P.A.Z., Dullius, J.E.L., Einloft, S. et al. (1996). *Polyhedron* 15: 1217–1219.

5 Dupont, J. (2004). *J. Braz. Chem. Soc.* 15: 341–350.

6 Antonietti, M., Kuang, D., Smarsly, B., and Zhou, Y. (2004). *Angew. Chem. Int. Ed.* 43: 4988–4992.

7 (a) Dupont, J. (2011). *Acc. Chem. Res.* 44: 1223–1231. (b) Canongia Lopes, J.N.A. and Pádua, A.A.H. (2006). *J. Phys. Chem. B* 110: 3330–3335.

8 He, Z. and Alexandridis, P. (2015). *Phys. Chem. Chem. Phys.* 17: 18238–18261.

9 Scholten, J.D., Leal, B.C., and Dupont, J. (2012). *ACS Catal.* 2: 184–200.

10 Dupont, J., Fonseca, G.S., Umpierre, A.P. et al. (2002). *J. Am. Chem. Soc.* 124: 4228–4229.

11 (a) Ott, L.S., Cline, M.L., Deetlefs, M. et al. (2005). *J. Am. Chem. Soc.* 127: 5758–5759. (b) Schrekker, H.S., Gelesky, M.A., Stracke, M.P. et al. (2007). *J. Colloid Interface Sci.* 316: 189–195.

12 Pensado, A.S. and Pádua, A.A.H. (2011). *Angew. Chem. Int. Ed.* 50: 8683–8687.

13 (a) Scheeren, C.W., Machado, G., Dupont, J. et al. (2003). *Inorg. Chem.* 42: 4738–4742. (b) Scheeren, C.W., Machado, G., Teixeira, S.R. et al. (2006). *J. Phys. Chem. B* 110: 13011–13020.

14 (a) Cassol, C.C., Umpierre, A.P., Machado, G. et al. (2005). *J. Am. Chem. Soc.* 127: 3298–3299. (b) Prechtl, M.H.G., Scariot, M., Scholten, J.D. et al. (2008). *Inorg. Chem.* 47: 8995–9001. (c) Qadir, M.I., Weilhard, A., Fernandes, J.A. et al. (2018). *ACS Catal.* 8: 1621–1627.

15 (a) Redel, E., Thomann, R., and Janiak, C. (2008). *Inorg. Chem.* 47: 14–16. (b) Yang, M., Campbell, P.S., Santini, C.C., and Mudring, A.-V. (2014). *Nanoscale* 6: 3367–3375.

16 (a) Redel, E., Krämer, J., Thomann, R., and Janiak, C. (2009). *J. Organomet. Chem.* 694: 1069–1075. (b) Vollmer, C., Redel, E., Abu-Shandi, K. et al. (2010). *Chem. Eur. J.* 16: 3849–3858.

17 (a) Zhu, J., Shen, Y., Xie, A. et al. (2007). *J. Phys. Chem. C* 111: 7629–7633. (b) Höfft, O. and Endres, F. (2011). *Phys. Chem. Chem. Phys.* 13: 13472.

18 (a) Torimoto, T., Okazaki, K.-i., Kiyama, T. et al. (2006). *Appl. Phys. Lett.* 89: 243117. (b) Okazaki, K.-i., Kiyama, T., Hirahara, K. et al. (2008). *Chem. Commun.* 6: 691–693.

19 (a) Wender, H., Migowski, P., Feil, A.F. et al. (2011). *Phys. Chem. Chem. Phys.* 13: 13552–13557. (b) Suzuki, S., Suzuki, T., Tomita, Y. et al. (2012). *Crist. Eng. Commun.* 14: 4922–4926. (c) Oda, Y., Hirano, K., Yoshii, K. et al. (2010). *Chem. Lett.* 39: 1069–1071. (d) Hamm, S.C., Shankaran, R., Korampally, V. et al. (2012). *ACS Appl. Mater. Interfaces* 4: 178–184.

20 (a) Kauling, A., Ebeling, G., Morais, J. et al. (2013). *Langmuir* 29: 14301–14306. (b) Qadir, M.I., Kauling, A., Calabria, L. et al. (2018). *Nano-Struct. Nano-Objects* 14: 92–97. (c) Qadir, M.I., Kauling, A., Ebeling, G. et al. (2019). *Aust. J. Chem.* 72: 49–55.

21 (a) Luza, L., Rambor, C.P., Gual, A. et al. (2016). *ACS Catal.* 6: 6478–6486. (b) Luza, L., Rambor, C.P., Gual, A. et al. (2017). *ACS Catal.* 7: 2791–2799.

22 (a) Dupont, J. and Scholten, J.D. (2010). *Chem. Soc. Rev.* 39: 1780–1804. (b) Janiak, C. (2015). Metal nanoparticle synthesis in ionic liquids. *Ionic Liquids (ILs) in Organometallic Catalysis* (eds. J. Dupont and L. Kollár), 17–53. Berlin, Heidelberg: Springer Berlin Heidelberg https://doi.org/10.1007/3418_2013_70.

23 Zeng, Y., Wang, Y., Xu, Y. et al. (2012). *Chin. J. Catal.* 33: 402–406.

24 Ullmann, F., Gerhartz, W., Yamamoto, Y.S. et al. (1995). *Ullmann's Encyclopedia of Industrial Chemistry*. Wiley-VCH.

25 Foppa, L. and Dupont, J. (2015). *Chem. Soc. Rev.* 44: 1886–1897.

26 Morin, C., Simon, D., and Sautet, P. (2006). *Surf. Sci.* 600: 1339–1350.

27 (a) Su, X., Kung, K., Lahtinen, J. et al. (1998). *Catal. Lett.* 54: 9–15. (b) Koel, B.E., Blank, D.A., and Carter, E.A. (1998). *J. Mol. Catal. A: Chem.* 131: 39–53.

28 Silveira, E.T., Umpierre, A.P., Rossi, L.M. et al. (2004). *Chem. Eur. J.* 10: 3734–3740.

29 Weilhard, A., Abarca, G., Viscardi, J. et al. (2017). *ChemCatChem* 9: 204–211.

30 Rossi, L.M., Dupont, J., Machado, G. et al. (2004). *J. Braz. Chem. Soc.* 15: 904–910.

31 Wegner, S., Rutz, C., Schutte, K. et al. (2017). *Chem. Eur. J.* 23: 6330–6340.

32 Dykeman, R.R., Yan, N., Scopelliti, R., and Dyson, P.J. (2011). *Inorg. Chem.* 50: 717–719.

33 Léger, B., Denicourt-Nowicki, A., Olivier-Bourbigou, H., and Roucoux, A. (2008). *Inorg. Chem.* 47: 9090–9096.

34 Hu, Y., Yu, Y., Hou, Z. et al. (2008). *Adv. Synth. Catal.* 350: 2077–2085.

35 Jiang, H.-Y. and Zheng, X.-X. (2015). *Catal. Sci. Technol.* 5: 3728–3734.

36 Chacón, G. and Dupont, J. (2019). *ChemCatChem* 11: 333–341.

37 Trzeciak, A.M. and Augustyniak, A.W. (2019). *Coord. Chem. Rev.* 384: 1–20.

38 Wu, X.-F., Anbarasan, P., Neumann, H., and Beller, M. (2010). *Angew. Chem. Int. Ed.* 49: 9047–9050.

39 Cotugno, P., Monopoli, A., Ciminale, F. et al. (2012). *Org. Biomol. Chem.* 10: 808–813.

40 Calò, V., Nacci, A., Monopoli, A., and Cotugno, P. (2009). *Angew. Chem. Int. Ed.* 48: 6101–6103.

41 Gaikwad, D.S., Undale, K.A., Patil, D.B., and Pore, D.M. (2019). *J. Iran. Chem. Soc.* 16: 253–261.

42 Gaikwad, D.S., Park, Y., and Pore, D.M. (2012). *Tetrahedron Lett.* 53: 3077–3081.

43 Silarska, E., Trzeciak, A.M., Pernak, J., and Skrzypczak, A. (2013). *Appl. Catal., A Gen.* 466: 216–223.

44 Górna, M., Szulmanowicz, M.S., Gniewek, A. et al. (2015). *J. Organomet. Chem.* 785: 92–99.

45 Xu, Y., Wang, Y., Zeng, Y. et al. (2012). *Catal. Lett.* 142: 914–919.

46 (a) Bezemer, G.L., Bitter, J.H., Kuipers, H.P.C.E. et al. (2006). *J. Am. Chem. Soc.* 128: 3956–3964. (b) de Smit, E. and Weckhuysen, B.M. (2008). *Chem. Soc. Rev.* 37: 2758–2781. (c) Jacobs, G., Das, T.K., Zhang, Y. et al. (2002). *Appl. Catal., A* 233: 263–281. (d) Ojeda, M., Nabar, R., Nilekar, A.U. et al. (2010). *J. Catal.* 272: 287–297.

47 Silva, D.O., Scholten, J.D., Gelesky, M.A. et al. (2008). *ChemSusChem* 1: 291–294.

48 (a) Posada-Pérez, S., Ramírez, P.J., Evans, J. et al. (2016). *J. Am. Chem. Soc.* 138: 8269–8278. (b) Roiaz, M., Monachino, E., Dri, C. et al. (2016). *J. Am. Chem. Soc.* 138: 4146–4154.

49 Kattel, S., Yan, B., Yang, Y. et al. (2016). *J. Am. Chem. Soc.* 138: 12440–12450.

50 Prieto, G. (2017). *ChemSusChem* 10: 1056–1070.

51 (a) Gnanamani, M.K., Jacobs, G., Hamdeh, H.H. et al. (2013). *Catal. Today* 207: 50–56. (b) Landau, M.V., Vidruk, R., and Herskowitz, M. (2014). *ChemSusChem* 7: 785–794. (c) Owen, R.E., O'Byrne, J.P., Mattia, D. et al. (2013). *ChemPlusChem* 78: 1536–1544. (d) Dorner, R.W., Hardy, D.R., Williams, F.W., and Willauer, H.D. (2010). *Energy Environ. Sci.* 3: 884.

52 Newsome, D.S. (1980). *Cat. Rev. - Sci. Eng.* 21: 275–318.

53 Melo, C.I., Szczepańska, A., Bogel-Łukasik, E. et al. (2016). *ChemSusChem* 9: 1081–1084.
54 Qadir, M.I., Bernardi, F., Scholten, J.D. et al. (2019). *Appl. Catal., B* 252: 10–17.
55 Agnelli, M., Kolb, M., and Mirodatos, C. (1994). *J. Catal.* 148: 9–21.

Part II

Supported Nanoparticles

7

Nanocellulose in Catalysis: A Renewable Support Toward Enhanced Nanocatalysis

Tony Jin[1,2] *and Audrey Moores*[1,2]

[1] *McGill University, Centre for Green Chemistry and Catalysis, Department of Chemistry, 801 Sherbrooke Street West, Montreal, QC H3A 0B8, Canada*
[2] *McGill University, Department of Materials Engineering, 3610 University Street, QC H3A 0C5, Canada*

7.1 Introduction

Although nanocellulose and its properties have been studied for over six decades, the last two have witnessed an explosion of discoveries and applications of this nanomaterial into a variety of domains, including biomedical and drug delivery applications [1], Pickering emulsions [2], environmental remediation [3, 4], oil and gas [5], adhesives [6], cement [7], plastics [8], supercapacitors [9], energy applications [10], catalysis [11], paint and coating [12], and sensing [13]. Nanocellulose defines a family of cellulosic materials with at least one dimension at the nanoscale [14]. Within nanocelluloses, two forms have been the focus of researcher's attention: cellulose nanocrystals (CNCs) and cellulose nanofibers (CNFs). CNCs are whisker-shaped nanomaterials made of crystalline cellulose, with a typical size of 5–20 nm in width and 100 nm up to several microns in length, depending on the cellulose source (Scheme 7.1). Cellulose may be sourced either from plants, with typical examples being wood, cotton, or bamboo or from bacterial sources [14]. CNCs are typically produced from cellulose pulp – after separation from lignin and hemicellulose by traditional pulping methods – by applying a strong inorganic acid hydrolysis process, following a widely accepted mechanism, which implies the dissolution of amorphous regions in native cellulose to liberate the nanocrystalline regions. This material is commercially produced in several countries by companies such as CelluForce (Canada, 1000 $kg\,d^{-1}$), American Process (USA, 500 $kg\,d^{-1}$), Melodea (Israel, 100 $kg\,d^{-1}$), Melodea/Holmen (Sweden, 100 $kg\,d^{-1}$), and Alberta Innovates (Canada, 20 $kg\,d^{-1}$) [15]. CNCs have a number of advantages in terms of sustainable development [16]. They are non-toxic and readily available from biomass, a feature of great interest for modern materials [17]. They also have a host of unique properties, including photonics, self-assembly, thixotropy, rheology, and liquid crystal properties, which make them very well suited for the development of new materials in a clean way. Conversely, CNFs are typically produced by mechanical treatment of cellulose pulp that features larger domains than CNCs

Nanoparticles in Catalysis: Advances in Synthesis and Applications, First Edition.
Edited by Karine Philippot and Alain Roucoux.

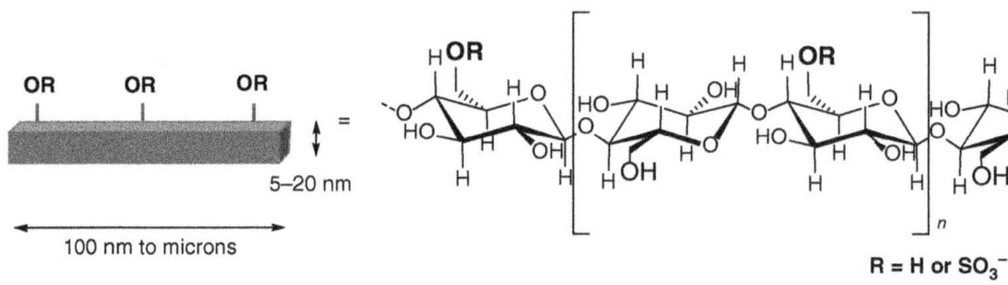

Scheme 7.1 Nano (left) and chemical structure (right) of CNCs produced from native cellulose by sulfuric acid hydrolysis. Additional oxidative treatments afford CNCs with carboxylated surfaces.

[18]. They do not suspend in solution, which allows using them as solid supports in catalysis. They have been applied to the fabrication of scaffolds and reinforced materials, for instance [18].

A large number of publications testify for the scientific excitement around this emerging material, with over 6500 publications in the field of CNCs alone published in the past 25 years. Figure 7.1 illustrates the publication trend in the field and demonstrates a steady and constant progress. Interestingly, application fields concerned with this research are very broad, including all corners of chemistry and chemical engineering specialties. In particular, polymer, nanoscience, biological, and sustainability research are well featured in these developments. Focusing on the past three years, the reader is directed to the general reviews by Sanchez and coworkers [16], Sheikhi [19], Tam and coworkers [20], Trache et al. on CNCs production [15], van Zyl and coworker on application of CNCs for surface-enhanced Raman spectroscopy [21], Shak et al. on CNCs in remediation applications [4], and also Tam and coworkers on hybrid structures with inorganic materials [22].

Among the different areas of application of CNCs, catalysis stands as an exciting opportunity, mostly developed over the past decade [11]. The use of biomass as a resource for catalyst fabrication, either as a support or as a catalytically active species, is a powerful means to diverge from mined and fossil resources and constitutes an interesting sustainable solution [17]. Also, CNC surface features hydroxyl, carboxylate, and/or sulfate half ester functionalities [23], stemming from cellulose itself (–OH), oxidative treatment ($-COO^-$), or from using sulfuric acid ($-OSO_3^-$), respectively (Scheme 7.1) [23]. These functionalities play a key role in the colloidal stability of these materials in water and are also important when considering interaction of metal species for catalyst synthesis. CNC surface chemistry, allied with their crystallinity, has also been exploited for enantioselective catalysis, for example, in the reduction of aromatic ketones with Pd(0) species on the surface of CNCs [24].

The use of nanocellulose in catalysis has been extensively reviewed by us in 2016 [11]. As this is an active field of research, a number of applications of CNCs for catalysis were published in the past four years. This book chapter will specifically focus on these recent developments, while providing a bit of context when needed to help the reader gauge the significance of the work. Two major aspects of these recent advances will be developed in the next sections: the catalyst design and synthesis (Section 7.2) as well as the reactions catalyzed using CNCs as supports (Section 7.3).

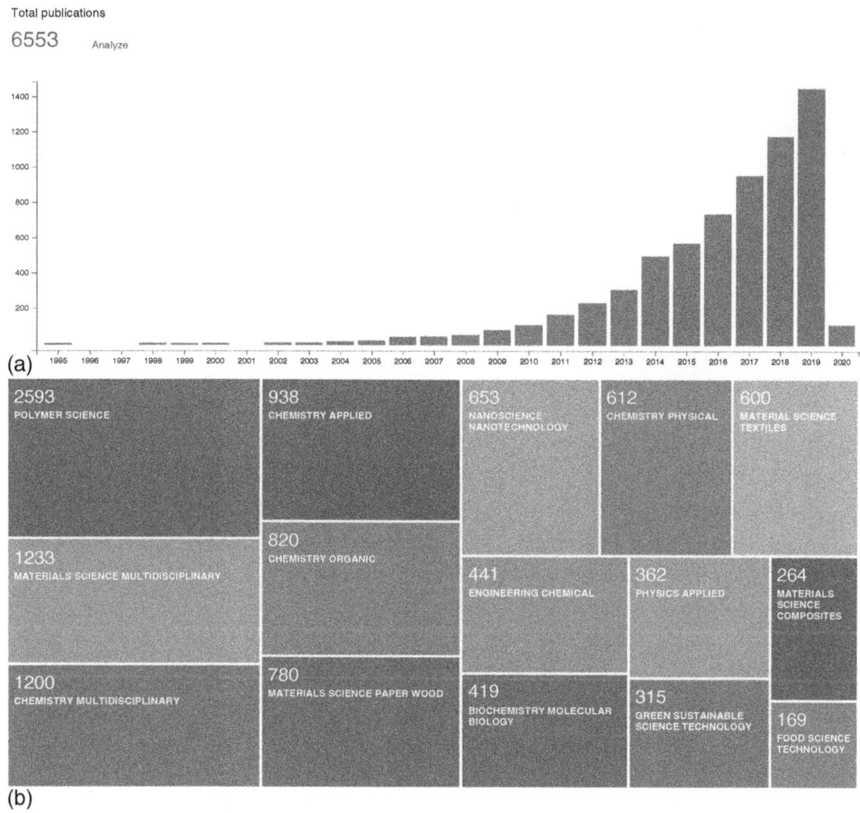

Figure 7.1 Number of publications per year on the topic of cellulose nanocrystals (CNCs) or nanocrystalline cellulose published in the past 25 years (a). Number of publications per research area on the same topic published since 1995 (b). Source: Web of Science, Clarivate Analytics, search performed on 5 February 2020 – reference group: 6553 articles.

Following the scope of this book, only catalysis involving metal nanoparticles is covered in this chapter. Recent developments in the area of organocatalysis [25–27] featuring nanocellulose as the core of the catalytic activity will not be discussed. Interested readers are directed to the elegant work of the Jones and Kitaoka groups on this topic, applied to aldol condensation [28, 29], including in the context of biomass conversion [30], and Michael addition [31].

7.2 Nanocellulose-Based Catalyst Design and Synthesis

In this section, we will review the design and synthesis of hybrid catalysts made of metal nanoparticles supported on nanocellulose or nanocellulose-derived materials. First, we will present catalysts that are suspendable and thus are separable via biphasic segregation (Section 7.2.1). In the second part, we will cover catalysts that are supported on solids, either made of nanocelluloses, of their hybrid derivatives, or of materials for which nanocellulose was used at some point in the synthesis of the catalysts (Section 7.2.2). Additionally, CNCs and CNFs are amenable to surface modifications that can be beneficial for the stabilization of catalytic nanospecies. Below, we will distinguish unmodified vs. modified nanocelluloses in catalysis design, for both suspendable and solid supports. Additionally, it is important to note that the use of monosaccharide "ligand" systems will not be included in this chapter. Indeed, the monosaccharide biopolymer is in itself used as a "ligand" to stabilize metal catalysts, such as for glucose, sucrose, or glucosamine. This does not constitute a physical "support" as with CNCs and will thus not be covered here. For this approach, the reader is directed to a recent review on the use of monosaccharides in metal-catalyzed coupling reactions by Kyne and Camp [32].

7.2.1 Synthesis of Suspendable, CNC-Based Nanocatalysts

CNCs are well suited in the context of catalysis because of their ability to suspend well in water. Catalytic conversions of substrates soluble in solvents that provide a biphasic liquid–liquid system with water are greatly simplified because it opens opportunities for biphasic separation. This approach allows an easy recycling of the catalyst, which should not be seen as a hard support but rather as a suspension bearing the catalyst (Figure 7.2) [33, 34].

7.2.1.1 Unmodified CNCs as a Support for Metal NPs

Owing to their surface functionalities, namely, hydroxyl groups and sulfonate half ester ones, CNCs are well adapted for the direct deposition and anchoring of metal nanoparticles. A typical method to achieve this consists of mixing a colloidal aqueous suspension of CNCs with metal salts chosen as NP precursors, followed by a reduction step of the resulting material, which can be performed through the use of different chemical reductants. Sodium borohydride is a typical reagent to access Au [35], Ag [36], or Pd [37] NPs deposited onto CNCs. Hydrazine hydrate and dihydrogen have also been successfully used [33, 38]. A variation of this protocol was reported in 2019 by the Foster group. They reported a pulsed method, which consists of alternating additions of Ag salts combined with two reducing agents, $NaBH_4$ and citrate sodium salt, over up to eight cycles, to grow Ag NPs onto CNCs with good size control in the range of 10 nm (Figure 7.3) [39]. Interestingly, CNCs themselves are able to reduce metal salts into NPs. This redox behavior derives from the

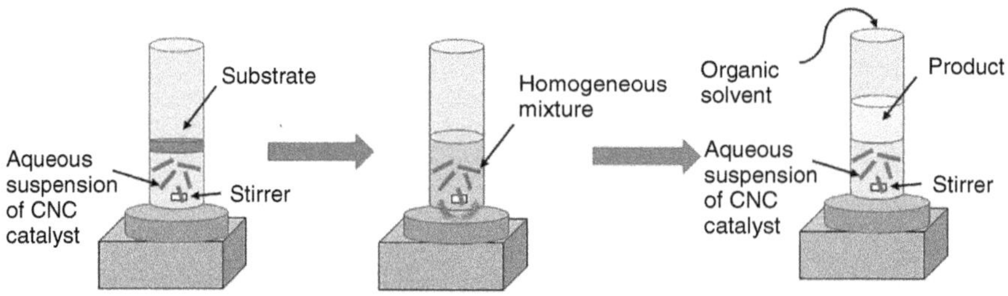

Figure 7.2 Schematic view of the use of CNC-based catalysts for the conversion of substrates showing the product separation by biphasic approach. Source: Cirtiu et al. [33]; Kaushik et al. [34].

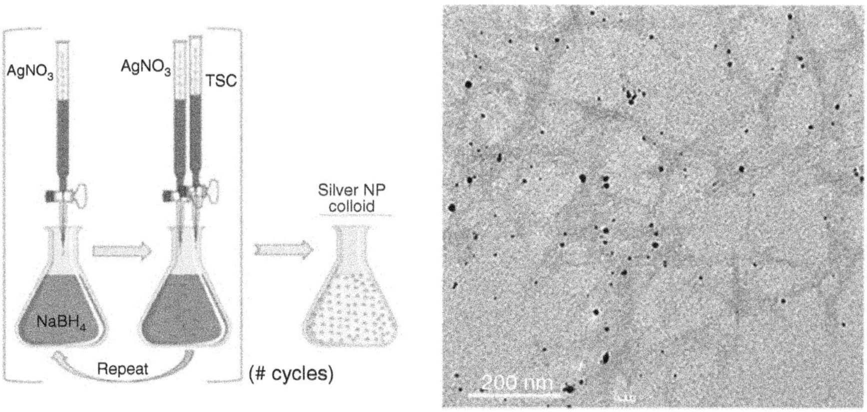

(a) (b)

Figure 7.3 Schematic view of the pulsed synthesis method for the generation of Ag NPs onto CNCs (a) and the resulting materials imaged by transmission electron microscopy (b). TSC means trisodium citrate. Source: Reprinted with permission from Stinson-Bagby et al. [39]. Copyright 2019, American Chemical Society.

presence of primary hydroxyl functionalities at the CNC surface, which can act as a reducing agent for H_2PtCl_6, in a mixture of water and supercritical CO_2 (100 °C, 100 bar) to form Pt NPs (5–30 nm in diameter) [40], or Ag salts in a procedure where ultrasmall Ag NPs (1.3 ± 0.3 nm) were generated *in situ* from the oxidation of bulk Ag wire in water, and subsequent reduction into Ag NPs via CNCs [41].

The most synthetic methods for CNC functionalization are aqueous solution based, as water is the solvent of choice to suspend CNCs [1]. Nevertheless, CNC functionalization to achieve suspendability in organic media has also been reported [20, 42], yet it requires significant synthetic effort. Mechanochemistry, defined as the study of "chemical transformations initiated or sustained by mechanical force and generally taking place in the solid phase" [43, 44], has been explored as a simplified way to functionalize the surface of CNCs. In 2018, the Eisa group pioneered this technique, relying on dry milling to introduce metal NPs at the surface of CNCs [45]. Specifically, they used ascorbic acid in order to *in situ* reduce Ag and Au salts at the surface of CNCs, in the solid phase, allowing the formation of AuNPs and AgNPs attached to the surface via strong hydrogen bonding with surface hydroxyl groups in the diameter size range of 6–35 nm and 18–24 nm, respectively. Our group reported a CNC surface functionalization method by mechanochemistry as well, allowing the synthesis of phosphorylated CNCs, with application as flame retardants [46]. These developments are promising in the prospect of scaled up catalyst syntheses.

7.2.1.2 Functionalized CNCs as a Support for Metal NPs

Native nanocelluloses feature hydroxy groups as their main surface functionalities. Oxidation methodologies allow the introduction of carboxylate groups, which are even better anchoring groups for metal ions and NPs. The Luong and Lam group first treated microcrystalline cellulose with ammonium persulfate, which afforded highly carboxylated CNCs, with dense negative surface charges [47]. They coated these carboxylated CNCs with diallyldimethylammonium chloride, a positively charged polymer that allowed a facile deposition of as-made carbonate-stabilized Au NPs onto the CNCs via electrostatic interactions. A more traditional 2,2,6,6-tetramethylpiperidine-1-oxyl (TEMPO) oxidation method was used by the Kitaoka group in order to stabilize bimetallic Au–Pd NPs via the thus-created carboxylate groups [48]. The Wang team further developed this approach by first oxidizing CNFs with TEMPO to access carboxylated species, before subsequent oxidation with periodate, which allowed the opening of the sugar rings and access to dialdehyde functionalities. This material then served as both a support and a reducing agent for the *in situ* growth of Pd NPs (diameter size c. 19 nm) and Au NPs (diameter size from 15 to 30 nm) from $PdCl_2$ and $HAuCl_4$, respectively [49]. In another example, Li and coworkers used 3-mercaptopropyltrimethoxysilane in order to introduce thiol groups onto CNCs, following a procedure developed by MacLachlan and coworkers [50] and stabilize Au NPs [51]. Beside these examples of CNC covalent functionalizations, noncovalent modifications with surfactants have also been explored. For instance, Ni and coworkers used cetyltrimethylammonium bromide (CTAB) to functionalize CNCs [52], following a strategy pioneered by Stanciu and coworkers [53], before an *in situ* chemical reduction of Ag ions with $NaBH_4$. CTAB, as a cation, was attracted to the negatively charged CNC surface and served as a stabilizer for the Ag NPs, although with limited size control. The obtained Ag NP featured a diameter ranging from 1 to 20 nm.

7.2.2 Nanocellulose-Based Solid Supports for Metal NPs

Although the suspendability of CNCs can be an advantage in a number of circumstances, most industrial catalytic systems rely on solid supports that are easy to separate from the reactive mixtures by simple filtration. Nanocellulose has been exploited to generate such solid supports, according to several major schemes: (i) CNCs are embedded into other carrier materials to combine their unique surface chemistry with the solid nature of the carrier materials; (ii) CNFs being fibrous materials can be used as solid supports; (iii) nanocellulose may be transformed, for instance, by carbonization to ultimately afford solid supports; and (iv) CNCs can be used as a template to guide the shape of another material, ultimately serving as the catalyst support.

7.2.2.1 CNC-Embedded Supports

Based on a reported method, the Nicolau group used (3-Aminopropyl)triethoxysilane (APES) in order to introduce amino groups at the surface of CNCs [54]. Ag and Pt NPs were grown *in situ* onto these amine-functionalized CNCs by $NaBH_4$ reduction of corresponding salts and the resulting hybrid materials were embedded with a

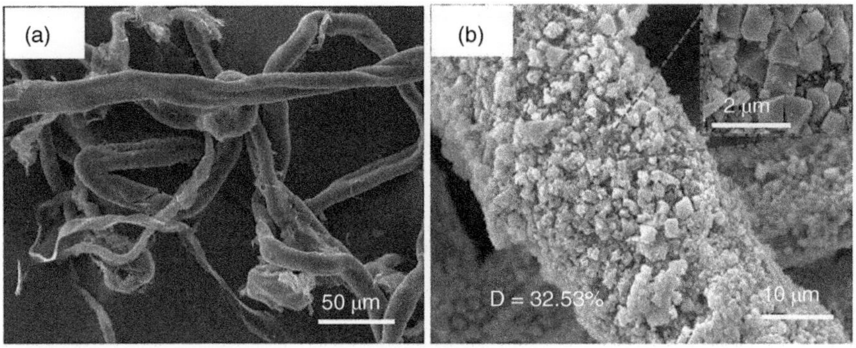

Figure 7.4 Scanning electron microscopy (SEM) images of carboxymethylated CNFs (a) and MOF-119s-functionalized carboxymethylated CNFs (b) (D = deposition ratio of MOF). Source: Reprinted and adapted with permission from Duan et al. [58]. Copyright 2019, Wiley VCH.

polyester membrane with excellent water cleaning properties [55]. In the same way, Dichiara and coworkers used an aerogel structure in order to build heterogeneous catalysts. Nanocelluloses are easily amenable to the synthesis of aerogels, and these materials were explored for a wide array of applications, including water treatment, depollution, or electronics [56]. Specifically, Pd NPs were deposited onto TEMPO-oxidized CNFs, which were turned into an aerogel by freeze drying the resulting suspensions. This material was applied to dye decolorization [57].

7.2.2.2 Functionalized CNFs as a Support for Metal NPs

CNFs feature long fiber-like structures, which can serve as supports to stabilize nanocatalysts with a facile recovery scheme. The Duan and Ni groups have explored this route using carboxymethylated CNFs, onto which they grew Cu-based metal organic framework (MOF)-199s nanostructures. They then used a classic $NaBH_4$ reduction approach to reduce $AgNO_3$ in order to functionalize the carboxymethylated CNFs with Ag NPs in the range of 6–20 nm (Figure 7.4) [58].

7.2.2.3 Use of CNCs as a Source for Carbon Supports

Another strategy lies on the use of nanocellulose as a source of carbon for the development of materials by thermal treatment. For instance, Qing and coworkers prepared nanomaterials based on graphitic carbon nitride (g-CN) using urea as the nitrogen and carbon source, with CNFs serving as an additional carbon source. The system was used to anchor Ag NPs for the enhanced photocatalytic reduction of rhodamine B and tetracycline, although extensive NP characterization is lacking (Figure 7.5) [59]. Specifically, the synthesis consisted of mixing the starting materials; urea, CNFs, and/or $AgNO_3$ in water; drying this mixture at 70 °C; and calcining the resulting powder at 550 °C. The presence of CNF in the mix allowed the tuning of the carbon content in the resulting material. The Ag NPs acted as a visible light harvester for the photocatalytic g-CN, while the additional carbon content amplified the photocatalytic effect.

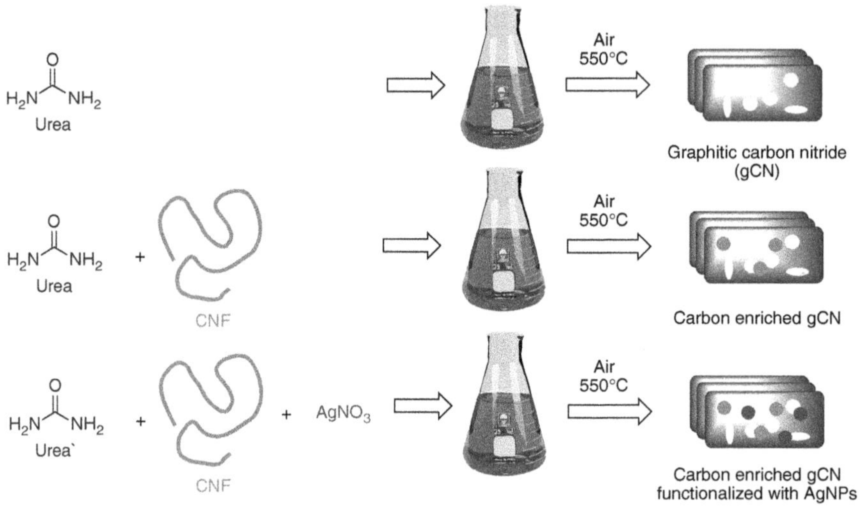

Figure 7.5 Comparative schematic illustration of the preparation of graphitic carbon nitride from urea, CNF, and/or Ag NPs. Aqueous solutions are first prepared before drying in air at 70 °C and calcination at 550 °C. Source: Modified from Tian et al. [59].

7.3 Organic Transformations Catalyzed by Metal NP/nanocellulose Hybrids

As illustrated in Section 7.2, nanocellulose and analogs have a high impact as biomass-based nanomaterials. A array of techniques allows to access metal–nanocellulose hybrids. Besides the construction of these intricate nanocomposites, the recent literature has also unraveled how these advanced materials behave as catalysts toward a wide variety of reactions. Much work has been done in order to take advantage in catalysis of nanocellulose's unique functionalities, bioavailability, and aqueous solubility [11]. In many cases, functionalities such as sulfonates, carboxylates, and hydroxyls have been evidenced to provide binding sites for metal catalysts [1]. In this instance, nanocelluloses are considered as a support material or platform onto which the active metal catalyst can be immobilized. This is especially important when catalytic reactions require precious metal containing catalysts such as those in the platinum metal group (PMG). Indeed, the recyclability and catalytic lifetime of these metal species are crucial factors toward their development on a multigram scale.

7.3.1 C–C Coupling Reactions

Cross-coupling reactions have long been a mainstay in common pharmaceutical protocols in order to synthesize active pharmaceutical ingredients (APIs) and drug molecules [60]. These reactions can be traced back to the early works of Tsuji and the formation of allylmalonate through an allylpalladium analog [61], as well as the studies that Heck [62, 63] and Mizoroki [64] conducted in proving carbon—carbon bond formation with palladium catalysts in the late 1960s. These early achievements and many others led up to the 2010 Nobel Prize in Chemistry awarded to Suzuki, Negishi, and Heck for their work in using Pd-catalyzed cross-coupling reactions. Owing to these discoveries, industrial-scale catalytic reactions resulted in increased consumption of precious, platinum group metals, and renewed interest in recycling or flow-immobilization strategies [65, 66]. In this context, CNCs have become an attractive option in stabilizing such metal nanocatalysts. Although the number of publications in the use of CNCs in order to stabilize metal NPs for coupling reactions is high, we will describe a few case examples on how the unique properties of CNCs and modified CNCs can affect the overall catalytic activity of the supported nanocatalysts.

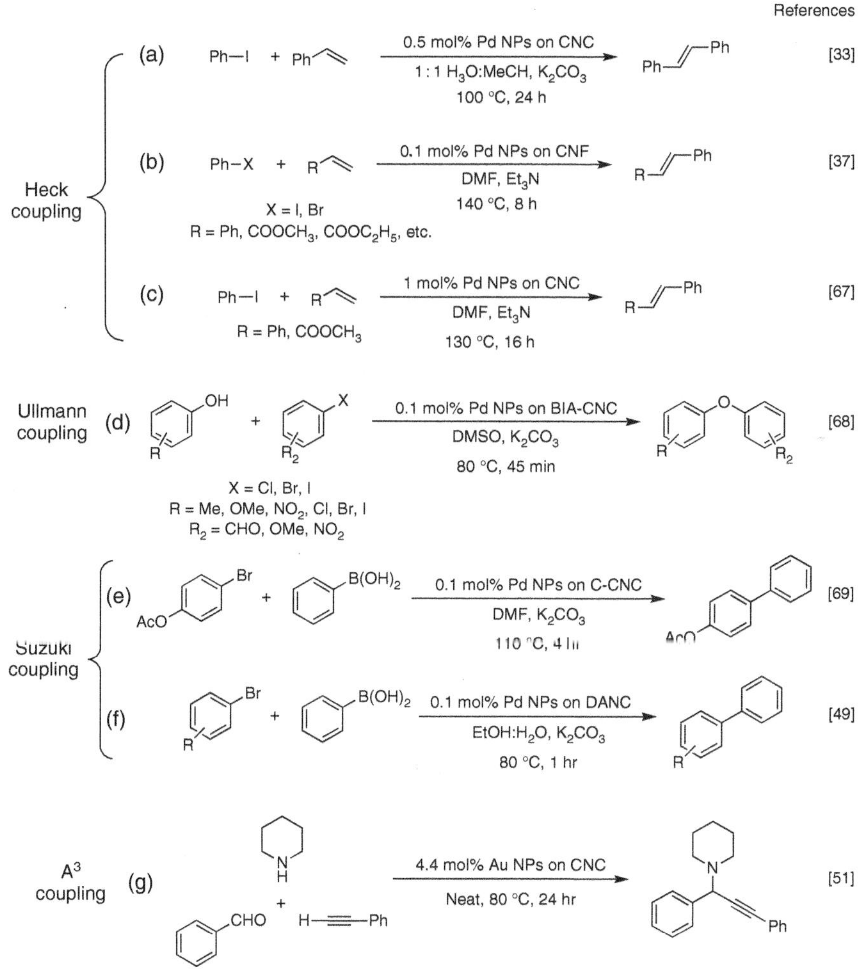

Scheme 7.2 Examples of C–C coupling reactions catalyzed by nanocellulose-supported nanocatalysts. Examples include Heck-coupling (a–c), Ullman coupling (d), Suzuki-coupling (e,f), and A^3 coupling (g). Source: Based on Refs. [33, 37, 49, 51, 67–69].

Scheme 7.2 illustrates the coupling reactions performed with metal NPs deposited onto CNCs. A large variety of coupling reactions have been performed especially with Pd NPs, and many of such studies have involved a multitude of recycling experiments to effectively prove the advantage of CNCs as a support for the recovery of the catalytic species.

Heck coupling has been a frequently used reaction in testing the catalytic activity of metal NPs deposited onto CNCs. Our group demonstrated the potency of Pd NPs deposited onto CNCs via an *in situ* hydrogen reduction synthetic method; reductive treatment of $PdCl_2$ under 4 bar of H_2 pressure for two hours in the presence of the CNCs led to 3.6 ± 0.8 nm CNC-supported Pd NPs (Scheme 7.2a) [33]. Heck coupling was performed using styrene and iodobenzene as model substrates. In a 1 : 1 mixture of water–acetonitrile as a solvent, K_2CO_3 as a base, and 24 hours of reaction time at 100 °C, 75% conversion into stilbene was obtained using only 0.5 mol% Pd loading. Zhou et al. demonstrated the use of bacterial CNFs as a support of Pd NPs for the Heck coupling reaction (Scheme 7.2b) [37]. This system yielded close to a complete conversion (99%) at 140 °C in eight hours, using only 0.1 mol% Pd loading. Moreover, they furthered the scope of the reaction by exploring various aryl halide (Br and I) and acrylate derivatives. Throughout the catalytic recycling experiments, they observed that no Pd contamination was detected in the products through inductively coupled plasma (ICP) emission spectroscopy measurements. Rezayet et al. fabricated Pd NPs (6–13 nm) immobilized onto CNCs in subcritical or supercritical CO_2 conditions that were then used as a catalyst for the Heck reaction (Scheme 7.2c), evidencing good yields in (*E*)-stilbene (96%) and *trans*-cinnamate production (87%) [67].

Recently, Venditti and coworkers reported a successful catalytic C–O Ullmann coupling reaction using Pd NPs supported on CNCs modified with 2-(1*H*-Benzo[*d*]imidazol-2-yl) (BIA) (Scheme 7.2d), hypothesizing that the nitrogen-containing functionalities present on the support can aid in "binding" Pd(II) [68]. Using phenol and 4-nitroiodobenzene, the authors reported close to full conversion (98%) of the diphenyl ether derivative in dimethylsulfoxide (DMSO) solution with K_2CO_3 as a base, at 80 °C in 45 minutes, and 0.1 mol% Pd loading. They further expanded the scope of the reaction with various functionalized phenols and aryl halides.

Suzuki coupling has also been a frequently used reaction to model the catalytic activity of metal–nanocellulose-based nanocomposites. For example, Majdoub and coworkers used glycidyltrimethylammonium-modified CNCs, a cationic cellulose nanocrystal (C-CNC), as a support for Pd NPs (3–9 nm in diameter) (Scheme 7.2e) [69]. Very high yields (98%) were obtained at 110 °C and four hours of reaction time in dimethylformamide (DMF) solution with only 0.1 mol% Pd loading. Zhang et al. also obtained successful Suzuki coupling using Pd NPs supported on CNFs that were oxidized by periodate to form 2,3-dialdehyde ends, which were used to reduce Pd or Au salts *in situ* to form Pd NPs (c. 18 nm) and Au NPs (c.16 nm) supported by the dialdehyde nanocellulose (DANC) (Scheme 7.2f) [49]. They report close to full conversion (96%) with Pd NPs on DANC in one hour of reaction time at 80 °C, along with other aryl halide and phenylboronic acid derivatives.

Other coupling reactions include the aldehyde–amine–alkyne (A^3) multicomponent coupling reaction. Huang et al. used Au NPs supported on thiol-functionalized CNCs to conduct the A^3 coupling to high yield at 80 °C in 24 hours, using benzaldehyde, piperidine, and phenylacetylene as model substrates and 4.4 mol% catalyst loading (Scheme 7.2g) [51]. They report full yield in neat conditions, while acquiring lower yield using solvent such as water (87%), toluene (92%), and ethanol (56%).

7.3.2 Reduction Reactions

The catalytic reduction of 4-nitrophenol is frequently applied as a model reaction in order to gauge the catalytic ability of metal/nanocellulose composite catalysts [11]. Other reduction-type catalysis has also been explored. For example, reductive environmental waste treatments are prevalent in the literature. Reductive organic transformations are also of great interest, including the hydrogenation of olefins and carbonyls [41], along with the hydrogenation of arenes [34] and also the hydrodeoxygenation of vanillin [70]. Typically, the turnover frequency (TOF) provides accurate information on the catalysis performance toward the development of more efficient catalytic systems and enables direct comparison between catalytic systems.

Many studies based on the model catalytic reduction of 4-nitrophenol within the metal NP-CNC literature have been reported (Scheme 7.3a). The mechanism follows a standard Langmuir–Hinshelwood model, in which metal hydrides are formed through sodium borohydride or H_2 bond cleavage [11]. Subsequent electron transfer to an adsorbed 4-nitrophenolate ion, which is the rate-determining step, leads to the desorption of 4-aminophenol. Because 4-nitrophenolate anion conveniently absorbs visible light, UV–vis spectroscopy is usually applied for *in situ* monitoring of the reaction evolution, leading to a plot absorbance vs. time, allowing to calculate TOF, rate constant, and activation energy reaction parameters. Reduction of 4-nitrophenol proved to be successful with Au [35, 47, 71–73], Ag [52, 58, 74], Pd [75], Au–Pd [48], Pt [76], Cu [77], Ni [78], and CuO [79] NPs supported onto either CNCs or CNFs. In many cases, the CNC-supported metal NPs are more effective and provide better TOFs than metal NPs supported onto other supports such as mesoporous silica or other polymers such as poly(methyl methacrylate) (PMMA). This is attributed to intrinsic properties of CNCs, specifically, their ability to suspend well and facilitate diffusion as compared to porous supports, while the intrinsic small nanodimensions of CNCs (5–10 nm in width) account for the stabilization of minute metal NPs [11]. Tam and coworkers published highest TOF for Ag NPs out of Au, Pd, and Cu NPs [74]. Huang and coworkers evidenced a NP size effect on the TOF, observing higher TOF values for c. 2 nm in size Pt NPs (2 nm) compared to c. 21 nm Pt NPs (21 nm), as an effect of both higher specific surface and higher defect content in the smaller NPs [76].

Recent literature has also focused on reductive environmental remediation of pollutant dyes, such as methylene blue. This model reaction has been explored using CNC-supported metal NPs as a catalyst, $NaBH_4$ as a reducer, and UV–vis as a monitoring technique (Scheme 7.3b) [82]. Both Pd [75] and Fe [80] NPs immobilized onto CNCs proved to catalyze this reaction effectively in terms of reaction yields.

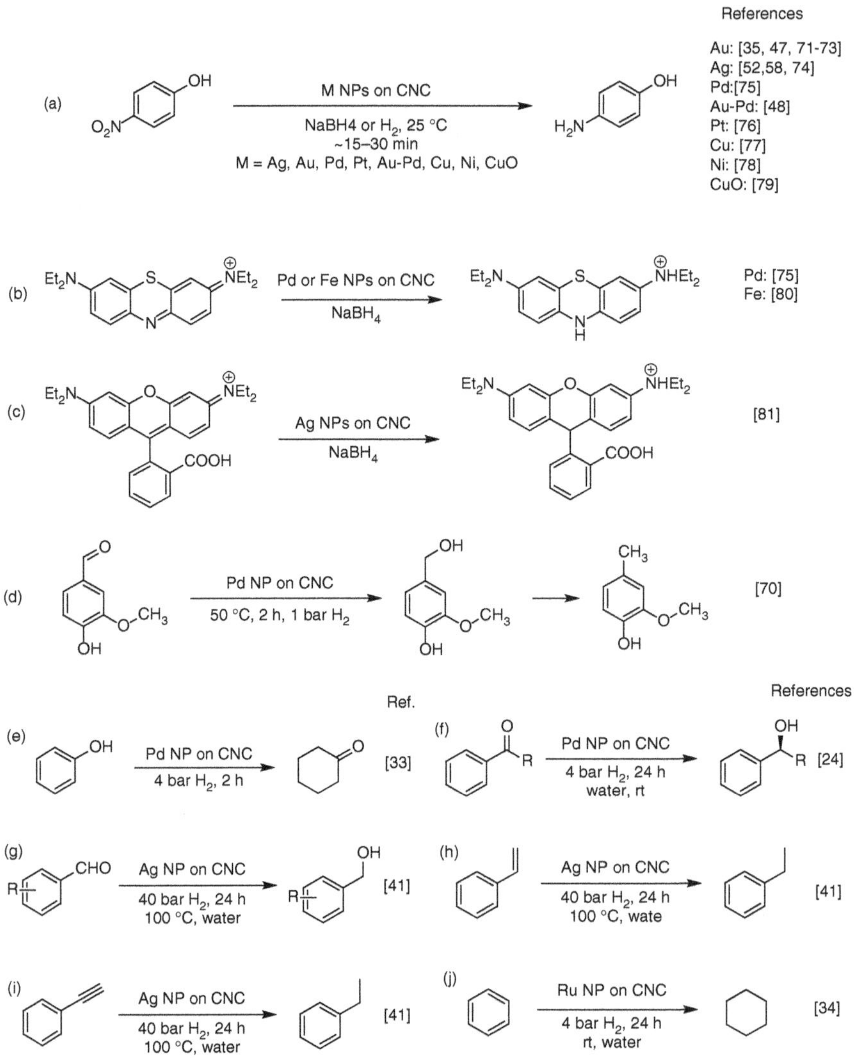

Scheme 7.3 Reduction reactions catalyzed by nanocellulose-supported nanocatalysts. Examples include aromatic nitro reduction using a variety of metal NPs (a), reduction of methylene blue using Pd or Fe NPs (b), reduction of methyl orange using Ag NPs (c), vanillin hydrodeoxygenation using Pd NPs (d), hydrogenation of phenol using Pd NPs (e), enantioselective reduction of benzaldehyde derivatives using Pd NPs (f), reduction of benzaldehyde using Ag NPs (g), alkene hydrogenation of styrene using Ag NPs (h), hydrogenation of alkyne using Ag NPs (i), and hydrogenation of benzene using Ru NPs (j). Source: Based on Refs [24, 34, 35, 41, 47, 48, 52, 58, 70–81].

The latter example established the potential applicability of CNC-stabilized Fe NPs as a material for water pollution remediation. The degradation of methyl orange was also catalyzed by a nanoporous cellulose membrane decorated by Ag NPs leading to very high yield (99%), even with over 10 catalytic recycling experiments (Scheme 7.3c) [81]. Further, Pd NPs supported on a CNC-based aerogel was used for the degradation of congo red, reaching at least 90% discoloration efficiency up to five successive catalytic cycles [57]. The reduction of nitrates for water denitrification is also reported with Cu–Pd NPs onto bacterial CNFs, with a nitrate conversion of 87%, along with similar reaction yields for up to five consecutive catalytic cycles [83]. Other works describe the valorization of lignocellulosic resources by hydrodeoxygenation reaction into valuable materials for biofuels. Vanillin is often used as a model molecule of lignin transformation [84, 85]. Examples include the use of Pd NPs supported onto modified CNCs as nanocatalysts for this reaction, as published by Cai and coworkers (Scheme 7.3d) [70]. In their study, they report full conversion of vanillin to 2-methoxy-4-methylphenol under relatively mild conditions (atmospheric H_2 pressure and 50 °C) in two hours of reaction time. They functionalized CNCs with trifluoromethyl groups in order to increase the interaction of the nanocatalyst with the organic substrates such as vanillin.

The reduction of olefins, carbonyls, and arenes are an essential reaction in both academy and industry laboratories, with the goal to boost atom economy by shifting away from stoichiometric quantities of reductant and find "greener" reaction conditions (according to the 12 principles of green chemistry) [86]. The use of a metal catalyst and of hydrogen gas as a reductant provides an outlet to meet these concerns, with CNC and CNF supports offering recyclability and in some cases superior enantioselectivity [24]. For example, Pd NPs supported onto CNCs were shown to be very active for the hydrogenation of phenols in mild reaction conditions (4 bar H_2, room temperature, and two hours), with up to 90% yield in 24 hours (Scheme 7.3e) [33]. The obtained data were compared to other supported Pd NP catalysts such as Pd NPs onto alumina or carbonaceous supports, which showed that even with half the Pd loading, similar high yields were obtained, outlining the effectiveness CNCs have as a support. Further work reported the enantioselective reduction of prochiral aromatic ketones in water, with 4 bar H_2 in two hours (Scheme 7.3f) [24]. The CNCs were found to be a noninnocent support, as suggested by cryo-TEM and tomography, which acted as chiral inducers allowing enantiomeric excess (ee) yields up to 65% owing to the local chiral environment around the active catalytic centers. Although this enantioselectivity result was low compared to values obtained from grafted chiral inducers, it constituted at the time the highest ee obtained with an unmodified biomass-based support as sole chiral source [87]. In addition to the Pd NP-based catalysts, CNC-supported Ag NPs were also shown to be active for the reduction of aromatic aldehydes (Scheme 7.3g), alkenes (Scheme 7.3h), and alkynes (Scheme 7.3i) but in more drastic conditions, namely, under 40 bar of H_2 and at 100 °C for 24 hours [41]. Ru NPs were also used for the very unfavorable hydrogenation of arenes with H_2, using mild conditions (4 bar of H_2, room temperature, and 24 hours) to afford full conversion into the saturated ring product (Scheme 7.3j) [34]. Other substrates such as styrene and acetophenone were also investigated in the same conditions to expand the scope of the reaction, with relatively good yields of 96% and 76%, respectively toward the fully reduced ethylcyclohexane product. Interestingly, a heterocyclic aromatic compound (2-acetylfuran) was also tested, producing excellent yield (90%).

7.4 Conclusions

As it continues to grow, the field of nanocellulose-supported catalysis is revealing the potential to deliver on its promise to afford more active, more selective, and more benign solutions. Nanocellulose affords catalytic systems that may be suspendable, typically in aqueous environments, or insoluble. The surface chemistry of nanocellulose is well suited for their direct functionalization with metal nanoparticles, yet prefunctionalization to introduce anchoring groups is also reported. The resulting systems have been reported as active catalysts for a wide range of reactions. Many coupling reactions such as the Heck, Ullmann, Suzuki, and A^3 couplings proceed with high yields, typically with Pd-based catalysts. Reduction reactions have also been successfully explored with Au-, Ag-, Pd-, Pt-, and Cu-based catalysts. The success of nanocellulose-based catalysts lies in the ability for such supports to sustain very small metal NPs, while favoring good mass diffusion. These systems have enabled high TOFs and even enantioselective induction. In the future, it is expected that emerging nanoparticles of polysaccharides will also be considered in the field. In particular, nanocrystalline chitosan offers the opportunity to complement the chemistry of cellulose by providing access to the amenable nitrogen manifold [88]. Recent examples are paving the way to this new type of nanocatalysts [89, 90].

References

1 Lam, E., Male, K.B., Chong, J.H. et al. (2012). *Trends Biotechnol.* 30 (5): 283–290.

2 Kalashnikova, I., Bizot, H., Cathala, B., and Capron, I. (2012). *Biomacromolecules* 13 (1): 267–275.

3 Bossa, N., Carpenter, A.W., Kumar, N. et al. (2017). *Environ. Sci. Nano* 4 (6): 1294–1303.

4 Shak, K.P.Y., Pang, Y.L., and Mah, S.K. (2018). *Beilstein J. Nanotechnol.* 9 (1): 2479–2498.

5 Hall, L.J., Deville, J.P., Rojas, O.J. et al. (2019). Additive of chemically-modified cellulose nanofibrils or cellulose nanocrystals. US Patent 10, 294,403, filed 19 December 2014 and issued 21 May 2019.

6 Kaboorani, A., Riedl, B., Blanchet, P. et al. (2012). *Eur. Polym. J.* 48 (11): 1829–1837.

7 Cao, Y., Tian, N., Bahr, D. et al. (2016). *Cem. Concr. Compos.* 74: 164–173.

8 Li, Y., Zhu, L., Wang, B. et al. (2018). *ACS Appl. Mater. Interfaces* 10 (33): 27831–27839.

9 Liew, S.Y., Walsh, D.A., and Thielemans, W. (2013). *RSC Adv.* 3 (24): 9158.

10 An, X., Wen, Y., Almujil, A. et al. (2016). *RSC Adv.* 6 (92): 89457–89466.

11 Kaushik, M. and Moores, A. (2016). *Green Chem.* 18 (3): 622–637.

12 Gu, M., Jiang, C., Liu, D. et al. (2016). *ACS Appl. Mater. Interfaces* 8 (47): 32565–32573.

13 Kelly, J.A., Shukaliak, A.M., Cheung, C.C.Y. et al. (2013). *Angew. Chem. Int. Ed.* 52 (34): 8912–8916.
14 Brinchi, L., Cotana, F., Fortunati, E., and Kenny, J.M. (2013). *Carbohydr. Polym.* 94 (1): 154–169.
15 Trache, D., Hussin, M.H., Haafiz, M.K.M., and Thakur, V.K. (2017). *Nanoscale* 9 (5): 1763–1786.
16 Thomas, B., Raj, M.C., Athira, K.B. et al. (2018). *Chem. Rev.* 118 (24): 11575–11625.
17 Varma, R.S. (2019). *ACS Sustainable Chem. Eng.* 7 (7): 6458–6470.
18 Kargarzadeh, H., Mariano, M., Huang, J. et al. (2017). *Polymer* 132: 368–393.
19 Sheikhi, A. (2019). Emerging cellulose-based nanomaterials and nanocomposites. In: *Nanomaterials and Polymer Nanocomposites*, 307–351. Elsevier.
20 Tang, J., Sisler, J., Grishkewich, N., and Tam, K.C. (2017). *J. Colloid Interface Sci.* 494: 397–409.
21 Ogundare, S.A. and van Zyl, W.E. (2019). *Cellulose* 26 (11): 6489–6528.
22 Islam, M.S., Chen, L., Sisler, J., and Tam, K.C. (2018). *J. Mater. Chem. B* 6 (6): 864–883.
23 Johnston, L.J., Jakubek, Z.J., Beck, S. et al. (2018). *Metrologia* 55 (6): 872–882.
24 Kaushik, M., Basu, K., Benoit, C. et al. (2015). *J. Am. Chem. Soc.* 137 (19): 6124–6127.
25 Nagorny, P. and Sun, Z. (2016). *Beilstein J. Org. Chem.* 12: 2834–2848.
26 Qin, Y., Zhu, L., and Luo, S. (2017). *Chem. Rev.* 117 (13): 9433–9520.
27 van der Helm, M.P., Klemm, B., and Eelkema, R. (2019). *Nat. Rev. Chem.* 3 (8): 491–508.
28 Ellebracht, N.C. and Jones, C.W. (2018). *Cellulose* 25 (11): 6495–6512.
29 Kanomata, K., Tatebayashi, N., Habaki, X., and Kitaoka, T. (2018). *Sci. Rep.* 8 (1): 4098.
30 Ellebracht, N.C. and Jones, C.W. (2019). *ACS Catal.* 9 (4): 3266–3277.
31 Ranaivoarimanana, N., Kanomata, K., Kitaoka, T. et al. (2019). *Molecules* 24 (7): 1231.
32 Kyne, S.H. and Camp, J.E. (2017). *ACS Sustainable Chem. Eng.* 5 (1): 41–48.
33 Cirtiu, C.M., Dunlop-Brière, A.F., and Moores, A. (2011). *Green Chem.* 13 (2): 288–291.
34 Kaushik, M., Friedman, H.M., Bateman, M., and Moores, A. (2015). *RSC Adv.* 5 (66): 53207–53210.
35 Chen, L., Cao, W., Quinlan, P.J. et al. (2015). *ACS Sustainable Chem. Eng.* 3 (5): 978–985.
36 Lokanathan, A.R., Uddin, K.M.A., Rojas, O.J., and Laine, J. (2014). *Biomacromolecules* 15 (1): 373–379.
37 Zhou, P., Wang, H., Yang, J. et al. (2012). *Ind. Eng. Chem. Res.* 51 (16): 5743–5748.
38 Dutta, A., Chetia, M., Ali, A.A. et al. (2019). *Catal. Lett.* 149 (1): 141–150.
39 Stinson-Bagby, K.L., Owens, J., Rouffa, A. et al. (2019). *ACS Appl. Nano Mater.* 2 (4): 2317–2324.

40 Benaissi, K., Johnson, L., Walsh, D.A., and Thielemans, W. (2010). *Green Chem.* 12 (2): 220–222.

41 Kaushik, M., Li, A.Y., Hudson, R. et al. (2016). *Green Chem.* 18 (1): 129–133.

42 Viet, D., Beck-Candanedo, S., and Gray, D.G. (2007). *Cellulose* 14 (2): 109–113.

43 James, S.L., Adams, C.J., Bolm, C. et al. (2012). *Chem. Soc. Rev.* 41 (1): 413–447.

44 Do, J.-L. and Friščić, T. (2017). *ACS Cent. Sci.* 3 (1): 13–19.

45 Eisa, W.H., Abdelgawad, A.M., and Rojas, O.J. (2018). *ACS Sustainable Chem. Eng.* 6 (3): 3974–3983.

46 Fiss, B.G., Hatherly, L., Stein, R.S. et al. (2019). *ACS Sustainable Chem. Eng.* 7 (8): 7951–7959.

47 Lam, E., Hrapovic, S., Majid, E. et al. (2012). *Nanoscale* 4 (3): 997.

48 Azetsu, A., Koga, H., Isogai, A., and Kitaoka, T. (2011). *Catalysts* 1 (1): 83–96.

49 Zhang, K., Shen, M., Liu, H. et al. (2018). *Carbohydr. Polym.* 186: 132–139.

50 Shopsowitz, K.E., Qi, H., Hamad, W.Y., and MacLachlan, M.J. (2010). *Nature* 468 (7322): 422–426.

51 Huang, J.L., Gray, D.G., and Li, C.J. (2013). *Beilstein J. Org. Chem.* 9: 1388–1396.

52 An, X., Long, Y., and Ni, Y. (2017). *Carbohydr. Polym.* 156: 253–258.

53 Padalkar, S., Capadona, J.R., Rowan, S.J. et al. (2010). *Langmuir* 26 (11): 8497–8502.

54 Koga, H., Kitaoka, T., and Isogai, A. (2011). *J. Mater. Chem.* 21 (25): 9356–9361.

55 Cruz-Tato, P., Ortiz-Quiles, E.O., Vega-Figueroa, K. et al. (2017). *Environ. Sci. Technol.* 51 (8): 4585–4595.

56 De France, K.J., Hoare, T., and Cranston, E.D. (2017). *Chem. Mater.* 29 (11): 4609–4631.

57 Gu, J., Hu, C., Zhang, W., and Dichiara, A.B. (2018). *Appl. Catal., B* 237: 482–490.

58 Duan, C., Liu, C., Meng, X. et al. (2019). *Appl. Organomet. Chem.* 33 (5): e4865.

59 Tian, C., Tao, X., Luo, S. et al. (2018). *Environ. Sci. Nano* 5 (9): 2129–2143.

60 Busacca, C.A., Fandrick, D.R., Song, J.J., and Senanayake, C.H. (2011). *Adv. Synth. Catal.* 353 (11–12): 1825–1864.

61 Tsuji, J., Takahashi, H., and Morikawa, M. (1965). *Tetrahedron Lett.* 6 (49): 4387–4388.

62 Heck, R.F. (1968). *J. Am. Chem. Soc.* 90 (20): 5518–5526.

63 Heck, R.F. (1968). *J. Am. Chem. Soc.* 90 (20): 5531–5534.

64 Mizoroki, T., Mori, K., and Ozaki, A. (1971). *Bull. Chem. Soc. Jpn.* 44 (2): 581–581.

65 Cue, B.W. and Zhang, J. (2009). *Green Chem. Lett. Rev.* 2 (4): 193–211.

66 Magano, J. and Dunetz, J.R. (2011). *Chem. Rev.* 111 (3): 2177–2250.

67 Rezayat, M., Blundell, R.K., Camp, J.E. et al. (2014). *ACS Sustainable Chem. Eng.* 2 (5): 1241–1250.

68 Seyednejhad, S., Khalilzadeh, M.A., Zareyee, D. et al. (2019). *Cellulose* 26 (8): 5015–5031.

69 Jebali, Z., Granados, A., Nabili, A. et al. (2018). *Cellulose* 25 (12): 6963–6975.

70 Li, D., Zhang, J., and Cai, C. (2018). *J. Organomet. Chem.* 83 (14): 7534–7538.

71 Chen, Y., Chen, S., Wang, B. et al. (2017). *Carbohydr. Polym.* 160: 34–42.

72 Koga, H., Tokunaga, E., Hidaka, M. et al. (2010). *Chem. Commun.* 46 (45): 8567–8569.

73 Wu, X., Lu, C., Zhou, Z. et al. (2014). *Environ. Sci. Nano* 1 (1): 71–79.

74 Tang, J., Shi, Z., Berry, R.M., and Tam, K.C. (2015). *Ind. Eng. Chem. Res.* 54 (13): 3299–3308.

75 Wu, X., Lu, C., Zhang, W. et al. (2013). *J. Mater. Chem. A* 1 (30): 8645–8652.

76 Lin, X., Wu, M., Wu, D. et al. (2011). *Green Chem.* 13 (2): 283–287.

77 Bendi, R. and Imae, T. (2013). *RSC Adv.* 3 (37): 16279–16282.

78 Prathap, K.J., Wu, Q., Olsson, R.T., and Dinér, P. (2017). *Org. Lett.* 19 (18): 4746–4749.

79 Zhou, Z., Lu, C., Wu, X., and Zhang, X. (2013). *RSC Adv.* 3 (48): 26066–26073.

80 Dhar, P., Kumar, A., and Katiyar, V. (2015). *Cellulose* 22 (6): 3755–3771.

81 Yang, Y., Chen, Z., Wu, X. et al. (2018). *Cellulose* 25 (4): 2547–2558.

82 Ramaraju, B., Imae, T., and Destaye, A.G. (2015). *Appl. Catal., A* 492: 184–189.

83 Sun, D., Yang, J., Li, J. et al. (2010). *Appl. Surf. Sci.* 256 (7): 2241–2244.

84 Behling, R., Valange, S., and Chatel, G. (2016). *Green Chem.* 18 (7): 1839–1854.

85 Strassberger, Z., Tanase, S., and Rothenberg, G. (2014). *RSC Adv.* 4 (48): 25310–25318.

86 de Vries, J.G. and Elsevier, C.J. (2007). *Introduction, Organometallic Aspects and Mechanism of Homogeneous Hydrogenation*. Weinheim, Germany: Wiley VCH.

87 Yasukawa, T., Miyamura, H., and Kobayashi, S. (2015). *Chem. Sci.* 6 (11): 6224–6229.

88 Fan, Y., Saito, T., and Isogai, A. (2008). *Biomacromolecules* 9 (1): 192–198.

89 Jin, T., Kurdyla, D., Hrapovic, S. et al. (2020). *Biomacromolecules* 6 (6): 2236–2245.

90 Jin, T., Hicks, M., Kurdyla, D. et al. (2020). *Beilstein J. Org. Chem.* 16 (1): 2477–2483.

8

Magnetically Recoverable Nanoparticle Catalysts

Liane M. Rossi[1], *Camila P. Ferraz*[1,2], *Jhonatan L. Fiorio*[1], *and Lucas L. R. Vono*[1]

[1] *Universidade de São Paulo, Instituto de Química, Departamento de Química Fundamental, Avenida Prof. Lineu Prestes 748, São Paulo 05508-000, Brazil*
[2] *Univ. Lille, CNRS, Centrale Lille, ENSCL, Univ. Artois, UMR 8181 - UCCS - Unité de Catalyse et Chimie du Solide, Lille 59000, France*

8.1 Introduction

Supported catalysts have the advantage of easy separation compared to their homogeneous counterparts, and this is also valid for colloidal metal nanoparticles. However, even heterogeneous catalysts can be harsh to separate from liquid products depending on the grain size and density of the catalyst support. Separation processes that involve centrifugation, filtration, decantation, and distillation are time or energy consuming. On the other hand, magnetic separation allows a fast removal of catalysts from crude samples avoiding mass loss and catalyst oxidation because the whole separation process occurs without removing the catalysts from the reaction vessel. The catalyst recycling can be easily performed by washing it directly in the reaction apparatus and by adding new portions of the substrate. In comparison with usual separation, the magnetic separation is faster, easier, minimizes the use of solvents, and the generation of waste; thus can be considered more efficient and green (BOX 8.1).

Box 8.1 Magnetic separation as a green technology for recovering and recycling of catalysts

- Fast and efficient
- Specific for separation of magnetic components
- Low energy consumption (magnet and electromagnet)
- Easy catalyst recycling through external magnet use – the catalyst is kept inside the reactor

(Continued)

Nanoparticles in Catalysis: Advances in Synthesis and Applications, First Edition.
Edited by Karine Philippot and Alain Roucoux.

Box 8.1 (Continued)

- Minimizes catalyst loss
- Easy catalyst handling under inert condition
- Easy sampling and product isolation under inert condition
- Minimizes exposure to air
- Solvent-free process
- Minimizes the production of waste
- Applicable for any volume – possible to scale up
- Simple design
- Applicable for any kind of catalysts (molecular, enzyme, metal, acid, etc.)

Magnetism is a powerful tool to separate magnetic from nonmagnetic components of a mixture. For example, it has been used for removal of ferromagnetic impurities from large volumes of boiler water in both conventional and nuclear power plants, removal of weakly magnetic colored impurities from kaolin clay, the enrichment of low-grade iron ore, or to select and sort magnetic materials [1, 2]. Magnetic separation can find use in environmental applications, mining, large-scale water purification, sewage treatment, food production, pharmaceutical and biochemical activities, and medical sciences [1]. The use of magnetism for separations is not always the first choice, unless to separate or sort an intrinsically magnetic component of a mixture. However, the development of strategies for attaching "nonmagnetic" components to magnetic carriers allows expanding this separation concept. In this case, the best candidate materials for magnetic carriers are those with a superparamagnetic behavior. The superparamagnetism is a size-dependent property that is characterized by high saturation magnetization (M_s) in the presence of a magnetic field and the absence of residual magnetization when ceased the application of a magnetic field [3, 4]. Bulk magnetic materials present a multidomain structure, but when the material is reduced below a critical diameter, the formation of single-domain nanoparticles is energetically favorable, characterizing a superparamagnetic state. The absence of residual magnetization of a superparamagnetic material is a consequence of the easy magnetic reorientation at temperatures higher than the blocking temperature (T_b). T_b is the temperature where the thermal energy is higher than the barrier energy, which allows the magnetic reorientation. At temperature lower than T_b, the material is in a blocked condition showing residual magnetization. Above T_b, the material can be separated (high magnetization in the presence of an applied magnetic field) and easily redispersed (no residual magnetization) after removal of magnetic field. The absence of residual magnetization of superparamagnetic supports is a key parameter for an efficient catalyst separation, allowing the catalyst to be dispersed in reaction media without support aggregation after the separation step [5, 6].

Even if they represent only a very small portion of the existing catalysts, the magnetic materials themselves can play a dual role in serving both as catalysts and

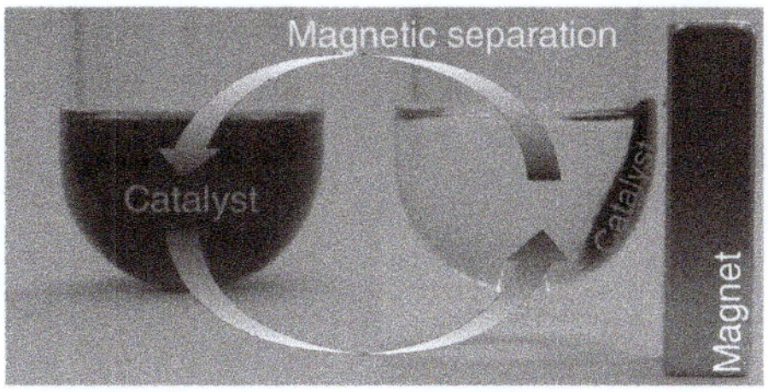

Figure 8.1 Principle of magnetic separation of a catalyst by simple application of a magnet. Source: Reproduced with permission from Rossi et al. [16]. Copyright 2012 Brazilian Chemical Society.

as magnetic supports. Catalysts based on iron [7], iron-based alloys [8], cobalt [9], nickel, and the corresponding oxides [10], such as magnetite (Fe_3O_4), maghemite (Fe_2O_3), and other ferrites ($MnFe_2O_4$ and $CoFe_2O_4$) [1, 11, 12], can be directly magnetically separated owing to their intrinsic magnetic properties [13]. All these materials can also be applied as support materials for the immobilization of "nonmagnetic" catalytically active species, rendering magnetically recoverable catalysts [14, 15]. Therefore, magnetic separation can also be straightforwardly used for the recovery of a large variety of catalysts such as metal complexes, metal nanoparticles, biocatalysts, or organocatalysts, which are immobilized onto magnetic supports and then can be magnetically recovered with a magnet (Figure 8.1).

8.2 Magnetic Support Material

Metallic Fe, Ni, or Co are very sensitive to oxidation and agglomeration and can also exhibit significant chemical reactivity, which are undesirable characteristics for a catalyst support. The use of the corresponding oxides, which present lower saturation magnetization, but are easily handled, stored, and synthesized can be a good alternative. Iron oxides are less toxic, cheaper, and easier to produce than Fe, Co, and Ni and are thus preferred as catalyst supports. Magnetite has been the most used magnetic support because of the higher saturation magnetization compared to the Co and Ni oxides and ferrites (γ-Fe_2O_3, $MnFe_2O_3$, $CoFe_2O_4$, $NiFe_2O_4$, and $CuFe_2O_4$) [17]. For the immobilization of the catalytic component, there are two possible strategies: (i) deposition, precipitation, or coprecipitation of the catalyst directly on the magnetic material, similarly to the preparation of supported heterogeneous catalysts onto inorganic oxides or carbon (Figure 8.2a) or (ii) anchoring the catalyst using functionalized surfaces (Figure 8.2b). The surface functionalization can be performed on a naked magnetic material or after coating with a shell of a protective material (Figure 8.2c) to preserve the magnetic cores and also facilitate the interaction and modification with desired functional groups. The support functionalization is accomplished with anchoring agents such as carboxylic acid, phosphonic acid, organoalkoxysilanes, and dopamine derivatives [18, 19].

The magnetic separation efficiency depends on the homogeneity and overall quality of the magnetic material used as a support, which reflects on the magnetic properties of the sample as a whole. Moreover, for an efficient separation, the

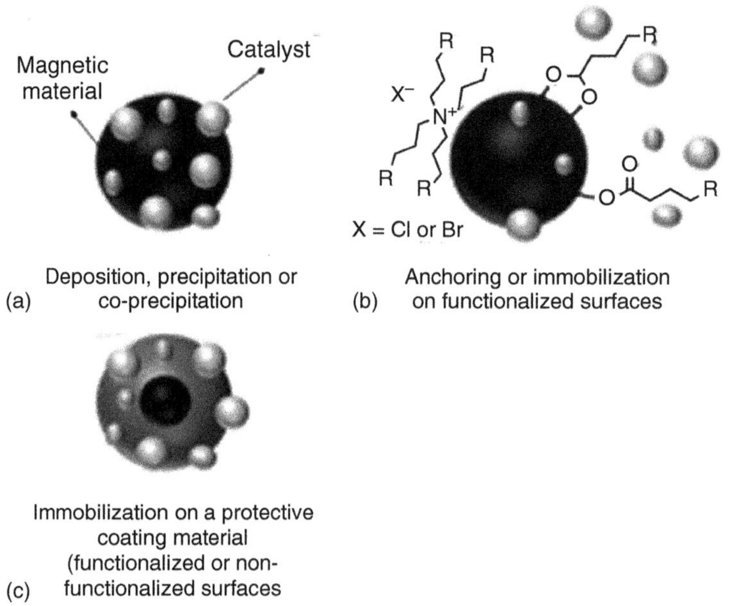

Figure 8.2 Strategies for magnetically recoverable catalysts: (a) deposition, (b) binding through functional groups, or (c) coating with a protective shell material.

interaction between the "nonmagnetic" catalytic component and the supporting magnetic material has to be strong and stable [6, 16]. Thus, the coating of the magnetic cores with a protective material, e.g. silica, polymer, and carbon, is a common strategy to improve the catalyst–support interaction and produce more stable magnetic support materials. The coating also presents a benefit for the catalyst preparation such as offering a surface that can be easily functionalized before the immobilization process. Silica and carbon are by far the most explored coatings for magnetic nanoparticles [20, 21]. The silica coating is (i) the most versatile because it is easily performed, (ii) chemically and thermally stable under typical reaction conditions including a resistance to organic solvents, and (iii) provides an easy functionalization. The most known methods to coating iron oxide with a protective shell of silica (Figure 8.3) are the Stöber [22], microemulsion [23], reverse microemulsion [24], aerosyl pyrolysis [25] etc. All these methods have advantages and disadvantages. For example, the Stöber method is simple and of low cost, but it can easily result in pearl-shaped necklace structures instead of single particles [26]. The aerosyl pyrolysis method can produce large amounts of materials, but requires sophisticated equipment. The main disadvantage of microemulsion and reverse microemulsion methods is the higher cost compared to the sol–gel Stöber method. However, the formation of micelles contributes to both the confinement of the particles and the controlled growth of a silica shell [27], providing high-quality materials [28]. The carbon coating is resistant in acid and basic conditions, but requires more sophisticated methods

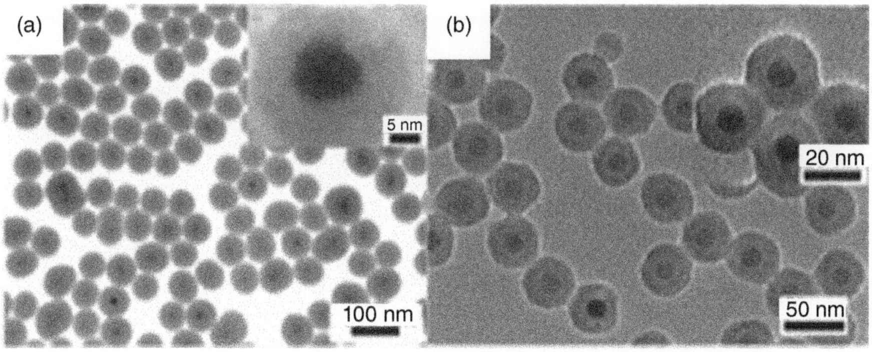

Figure 8.3 Silica-coated magnetic nanoparticles prepared by different methods: (a) Stöber. Source: Reproduced with permission from Lu et al. [22]. Copyright 2002 American Chemical Society. Reverse microemulsion. Source: Reproduced with permission from Ding et al. [24]. Copyright 2006 American Chemical Society.

and is usually applied to bigger magnetic cores. The most known methods to coating magnetic materials with a protective shell of carbon are template [29], chemical vapor deposition (CVD) [30], and hydrothermal/solvothermal methods [31]. The careful selection of the coating process is fundamental to obtain a support material with uniform morphology, loading, and distribution of magnetic cores into the coating matrix. The ideal support material should be composed of a magnetic core surrounded by a uniform shell (controlled/uniform thickness), such as a core–shell-like structure. Any heterogeneity among the individual particles, for example, particles of the matrix material without any magnetic core, or polydispersity (various sizes and shell thickness) will mean different responses to the applied magnetic field and thus inefficient separation processes.

8.2.1 Magnetite Coated with Silica

Silica is often used as a support in the field of catalysis. It is an excellent coating material for the preparation of magnetic supports because it is quite stable and inert, and the surface silanol groups are easy to functionalize. The literature contains plenty of methodologies in order to get silica-coated magnetic nanoparticles [22, 32–40], with different levels of success. Many reported protocols are difficult to reproduce and show a low level of success in the deposition of a uniform layer of silica around each nanoparticle (a core–shell-like structure is desired). The reverse microemulsion methodology is more reproducible than other methods and also guarantees a higher quality core–shell-like material [16, 28]. The micelles or reverse micelles work as nanoreactors, controlling the spherical morphology of the material, leading to better-defined core–shell-like structures [32, 33, 41, 42]. Excellent materials have been obtained by different research groups via a reverse microemulsion method from an organic phase containing the magnetite nanoparticles prepared by coprecipitation and stabilized with oleic acid, cyclohexane, tetraethyl orthosilicate (TEOS),

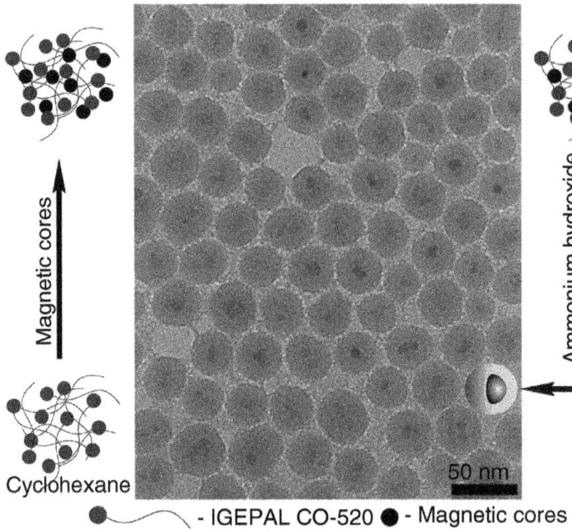

Figure 8.4 Silica-coated magnetic nanoparticles prepared by reverse microemulsion.

and the surfactant Igepal® CO-520 (Figure 8.4). The separation of the silica-covered material from the microemulsion is a critical step to obtain a high-quality material. Wang et al. [44] reported a comparative work where they applied different methods to separate and wash silica-coated Ag nanoparticles, all prepared by a similar reverse microemulsion. The authors observed that the sedimentation washing procedure resulted in aggregation and that washing by Soxhlet extraction caused dissolution of the silica shell. Even if ethanol (the solvent used in washing by Soxhlet extraction) is not expected to cause silica corrosion, the authors indicated that this washing procedure gives the best result. In the case of silica-coated magnetite, the work of Rossi's group suggests that precipitation of the microemulsion with methanol is a key step to avoid the condensation of excess of TEOS and agglomeration, followed by centrifugation and washing with ethanol [43]. The reaction time is also an important parameter to obtain a uniform silica shell. The material obtained after six hours of reaction comprised well-defined spherical core–shell magnetite–silica nanospheres with a final size of ~28 nm, and the silica shell was shown to grow slightly with the reaction time up to a size of ~35 nm after 48 hours of reaction (Figure 8.4). The reaction typically takes 16 hours to get a very homogeneously coated material [43]. The reverse microemulsion method is efficient, reproducible, and easy to scale up, while maintaining a well-defined core–shell-like morphology. Nevertheless, this synthesis method generates a considerable amount of chemical waste, mainly because of the solvents and the surfactant used. In order to have an economical and environmental synthesis method, a separation and recuperation protocol was investigated by the same group [45].

The silica coating contributed to wide spread the range of applications of magnetic materials in catalysis and biomedicine because it has several advantages, including high stability of the material in aqueous media and easy surface functionalization.

This was the case of the developed biologically inspired magnetically recoverable metalloporphyrin-based catalysts reported for the oxidation of hydrocarbons with molecular oxygen [46] and for sequential transformation of alkenes and CO_2 into cyclic carbonates [47]. In addition, a magnetically recoverable enzyme catalyst Fe_3O_4@silica–αCT and a magnetically recoverable metal nanocatalyst Fe_3O_4@silica–Pd were used together for the synthesis of a bitter-taste dipeptide. This represents an inspiring combination of technologies by allying biocatalysis, heterogeneous metal catalysis, and magnetic separation [48]. More examples of magnetically recoverable nanoparticle catalysts will be discussed later on (part 3).

It is worth to mention that the silica coating is able to protect the iron oxide (Fe_3O_4) cores from oxidation and sintering, as shown by the following set of experiments. A comparative study with uncoated and silica-coated magnetite nanoparticles revealed that the thermal stability of the magnetite cores significantly increased after the coating with silica [43]. The sample containing uncoated magnetite nanoparticles oxidized into α-Fe_2O_3 and lost the magnetic properties after thermal treatment, while the silica-coated magnetite nanoparticles preserved their superparamagnetic behavior and saturation magnetization. Additionally, transmission electron microscopy (TEM) images of the silica-coated magnetite material before and after calcination revealed that the core–shell-like morphology and mean particle size were maintained. As expected, the surface area of the silica-coated magnetite material increased after calcination (from c. 20 to 110 $m^2\ g^{-1}$) because of the removal of organic matrix, thus offering more surface available for functionalization. Morevoer, the morphology, size, crystal structure, and magnetic behavior of the silica-coated sample did not change much after calcination at 500 °C for 2 h in air. No significant changes were observed in M(H) and ZFC/FC curves for the silica-coated magnetite material before and after calcination. The two samples show typical characteristics of superparamagnetic particles, and the saturation magnetization (Ms) was 82 and 79 $emu\ g^{-1}$ for as-prepared and calcined support samples, respectively. Thus, no damage to the magnetic material by heating was observed. On the contrary, the uncoated magnetite nanoparticles calcined under similar conditions agglomerated and oxidized into α-Fe_2O_3, which lead to important changes in the magnetic properties. After calcination, the saturation magnetization (Ms) of the resulting material dropped by nearly 80% of the original value. Hence, the silica shell protects the magnetic core, avoiding both the oxidation process and particle agglomeration and ultimately preserving the superparamagnetic properties. In summary, there is a great advantage of using the magnetic cores coated by silica for application in catalysis, which may involve high temperatures. Therefore, calcination of the silica-coated magnetite material before catalyst preparation is strongly recommended.

8.2.2 Magnetite Coated with Ceria, Titania, and Other Oxides

The support material plays an important role in heterogeneous catalysis and is generally chosen based on the desired application or property (inert, redox, acid, etc.). Besides carbon and silica, other inorganic oxides, such as titania, alumina, zirconia, and ceria, have widespread use in catalysis but are rarely reported as coatings for magnetic supports. Most methodologies for coating iron oxides with alumina results in a poor morphology control, but some improvement was observed when the material was precoated with silica [49, 50]. The controlled addition of tetrabutyltitanate resulted in the preparation of a material with spherical to "peapod"-like morphology. The precipitation of $Ti(SO_4)_2$ in the presence of magnetic NPs and urea resulted in the direct coating of magnetite with titania [51] because of the slow release of ammonia and the control of the hydrolysis and condensation rate of the TiO_2 precursor. The hydrolysis of titanium(IV) butoxide over magnetic particles in an ultrasonic bath at 90 °C for 1.5 hours using a mixture of ethanol/water [52] or over silica-coated magnetite using glacial acetic acid [53] resulted in a magnetic material tentatively coated with a layer of titania. The hydrolysis of aluminum isopropoxide over magnetite in water/ethanol mixture was not successful in terms of the quality of the alumina coating [54]. Many studies in the literature report on the encapsulation of metal nanoparticles by other oxides than silica, even in a core–shell structure, but the conditions of the synthesis (dilution required) do not allow the isolation of a sufficient amount of material to make their use as a suitable catalyst support [55, 56]. After many attempts to prepare titania-, ceria-, and alumina-coated magnetite by the substitution of TEOS by different Ti, Ce, and Al precursors in the reverse microemulsion process as described for the silica coating (see Section 8.2.1) have failed, the postcoating of silica-coated magnetite with different oxides was explored by Rossi's group. The main problem observed was the lack of control of the rate of hydrolysis and condensation of the precursors, leading to a fast precipitation process in the microemulsion. Moreover, the crystallization of ceria and other oxides occurs at high temperature, and it could have been a problem for uncoated magnetite nanoparticles. Considering the thermal stability of the silica-coated magnetite nanoparticles, part of the problem was solved by developing a postcoating process. This adapted process allowed decorating the silica shell at $Fe_3O_4@SiO_2$ with different oxides by wetness impregnation method, followed by thermal decomposition of the precursors into oxides. The preparation and characterization of these materials were reported elsewhere [43].

8.2.3 Magnetite Coated with Carbon-Based Materials

Carbon is also a very exploited material as a support owing to its properties such as large specific surface area, high porosity, excellent electron conductivity, chemical and thermal stability, relative chemical inertness, and also its versatility [57]. Different carbon materials can be formed, such as graphene, carbon nanotubes, carbon nanowires, and carbon monoliths, and they can also be chemically functionalized and/or decorated with metallic nanoparticles, metal complexes, and enzymes. Thus, coating magnetic cores with a carbon-based layer is a good alternative to access magnetic supports. Well-developed graphitic carbon layers provide an effective barrier against oxidation or acid erosion and improve novel catalytic activity through surface functionalization (e.g. sulfonated porous carbon materials). Magnetite coated with carbon can be obtained by a hydrothermal treatment in the presence of glucose, polyethylene glycol, or citric acid [58–60]. The most interesting magnetic supports based on carbon consist of cobalt particles surrounded by a protective layer of carbon, which allows the utilization of high saturation magnetization material while avoiding oxidation of the cobalt cores [61–64]. These examples will not be discussed here in detail, but information can be easily found in the literature [46, 65–67].

8.3 Preparation of Magnetically Recoverable Metal Nanoparticle Catalysts

Magnetic separation has become a subject of intense investigation in the past 15 years as illustrated by review papers from the literature [68–70]. Rossi's group has used magnetic supports for the development of magnetically recoverable metal nanoparticle (NP) catalysts, especially to overcome the difficulty of filtering colloidal nanoparticle solutions. The immobilization of metal NPs onto a magnetic support may be tricky, and its efficiency is a key parameter for the ulterior separation process. As a consequence, a strong metal–support interaction is required in order to avoid any metal leaching during the catalyst separation process. Heterogeneous catalysts are typically prepared by deposition (incipient wetness impregnation) – reduction steps, which can be improved by changing the reaction parameters (concentration, pH, solvent, temperature, metal precursors, reducing agents, etc.), wherein usually the goal is to improve metal dispersion as well as the metal–support interaction. The support surface can be modified with ligands or binding sites, which provides a very effective strategy to enhance the metal–support interactions. The ligands grafted on the solid support act as coordination sites, thus allowing the capture of metal precursors before reduction but also of preformed colloidal NPs (Figure 8.5) [71]. The ligand-assisted immobilization methods were also referred as "coordination capture methods" because the recovery of metal is promoted by the coordinating ligands grafted on the support surface [72, 73]. The pros and cons of these preparation methods will be summarized in the following sections.

8.3.1 Immobilization of Metal Precursors Before Reduction

Magnetically recoverable catalysts have been prepared by standard wet impregnation methods, but usually, the functionalization of silica is required to improve the adsorption of the molecular metal precursor [74]. The simplest functional group is aminopropyl obtained by reaction of the silica surface with 3-aminopropyltriethoxysilane, which has shown advantages in the impregnation of metal precursors for the synthesis of diverse supported metal nanoparticle catalysts. Comparatively, the loading of metal ions on the amino-functionalized silica support was always higher (at least 10 times) than the metal species loading

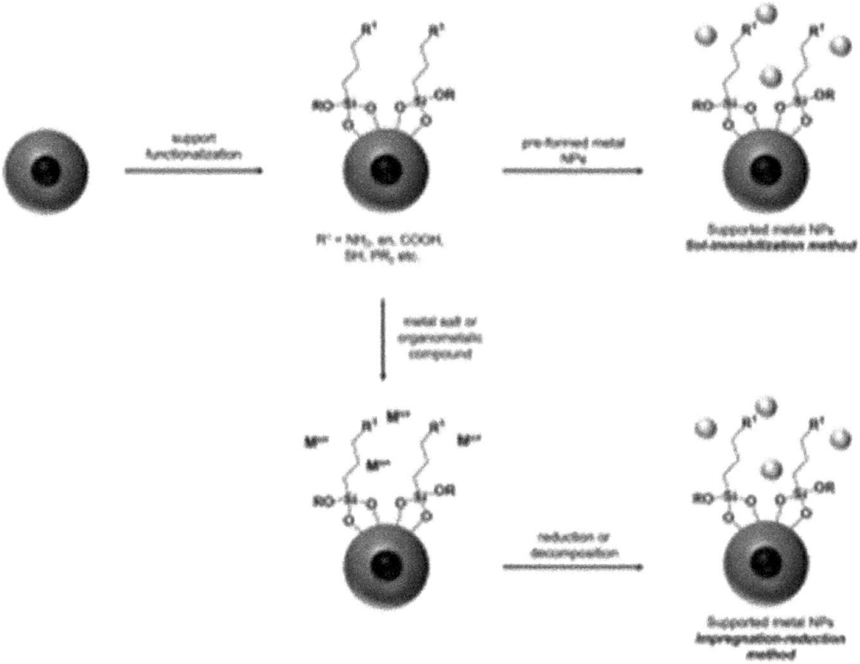

Figure 8.5 Ligand-assisted preparation of supported metal NP catalysts.

on nonfunctionalized silica. This behavior was observed during the immobilization of Rh(III) [28], Pt(II) [75], Ir(III) [76], and Au(III) molecular precursors [77]. After their loading on the silica-coated magnetic support, the metal precursors were easily reduced either under 6 atm of H_2 and 75–100 °C or $NaBH_4$, resulting in both cases in well-dispersed nearly 2–5 nm supported metal NPs (Figure 8.6). A comparative X-ray absorption near-edge spectroscopy (XANES) study confirmed a very low affinity of gold(III) ions to silica surfaces, but an enhanced interaction favored by coordination of gold ions to amine groups grafted on the functionalized silica surfaces [79]. After the reduction step, the gold NPs leached from the non-functionalized support, indicating the weak interaction with the silanol groups. In contrast, the gold NPs were strongly attached to the amine-functionalized support. All the supported metal nanoparticle catalysts mentioned here could be recovered with the assistance of an external magnet, which greatly simplifies the workup procedure and purification of products, also minimizing the use of solvents, costly consumables, energy and time, and were reused in successive hydrogenation reactions.

Besides serving as anchoring groups for the metal precursors (improving metal loading), the ligands grafted on the support surface had an influence on the particles' size during the preparation of Pd catalysts. When the support surface was functionalized with different organotrialkoxysilanes (amine and ethylenediamine), a clear effect on the particle size and dispersion was observed for the Pd NPs

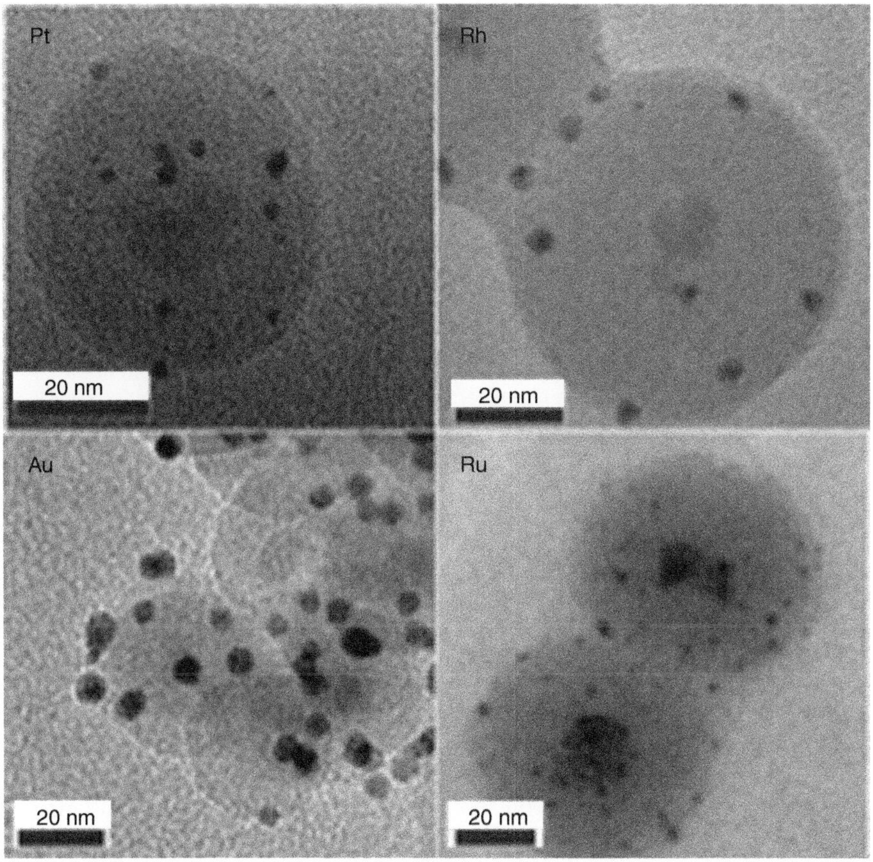

Figure 8.6 TEM micrographs of (a) Pt NPs (Source: Reprinted with permission from Jacinto et al. [75]. Copyright 2009 Elsevier) (b) Rh NPs (Source: Reprinted with permission from Jacinto et al. [28]. Copyright 2008 Elsevier) (c) Au NPs (d) Ru NPs (Source: Reprinted with permission from Jacinto et al. [78]. Copyright 2009 Elsevier) supported on silica-coated magnetite functionalized with amino groups.

obtained after reduction of adsorbed metal precursors (as exemplified by $[PdCl_4]^{2-}$) with molecular hydrogen [80]. Amine and ethylenediamine groups grafted on the surface of the silica support assisted the preparation of magnetically recoverable Pd NPs of ~6 and ~1 nm, respectively (Figure 8.7), which is significantly different from the metal aggregates obtained when using nonfunctionalized surfaces (a few isolated Pd NPs of ~5 nm and very irregular metal aggregates >20 nm). The obtained catalysts also exhibited different catalytic activities when tested in the hydrogenation of cyclohexene model reaction. In order to understand the influence of the ligands grafted on the support surfaces on the catalytic activity, complementary experiments were performed with the immobilization of preformed Pd NPs of similar size on supports functionalized with amine and ethylenediamine groups (see Section 8.3.4).

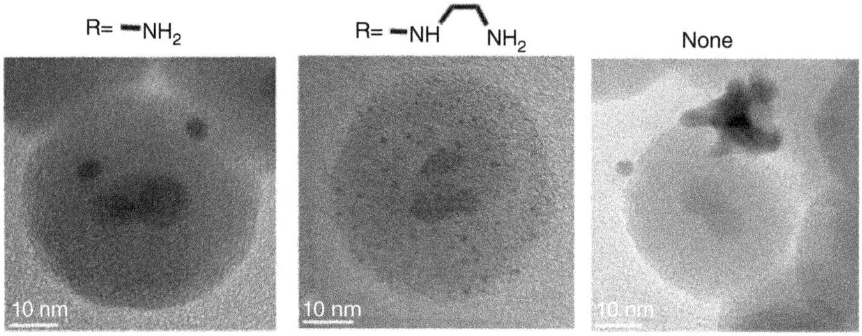

Figure 8.7 TEM micrographs of Pd NPs. Source: Reprinted with permission from Rossi et al. [80]. Copyright 2009 American Chemical Society.

8.3.2 Decomposition of Organometallic Precursors

Despite disadvantages related to the tedious synthesis of the organometallic precursors and their frequent sensitivity to air and moisture, the organometallic method is known as an advantageous way for the reproducible preparation of metal NPs with well-controlled size, dispersion, chemical composition, and morphology, thus offering nanomaterials of high quality. In addition, when applied in catalytic reactions, the so-obtained nanoparticles present remarkable activity because of a cleaner surface state [32, 33, 81]. Typically, when decomposed under H_2 atmosphere, olefinic organometallic complexes lead to the release of reduced olefinic ligands (alkanes) as the only by-products. The result is the obtention of free metal(0) atoms that tend to aggregate to form metal bulk. However, if the decomposition is accomplished in the presence of a support or a protective agent (polymer and ligand), the aggregation is controlled, and populations of nanoparticles with narrow size distributions are formed [32].

The organometallic approach proved to be particularly interesting for the *in situ* synthesis of supported Ni NPs. A robust, oxidation-resistant, and very active nickel catalyst was prepared by direct decomposition of the organometallic precursor [bis(1,5-cyclooctadiene)nickel(0)], $Ni(COD)_2$, over the silica-coated magnetite support [82]. The recovered sample contained mostly Ni(0) with only partial surface oxidation after storage in air as confirmed by XANES and XPS. The catalytic results obtained in the hydrogenation of cyclohexene were surprising because these surface-oxidized nickel species could be reduced into Ni(0)-active catalyst under very mild conditions (1 bar of H_2 and 75 °C). This contrasted with the results obtained with NiO bulk that was unreactive under the same conditions. The catalyst exhibited very promising activity in the hydrogenation of cyclohexene converting 4500 mol substrate per mole of catalyst (TOF [at 20% conversion] up to 1500 h^{-1} – 15 recycles without deactivation) under the conditions studied (75 °C and 6 atm H_2). Metal leaching was not detected by ICP OES analysis, demonstrating the efficiency of Ni–support interaction. In contrast, Raney nickel catalyst was tested under similar conditions, giving only 29% conversion after 48 hours of reaction, which means that Raney Ni was also not activated in the applied reaction conditions. An important feature of the Ni catalyst prepared from the organometallic precursor is that it can be activated *in situ* in the hydrogenation reaction under very mild condition.

Ligands grafted on the support surface also influence the size and morphology of supported metal NPs prepared by decomposition of an organometallic precursor. Supported Pd NPs synthesized by the decomposition of $Pd_2(dba)_3$ (dba = dibenzylideneacetone) with molecular hydrogen (3 bar; room temperature) on a silica-coated magnetic support functionalized with aminopropyl or terpyridine ligands appeared well dispersed on the support and with a mean size of ~2 nm. Only large aggregates of palladium were found for samples prepared on bare silica or chloropropyl-functionalized silica surfaces. The activity and selectivity of on terpyridine-functionalized support was largely increased when compared to a similar nanocatalyst prepared on amino-modified magnetic support or to Pd/C [83].

The organometallic approach was also successfully applied for the immobilization of colloidal NPs onto a support in a two-step procedure. A much higher control over particle size and composition on the preparation of Ni, Pd, and NiPd NPs was achieved by the preparation of colloidal NPs using $Pd_2(dba)_3$ and $Ni(COD)_2$ in different molar ratios and hexadecylamine (HDA) as a stabilizer, followed by their immobilization on the magnetic support [84–86] when compared to the direct decomposition of Ni precursor onto the magnetic support [59]. The preparation of the monometallic NPs by the decomposition of $Ni(COD)_2$ [87] or $Pd_2(dba)_3$ [88], after optimization, was done in toluene under 3 bar of dihydrogen at 110 °C, with a molar ratio metal/hexadecylamine (HDA) of 1 : 10. The same reaction conditions were applied to prepare bimetallic NiPd nanoparticles by codecomposition of the two organometallic precursors. The metal particles' size ranged from 4 to 6 nm, and the immobilization was only possible after functionalization of the support with aminopropyl groups. The catalyst containing a molar ratio Ni/Pd of 1 : 9 was the active in a comparative hydrogenation reaction of cyclohexene, which is economically important, because it provides a catalyst with less Pd but activity than the Pd monometallic counterpart. The magnetic separation was crucial to the performance of the catalysts because it provided excellent separation of the catalyst from the liquid products without metal leaching exposure to air, allowing for efficient recycling [84]. Rh, Pd, and Ru NPs were also synthesized by the decomposition of organometallic precursors $Rh(C_3H_5)_3$, Ru(COD)(COT), and $Pd_2(dba)_3$ in the presence of H_2 and polyvinylpyrrolidone (PVP) as the protective agent [85, 86, 89]. The metal NPs were obtained with high morphological and size control with average 1–2 nm and successfully immobilized on magnetic supported coated by silica, ceria, and titania (Section 8.2.2). It is important to note that differently from the previous results, the immobilization of the metal NPs on nonfunctionalized solid was efficiently achieved using THF colloidal solutions of preformed NPs.

8.3.3 Immobilization of Colloidal Nanoparticles

Multiple strategies and possibilities are available to achieve the desired size, shape, and composition control during the synthesis of NPs, such as liquid phase (aqueous, organic solvents, ionic liquids, etc.) and metal precursors (metals salts, complexes, or organometallic compounds), stabilizers, and reducing process. The immobilization of presynthesized nanoparticles on solid supports has received attention as a versatile strategy to prepare supported catalysts [90]. This approach, also called sol-immobilization (SI) method [91], improves the size and morphology control of supported nanoparticles; the same control is usually not possible by the direct deposition of metal salts followed by reduction. The properties of the colloidal NPs, such as size and size distribution, are retained when the NPs are immobilized on the support. However, the organic molecules, polymers, and surfactants that are used to control the particle growth and protect them against agglomeration may be a poison for the active surface sites. The selective poisoning of metal surface atoms is not always detrimental, and it can be used as a strategy to prepare selective catalysts. However, the ligands can also be removed to clean the catalyst surface and recover activity.

The aqueous-phase reduction of metal salts by $NaBH_4$ in the presence of stabilizing agents, such as polyvinyl alcohol (PVA) or citrate, has been very often used for the preparation of colloidal nanoparticles and then supported catalysts, with controlled size, by the solimmobilization method [91]. This very simple method provides metal NPs (Au, Pd, Rh, Ru, etc.) suitable for catalytic application and, if necessary, the excess of stabilizers can be removed by several methods (calcination under an inert or reactive atmosphere, solvent extraction, chemical-assisted removal of ligands at low temperatures, UV ozone, or plasma cleaning procedures) [92, 93]. The immobilization of colloidal nanoparticles occurs typically by the interaction of the support with metal surface or with nanoparticle's stabilizing agent. The interaction between the support and colloidal nanoparticles can be tuned by the support surface functionalization [71] or changing the characteristics of the nanoparticle's stabilizer. Colloidal Au-PVA NPs did not immobilize on nonfunctionalized silica but could be firmly attached onto silica functionalized with amino groups [77]. Further studies showed that besides allowing immobilization of metal NPs, the presence of organic ligands on the surface of silica-coated magnetite also affected the catalytic activity [73, 94]. For instance, colloidal AuPd-PVA NPs immobilized onto magnetic supports functionalized with amino and thiol groups exhibited different catalytic activities [73]. The catalyst supported on amino-silica exhibited higher activity for oxidation of benzyl alcohols and stability toward catalyst recycling than the catalyst supported on thiol-silica. The catalyst prepared on nonfunctionalized support was the least active and exhibited metal leaching during recycling. The presence of thiol groups, known to poison gold catalyst surfaces, may have been responsible for blocking the activity of supported AuPd NPs causing a decrease of activity upon recycling.

Stabilized Au NPs were immobilized on magnetic magnesium ferrite for comparison studies in the Au-based reactions where the catalytic activity is highly dependent on the presence of a base. For example, the basicity of magnesium ferrite/oxide supports ($MgFe_2O_4$ and $MgO/MgFe_2O_4$) was explored as an alternative to the use of strong bases, such as NaOH, in the oxidation of benzyl alcohol [95]. The presence of Mg^{2+} ions in the ferrite structure improved the catalytic activity of supported Au-PVA NPs to 35% conversion in the absence of any additional base and to 50% conversion after modifying the support with MgO. However, the catalyst deactivated after successive recycling tests. In the presence of a substoichiometric amount of K_2CO_3, the catalyst became more active and remained stable upon recycling with no loss of activity and selectivity. The effect of the nature and concentration of the base on the activity and selectivity for the oxidation of benzyl alcohol into benzaldehyde versus benzoic acid catalyzed by Au catalysts was further investigated [96]. The concentration of weak bases (K_2CO_3, $Na_2B_4O_7$, and NaAcO) had little effect on conversion because of the buffer effect, but strongly affected selectivity. The bases with higher pK_a values provided higher conversions and increased production of benzoic acid. For the strong base NaOH, benzoic acid was always the major product, although conversion decreases in excess of base. In organic media, as cyclohexane and benzyl alcohol (solvent-free conditions), the overoxidation into benzoic acid or benzyl benzoate is avoided when using low amount of base. Another strategy to avoid the use of base in Au-catalyzed reactions is the alloying with a second metal (e.g. Pd, Pt, and Cu) [94, 97]. The variation in the Au:Pd molar ratio has shown to affect activity. The key to increase catalytic activity is a right balance between the number of active sites and the ease of product desorption; however, both parameters are extremely sensitive to the Pd content resulting in the volcano-like activity [94].

8.3.4 Influence of Ligands on Catalytic Properties

Although common in enzymatic reactions and also present in homogeneous catalysts [98, 99], the cooperative action of ligands has only been recently investigated in heterogeneous catalysis. When properly applied, catalysis at metal–ligand interfaces can lead to a very significant increase in activity and selectivity [100–103]. Organic modifiers (ligands) adsorbed on metal surfaces of heterogeneous catalysts have been used to control parameters such as reactive intermediates and the electronic properties of the surface of the heterogeneous catalysts and, in some cases, enhancing selectivity and catalytic activity [104]. A number of recent studies have demonstrated the tunability in the catalytic activity when amines [102, 105], thiols [104, 106], phosphorous-based [107–112], and N-heterocyclic carbenes [113–115] are combined with metallic catalysts.

In this respect, combining gold nanoparticles (Au NPs) with ligands allowed the discovery of a new reactivity pattern of gold catalysts, leading to highly selective hydrogenation reactions (Figure 8.8). During screening the use of amine ligands as possible activating ligands, it was found that there is a

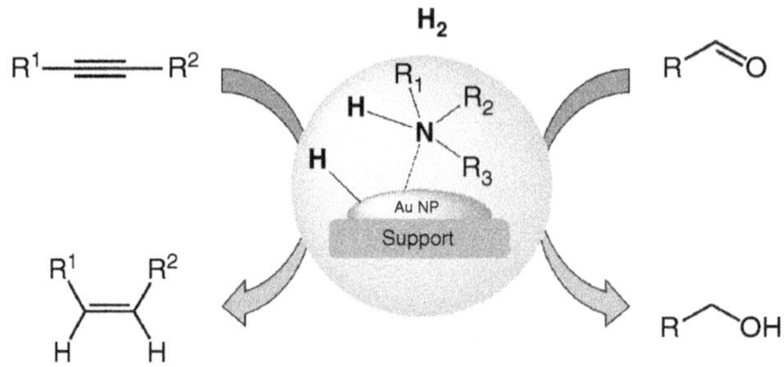

Figure 8.8 Ligand-assisted activation of supported Au NPs in hydrogenation reactions.

correlation between the catalytic activity for the selective semihydrogenation of alkynes and the computed activation energies for heterolytic H_2 dissociation at the N ligand–Au(111) interface for structurally different amines [116, 117]. Piperazine–gold interface showed a unique catalytic activity, attributed to the favorable dissociation of H_2, and was successfully used for the hydrogenation of a variety of terminal and internal alkynes. Also, the combination of Au NPs with the appropriate N-ligand (e.g. 2,4,6-trimethylpyridine) furnishes a catalytic system able to selectively reduce aldehyde carbonyl group, without further reduction of reducible moieties such as alkene or furan ring [117]. The group of van Leeuwen demonstrated the selective hydrogenation of a wide variety of unsaturated and functionalized aldehydes, such as cinnamaldehyde and furaldehyde, when using secondary phosphine oxides adsorbed on Au NPs [110, 111]. The nanoparticles prepared using aryl secondary phosphine oxides as a stabilizer were highly selective for aldehyde hydrogenation into alcohol, and a crucial ligand effect was noticed in the heterolytic hydrogenation mechanism.

The ligand can also play a major role in engineering the heterogeneous catalysts in order to increase selectivity and overcome the usual promiscuity associated with it [118–120]. Taking advantage of this ligand effect, preformed Pd NPs (PVA-stabilized Pd NPs) were supported on ligand-modified magnetic silica surfaces bearing amino, ethylenediamine, and diethylenetriamine groups [121]. The organic moiety allows a strong metal–support interaction and an efficient NP immobilization process. Higher activity in the hydrogenation of cyclohexene was observed for Pd NPs supported on amino-functionalized support, while the ethylenediamine and diethylenetriamine groups were detrimental to activity (Figure 8.9). Because of the observed poisoning effect, the addition of ethylenediamine was explored to tune the catalytic activity and selectivity of Pd NPs. Incremental addition of ethylenediamine to the reaction medium led to a decrease in the catalytic activity, until deactivation of the catalyst, suggesting the poisining of catalytic sites. Nonetheless, the supported Pd NPs assisted by the ligand was found to be able to suppress the overhydrogenation of the alkene product in the semihydrogenation of alkynes.

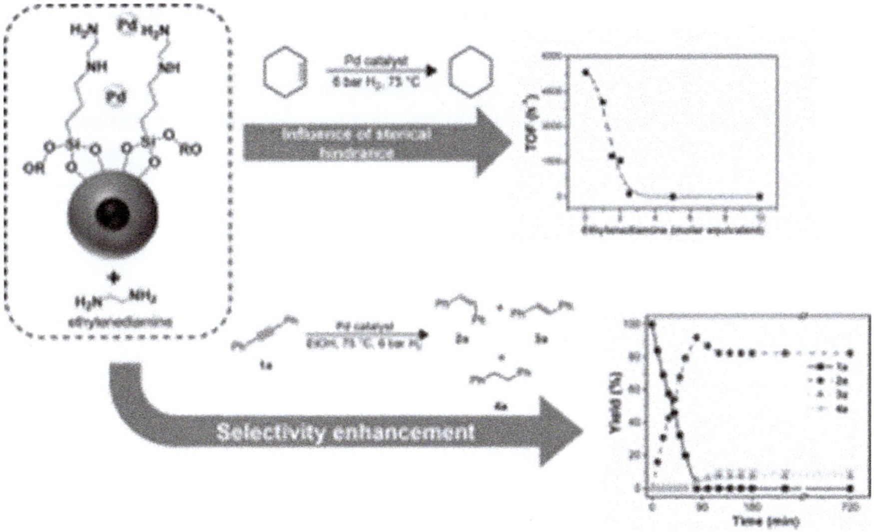

Figure 8.9 Impact of the addition of ethylenediamine on Pd NP activity. Source: Reprinted with permission from [121]. Copyright 2017 American Chemical Society.

Besides palladium ligand-modified catalysts, supported copper nanocatalyst has shown to be a promising catalyst in chemoselectivity and semihydrogenation of alkynes when combined with phosphorous ligand. For instance, a PCy_3(tricyclohexylphosphine)-modified Cu catalyst was capable to semihydrogenate a wide range of alkynes with a turnover number and a turnover frequency over 540 and 1.9 min^{-1}, respectively [107]. Moreover, this catalytic system was also successfully applied in the semihydrogenation of alkynes in flow conditions [122]. In the same direction, the introduction of an NHC (N-heterocyclic carbene) ligand (IMes = 1,3-bis(2,4,6-trimethylphenyl) imidazol-2-ylidene) to Cu NPs proved to greatly improve the selectivity of these particles for semihydrogenation of internal alkynes. The generated catalyst readily hydrogenates internal alkynes to the corresponding *cis*-olefins with very high selectivities (>95%) at full conversion [123]. NHC ligands have also been applied to electronically activate heterogeneous catalysts [114]. A Pd/Al_2O_3 catalyst modified with NHCs promptly catalyzed the Buchwald–Hartwig amination reaction, the a priori inactive catalyst becomes active because of ligand binding to the catalyst surface, which increases electron density. Therefore, the NHCs bind to the palladium surface, leading to a decreased ionization potential of the Pd NPs and a lower energy barrier for the cleavage of the C—Br bond of bromobenzene, as confirmed by DFT [113].

Despite the great advances already reached in the field, it is still unclear how the ligand acts either improving the catalytic activity or tuning the selectivity [93, 124]. To untangle the role of the ligands and get an in-depth understanding of the resulting catalytic systems, the combination of systematic experimental supported by *in situ* characterizations and theoretical studies will be crucial.

8.4 Summary and Conclusions

This chapter provides an overview on the preparation of magnetically recoverable supported metal nanoparticle catalysts. In the past few decades, different research groups have shown the applicability of magnetic supports for the preparation of catalysts (molecular, enzymatic, or heterogeneous) that can be recovered simply by applying an external magnetic field, which saves time, consumables, and energy when compared to conventional separation techniques, for example, filtration and centrifugation. The support material varies from simple iron oxide nanoparticles, e.g. magnetite, to composite materials containing iron oxides coated with a protective layer (mainly, carbon, silica, and other oxides). The protective material improves the thermal and chemical stability and facilitates the immobilization process through functionalization. A reverse microemulsion system was optimized to prepare a core–shell-type composite of magnetic nanoparticles spherically coated with silica. The material obtained exhibits excellent magnetic properties, high thermal stability, and has been used as a support for many types of catalysts. We have shown here the preparation of monometallic, bimetallic, or metal oxide nanoparticle catalysts following three methodologies: immobilization of metal precursors followed by reduction, decomposition of organometallic precursors, or immobilization of preformed colloidal metal nanoparticles. Moreover, these immobilization methods can be assisted by the functionalization of the support with ligands containing different functional groups. In principle, the role of the ligands grafted on the support surface was to improve the metal–support interactions to avoid metal leaching, but it was also demonstrated that the ligands can affect both particles sizes and ultimately the catalytic activity and selectivity of the supported metal catalyst. The ligands are most often used as selectivity enhancers, but they can also act cooperatively with the metal to enhance reaction rates. There is a huge number of examples in the scientific literature that illustrate well the interest of using magnetic separation in catalysis and its different advantages. The field has still potential to move to the next step, especially to build prototypes in order to transfer this technology to industry.

References

1 Iranmanesh, M. and Hulliger, J. (2017). *Chem. Soc. Rev.* 46: 5925–5934.

2 Yavuz, C.T., Prakash, A., Mayo, J.T., and Colvin, V.L. (2009). *Chem. Eng. Sci.* 64: 2510–2521.

3 Lu, A.-H., Salabas, E.L., and Schüth, F. (2007). *Angew. Chem. Int. Ed.* 46: 1222–1244.

4 Wills, B.A. and Napier-Munn, T. (2016). Magnetic and electrical separation in Wills' Mineral Processing Technology, an introduction to the practical aspects of ore treatment and mineral recovery (8th ed.), Butterworth-Heinemann, Elsevier, Oxford, UK, pp. 381–406.

5 Nyholm, R.S. (1953). *Q. Rev. Chem. Soc.* 7: 377.
6 Rossi, L.M., Costa, N.J.S., Silva, F.P., and Wojcieszak, R. (2014). *Green Chem.* 16: 2906.
7 Stein, M., Wieland, J., Steurer, P. et al. (2011). *Adv. Synth. Catal.* 353: 523–527.
8 Astinchap, B., Moradian, R., Ardu, A. et al. (2012). *Chem. Mater.* 24: 3393–3400.
9 Yao, Y., Xu, C., Qin, J. et al. (2013). *Ind. Eng. Chem. Res.* 52: 17341–17350.
10 Bhattacharjee, D., Sheet, S.K., Khatua, S. et al. (2018). *Bioorg. Med. Chem.* 26: 5018–5028.
11 Tripathi, V.K. and Nagarajan, R. (2016). *Adv. Powder Technol.* 27: 1251–1256.
12 Zampiva, R.Y.S., Kaufmann Junior, C.G., Pinto, J.S. et al. (2017). *Appl. Surf. Sci.* 422: 321–330.
13 Zhu, M. and Diao, G. (2011). *J. Phys. Chem. C* 115: 18923–18934.
14 Lee, Y., Garcia, M.A., Frey Huls, N.A., and Sun, S. (2010). *Angew. Chem. Int. Ed.* 49: 1271–1274.
15 Sun, W., Li, Q., Gao, S., and Shang, J.K. (2013). *J. Mater. Chem. A* 1: 9215.
16 Rossi, L.M., Garcia, M.A.S., and Vono, L.L.R. (2012). *J. Braz. Chem. Soc.* 23: 1959–1971.
17 Cullity, B.D. and Graham, C.D. (2008). *Introduction to Magnetic Materials*, 2e. Hoboken, NJ: Wiley.
18 Sperling, R.A. and Parak, W.J. (2010). *Philos. Trans. R. Soc. London, A* 368: 1333–1383.
19 Mazur, M., Barras, A., Kuncser, V. et al. (2013). *Nanoscale* 5: 2692.
20 Gawande, M.B., Monga, Y., Zboril, R., and Sharma, R.K. (2015). *Coord. Chem. Rev.* 288: 118–143.
21 Zhu, M. and Diao, G. (2011). *Nanoscale* 3: 2748.
22 Lu, Y., Yin, Y., Mayers, B.T., and Xia, Y. (2002). *Nano Lett.* 2: 183–186.
23 Shao, H., Zhou, Y., Qi, J. et al. (2019). *J. Supercond. Novel Magn.* 32: 247–252.
24 Ding, H.L., Zhang, Y.X., Wang, S. et al. (2012). *Chem. Mater.* 24: 4572–4580.
25 Li, Y., Hu, Y., Jiang, H., and Li, C. (2013). *Nanoscale* 5: 5360.
26 Tada, D.B., Vono, L.L.R., Duarte, E.L. et al. (2007). *Langmuir* 23: 8194–8199.
27 Vogt, C., Toprak, M.S., Muhammed, M. et al. (2010). *J. Nanopart. Res.* 12: 1137–1147.
28 Jacinto, M.J., Kiyohara, P.K., Masunaga, S.H. et al. (2008). *Appl. Catal., A* 338: 52–57.
29 Dong, X., Chen, H., Zhao, W. et al. (2007). *Chem. Mater.* 19: 3484–3490.
30 Wang, Z.H., Choi, C.J., Kim, B.K. et al. (2003). *Carbon.* 41: 1751–1758.
31 Wang, Z., Xiao, P., and He, N. (2006). *Carbon.* 44: 3277–3284.
32 Yi, D.K., Lee, S.S., Papaefthymiou, G.C., and Ying, J.Y. (2006). *Chem. Mater.* 18: 614–619.
33 Yi, D.K., Selvan, S.T., Lee, S.S. et al. (2005). *J. Am. Chem. Soc.* 127: 4990–4991.

34 Im, S.H., Herricks, T., Lee, Y.T., and Xia, Y. (2005). *Chem. Phys. Lett.* 401: 19–23.

35 Narita, A., Naka, K., and Chujo, Y. (2009). *Colloids Surf., A* 336: 46–56.

36 Ohmori, M. and Matijević, E. (1992). *J. Colloid Interface Sci.* 150: 594–598.

37 Ohmori, M. and Matijević, E. (1993). *J. Colloid Interface Sci.* 160: 288–292.

38 Zhang, M., Cushing, B.L., and O'Connor, C. (2008). *J. Nanotechnology* 19: 085601.

39 Park, J.C., Gilbert, D.A., Liu, K., and Louie, A.Y. (2012). *J. Mater. Chem.* 22: 8449.

40 Abramson, S., Safraou, W., Malezieux, B. et al. (2011). *J. Colloid Interface Sci.* 364: 324–332.

41 Tartaj, P. and Serna, C.J. (2003). *J. Am. Chem. Soc.* 125: 15754–15755.

42 Yang, H.-H., Zhang, S.-Q., Chen, X.-L. et al. (2004). *Anal. Chem.* 76: 1316–1321.

43 Vono, L.L.R., Damasceno, C.C., Matos, J.R. et al. (2018). *Pure Appl. Chem.* 90: 133–141.

44 Wang, J., White, W.B., and Adair, J.H. (2006). *J. Am. Ceram. Soc.* 89: 2359–2363.

45 Garcia, M.A.S., Teruya, L.C., Yoshimoto, K.M. et al. (2017). *Sep. Sci. Technol.* 52: 504–511.

46 Henriques, C.A., Fernandes, A., Rossi, L.M. et al. (2016). *Adv. Funct. Mater.* 26: 3359–3368.

47 Dias, L.D., Carrilho, R.M.B., Henriques, C.A. et al. (2018). *ChemCatChem* 10: 2792–2803.

48 Ungaro, V.A., Liria, C.W., Romagna, C.D. et al. (2015). *RSC Adv.* 5: 36449–36455.

49 Álvarez, P.M., Jaramillo, J., López-Piñero, F., and Plucinski, P.K. (2010). *Appl. Catal., B* 100: 338–345.

50 Ye, M., Zorba, S., He, L. et al. (2010). *J. Mater. Chem.* 20: 7965.

51 He, Q., Zhang, Z., Xiong, J. et al. (2008). *Opt. Mater.* 31: 380–384.

52 Beydoun, D., Amal, R., Low, G.K.-C., and McEvoy, S. (2000). *J. Phys. Chem. B* 104: 4387–4396.

53 Liu, H., Jia, Z., Ji, S. et al. (2011). *Catal. Today* 175: 293–298.

54 Sun, L., Zhang, C., Chen, L. et al. (2009). *Anal. Chim. Acta* 638: 162–168.

55 De Rogatis, L., Cargnello, M., Gombac, V. et al. (2010). *ChemSusChem* 3: 24–42.

56 Ghosh Chaudhuri, R. and Paria, S. (2012). *Chem. Rev.* 112: 2373–2433.

57 Lam, E. and Luong, J.H.T. (2014). *ACS Catal.* 4: 3393–3410.

58 Sun, Z., Yang, J., Wang, J. et al. (2014). *J. Mater. Chem. A* 2: 6071–6074.

59 Li, Y., Zhang, Z., Shen, J., and Ye, M. (2015). *Dalton Trans.* 44: 16592–16601.

60 Liu, Y., Li, L., Xie, C. et al. (2016). *Chem. Eng. J.* 303: 31–36.

61 Schätz, A., Grass, R.N., Stark, W.J., and Reiser, O. (2008). *Chem. – A Eur. J.* 14: 8262–8266.

62 Wittmann, S., Schätz, A., Grass, R.N. et al. (2010). *Angew. Chem. Int. Ed.* 49: 1867–1870.

63 Linhardt, R., Kainz, Q.M., Grass, R.N. et al. (2014). *RSC Adv.* 4: 8541.

64 Kainz, Q.M., Linhardt, R., Grass, R.N. et al. (2014). *Adv. Funct. Mater.* 24: 2020–2027.

65 Yoon, H., Ko, S., and Jang, J. (2007). *Chem. Commun.*: 1468.

66 Jang, J. and Yoon, H. (2003). *Adv. Mater.* 15: 2088–2091.

67 Jang, J. and Yoon, H. (2005). *Small* 1: 1195–1199.

68 Polshettiwar, V., Luque, R., Fihri, A. et al. (2011). *Chem. Rev.* 111: 3036–3075.

69 Wu, W., Wu, Z., Yu, T. et al. (2015). *Sci. Technol. Adv. Mater.* 16: 023501.

70 Zhu, N., Ji, H., Yu, P. et al. (2018). *Nanomaterials* 8: 810.

71 Costa, N.J.S. and Rossi, L.M. (2012). *Nanoscale* 4: 5826.

72 Wang, Y. and Liu, H. (1991). *Polym. Bull.* 25: 139–144.

73 Silva, T.A.G., Landers, R., and Rossi, L.M. (2013). *Catal. Sci. Technol.* 3: 2993–2999.

74 Rossi, L.M., Costa, N.J.S., Silva, F.P., and Gonçalves, R.V. (2013). *Nanotechnol. Rev.* 2: 597–614.

75 Jacinto, M.J., Landers, R., and Rossi, L.M. (2009). *Catal. Commun.* 10: 1971–1974.

76 Jacinto, M.J., Silva, F.P., Kiyohara, P.K. et al. (2012). *ChemCatChem* 4: 698–703.

77 Oliveira, R.L., Kiyohara, P.K., and Rossi, L.M. (2010). *Green Chem.* 12: 144–149.

78 Jacinto, M.J., Santos, O.H.C.F., Jardim, R.F. et al. (2009). *Appl. Catal., A* 360: 177–182.

79 Oliveira, R.L., Zanchet, D., Kiyohara, P.K., and Rossi, L.M. (2011). *Chem. – A Eur. J.* 17: 4626–4631.

80 Rossi, L.M., Nangoi, I.M., and Costa, N.J.S. (2009). *Inorg. Chem.* 48: 4640–4642.

81 Amiens, C., Ciuculescu-Pradines, D., and Philippot, K. (2016). *Coord. Chem. Rev.* 308: 409–432.

82 Costa, N.J.S., Jardim, R.F., Masunaga, S.H. et al. (2012). *ACS Catal.* 2: 925–929.

83 Hatch, G.P. and Stelter, R.E. (2001). *J. Magn. Magn. Mater.* 225: 262–276.

84 Costa, N.J.S., Guerrero, M., Collière, V. et al. (2014). *ACS Catal.* 4: 1735–1742.

85 Pellegatta, J.L., Blandy, C., Choukroun, R. et al. (2003). *New J. Chem.* 27: 1528–1532.

86 Philippot, K. and Chaudret, B. (2003). *C.R. Chim.* 6: 1019–1034.

87 Cordente, N., Respaud, M., Senocq, F. et al. (2001). *Nano Lett.* 1: 565–568.

88 Ramirez, E., Jansat, S., Philippot, K. et al. (2004). *J. Organomet. Chem.* 689: 4601–4610.

89 Pan, C., Pelzer, K., Philippot, K. et al. (2001). *J. Am. Chem. Soc.* 123: 7584–7593.

90 Jia, C.-J. and Schüth, F. (2011). *Phys. Chem. Chem. Phys.* 13: 2457.

91 Villa, A., Wang, D., Veith, G.M. et al. (2013). *Catal. Sci. Technol.* 3: 3036.

92 Lopez-Sanchez, J.A., Dimitratos, N., Hammond, C. et al. (2011). *Nat. Chem.* 3: 551–556.

93 Rossi, L.M., Fiorio, J.L., Garcia, M.A.S., and Ferraz, C.P. (2018). *Dalton Trans.* 47: 5889–5915.

94 Silva, T.G., Teixeira-Neto, E., López, N., and Rossi, L.M. (2014). *Sci. Rep.* 4: 5766.

95 De Moura, E.M., Garcia, M.A.S., Gonçalves, R.V. et al. (2015). *RSC Adv.* 5: 15035–15041.

96 Ferraz, C.P., Garcia, M.A.S., Teixeira-Neto, É., and Rossi, L.M. (2016). *RSC Adv.* 6: 25279–25285.

97 Silva, T.A.G., Ferraz, C.P., Gonçalves, R.V. et al. (2019). *ChemCatChem* 11: 4021–4027.

98 Liu, Z.-P. and Hu, P. (2002). *J. Chem. Phys.* 117: 8177–8180.

99 Noyori, R. (2002). *Angew. Chem. Int. Ed.* 41: 2008.

100 Marshall, S.T., O'Brien, M., Oetter, B. et al. (2010). *Nat. Mater.* 9: 853–858.

101 Teschner, D., Borsodi, J., Wootsch, A. et al. (2008). *Science* 320: 86–89.

102 Chen, G., Xu, C., Huang, X. et al. (2016). *Nat. Mater.* 15: 564–569.

103 Xiao, B., Niu, Z., Wang, Y.-G. et al. (2015). *J. Am. Chem. Soc.* 137: 3791–3794.

104 Schoenbaum, C.A., Schwartz, D.K., and Medlin, J.W. (2014). *Acc. Chem. Res.* 47: 1438–1445.

105 Li, G. and Jin, R. (2014). *J. Am. Chem. Soc.* 136: 11347–11354.

106 Makosch, M., Lin, W.-I., Bumbálek, V. et al. (2012). *ACS Catal.* 2: 2079–2081.

107 Fedorov, A., Liu, H.-J., Lo, H.-K., and Copéret, C. (2016). *J. Am. Chem. Soc.* 138: 16502–16507.

108 Lari, G.M., Puértolas, B., Shahrokhi, M. et al. (2017). *Angew. Chem. Int. Ed.* 56: 1775–1779.

109 Almora-Barrios, N., Cano, I., van Leeuwen, P.W.N.M., and López, N. (2017). *ACS Catal.* 7: 3949–3954.

110 Cano, I., Chapman, A.M., Urakawa, A., and van Leeuwen, P.W.N.M. (2014). *J. Am. Chem. Soc.* 136: 2520–2528.

111 Cano, I., Huertos, M.A., Chapman, A.M. et al. (2015). *J. Am. Chem. Soc.* 137: 7718–7727.

112 Liu, C., Abroshan, H., Yan, C. et al. (2016). *ACS Catal.* 6: 92–99.

113 Ernst, J.B., Schwermann, C., Yokota, G. et al. (2017). *J. Am. Chem. Soc.* 139: 9144–9147.

114 Ernst, J.B., Muratsugu, S., Wang, F. et al. (2016). *J. Am. Chem. Soc.* 138: 10718–10721.

115 Ye, R., Zhukhovitskiy, A.V., Kazantsev, R.V. et al. (2018). *J. Am. Chem. Soc.* 140: 4144–4149.

116 Fiorio, J.L., López, N., and Rossi, L.M. (2017). *ACS Catal.* 7: 2973–2980.

117 Silva, R., Fiorio, J., Vidinha, P., and Rossi, L. (2019). *J. Braz. Chem. Soc.* 00: 1–8.

118 Kwon, S.G., Krylova, G., Sumer, A. et al. (2012). *Nano Lett.* 12: 5382–5388.

119 Wu, B., Huang, H., Yang, J. et al. (2012). *Angew. Chem. Int. Ed.* 51: 3440–3443.

120 Vilé, G., Albani, D., Almora-Barrios, N. et al. (2016). *ChemCatChem* 8: 21–33.

121 Da Silva, F.P., Fiorio, J.L., and Rossi, L.M. (2017). *ACS Omega* 2: 6014–6022.

122 Salnikov, O.G., Liu, H.-J., Fedorov, A. et al. (2017). *Chem. Sci.* 8: 2426–2430.

123 Kaeffer, N., Liu, H.J., Lo, H.K. et al. (2018). *Chem. Sci.* 9: 5366–5371.

124 Niu, Z. and Li, Y. (2014). *Chem. Mater.* 26: 72–83.

9

Synthesis of MOF-Supported Nanoparticles and Their Interest in Catalysis

Guowu Zhan[1] *and Hua C. Zeng*[2]

[1] *Huaqiao University, Department of Chemical and Pharmaceutical Engineering, College of Chemical Engineering, 668 Jimei Boulevard, Xiamen, Fujian 361021, P.R. China*
[2] *National University of Singapore, Department of Chemical and Biomolecular Engineering, Faculty of Engineering, 10 Kent Ridge, Crescent 119260, Singapore*

9.1 Introduction

Metal–organic frameworks (MOFs) built by coordination between metal ions (or clusters) and organic ligands are an emerging class of porous materials with high surface area, versatile structure, and designable porosity [1]. The acronym "MOF" was firstly used by Yaghi et al. in 1999 [2]. Among various existing MOFs, MOF-5 is recognized as the first framework with permanent porosity after guest solvent molecule evacuation. MOF materials are also termed as porous coordination polymers (PCPs) in many publications by Kitagawa et and coworkers [3]. Nomenclatures about MOFs are based on either topology/structure type (e.g. ZIF-8, ZIF-67, and MOF-5) or the associated educational or research institutions (e.g. MIL-101, UiO-66, and HKUST-1). Sometimes, the direct chemical formulas are also used to name a specific MOF. MOFs are well known for their extremely high specific surface area, which has exceeded other traditional porous materials (e.g. mesoporous silica/silicate, zeolites, and activated carbon). In particular, by functionalization with the "space-efficient" organic linker, the computed maximum Brunauer–Emmett–Teller (BET) surface area for MOF materials can be higher than 14 600 m^2/g [4]. In addition, MOFs have rich topological diversity, and even metal ion nodes can copolymerize with two more organic linkers to generate different cages in the same structure (i.e. mixed-ligand MOFs). For example, Zn(II) ions coordinate with both dicarboxylic and tricarboxylic acids in UMCM-2 structure forming three types of cages in the framework (including two microporous cages and one mesoporous cage) [5]. Benefiting from such extraordinary compositional, topological, and structural variations, the number of reported MOFs in the literature is higher than 20 000 in the past decades, whereas the types of zeolites are about 300 only. Accordingly, the majority of work has been focused on the synthesis and characterizations of porous MOFs over the past years. Also, research attention turns to seek prominent applications of MOFs such as catalysis, gas storage,

Nanoparticles in Catalysis: Advances in Synthesis and Applications, First Edition.
Edited by Karine Philippot and Alain Roucoux.

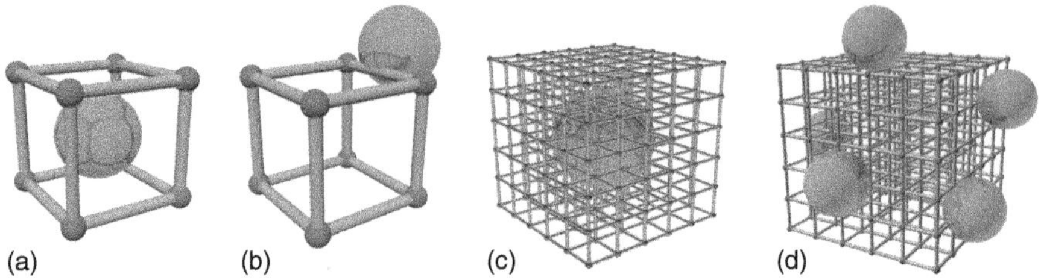

Figure 9.1 Schematic illustration showing the spatial locations of MNPs in the MOF matrix (a) inside the pore cavity, (b) outside the pore cavity, (c) inside the MOF particle, and (d) outside the MOF particle. The spheres represent MNPs and the frameworks represent MOFs. Note that the depicted objects are not proportional to real dimensions.

membrane separation, light harvesting, drug delivery, ion conduction, and so forth. Within these promising applications, in this chapter, we have limited ourselves to summarizing the state-of-the-art achievements of MOFs in catalysis.

In general, MOFs can not only serve as pristine catalysts but also be used as host materials to load catalytic metal nanoparticles (MNPs). Regarding MNPs, especially for the noble MNPs with tailored size and shape, research efforts have been dedicated to their catalytic properties [6]. However, MNPs are thermodynamically unstable because of high surface energy; after removing their surface capping, they are prone to aggregate during catalytic reactions resulting in poor cycling stability. In this regard, MOFs would be an advantageous support to load MNPs to prevent them from leaching and aggregation during reactions, by virtue of high porosity, high surface area, and strong interactions with MNPs [7, 8]. It becomes obvious that more research studies have been focused on integrated nanocatalysts (INCs) with the inclusion of two or more catalytically active components with compositional, size, and shape controls into a larger material system [9–11]. Similarly, controllable integration of MNPs and MOFs (see Figure 9.1) will create novel features, multifunctionality, and synergistic effects, which will greatly enhance catalysis performance or new avenues to multistep reactions [12]. For instance, synergistic effects between the acidity and basicity of MOFs and MNPs will yield bifunctional catalysts, which could be used to perform tandem reactions [13]. Recently, one-pot tandem catalysis involving multistep chemical transformations has been carried out with MNP/MOF composite with two or more active sites, which has gained increasing research attention [14].

Moreover, the usage of MOFs will further boost size-selective catalysis because of the molecular sieving effect of their controllable porosity [15]. For instance, Huo and coworkers have investigated the catalytic activity of Pt@UiO-66 for liquid-phase olefin hydrogenation [16]. Because the pore window of the UiO-66 frame is 6 Å, Pt@UiO-66 core–shell exhibited significantly different conversions for size-dependent alkenes: hexene (size of 2.5 Å) with 100% conversion, cyclooctene (5.5 Å) with 66% conversion, trans-stilbene (5.6 Å) with 35% conversion, triphenyl ethylene (5.8 Å) with 8% conversion, and tetraphenyl ethylene (6.7 Å) with 0% conversion. Because the spatial locations of MNPs in the MOF matrix varied, such as inside the pore cavity (Figure 9.1a), outside the pore cavity (Figure 9.1b), inside the MOF particles (Figure 9.1c), and outside the MOF particles (Figure 9.1d), for easy representation, herein, we refer to all composites consisting of MOFs and MNPs as "MNP/MOF," which do not specifically describe a detailed structure.

To date, there have been a rapidly increasing number of publications devoted to the synthesis, characterization, and catalytic applications of integrated monometallic and multimetallic MNPs in MOF matrixes [17, 18]. One early work on combining MNPs and MOFs was published in 2005 by Fischer and coworkers, who have successfully introduced several MNPs (M = Pd, Au, and Cu) in MOF-5 [19]. Nowadays, the development of synthetic strategy for MNP/MOF composites has become an active research area in heterogeneous catalysis. It is seemingly urgent to develop MNP/MOF composites by a rational design to attain more collective effects and to achieve higher catalytic performances. On the other hand, significant progress on exploring catalytic reactions using such integrated composites has also been made. In addition to the model reactions such as 4-nitrophenol reduction in the liquid phase or CO oxidation in the gas phase, which have been often used to demonstrate the workability of MNP/MOF materials, significant research efforts have recently been extended to catalytic applications with industrial potential. Specifically, a myriad of tandem catalytic reactions for synthesis of intermediates in fine chemical and pharmaceutical industries have been investigated [20]. Thus, the catalytic reactions using MNP/MOF composites will be shown throughout this chapter along with introducing their synthetic strategies.

In the following parts of this chapter, we aim at providing a general synthesis of MNP/MOF materials for their application in catalysis. First, we give brief backgrounds of MOFs and MNPs in catalysis and the merits of their integration compared to their single-component counterparts. Next, we introduce various synthetic approaches for single MNP and single MOF, respectively, and integrative approaches of these two classes of functional materials. The effects of synthesis parameters on the integration efficiency and spatial distributions of MNPs in MOF matrixes are systematically discussed. In view of diverse compositions and configurations of MNPs and MOFs, which lead to endless combinatory possibilities, we try to classify them according to dimensional variation of MOFs in the composites, e.g. zero-dimensional (0D), one-dimensional (1D), two-dimensional (2D), and three-dimensional (3D) structures. Afterward, some representative MNP/MOF composites with well-defined structures in heterogeneous catalysis are also discussed with respect to their structure–function relationships. Finally, we propose a critical outline of future perspectives and the challenges in this field.

9.2 General Synthetic Methodologies

9.2.1 Catalytic Properties of Metal Nanoparticles

The mass of metals takes up ~24% of Earth's crust, and metals are incredibly important for industrial use in modern society. Generally, the term MNPs is referred to spherical or faceted particles having a spatial dimension between 1 and 100 nm, although sometimes, it is also loosely used for those in submicron regime. Compared to their bulk counterparts, MNPs are highly active in catalysis because of their large surface-to-volume ratio, quantum confinement, and surface effects (viz., surface atoms have fewer neighbors and thus more unsaturated sites) [21]. The most famous example is that nanogold is highly active for CO oxidation to CO_2, while gold in bulk state is extremely unreactive as a catalyst [22]. Therefore, the catalytic properties of MNPs are size dependent. For instance, Au NPs exhibit the unusual size dependence activity in the low-temperature CO catalytic oxidation [22]. However, smaller gold particles cannot guarantee higher catalytic activities. Goodman and coworkers have demonstrated that the turnover frequency (TOF) increases when the diameter of the Au NPs is decreased below 3.5 nm, and a further downsizing below 3 nm leads to a decrease of Au activity, yielding a volcano-shaped plot between TOF and size [23]. Therefore, the controlled growth of MNPs is strategic to obtain active catalysts; however, it may require more sophisticated ways of synthesis.

Apart from the effects of size and size distribution, the shape of MNPs may also have a strong impact on catalytic performance, which is often known as "structure dictates function" principle. In the case of Ag NPs for styrene oxidation, for example, the reaction rate over Ag nanocubes was 18 times higher than that over Ag nanoplates and 4 times higher than that over the Ag nanospheres. The reason can be ascribed to different crystal faces on the different crystal morphologies. For instance, Ag nanocubes own mainly {100} terminated surface, whereas the relatively inactive {111} crystal faces were predominantly exposed on Ag nanoplates and Ag nanospheres [24]. Accordingly, there is a significant amount of work on the shape control of MNPs via regulation of nucleation and growth rates at the atomic level. For example, Au NPs can be grown into many anisotropic shapes (e.g. nanorods, nanoplates, platonic solids, and branched crystals) via "seed-mediated growth" strategies [25]. Furthermore, the catalytic activities of MNPs are also surface and interface sensitive. Therefore, the controlled growth of MNPs is very important which can be realized by using facet-selective capping agents such as surfactants, polymers, halide anions, formaldehyde, CO, amines, biomolecules, etc., to direct the growth of MNP crystals with specific exposure surfaces [26].

Besides the monometallic state, the fabrication of bimetallic alloy-type MNPs will further enhance their catalytic activity and durability. For instance, alloyed PtCo NPs were highly active, selective, and stable for catalyzing the hydrogenation of nitroarenes to anilines [27]. Likewise, alloyed AuPd NPs were excellent catalysts for selective oxidation of benzyl alcohol to benzaldehyde [28]. In this chapter, composites with alloyed MNPs loaded on MOFs are also discussed. In addition, metal nanoclusters (MNCs), consisting of just tens of atoms with a size typically <1 nm, have also been proven to be a class of excellent catalysts [29]. They own quantum electronic state, molecule-like behavior, and even higher surface-to-volume ratios than the conventional MNPs. However, this chapter will not describe the synthetic methods for MNCs and their combination with MOFs. Researchers who are interested in the topic can refer to the review paper of catalytic applications of MNCs provided by Schmid et al. [30].

9.2.2 Synthetic Strategies of Metal Nanoparticles

In general, the selection of a preparation method for MNPs depends on the nature of the metal precursors, which differs from one to another. Preparation methods can be classified into two main categories, namely, "top-down" and "bottom-up." The former starts from a large bulk material via physical (e.g. sonication, laser ablation, and lithography), mechanical (e.g. grinding), chemical, or thermal etching processes [31–33], while the latter refers to the assembly of ionic, atomic, or molecular species to form MNPs in nanoscale through chemical reactions. For instance, according to the "top-down" strategy, micromechanical cleavage (scotch tape exfoliation) and liquid-phase exfoliation have been frequently used for 2D nanomaterials [34]. However, the "top-down" strategy might suffer from a relatively broad size distribution and varied shapes of the obtained MNPs. The chemical transformations involved in the "bottom-up" strategy include sol–gel processes, chemical vapor deposition (CVD), plasma or flame spraying synthesis, laser pyrolysis, atomic or molecular condensation, and so forth. In addition, it should be mentioned that the syntheses of MNPs from the above two strategies can be conducted in both the liquid phase and the gas phase [35]. In the following section, we will focus on the advantages/disadvantages of the most common methods based on the "bottom-up" strategy to produce different types of colloidal MNPs.

9.2.2.1 Wet Chemical Reduction Method

Wet chemical reduction method is widely used for the synthesis of various MNP colloidal suspensions, especially for the noble metals, by the reduction of corresponding metallic salts in the solution phase. Different types of reducing agents have been explored, such as sodium citrate, $NaBH_4$, ascorbic acid, alcohol, sugar (glucose, fructose, and sucrose), and so forth [36]. The pioneering example was demonstrated by Faraday who prepared gold colloids by reduction of $HAuCl_4$ with phosphorus in 1857. In some reports, reducing agents play additional roles of solvents and stabilizers (i.e. capping agents) [37]. Moreover, the quality, size, shape, and crystallinity of MNPs can be well controlled by fine-tuning of the reaction conditions (e.g. temperature, concentration, pH, and type of precursor). To avoid batch-to-batch variations, the wet chemical reduction can be realized via flow synthesis, which can afford reproducible and high-quality MNPs with large-scale production [38].

9.2.2.2 Metal Vapor Condensation/Deposition Method

MNPs can be continuously synthesized by the condensation of metal vapors under flowing conditions, wherein inert gas pressure, evaporation temperature, vapor quenching rate, and residence time are four critical factors determining the final particle size of MNPs. Pratsinis and coworkers have designed an aerosol flow apparatus with the evaporation zone and quenching zone to synthesize Bi NPs with an average size from 12 to 37 nm [39]. MNPs with colloidal suspensions of nanosheets/nanofilms can also be prepared by CVD and aerosol-assisted chemical vapor deposition (AACVD) [40]. Considering the difference between them, AACVD relies on the solubility of metal precursor in a solvent to form an aerosol, while conventional CVD requires high volatility of the metal precursor [40].

9.2.2.3 Electrochemical Method

Electrochemical techniques avoid the use of reducing agents to produce MNPs with high purity and short durations. For instance, Ag NPs can be electrochemically deposited onto an atomically smooth surface with overpotentials of 100–500 mV for durations of either 10 or 50 ms [41]. Sun and coworkers have reported a unique electrochemical method to synthesize Pt NPs with the exposed high Miller index surface facets, which exhibited higher activity than commercial Pt/C catalysts in formic acid/ethanol electrooxidation reaction [42]. Moreover, shape-controlled synthesis of MNPs can also be achieved by electrochemical method. Wang and coworkers have reported the preparation of Au nanorods with tunable aspect ratios (between 1 and 7) in a two-electrode-type electrochemical cell [43].

9.2.2.4 Biosynthesis Method

Biosynthesis is a newer environmentally benign approach to produce MNPs by using biomaterials (such as proteins, DNA molecules, viruses, fungi, bacteria, and plants) as reducing agents [44]. The basic mechanism of this method relies on the abundant functional groups on the biomaterials, which not only have high affinity to bind metal ions but also provide nucleation sites for MNP formation via self-reducing or auxiliary reductants. Although MNP nucleation involved in these synthetic processes is complex, biosynthesis method has recently received more research attention and is viewed as a suitable alternative to chemical and physical procedures [45]. However, as compared to the physical/chemical synthetic methods, the reduction rate and reproducibility of MNPs via biosynthesis are usually low. Accordingly, synthetic parameters such as solution pH, incubation temperature, and reaction duration; the concentrations of metal ions; and the amount of biomaterials have to be optimized in order to promote the approach in a large-scale production [46].

9.2.3 Catalytic Activity and Catalytic Sites of MOFs

As mentioned earlier, MOFs are an interesting class of materials for catalysis because of their chemical tunability and high internal surface area. The catalytic activities of MOFs can be ascribed to their abundant framework nodes (metal ion sites or terminal ligands) and cavity modifiers, which usually serve as good Lewis acids or Lewis bases in catalysis reactions [47]. The heterogeneity nature of MOF catalysts can be confirmed by either substrate shape and size selectivity or ceasing the reaction by removing solid MOFs with filtration, which is significantly different from the homogeneous catalysis [48].

Coordinatively unsaturated metal sites (viz., coordinative defect sites) are viewed as Lewis acid sites, which can be obtained by removing coordinated solvent molecules in MOFs, e.g. heating HKUST-1 leads to obtaining open Cu(II) sites [49]. In a catalytic reaction, even substrates could displace the labile, axially coordinated solvent molecules on the initially coordinatively saturated metal ion nodes in a MOF structure, while the equatorially bound organic ligands in the framework are relatively difficult to be replaced by the substrates during the reactions [50]. Then, the obtained coordinatively unsaturated metal ion nodes would work as catalytically active sites. Moreover, such metal ion active sites are majorly distributed on the exterior solid surface of MOFs, which are easily accessible to reaction substrates. For instance, In(III) MOF has been used as an efficient heterogeneous catalyst for acetalization of aldehydes because of the highly Lewis acidic In(III) sites [51]. Solid solution MOFs (e.g. $[In_xGa_{1-x}(O_2C_2H_4)_{0.5}(hfipbb)]$) with more than one metal ion node (e.g. In^{3+} or Ga^{3+}) in equivalent crystallographic sites can be utilized for a heterogeneous catalyst in one-pot multicomponent reactions, and their catalytic activity can be tuned by changing the ratio of the two different metals [52]. It is reported that both crystalline and amorphous MOFs can serve as good catalysts if there is enough amount of unsaturated coordination sites in the frameworks [53]. Moreover, the secondary building units (SBUs) in some MOFs can play a similar role as catalytic inorganic metal oxides (such as TiO_2 and CeO_2). For example, PCN-415 consisting of inorganic $[Ti_8Zr_2O_{12}(MeCOO)_{16}]$ clusters as SBUs is highly photoactive and exhibits good photocatalytic activity toward water reduction [54]. Reviews in Ti-bearing MOFs with photoactive SBUs for advanced photocatalysts have been presented by Zou and coworkers [55].

Compared with metal ion nodes, the organic linkers of MOFs have been largely utilized as base catalysts. For instance, 3D porous $\{[Cd(4\text{-btapa})_2(NO_3)_2]\cdot 6H_2O\cdot 2DMF\}_n$ owning amide groups in the frameworks was used to catalyze Knoevenagel condensation reactions of benzaldehyde with malononitrile with a conversion of 98% [56]. Furthermore, the organic nodes can be further functionalized with amino or sulfo groups to expand their scape of catalytic applications. For instance, MIL-101 was dually functionalized with basic amino and acidic sulfo groups, so as to catalyze tandem catalysis (i.e. one-pot deacetalization–Knoevenagel condensation). Importantly, the acid-to-base ratio can be fine-tuned to achieve optimal catalytic performance [57]. After loading Pd NPs and/or doping other metal ions with a lower electron affinity, the pendant carboxylate groups in HKUST-1-R can be engineered as a new type of Lewis basic sites because of the high electron density of the oxygen atoms [58, 59].

In addition, the catalytical features of MOFs can also come from cavity modifiers. Sun et al. have designed a strong base MOF by using guest–host redox interaction between KNO_3 and low-valence metal centers (i.e. Cr^{3+}) at 300 °C, which exhibited excellent catalytic performance in Knoevenagel condensation and transesterification reactions [60]. It should be noted that in many other reports, MOFs have transformed to "Quasi-MOF" or transition metal/nitrogen-doped carbon materials or other metal derivatives (oxides and sulfides) through pyrolysis before the usage as catalysts [61, 62]; however, this chapter will not focus on this type of works, as the chemical nature of the transformed products is similar to that of conventional catalysts.

9.2.4 Porosity of MOFs for Catalysis Applications

The ultrahigh areas of MOFs are generally attained and measured after activation treatments to evacuate the occupied solvents without pore collapse or channel blockage. If the pore window size is too small, the guest solvent molecules are difficult

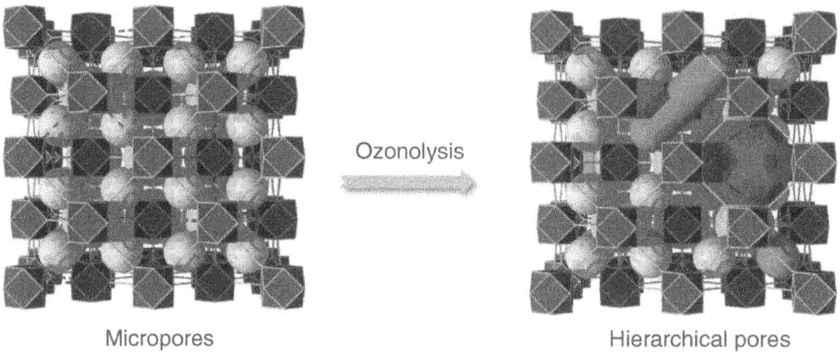

Figure 9.2 Schematic illustration of the modification of Zr-fcu-MOFs through ligand cleavage method by solid–gas ozonolysis leading to hierarchical porosity. Source: Reproduced with permission from Ref. [66]. Copyright 2018, American Chemical Society.

to be depleted or the frameworks are easily collapsing during the activation process. It has been observed that the pore elimination during removing guest solvent molecules frequently leads to significant discrepancies between the experimentally measured and the calculated surface areas of MOFs. To tackle this issue, Hupp et al. have adopted a supercritical CO_2 technique to activate NU-109E and NU-110E without pore collapse or channel blockage, which have large BET surface areas of 7010 and 7140 $m^2\ g^{-1}$, respectively [4].

Traditionally, MOFs are microporous materials with pore sizes <2 nm, although the pore aperture can be expanded up to 9.8 nm by an elaborate design with longer organic linkers [63]. Therefore, the molecular accessibility in MOFs is critically important for catalysis applications, which need to be carefully addressed. Considering the slow mass diffusion of large molecules in MOFs in transport processes, mesoporous MOFs (mesoMOFs) with mesopore sizes ranging from 2 to 50 nm have received intensive attention because of the more efficient reactant and product transports [64]. Zhang and coworkers have used CO_2-expanded liquids as switchable solvents for the synthesis of mesoporous HKUST-1 with large mesopores (13–23 nm) [65]. Very recently, as displayed in Figure 9.2, Maspoch and coworkers have developed a novel ozonolysis post-treatment method to create mesopores in MOFs by selectively cleaving ligand and fusion of micropores into mesopores [66]. The success was based on designing a mixed ozone inert/active ligands in the same MOF structure.

9.2.5 Synthetic Strategies of MOFs

Currently, the discovery of new MOF compounds remains largely an art. To address this issue, Yaghi et al. have defined a "reticular synthesis" conceptual approach to synthesize new MOF compounds with the assembly of SBUs to default networks [67]. Besides chemical compositions and topological structures, similar to MNPs, size and shape also affect the physicochemical properties of MOFs [68, 69]. In terms of catalytic activity and material utilization, nano-sized MOF materials are particularly useful in heterogeneous catalysis. A recent review by Mirkin and coworkers has been devoted on the size- and shape-controlled synthesis of MOFs [70]. Through a "bottom-up" strategy, the following methods have been explored to tune nucleation and growth rates in reaction systems: (i) adjusting solvent composition, reaction temperature, or concentrations of metal ions and organic linkers; (ii) confinement synthesis by using reverse micelles and microemulsions; (iii) coordination modulation (e.g. adding block copolymer ligands, surfactants, or competitive ligands in the synthetic solution); and (iv) deprotonation of organic linkers to produce more reactive sites by triethylamine [71, 72]. Zhou et al. have also suggested that MOF formation reactions can be simplified as ligand substitution reactions on the metal clusters [73]. Therefore, manipulation of the kinetics of the coordination reactions eventually controls the particle size of the resultant MOFs.

Regarding their general synthetic methods, most MOFs are obtained prevailingly from homogeneous solution reactions between soluble metal salts with organic ingredients (i.e. ligands, and additives) in one or mixed polar solvents (e.g. water, alcohol, or other organic solvents) [74]. Besides conventional heating methods (e.g. hydrothermal or solvothermal methods), some other newer methods for MOF synthesis are summarized below to highlight their advantages/disadvantages.

9.2.5.1 Electrochemical Method

The electrochemical method is an alternative route to the hydrothermal synthesis of MOFs, which is advantageous for the synthesis of MOF thin films on any given substrate without pretreatments because of the remarkably fast reaction rate, mild preparation condition, and easy controls over product film morphology and thickness [75]. Regarding the reaction setting, both two-electrode and three-electrode systems have been used for electrochemical deposition of MOF thin films, wherein the coordination reaction occurs in electrolytic cells containing metal ions, organic ligands, and electrolytes. Kulandainathan and coworkers have fabricated Cu-BTC with controlled size by adjusting supporting electrolyte and current density, which could be used as an excellent catalyst for the reduction of 4-nitrophenol to 4-aminophenol [76].

9.2.5.2 Sonochemical Method

Different from the conventional hydrothermal and solvothermal methods with electric heating, ultrasound power has been used for the synthesis of MOFs in order to reduce reaction duration and increase energy efficiency. Jhung and coworkers have studied crystallization behaviors of MOFs by irradiation under ultrasound, and they found that the reaction time for producing MOFs under ultrasounds (0.5–1 hour) was much shorter than that using conventional methods (1.5–3 days) [77]. The increased reaction rate was due to the larger pre-exponential factors (A) in the Arrhenius equations. Moreover, the ultrasound-assisted synthesis also affects the particle size of MOFs because of the increased reaction rate. For instance, Abbasi and Rizvandi found that the average particle size for Cu-BTC from ultrasound method was around 80 nm, while those obtained from electrical heating were larger than 140 nm [78]. A comprehensive review on ultrasound-assisted preparations of MOF crystals has been made and published by Argirusis and coauthors [79].

9.2.5.3 Microwave Irradiation Method

Microwave irradiation technique has also been widely used for the synthesis of MOFs via fast crystallization and the control over crystal morphology. Microwave irradiation acting as high-frequency electric fields directly accelerates the heating process of polar molecules containing mobile electric charges by rotating with the field. A kinetic study on MIL-53(Fe) synthesis under microwave irradiation showed that the acceleration in crystal growth is more important than the acceleration in nucleation [77]. Similar results were also obtained by Tian et al. who fabricated medi-MOF-1 with the aid of microwave heating [80].

9.2.5.4 Mechanochemical Method

Mechanochemical synthesis can significantly reduce the amount of solvent in MOFs fabrication, which is a promising alternative method for the large-scale production of MOFs without considering the solubility issue of reactants in the solvent. Kitagawa and coworkers have prepared CPL-1 with 1D microporous channels, by mechanochemical mixing of copper nitrate, disodium pyrazine-2,3-dicarboxylate, and pyrazine in the air using a mortar and pestle [81]. It was found that no CPL-1 was produced if the environmental humidity was lower than that of ambient air. In addition, the intermediate phase of MOFs can be observed by using mechanochemical synthesis, which is useful for studying the formation mechanism of MOFs. However, it should be noted that the salt by-product from the coordination reactions cannot be removed during the mechanochemical process, which might occupy the pores and therefore should be further treated by postgrowth washing.

9.2.5.5 Synthesis of MOFs in Green Solvents

Organic solvents are commonly used in the synthesis of MOFs, which might hinder their industrial applications and scale-up production. In this regard, both ionic liquids (ILs) and supercritical carbon dioxide ($scCO_2$), considered as often environmentally friendly solvents with unusual properties, have been used as alternative media for MOF synthesis [82, 83]. For instance, chiral Ni-MOFs were fabricated by constructing chiral ILs, wherein chiral information from the ILs can be efficiently transferred to MOFs during the ionothermal synthesis [84]. Furthermore, ionothermal synthesis with ILs as solvents can be accelerated by combining microwave-assisted synthesis to reduce the reaction time of MOFs [85]. Simultaneous utilization of ILs and $scCO_2$ generates CO_2-in-IL microemulsion, which could also be used for synthesizing meso-MOFs with a hierarchical porous structure (mesopore size of 20–50 nm and micropore size of 2–3 nm) [86]. Regarding the formation mechanisms of the tiered pores, Xu and coworkers indicated that CO_2 droplets (18–50 nm) were used as the template for the cores, while CO_2/IL micelles were used as the template of mesopores (3–6 nm), and the micropores were derived from the intrinsic frameworks of MOFs [87]. Interestingly, the MOFs with the hierarchical porous structure have been demonstrated as effective and reusable heterogeneous catalysts in aerobic oxidation of benzylic alcohols because of their advantages of the low diffusion barrier [88].

9.2.5.6 Microemulsion Method

A thermodynamically stable microemulsion is a dispersion of two immiscible solvents stabilized by surfactants or NPs (namely, Pickering emulsion). The controlled synthesis of MOFs with different sizes and morphologies can be achieved by performing the reactions in the confined microemulsions. In particular, Han and coworkers have demonstrated that Ln-MOFs with various shapes (spherical, lamellar, and cylindrical) can be fabricated in water-in-IL microemulsions [89]. Similarly, Boyes and coworkers have also reported a reverse microemulsion method to synthesize Gd-MOF [90]. They found that the shape, size, and size distribution of the MOFs could be manipulated by changing the micelle rigidity with the addition of hydrotrope.

9.2.5.7 Transformation from Solid Matters to MOFs

Solid matters have been exploited as precursors to the fabrication of MOFs, which not only serve as metal ion sources but also as templates/supports to regulate the nucleation and accelerate the growth of MOFs [49, 91]. As displayed in Figure 9.3, our group has fabricated mesoporous HKUST-1 and 2D Cu(HBTC)-1 nanosheets by using Cu_2O nanocubes as a copper ion source in two different solutions containing organic linkers, wherein the dissolution rate of Cu_2O greatly affects the morphology of the produced MOFs [49, 92]. Falcaro et al. have positioned MOFs on a substrate

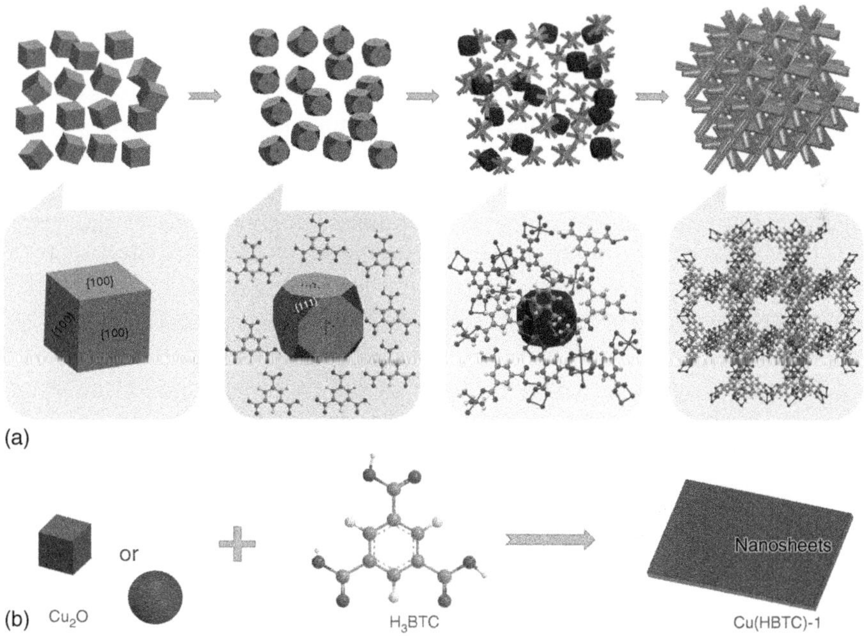

Figure 9.3 (a) Schematic illustration of the transformation of Cu_2O nanocubes to hierarchically macro–meso–microporous HKUST-1. Source: Reproduced with permission from Ref. [49]. Copyright 2016, Royal Society of Chemistry. (b) Transformation of Cu_2O nanocubes/or nanospheres to Cu(HBTC)-1 nanosheets. Source: From Zhan and Zeng [92]. Reprinted with permission of John Wiley and Sons.

by using solid matter as nucleating agents and templates, particularly vital to the fabrication of MOF-based device [93]. Importantly, it was found that the crystal growth rate by using solid matter was much faster compared to control samples. For example, Parsons et al. have designed an extremely rapid synthetic route for HKUST-1 (<1 minute) by using solid matter (such as hydroxy double salts) as intermediates [94].

9.2.6 Integration Methods of MNPs with MOFs

Although bare or so-called "unprotected" MNPs are also reported as stable catalysts in some literatures, solvent molecules or anionic species can actually adsorb on the external surface of MNPs as stabilizers, which might prohibit the diffusion of reactants and result in an induction period in a catalysis reaction [95]. Apart from commonly known organic anions or chelating agents, it was found that even inorganic anionic species such as S^{2-}, HS^-, Se^{2-}, HSe^-, Te^{2-}, HTe^-, $TeS_3{}^{2-}$, OH^-, and $NH_2{}^-$ can enhance the colloidal stability of MNPs in polar solvents [96]. Therefore, for most colloidal MNPs, they are not truly bare. In most cases, the freestanding MNPs are employed as heterogeneous catalysts only when surface-protecting agents are present. For instance, Pt_3Co alloy NPs with truncated octahedral shape were prepared with amine as a capping agent, and the hydrocarbon chains in the amine would affect the selectivity in catalytic hydrogenation of cinnamaldehyde [97]. However, in some cases, the added capping molecules may be deleterious to the catalytic performance. To circumvent this problem, MNPs have to be loaded on porous support (e.g. porous silica, zeolites, and MOFs) for construction of supported heterogeneous catalysts. As previously mentioned, the immobilization of MNPs on the exterior surface or inside pores of MOFs could greatly improve the stability of MNPs in their employed reaction conditions [98]. Moreover, sintering of MNPs, which is a major cause of catalyst deactivation, can be effectively suppressed. In this way, strong MNP–MOF interaction and new synergy between MNPs and MOFs could be expected, thus enhancing the performance of the catalysts.

In most of the MNP/MOF composites, the mentioned MNPs are separately fabricated and introduced to MOFs. However, Kitagawa et al. have prepared Ni@MOF by a novel method with partial thermal decomposition of Ni^{2+} ions in MOF to self-generate Ni NPs within the MOF [99]. More specifically, the Ni-MOF-74 was heated at 250–350 °C under vacuum, which led to Ni@MOF-74 with a mean size of Ni NPs at 4.3 ± 1.4 nm, wherein the entrapped Ni NPs were self-generated from the Ni ions in the starting material.

In general, the introduction of premade foreign MNPs onto MOF particle surface can be done with similar methods of loading MNPs on conventional porous supports (e.g. porous silica, zeolites, and carbon). Liu et al. have compared different ways to load Au NPs on MIL-101 (Cr). Their work showed that both the particle size of Au NPs and their corresponding catalytic performance in aerobic oxidations of primary or secondary aromatic alcohols are greatly influenced by the preparation methods (see Figure 9.4) [100]. Furthermore, the preparation methods also affect the spatial locations of MNPs in the MNP–MOF composites, which will directly dictate their

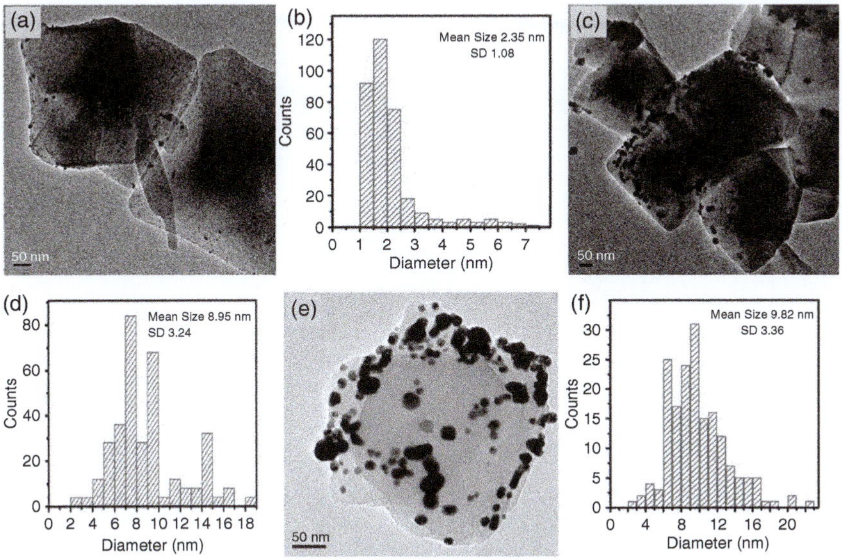

Figure 9.4 (a,b) TEM image of Au/MIL-101 prepared by colloidal deposition method and corresponding size distribution of Au NPs, (c,d) TEM image of Au/MIL-101 prepared by deposition–precipitation method and corresponding size distribution of Au NPs, and (e,f) TEM image of Au/MIL-101 prepared by impregnation–reduction method and corresponding size distribution of Au NPs. Source: (a, c, and e) Reproduced with permission from Liu et al. [100]. Copyright 2010, American Chemical Society.

catalytic properties. Jiang and coworkers have compared two types of catalysts for hydrogen production from water splitting, i.e. Pt@UiO-66-NH_2 and Pt/UiO-66-NH_2 [101]. The former catalyst had Pt NPs embedded in UiO-66-NH_2, while Pt NPs were supported on the external surface of UiO-66-NH_2 in the latter case. Interestingly, Pt@UiO-66-NH_2 greatly improved the electron–hole separation efficiency by shortening the transport distance, which has five times higher photocatalytic efficiency than that of Pt/UiO-66-NH_2.

In this section, we mainly focus on the incorporation of MNPs within the MOF matrixes, either embedded in MOF particles or inside their pore spaces. Usually, composites with MNPs confined by MOFs as host materials are named MNP@MOF or MOF⊃MNP [3]. According to the successive or concomitant steps for the synthesis of MNP and MOF, three known strategies can be further categorized (see Figure 9.5): (i) presynthesis of MNPs and then using them as seeds to induce the nucleation of MOFs, (ii) incorporation of metal precursors to the preformed MOFs and then *in situ* reduction to obtain MNPs, and (iii) simultaneous growth of MNPs and MOFs. The first two require a two-step route for the combination (called as *ex situ* incorporation), while the last type of integration belongs to one-pot synthesis (*in situ* incorporation).

9.2.6.1 Preformation of MNPs and Growth of MOFs

In a first approach, MNPs can be preformed and then used as seeds to induce the nucleation and growth of MOFs. However, there is no guarantee that the MOFs will

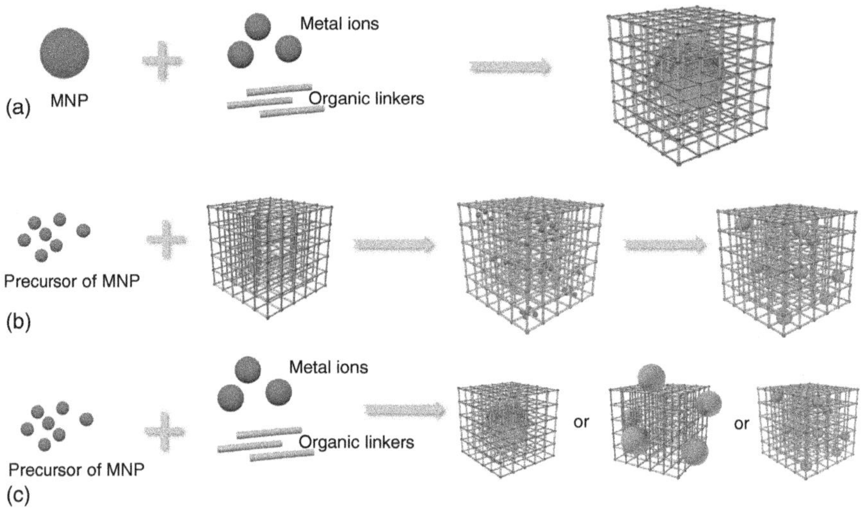

Figure 9.5 Schematic illustration showing the synthetic strategies of MNP/MOF nanocomposites, (a) presynthesis of MNPs and then coating of the MOF layer, (b) incorporation of metal precursors to the preformed MOFs and then *in situ* reduction to obtain MNPs, and (c) one-pot synthesis. Note that the depicted objects are not proportional to real dimensions.

follow heterogeneous nucleation and grow on the external surface of MNPs. Therefore, "binder agents" such as polydopamine, proteins (e.g. bovine serum albumin), amorphous alumina, and polyvinylpyrrolidone (PVP) are employed to enhance the affinity of MNPs toward MOFs and boost conformal coating of MOFs on nanoparticles [15, 102–104]. In the case of the harsh environment of MOF synthesis (e.g. low pH), MNP etching is a problem, which can be prevented by their stabilization with a protective layer of a second metal. For instance, when coating Pd NPs with UiO-66, PdPt core/shell NPs with Pt as the shell metal were used to prevent Pd NP agglomeration and leaching during the coating of UiO-66 layer [105]. Additionally, the environment for MOF growth might lead to serious agglomeration of MNPs. Therefore, Wang and coworkers have developed a crystalline ion-triggered growth strategy for embedding Au NPs in the ZIF-8 matrix without agglomeration [106], wherein lately added zinc ions would coordinate with the existing functional groups on the Au surface to enhance the interactions.

Under the above preparative conditions, the sizes of MNPs are not limited to the pore size of MOFs because of the fact that MNPs do not occupy the pore cavities, which is different from the cases to be discussed in Section 9.2.6.2 (*vide infra*). Accordingly, if the sizes of the preformed MNPs are larger than 2 nm, it is then difficult for them to enter the pore windows of MOFs. Usually, the sizes of MOF particles are much larger than those of MNPs; hence, several small MNPs can be encapsulated into a single MOF particle, as can be seen from the structure of Pd@MIL-101 displayed in Figure 9.6. Because the encapsulated MNPs

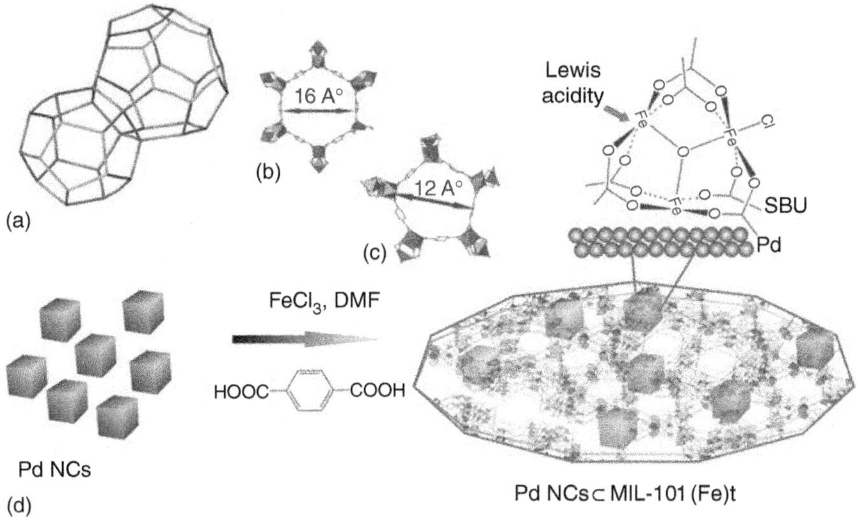

Figure 9.6 (a–c) Topology and hexagonal/pentagonal windows in the MIL-101(Fe) structure and (d) schematic illustrations of Pd@MIL-101 composite formation. Source: From Bakuru and Kalidindi [107]. Copyright 2017, John Wiley and Sons. Reprinted with permission of John Wiley and Sons.

are fully covered by the MOF shell, the resultant composite could be used for molecule-size-selective catalysis, which will be discussed more in the following sections. It should be noted that the notion of MNP@MOF is used not only for MNPs confined into the cavities or channels of a MOF matrix but also for those composites with MNPs enveloped by an MOF shell in forming core–shell or yolk–shell structures (*vide infra*).

9.2.6.2 Incorporation of Metal Precursors Followed by in Situ Reduction

The second strategy involves stepwise inclusion of a guest metal precursor (in the liquid phase or gas phase) in MOFs and then its *in situ* reduction or decomposition. In this process, the metal precursor should be introduced to the pores/channels of MOFs using specific techniques such as CVD, atomic layer deposition (ALD), solution impregnation (also called as capillary impregnation), surface grafting, and solid grinding. The reduction of the metal precursor to MNPs can be realized by using reductants such as $NaBH_4$ in the liquid phase, H_2 in the gas phase, electroreduction, or photoreduction. Tang et al. have employed protein (i.e. lysozyme) as a novel reductant to load Ag NPs and Au NPs into the nanopores of ZIF-8 following a protein-induced biomineralization process [108]. Compared to *in situ* reductions of metal precursors, colloidal deposition method offers better controllability over the size uniformity and morphology of the MNPs in the final composites. Nevertheless, the preformed MNPs are difficult to get into the nanopores of MOFs because of the presence of a large transportation barrier. As a result, an uneven distribution of MNPs is observed as they are located mainly in the surface regions or on the external surface of MOF particles.

Among the various infiltration methods, CVD, ALD, and solid grinding belong to dry chemical processes. Solid grinding involves grinding porous MOFs with volatile metal precursors, while CVD method involves exposure of porous MOFs to the vapor of volatile organometallic compounds. The self-limiting deposition technique ALD is able to introduce metal vapor phase to the micropores of MOFs, which could control the size of MNPs with atomic precision. One common requirement for such syntheses is that the metal precursors are volatile enough and can be adsorbed to the cavity space in MOFs [109]. Although it has been reported that the efficiency of loading metal precursor to MOFs via solution penetration is far less than through the gas phase, as previously discussed, the metal precursors used in CVD, ALD, and solid gridding methods are limited [19]. In addition, the CVD process faces the complexity of operation under stringent conditions (such as precise controls over pressure and temperature) and the instability of volatile organometallic precursors in the air. Haruta and coworkers have demonstrated that the obtained Au NPs on CPL-2 were much uniform and smaller when using the solid grinding method (2.2 ± 0.3 nm) than those obtained by using the CVD method (3.1 ± 1.9 nm) [110]. It should also be noted that a slow diffusion rate of metal precursor in MOFs will yield only MNP aggregates. Although the produced MNPs are confined in the nanopores, there are some reports of MNPs larger than the size of MOF pores, which could be probably resulted from the TEM-induced MOF degradation (by the electron beam) and the followed agglomeration of MNPs [111].

Solid grinding is suitable to introduce volatile organometallic complex (viz., the precursor of MNP) to the pore space in MOFs as well. Xu et al. have prepared Au@ZIF-8 by homogenously grinding ZIF-8 and $(CH_3)_2Au(acac)$ in an agate mortar in the air, followed by H_2 reduction at 230 °C [112]. The resultant Au@ZIF-8 presented good catalytic activity and high thermal stability in CO oxidation reaction. No solid grinding is required if the metal precursor can sublime at ambient or low-pressure conditions. For instance, Pt@MOF-177 was prepared by loading a volatile platinum precursor [Me_3PtCp'] in MOF-177 without solid grinding, followed by a reduction in a hydrogen atmosphere [113]. Moreover, the obtained Pt@MOF-177 (with Pt size between 2 and 3 nm) exhibited an excellent activity (with a turnover number up to 1000) toward the solvent- and base-free oxidation of allylic alcohols (both cinnamyl alcohol and benzyl alcohol) at room temperature with air as an oxidant. However, the recyclability of Pt@MOF-177 was poor because of the fact that the by-product water destroyed the MOF host lattice, leading to inaccessible Pt NPs and no catalytic activity even in the second run of the catalysis reaction.

Solution impregnation relies on the capillary action to draw the metal-containing solution into the pores of MOFs, which is similar to the incipient wetness impregnation to prepare traditional supported catalysts. One possible problem of solution impregnation is the existence of competition between guest and solvent molecules from the bulk solution to the pore, which might lead to a broad particle size distribution of MNPs [3]. Functionalization of MOFs with amino group provides coordination sites for metal precursors during the impregnation process [98]. For instance, it was calculated that the adsorption enthalpy of Pd atoms on

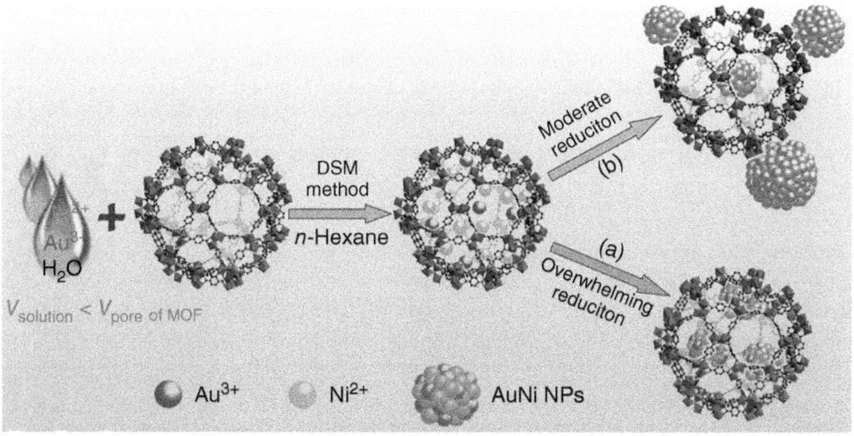

Figure 9.7 Schematic illustration of immobilization of the AuNi alloy NPs on MIL-101 by using a double-solvent method to avoid the deposition of the precursors on the outer surface of the MOF. Source: From Zhu et al. [115]. Copyright 2013, American Chemical Society. Reprinted with permission of American Chemical Society.

functionalized NH_2-UiO-66 is 187 kJ mol^{-1} higher than that on pristine UiO-66 (150 kJ mol^{-1}), indicating that NH_2-UiO-66 provides stronger adsorption sites for Pd during the embedment in the pores of MOF [114]. In order to exclusively deposit MNPs in the interior pores of MOFs without leaving MNPs on the external surface, Xu et al. have designed a double-solvent method to realize the full incorporation of metal precursors into the hydrophilic pores of MIL-101 by capillary action (as displayed in Figure 9.7) [115]. It should also be mentioned that the amount of aqueous metal precursor solution should be slightly less than the pore volume of the MOFs.

If the precursor solution contains two different metals, usually bimetallic NPs embedded in MOFs could be obtained. For example, simultaneous infiltration of Ni^{2+} and Mo^{6+} aqueous solution in MIL-101 generates NiMo@MIL-101, which has been used as an excellent photocatalyst for hydrogen evolution from water splitting with high transient photocurrent, lower overpotential, and longer fluorescence lifetime [116]. Solution impregnation process can also be coupled with electrochemical reduction method. Recently, Hupp and coworkers have applied an electrochemical reduction method to load Cu NPs in NU-1000 by electrochemical reduction of the adsorbed Cu(II) to metallic Cu, and the resultant composite exhibited promising electrocatalytic activity for CO_2 reduction [117].

9.2.6.3 One-pot Integration of MOFs and MNPs

The one-pot approach to integrating MOFs with MNPs means the simultaneous formation of both materials. Li and coworkers have designed a well-defined Pt@DUT-5 composite by directly mixing the Pt precursor, $Al(NO_3)_3$, and 4,4′-biphenyldicarboxylic acid in the *N*,*N*-dimethylformamide (DMF) solvent at 130 °C for 24 hours [118]. As shown in Figure 9.8, the reduction of H_2PtCl_6 to produce tiny Pt NPs (size between 1.5 and 2.0 nm) and the coordination between

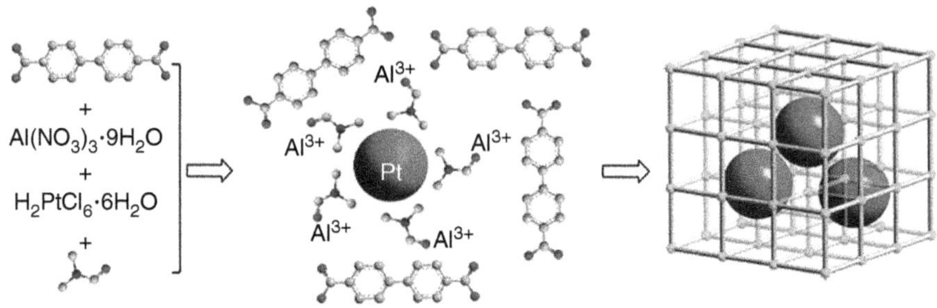

Figure 9.8 Schematic illustration of one-pot synthesis of Pt@DUT-5. Source: Reproduced with permission from Ref. [118]. Copyright 2015, Royal Society of Chemistry.

Al ions and organic linker to form MOFs (i.e. DUT-5) were conducted at the same time, wherein the oxophilic Al^{3+} cations will be adsorbed toward the C=O group on the Pt surface and promote the integration efficiency. Importantly, such method can be extended to prepare other MNP@MOF composites such as Pt@UiO-66 and Pt@MOF-253. Similarly, bimetallic PdNi NPs can be encapsulated in the UiO-67 shell via the one-pot incorporation method [119]. The heterogeneous PdNi@UiO-67 has significantly enhanced the catalytic activity in nitrobenzene hydrogenation reaction owing to the confinement effect offered by the pristine UiO-67 shell. Li and coworkers have found that the formation rate of MNPs was slower than that of MOF in the one-pot synthesis, leading to the distribution of MNPs in both the MOF matrix and the external surface. Therefore, they have further developed a kinetically modulated method to exclusively encapsulate MNPs with MOF shell by either decreasing the growth rate of the MOF with the use of acetic acid or increasing the reduction rate of MNPs using H_2 as an assistant reducing agent [120].

Solvents used for MOF synthesis can also serve as a reducing agent for the synthesis of MNPs under one-pot conditions. For example, DMF can act as a mild reducing agent for the preparation of Au NPs and Pt NPs as the temperature was increased to more than 120 °C [121]. Additionally, DMF can adsorb on the surface of MNPs to prevent them from agglomeration, wherein stabilizing agents (e.g. thiolates, surfactants, and polymers) are absent, resulting in "unprotected" MNPs discretely distributed on MOFs.

9.3 Architectural Designs and Catalytic Applications of MNP/MOF Nanocomposites

Along with the rapid development in nanotechnology, fashioning MOFs into desired shapes has recently appeared in the literature, ranging from 0D nanospheres, 1D nanorods/nanofibers, 2D nanosheets/nanofilms, and 3D monoliths/networks. According to the Gibbs–Curie–Wulff theorem, the morphology of an MOF is mainly determined by its crystal symmetry, such as unit cell parameters, axial relations, and relative specific surface energy of each facet in the crystal [122]. The crystal growth rates of MOFs are relatively faster in those directions that are associated with shorter unit cell axes [123]. For instance, ZIF-5 crystals are likely reflected on the cubic morphology because of the equal ***a–c*** unit cell axes in a cubic system, while ZIF-3 crystals perfectly exhibit a prismatic morphology because of one comparatively larger axis [124]. Surfactants such as PVP molecules have frequently been used in the shape control of MOFs for the formation of 2D nanosheets by inhibiting the crystal growth in certain directions [92]. Because of length constraint, in the following subsections, we will briefly introduce the catalytic applications of several typical MNP/MOF composites with different morphologies based on their dimensional properties, and the structure–property relationships of these catalysts will also be addressed.

9.3.1 Zero-Dimensional MNP/MOF Nanocomposites

On the basis of the structural classification, MOF particles with all the dimensions below 100 nm can be considered as 0D nanostructures. For instance, Yang have synthesized ultrasmall Zn-MOF-74 nanodots with an average size within 10 nm by using low zinc ion concentrations [125]. However, currently, there are rare reports about integrating such 0D MOF connected to MNPs forming dimers or trimers. In this regard, it was noted that the MOF-5 matrix is constructed by discrete 0D Zn_4O_{13} quantum dots (that is, SBUs) linked with terephthalate [126]. Therefore, the infiltration of MNPs with such Zn_4O_{13} quantum dots (QDs) could ideally form dimers. Aguilera-Sigalat and Bradshaw have reviewed the integrated MOF-quantum dot (QD@MOF) composites for applications in photocatalysis, energy, gas storage, and selective sensing [127]. However, the structural relationships between QDs and the 0D SBUs are not clearly described, which cannot fully accord with the notion of 0D MNP/MOF composites. The existing configurations of MNPs and 0D MOFs can be core–shell or yolk–shell structures (see below, Section 9.3.5.1). Lah and coworker have summarized templated and template-free fabrication strategies for 0D MOF hollow nanospheres via Ostwald ripening mechanism, surface-energy-driven mechanism, self-template mechanism, surface-protected selective etching mechanism, and other hard-templating strategies [128].

9.3.2 One-Dimensional MNP/MOF Nanocomposites

1D MOFs have one dimension that is outside the nanoscale, being mainly in the shape of nanotubes, nanorods, nanofibers, nanoribbons, nanowires, etc. Apparently, 1D MOFs can be obtained by using 1D template such as carbon nanotubes [28]. As displayed in Figure 9.9, Duan et al. have reported the synthesis of 1D wire-like MOF-545 crystallizing in the hexagonal space group *P6/mmm* [129]. Then, both Au NPs and Pt NPs were loaded in the pore cavity of 1D MOF-545 by the above-mentioned solution impregnation–reduction method. It was found that the solvent and the reducing agent were critical to the shapes of the MNPs confined in the pores of the 1D MOF. For example, as shown in Figure 9.9c,d, high-density Au nanodots were grown in the MOF crystal in the case of acetone solution, while a

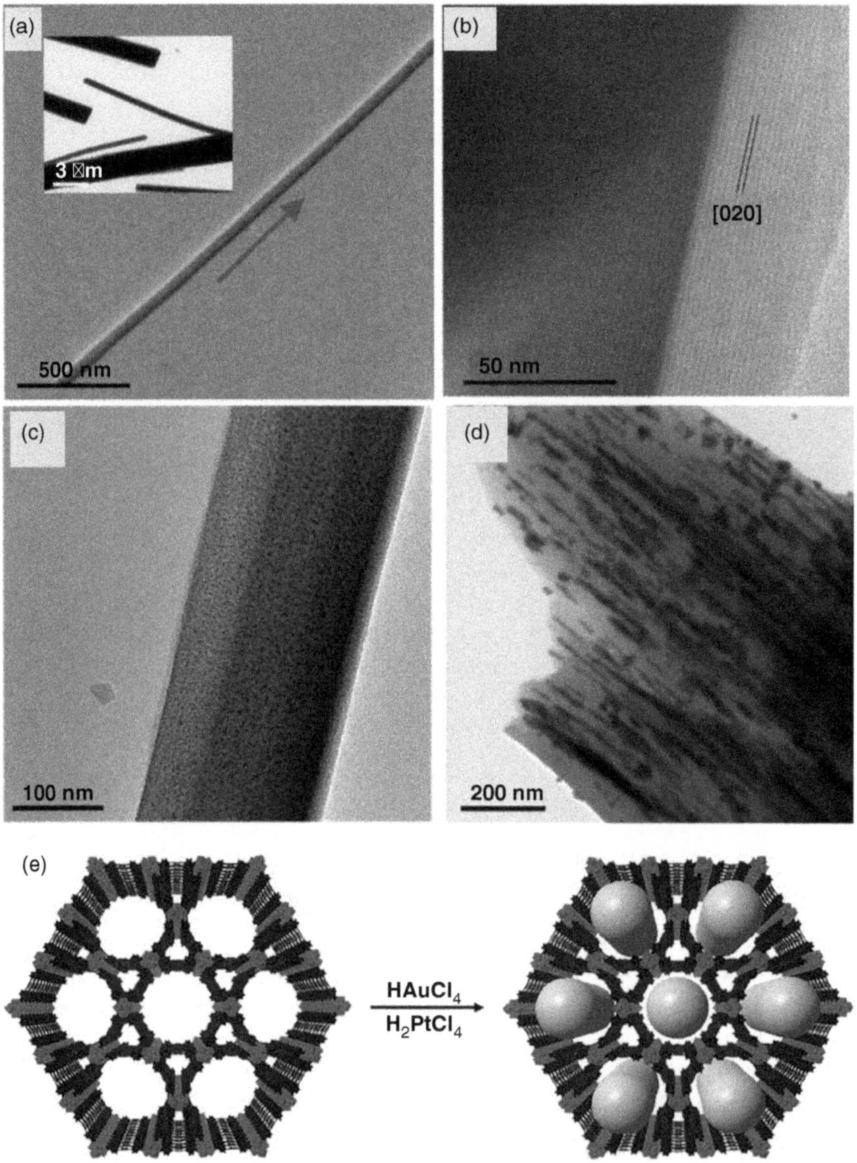

Figure 9.9 (a,b) Representative TEM images of the wire-like MOF-545, (c) representative TEM image of the integrated Au/MOF-545 composite with Au nanospheres, (d) representative TEM image of the integrated Au/MOF-545 composite with Au nanowires, and (e) structural model of MOF-545 with 1D pores as templates for the synthesis of MNP/MOF composites. Source: (a–d) Reproduced with permission from Volosskiy et al. [129]. Copyright 2015, American Chemical Society.

large number of elongated Au nanowires were grown inside the same MOF crystal if the reaction proceeded in water. *In situ* coating of MOF shell on 1D MNP can also lead to the fabrication of 1D MNP/MOF composite. For example, a ternary Au/ZnO@ZIF-8 flower-like nanocomposite was obtained by using 1D Au/ZnO as both morphological template and sacrificial precursor to provide zinc ions for construction of ZIF-8 [130]. During the epitaxial growth of ZIF-8 on the Au/ZnO surface, Au NPs were resident between the ZnO core and the ZIF-8 shell. The resulting 1D Au/ZnO@ZIF-8 showed high catalytic activity for the reduction of 4-nitrophenol to 4-aminophenol.

9.3.3 Two-Dimensional MNP/MOF Nanocomposites

The shapes of 2D MOFs include nanosheets, nanofilms, nanolayers, nanocoatings, even thin-shelled hollow spheres, etc. As compared to other morphologies, 2D MOFs with extended lateral dimensions and nanometer thickness have been less explored, probably because of the lack of easy-to-use techniques. 2D MOFs are more attractive for catalysis than bulk MOFs because of the unique physical, chemical, and electronic properties; minimal diffusion barriers; and more coordinatively unsaturated atoms on the external surface [131]. The surfactant-assisted method has been frequently used to fabricate MOFs with nanosheets by hindering the growth of a specific crystal facet [92, 132]. Of late, Lu and coworkers have prepared 2D Ni-based MOF nanosheets with PVP as a surfactant (Figure 9.10a) [132]. In contrast, they found that the bulk counterpart was obtained in the absence of PVP. Then, ultrasmall Au NPs (c. 1 nm) were confined in MOF pores through the infiltration–reduction method. Importantly, the resultant 2D Au@NMOF-Ni exhibits remarkable catalytic performance toward the conversion of MG 17 and 4-NP, where reaction rates were well correlated with the molecular size of substrates (as shown in Figure 9.10b) [132]. Wang's group has successfully synthesized bimetallic 2D MOFs based on the diffusion-mediated regulation of MOF growth [133]. Porphyrin-based MOFs are easy to be fabricated as 2D nanosheets [134]. In a recent work by Zhang and coworkers [135], Au NPs with the size of 2.1 nm were grown on 2D metalloporphyrinic MOF nanosheets (thickness < 10 nm). The integrated composite was used as a hybrid catalyst for mimicking the natural enzymatic cascade reaction. They concluded that Au NPs and 2D MOF nanosheets served as artificial glucose oxidase-like and peroxidase-like activities, respectively. Our group has reported a novel method to fabricate 2D copper-based MOFs with ultrathin thickness (~6 nm) by using Cu_2O as a copper ion source, and ultrafine MNPs (e.g. Au, Ag, Pt, Pd, $Au_{0.4}Pt_{0.6}$, $Au_{0.4}Pd_{0.6}$, and $Au_{0.3}Pt_{0.3}Pd_{0.4}$) were exclusively anchored on the external surface of 2D MOF nanosheets [92]. The catalytic performance of the derived 2D composite catalysts was also evaluated by using 4-nitrophenol reduction in the liquid phase.

9.3.4 Three-Dimensional MNP/MOF Nanocomposites

In principle, 3D MOFs have all dimensions not being confined or limited to the nanoscale; that is, they are just conventional as-prepared MOF crystals without

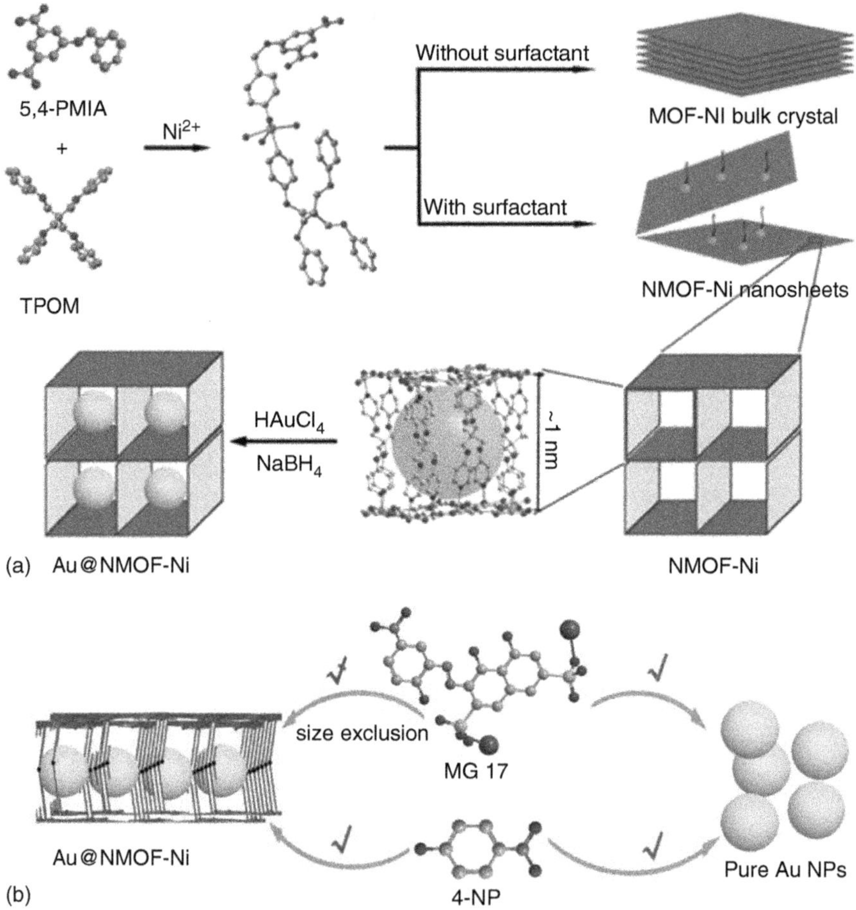

Figure 9.10 (a) Schematic illustration of preparations of bulk MOF, 2D MOF nanosheets, and Au@MOF composites, wherein the mentioned surfactant is polyvinylpyrrolidone (PVP). (b) Schematic illustration of the size-selective catalysis of Au@MOF, wherein the molecular sizes of 4-NP and Mordant Green 17 (MG 17) substrates were $2.4 \times 6.1\,Å^2$ and $7.2 \times 14.0\,Å^2$, respectively. Source: From Yan et al. [132] Copyright 2018, John Wiley and Sons. Reprinted with permission of John Wiley and Sons.

particular dimensional control. Besides the bulk powder forms of MOFs, 3D assemblies of MOFs can contain hierarchically porous monoliths, such as the incorporation of NPs, bundles of nanowires, and nanotubes as well as multi-nanolayers. Zhu et al. have reported the preparation of 3D hierarchical porous UiO-66 monoliths by using a Pickering high internal phase emulsion template (i.e. oil-in-water [o/w] emulsions) [136]. The monolith was very light, having a low density of c. 50 mg cm^{-3}. In addition, MOF monoliths can be fabricated by using the 3D printing method or mechanical compression method. For instance, Rezaei et al. have described MOF-74(Ni) and UTSA-16(Co) monoliths with uniform channels by a 3D printer [137]. Chahine et al. have synthesized MOF-177 monoliths by mechanically compressing with bulk densities three times more than its crystallographic density [138]. Furthermore, MOFs can be immobilized in a preformed monolith (e.g. silica) to form 3D MOFs. Likewise, there are several methods to incorporate both MOFs and MNPs in monolithic structures, such as *in situ* growth or postimmobilizations [139]. MNP/MOF composites with a rigid 3D monolith structure are highly desirable for large-scale applications because crystalline structures of unassembled MOFs are easily damaged under high backpressure in flow catalytic reactors. However, such 3D MNP/MOF composites have been rarely reported.

9.3.5 Other Representative Structures of MNP/MOF Composites

9.3.5.1 Core–Shell/Yolk–Shell Nanostructures

Core–shell structured MNP@MOF composites consist of an inner MNP core encapsulated by a porous MOF shell. As shown in Figure 9.11a–c, there are three different configurations between MNP and MOF in this type of core–shell structure, that is (a) one core coated by a shell, (b) multiple cores encapsulated in a matrix particle, and (c) yolk–shell structure with a hollow space between a movable core (i.e. yolk) and a shell. The main advantages of core–shell-structured MNP@MOF composites in catalysis include (i) providing a confined space for catalytic reaction, which might enhance the performance, (ii) enhancing the durability of the MNPs by the physical isolation, (iii) enriching the reactants on the MNP surface by MOF shell via coordination or π–π forces, and (iv) creating the interface for tandem catalysis [140]. In addition, because of the well-defined porous structure, MNP@MOF composites have also been used for size-selective catalysis by adjusting the pore size or thickness of the MOF shell. For instance, a kinetic analysis of the Knoevenagel condensation reactions with substrates of different sizes was examined under the same catalytic reactions, further confirming the size-selective catalysis of the MNP@MOF [142].

In order to facilitate the coating of MOFs on MNPs and eliminate the homogeneous growth, MNPs can be functionalized with organic groups/polymers that serve as a bridge layer. For example, Au NPs were functionalized with lipoic acid and modified by PEG-SH polymer before the growth of porphyrin MOFs, wherein the carboxyl groups on Au NP surface would adsorb zirconium cations for heterogeneous nucleation [143]. The control of the location of MNPs can be achieved by adjusting the addition time of MNPs. As shown in Figure 9.11d, Huo et al. have incorporated Au NPs (size of 13 nm) and Pt (size of 3.3 nm) in different locations in the ZIF-8 matrix by addition of the MNPs at the beginning (T_0) or after a certain time (T) during the MOF synthesis [141]. Furthermore, the spatial location of MNPs can directly dictate the catalytic properties of the final composites. In some reports, MNPs can be exploited as a self-template to grow porous MOF shell by partial dissolution of MNP cores, which serve as the metal ion source as well [12, 144]. For instance, Ni–Cu alloy NPs (Ni:Cu atomic ratio of 4.4/5.6) served as the metal ion sources when encapsulated into Ni–Cu–MOF in the absence of additional metal precursors. Importantly, the derived (Ni–Cu)@(Ni–Cu–MOF) with a high electrical conductivity exhibited enhanced catalytic performances and stability for oxygen evolution reaction (OER)

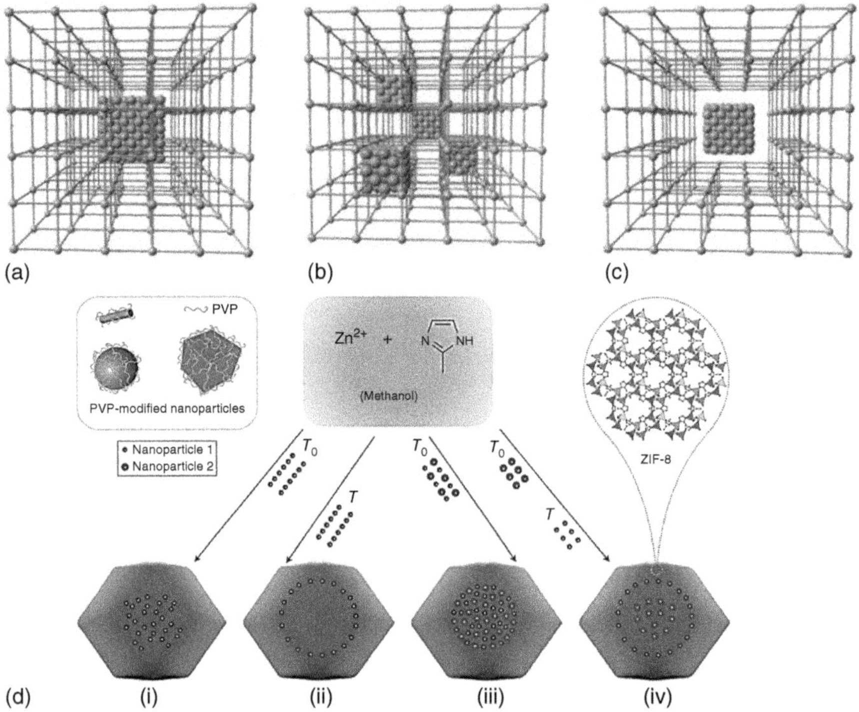

Figure 9.11 (a–c) Schematic illustration of three types of spatial configuration of MNPs and MOFs in core–shell structures: (a) single core–shell structure, (b) multiple cores in one shell structure, and (c) yolk–shell structure with void space. Source: Reproduced with permission from Ref. [140]. Copyright 2014, American Chemical Society. (d) Schematic illustrations of the controlled encapsulation of MNPs in ZIF-8 crystals with different spatial locations. Source: From Lu et al. [141]. Copyright 2012, Springer Nature. Reprinted with permission of Springer Nature

[144]. Sacrificial templates (such as silica) can be used to create a hollow space in core–shell-structured MNP@MOF forming a yolk–shell structure [145].

In addition, MNP@MOF composites can contain two or more different types of MNPs located in different parts of MOFs. For example, Pt/Au@Pd@UiO-66 contains Au@Pd NPs encapsulated in the center of UiO-66 crystals and Pt NPs dispersed well onto the external surface of UiO-66 [146]. These integrated catalysts were used as a high-performance catalysts for reverse water gas shift reaction to produce CO.

9.3.5.2 Sandwich-like Nanostructures

MOFs can serve as a support to immobilize MNPs or a coating shell to encapsulate them; both functions will prevent the detachment of the MNPs during particle collision. In sandwich-like nanostructures, MNPs are embedded between two MOF layers, wherein the inner and external MOF layers serve as a support and a coating shell, respectively. It was reported that metal–support interface sites are critically important for catalysis, especially involving the support as active sites in catalysis reaction [147]. Moreover, conformal coating supported MNP/MOF provides

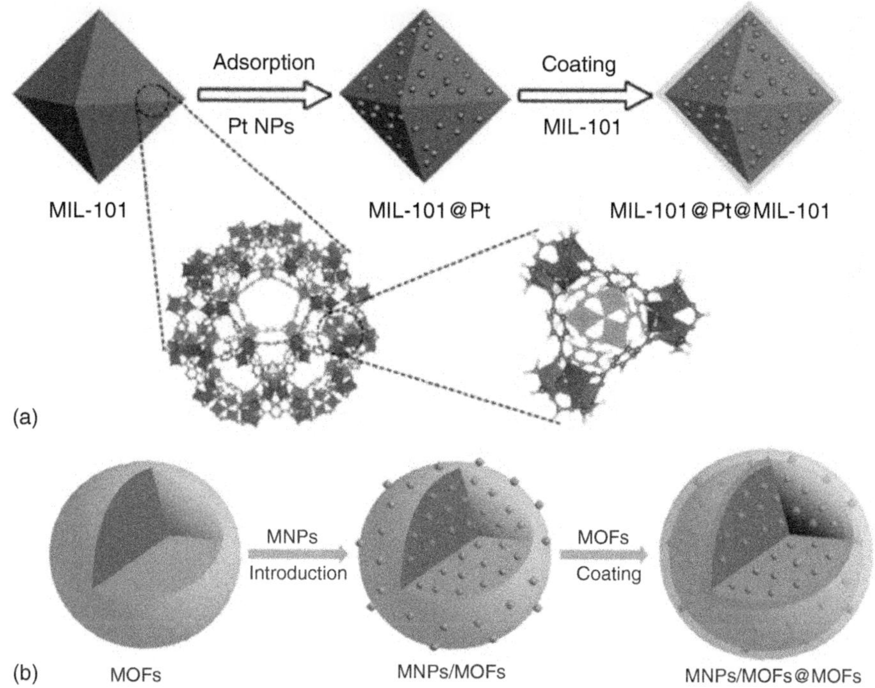

Figure 9.12 (a) Schematic illustration of preparation of MIL-101@Pt@MIL-101 with Pt NPs sandwiched between a core and a shell of MIL-101. Source: From Zhao et al. [148]. Copyright 2016, Springer Nature. Reprinted with permission of Springer Nature (b) Schematic illustrations of the stepwise preparation of MOF@Pt@MOF composites (MOF = MIL-100(Fe)). Source: From Liu et al. [149]. Copyright 2016, John Wiley and Sons. Reprinted with permission of John Wiley and Sons.

electron donation and steric restriction effects to the confined MNPs. Consequently, a lot of endeavors have been devoted to the fabrication of sandwich-structured MOF@MNP@MOF. As compared to the exposed MNP surface, sandwiching MNPs between two MOF layers leads to more abundant MOF/MNP interface sites. For instance, sandwich-like MIL-101@Pt@MIL-101 (see the preparation procedure in Figure 9.12a) was found as efficient catalysts for the selective hydrogenation of α,β-unsaturated aldehydes, which exhibited preferential hydrogenation of the C=O group over the C=C group [148].

It was suggested that the same type of MOFs in core and shell ensures a lattice match at the interface, which not only facilitates homoepitaxial growth during the catalyst fabrication but also enhances the substrate accessibility during the catalytic reaction. As shown in Figure 9.12b, Li and coworkers have prepared the MIL-100@Pt@MIL-100 composite for the selective hydrogenation of cinnamaldehyde to the desired cinnamyl alcohol [149]. Compared to single-side supported Pt/MOFs, the improved performance about activity and stability can be ascribed to both electron donation and confinement effects on the sandwiched Pt NPs. Density functional theory (DFT) calculation results by Tang and coworkers also reveal

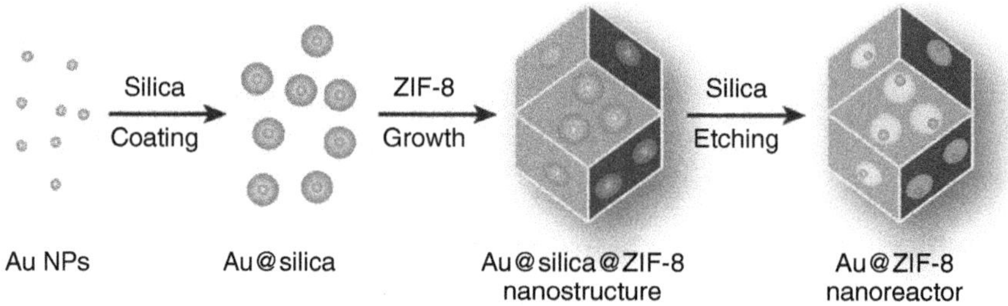

Figure 9.13 Schematic illustration of the synthesis of the Au@ZIF-8 nanoreactor. Source: From Wang et al. [150]. Copyright 2016, John Wiley and Sons. Reprinted with permission of John Wiley and Sons.

that MOF shells affect the interaction between substrates and the sandwiched MNPs, which will alter the thermodynamic equilibrium of the hydrogenation reaction [148].

9.3.5.3 Formation of Nanoreactors with a Central Cavity

Nanoreactor-like MNP/MOF composites consist of active MNPs as cores or yolks, porous MOFs as permeable shells, and hollow interior spaces if possible. The presence of void space will boost the diffusion of guest molecules from the bulk solution to the active phases and provide a venue for conducting selective confined catalytic reactions [9]. For instance, as shown in Figure 9.13, Su et al. have prepared Au@ZIF-8 nanoreactors by selectively etching silica layers with NaOH in the Au@silica@ZIF-8 template, wherein the size of Au NPs was 15 nm, the size of the reactor was around 800 nm, and the size of multiple hollow cavities was around 140 nm [150]. Importantly, Au@ZIF-8 nanoreactors showed much higher activity for the aerobic oxidation of alcohols with different molecular size than the core–shell Au@ZIF-8 nanostructures, further confirming the importance of void space between the Au NPs and the ZIF-8 shell. However, it is found that the MNPs inside a MOF-shelled nanoreactor are mostly unsupported; for this reason, we believe that the deliberately made supported MNPs that are simultaneously encapsulated in a MOF-shelled nanoreactor would be more desirable.

9.4 Summary and Conclusions

In summary, we have highlighted the recent studies on both synthetic strategies and catalytic applications of MNP/MOF composites. The general preparative methods for the MNP/MOF composites with various nanostructures are summarized. A fundamental goal of the combination between MNP and MOF within an integrated structure is manifesting the enhanced catalytic performance in a myriad of reactions, in terms of activity, selectivity, and stability. There is no doubt that integrated MNP/MOF composites will continue to attract much attention from researchers in the future. Therefore, the rational design of MNP/MOF composites should be carefully addressed in this field in order to maximize their collective effects, particularly the spatial configuration of MNP and MOF in the composites and the control of the interparticle spacing needs to be investigated more deeply. We believe that the issues addressed below are significant for the further developments of MNP/MOF composites in catalysis applications.

(i) Although a huge work has been done on the fabrication of MNP/MOF composites, it is still lacking a general protocol for accessing easily to MNP/MOF composites with global controls over size, shape, porosity, and surface properties, at multiple length scales. Especially, sporadic works have been focused on the preparation of 0D and 3D MNP/MOF composites. Moreover, some current techniques at laboratory scale require a tedious procedure and complex instrumentation. The issues about cost and related analysis have been largely ignored in most of the publications. Currently, it may be difficult to economically produce MNP/MOF in high quality and large scale. However, further research efforts toward the challenge are in urgent need, in order to promote this type of composite catalysts for industrial usage.

(ii) Some MNP/MOF composites may be catalytically inactive because of the lack of strong metal–support interaction [19, 61]. Few investigations have been devoted to understanding chemical interactions between MNPs and MOFs, which would be a key parameter to determining their synergistic effects. On the other hand, there are rare attempts on studying changes in the morphology, size, stability, electronic structure, and the chemical state of MNPs after immobilization on particle surface or within the pore spaces of MOFs, especially after their uses. These alternations have critical impacts on the catalytic activity and durability of MNPs because the MNPs might change surface charge and electronic structure of MOFs. Therefore, investigations of the chemical interactions and even the changes in the profound interactions during the catalytic reactions should be conducted in the future.

(iii) At the present time, the functionality of the composites mainly depends on the types of building blocks (viz., MNPs and MOFs). Integration of more than two active components in MNP/MOF, such as double MNPs or hybrid MNPs/metal oxide, will remarkably expand their catalytic applications and open the door to multistep reactions. For instance, both Cu and ZnO NPs can be integrated into MOF-5 by gas-phase infiltration method, which could be used as an excellent catalyst for methanol synthesis from the hydrogenation of CO/CO_2 [151]. The composites can be further assembled with other components, such as nucleic acids (DNAs or RNAs), to bring biocompatibility or bioaffinity, which could be applied to biocatalytic reactions in clinical diagnosis [152]. However, attention should be paid to the global control of spatial distributions of different MNPs in the MOF matrix in order to maximize their collective effects and realize the "make-on-demand" principle as designing INCs.

(iv) Last but not the least, compared to zeolite materials, MOFs have much lower stability, although their stability has been improved over the years. For instance, many reported MOFs are still suffering from hydrolysis of the weak metal–ligand coordination bonds even under an ambient atmospheric environment, which greatly hinders their large-scale applicability. It has been known that functionalization with hydrophobic surface and construction of MOFs with hydrophobic linkers are two main approaches to obtain robust MOFs. The addition of MNPs in MOF matrixes will also influence the surface properties and water resistance under their working conditions. To this end, the stability or alternation of these MNP/MOF composites including their compositional surface information, reacting particle size, and their structural configuration should be *in situ* characterized during their catalytic reactions in the future.

References

1 Howarth, A.J., Peters, A.W., Vermeulen, N.A. et al. (2017). *Chem. Mater.* 29: 26–39.
2 Li, H., Eddaoudi, M., O'Keeffe, M., and Yaghi, O.M. (1999). *Nature* 402: 276.
3 Kim, C.R., Uemura, T., and Kitagawa, S. (2016). *Chem. Soc. Rev.* 45: 3828–3845.
4 Farha, O.K., Eryazici, I., Jeong, N.C. et al. (2012). *J. Am. Chem. Soc.* 134: 15016–15021.
5 Koh, K., Wong-Foy, A.G., and Matzger, A.J. (2009). *J. Am. Chem. Soc.* 131: 4184–4185.
6 Burda, C., Chen, X., Narayanan, R., and El-Sayed, M.A. (2005). *Chem. Rev.* 105: 1025–1102.
7 Li, Z. and Zeng, H.C. (2014). *J. Am. Chem. Soc.* 136: 5631–5639.
8 Li, Z. and Zeng, H.C. (2013). *Chem. Mater.* 25: 1761–1768.
9 Zhan, G., Li, P., and Zeng, H.C. (2018). *Adv. Mater.* 30: 1802094.
10 Zeng, H.C. (2013). *Acc. Chem. Res.* 46: 226–235.
11 Zhan, G. and Zeng, H.C. (2016). *Coord. Chem. Rev.* 320–321: 181–192.
12 Li, G., Zhao, S., Zhang, Y., and Tang, Z. (2018). *Adv. Mater.* 30: 1800702.
13 Zhang, Y.-Y., Li, J.-X., Ding, L.-L. et al. (2018). *Inorg. Chem.* 57: 13586–13593.
14 Li, X., Guo, Z., Xiao, C. et al. (2014). *ACS Catal.* 4: 3490–3497.
15 He, L., Liu, Y., Liu, J. et al. (2013). *Angew. Chem. Int. Ed.* 52: 3741–3745.
16 Zhang, W., Lu, G., Cui, C. et al. (2014). *Adv. Mater.* 26: 4056–4060.
17 Liu, Y. and Tang, Z. (2013). *Adv. Mater.* 25: 5819–5825.
18 Yang, Q., Xu, Q., and Jiang, H.-L. (2017). *Chem. Soc. Rev.* 46: 4774–4808.
19 Hermes, S., Schröter, M.-K., Schmid, R. et al. (2005). *Angew. Chem. Int. Ed.* 44: 6237–6241.
20 Li, X., Zhang, B., Tang, L. et al. (2017). *Angew. Chem. Int. Ed.* 56: 16371–16375.
21 Roduner, E. (2006). *Chem. Soc. Rev.* 35: 583–592.
22 Haruta, M. (1997). *Catal. Today* 36: 153–166.
23 Valden, M., Lai, X., and Goodman, D.W. (1998). *Science* 281: 1647–1650.
24 Xu, R., Wang, D., Zhang, J., and Li, Y. (2006). *Chem. Asian J.* 1: 888–893.
25 Grzelczak, M., Pérez-Juste, J., Mulvaney, P., and Liz-Marzán, L.M. (2008). *Chem. Soc. Rev.* 37: 1783–1791.
26 Wu, B. and Zheng, N. (2013). *Nano Today* 8: 168–197.

27 Yang, H., Bradley, S.J., Chan, A. et al. (2016). *J. Am. Chem. Soc.* 138: 11872–11881.

28 Zhan, G. and Zeng, H.C. (2015). *Chem. Mater.* 27: 726–734.

29 Gates, B.C. (2000). *J. Mol. Catal. A* 163: 55–65.

30 Schmid, G., Bäumle, M., Geerkens, M. et al. (1999). *Chem. Soc. Rev.* 28: 179–185.

31 Zhang, X., Lai, Z., Liu, Z. et al. (2015). *Angew. Chem.* 127: 5515–5518.

32 Chen, J.S., Zhu, T., Yang, X.H. et al. (2010). *J. Am. Chem. Soc.* 132: 13162–13164.

33 Tian, J., Liu, Q., Asiri, A.M. et al. (2015). *Sens. Actuators, B* 216: 453–460.

34 Novoselov, K.S., Jiang, D., Schedin, F. et al. (2005). *Proc. Natl. Acad. Sci. U. S. A.* 102: 10451–10453.

35 Swihart, M.T. (2003). *Curr. Opin. Colloid Interface Sci.* 8: 127–133.

36 Panigrahi, S., Kundu, S., Ghosh, S. et al. (2004). *J. Nanopart. Res.* 6: 411–414.

37 Kelly, K.L., Coronado, E., Zhao, L.L., and Schatz, G.C. (2003). *J. Phys. Chem. B* 107: 668–677.

38 Nikam, A.V., Prasad, B.L.V., and Kulkarni, A.A. (2018). *CrystEngComm* 20: 5091–5107.

39 Wegner, K., Walker, B., Tsantilis, S., and Pratsinis, S.E. (2002). *Chem. Eng. Sci.* 57: 1753–1762.

40 Marchand, P., Hassan, I.A., Parkin, I.P., and Carmalt, C.J. (2013). *Dalton Trans.* 42: 9406–9422.

41 Zoval, J.V., Stiger, R.M., Biernacki, P.R., and Penner, R.M. (1996). *J. Phys. Chem.* 100: 837–844.

42 Tian, N., Zhou, Z.-Y., Sun, S.-G. et al. (2007). *Science* 316: 732–735.

43 Chang, S.-S., Shih, C.-W., Chen, C.-D. et al. (1999). *Langmuir* 15: 701–709.

44 Huang, J., Lin, L., Sun, D. et al. (2015). *Chem. Soc. Rev.* 44: 6330–6374.

45 Iravani, S. (2011). *Green Chem.* 13: 2638–2650.

46 Narayanan, K.B. and Sakthivel, N. (2010). *Adv. Colloid Interface Sci.* 156: 1–13.

47 Lee, J., Farha, O.K., Roberts, J. et al. (2009). *Chem. Soc. Rev.* 38: 1450–1459.

48 Horike, S., Dinca, M., Tamaki, K., and Long, J.R. (2008). *J. Am. Chem. Soc.* 130: 5854–5855.

49 Zhan, G. and Zeng, H.C. (2016). *Chem. Commun.* 52: 8432–8435.

50 Fujita, M., Kwon, Y.J., Washizu, S., and Ogura, K. (1994). *J. Am. Chem. Soc.* 116: 1151–1152.

51 Gándara, F., Gomez-Lor, B., Gutiérrez-Puebla, E. et al. (2008). *Chem. Mater.* 20: 72–76.

52 Aguirre-Díaz, L.M., Gándara, F., Iglesias, M. et al. (2015). *J. Am. Chem. Soc.* 137: 6132–6135.

53 Bhattacharjee, S., Choi, J.-S., Yang, S.-T. et al. (2010). *J. Nanosci. Nanotechnol.* 10: 135–141.

54 Yuan, S., Qin, J.-S., Xu, H.-Q. et al. (2018). *ACS Cent. Sci.* 4: 105–111.

55 Zhu, J., Li, P.-Z., Guo, W. et al. (2018). *Coord. Chem. Rev.* 359: 80–101.

56 Hasegawa, S., Horike, S., Matsuda, R. et al. (2007). *J. Am. Chem. Soc.* 129: 2607–2614.

57 Liu, H., Xi, F.-G., Sun, W. et al. (2016). *Inorg. Chem.* 55: 5753–5755.
58 Tan, Y.C. and Zeng, H.C. (2018). *Nat. Commun.* 9: 4326.
59 Tan, Y.C. and Zeng, H.C. (2017). *Adv. Funct. Mater.* 27: 1703765.
60 Liu, W., Zhu, L., Jiang, Y. et al. (2018). *Chem. Mater.* 30: 1686–1694.
61 Fan, L., Zhao, F., Huang, Z. et al. (2019). *Appl. Catal., A* 572: 34–43.
62 Tsumori, N., Chen, L., Wang, Q. et al. (2018). *Chem* 4: 845–856.
63 Deng, H., Grunder, S., Cordova, K.E. et al. (2012). *Science* 336: 1018–1023.
64 Xuan, W., Zhu, C., Liu, Y., and Cui, Y. (2012). *Chem. Soc. Rev.* 41: 1677–1695.
65 Peng, L., Zhang, J., Xue, Z. et al. (2014). *Nat. Commun.* 5: 4465.
66 Guillerm, V., Xu, H., Albalad, J. et al. (2018). *J. Am. Chem. Soc.* 140: 15022–15030.
67 Yaghi, O.M., O'Keeffe, M., Ockwig, N.W. et al. (2003). *Nature* 423: 705.
68 Ji, L., Jin, Y., Wu, K. et al. (2018). *Anal. Chim. Acta* 1031: 60–66.
69 Yang, J.-M., Liu, Q., and Sun, W.-Y. (2014). *Microporous Mesoporous Mater.* 190: 26–31.
70 Wang, S., McGuirk, C.M., d'Aquino, A. et al. (2018). *Adv. Mater.* 30: 1800202.
71 Guo, C., Zhang, Y., Guo, Y. et al. (2018). *Chem. Commun.* 54: 252–255.
72 Nordin, N.A.H.M., Ismail, A.F., Mustafa, A. et al. (2014). *RSC Adv.* 4: 33292–33300.
73 Park, J., Jiang, Q., Feng, D. et al. (2016). *J. Am. Chem. Soc.* 138: 3518–3525.
74 Stock, N. and Biswas, S. (2012). *Chem. Rev.* 112: 933–969.
75 Li, W.-J., Tu, M., Cao, R., and Fischer, R.A. (2016). *J. Mater. Chem. A* 4: 12356–12369.
76 Senthil Kumar, R., Senthil Kumar, S., and Anbu Kulandainathan, M. (2013). *Microporous Mesoporous Mater.* 168: 57–64.
77 Haque, E., Khan, N.A., Park, J.H., and Jhung, S.H. (2010). *Chem. Eur. J.* 16: 1046–1052.
78 Abbasi, A.R. and Rizvandi, M. (2018). *Ultrason. Sonochem.* 40: 465–471.
79 Vaitsis, C., Sourkouni, G., and Argirusis, C. (2018). *Ultrason. Sonochem.* https://doi.org/10.1016/j.ultsonch.2018.11.004.
80 Feng, X., Muhammad, F., Sun, F. et al. (2018). *Cryst. Growth Des.* https://doi.org/10.1021/acs.cgd.8b01442.
81 Sakamoto, H., Matsuda, R., and Kitagawa, S. (2012). *Dalton Trans.* 41: 3956–3961.
82 Zhang, B., Zhang, J., and Han, B. (2016). *Chem. Asian J.* 11: 2610–2619.
83 Parnham, E.R. and Morris, R.E. (2007). *Acc. Chem. Res.* 40: 1005–1013.
84 Lin, Z., Slawin, A.M.Z., and Morris, R.E. (2007). *J. Am. Chem. Soc.* 129: 4880–4881.
85 Lin, Z., Wragg, D.S., and Morris, R.E. (2006). *Chem. Commun.*: 2021–2023. https://doi.org/10.1039/B600814C.
86 Zhang, J., Han, B., Li, J. et al. (2011). *Angew. Chem. Int. Ed.* 50: 9911–9915.
87 Yu, H., Xu, D., and Xu, Q. (2015). *Chem. Commun.* 51: 13197–13200.
88 Sang, X., Zhang, J., Peng, L. et al. (2015). *ChemPhysChem* 16: 2317–2321.
89 Shang, W., Kang, X., Ning, H. et al. (2013). *Langmuir* 29: 13168–13174.

90 Hatakeyama, W., Sanchez, T.J., Rowe, M.D. et al. (2011). *ACS Appl. Mater. Interfaces* 3: 1502–1510.

91 Zhan, G. and Zeng, H.C. (2017). *Chem. Commun.* 53: 72–81.

92 Zhan, G. and Zeng, H.C. (2016). *Adv. Funct. Mater.* 26: 3268–3281.

93 Falcaro, P., Hill, A.J., Nairn, K.M. et al. (2011). *Nat. Commun.* 2: 237.

94 Zhao, J., Nunn, W.T., Lemaire, P.C. et al. (2015). *J. Am. Chem. Soc.* 137: 13756–13759.

95 Wang, Y., Ren, J., Deng, K. et al. (2000). *Chem. Mater.* 12: 1622–1627.

96 Nag, A., Kovalenko, M.V., Lee, J.-S. et al. (2011). *J. Am. Chem. Soc.* 133: 10612–10620.

97 Wu, B., Huang, H., Yang, J. et al. (2012). *Angew. Chem. Int. Ed.* 51: 3440–3443.

98 Guo, Z., Xiao, C., Maligal-Ganesh, R.V. et al. (2014). *ACS Catal.* 4: 1340–1348.

99 Mukoyoshi, M., Kobayashi, H., Kusada, K. et al. (2015). *Chem. Commun.* 51: 12463–12466.

100 Liu, H., Liu, Y., Li, Y. et al. (2010). *J. Phys. Chem. C* 114: 13362–13369.

101 Xiao, J.-D., Shang, Q., Xiong, Y. et al. (2016). *Angew. Chem. Int. Ed.* 55: 9389–9393.

102 Zhou, J., Wang, P., Wang, C. et al. (2015). *ACS Nano* 9: 6951–6960.

103 Khaletskaya, K., Reboul, J., Meilikhov, M. et al. (2013). *J. Am. Chem. Soc.* 135: 10998–11005.

104 Mao, H., Zhang, W., Zhou, W. et al. (2017). *ACS Appl. Mater. Interfaces* 9: 24649–24654.

105 Rösler, C., Dissegna, S., Rechac, V.L. et al. (2017). *Chemistry (Weinheim an der Bergstrasse, Germany)* 23: 3583–3594.

106 Gao, Q., Xu, S., Guo, C. et al. (2018). *ACS Appl. Mater. Interfaces* 10: 16059–16065.

107 Bakuru, V.R. and Kalidindi, S.B. (2017). *Chem. Eur. J.* 23: 16456–16459.

108 Tang, L., Shi, J., Wu, H. et al. (2017). *Nanotechnology* 28: 365604.http://stacks.iop.org/0957-4484/28/i=36/a=365604

109 Hermes, S., Schröder, F., Amirjalayer, S. et al. (2006). *J. Mater. Chem.* 16: 2464–2472.

110 Ishida, T., Nagaoka, M., Akita, T., and Haruta, M. (2008). *Chem. Eur. J.* 14: 8456–8460.

111 Houk, R.J.T., Jacobs, B.W., Gabaly, F.E. et al. (2009). *Nano Lett.* 9: 3413–3418.

112 Jiang, H.-L., Liu, B., Akita, T. et al. (2009). *J. Am. Chem. Soc.* 131: 11302–11303.

113 Proch, S., Herrmannsdörfer, J., Kempe, R. et al. (2008). *Chem. Eur. J.* 14: 8204–8212.

114 Coupry, D.E., Butson, J., Petkov, P.S. et al. (2016). *Chem. Commun.* 52: 5175–5178.

115 Zhu, Q.-L., Li, J., and Xu, Q. (2013). *J. Am. Chem. Soc.* 135: 10210–10213.

116 Zhen, W., Gao, H., Tian, B. et al. (2016). *ACS Appl. Mater. Interfaces* 8: 10808–10819.

117 Kung, C.-W., Audu, C.O., Peters, A.W. et al. (2017). *ACS Energy Lett.* 2: 2394–2401.

118 Liu, H., Chang, L., Chen, L., and Li, Y. (2015). *J. Mater. Chem. A* 3: 8028–8033.

119 Chen, L., Chen, X., Liu, H., and Li, Y. (2015). *Small* 11: 2642–2648.

120 Liu, H., Chang, L., Bai, C. et al. (2016). *Angew. Chem. Int. Ed.* 55: 5019–5023.

121 Yamamoto, H., Yano, H., Kouchi, H. et al. (2012). *Nanoscale* 4: 4148–4154.

122 Mullin, J.W. (2001). *Crystallization*, 4e. Boston: Butterworth Heinemann.

123 Seoane, B., Castellanos, S., Dikhtiarenko, A. et al. (2016). *Coord. Chem. Rev.* 307: 147–187.

124 Park, K.S., Ni, Z., Côté, A.P. et al. (2006). *Proc. Natl. Acad. Sci.* 103: 10186–10191.

125 Wang, J., Fan, Y., Lee, H.-w. et al. (2018). *ACS Appl. Nano Mater.* 1: 3747–3753.

126 Llabrés i Xamena, F.X., Corma, A., and Garcia, H. (2007). *J. Phys. Chem. C* 111: 80–85.

127 Aguilera-Sigalat, J. and Bradshaw, D. (2016). *Coord. Chem. Rev.* 307: 267–291.

128 Kim, H. and Lah, M.S. (2017). *Dalton Trans.* 46: 6146–6158.

129 Volosskiy, B., Niwa, K., Chen, Y. et al. (2015). *ACS Nano* 9: 3044–3049.

130 Song, X., Sun, H., Cao, X. et al. (2016). *RSC Adv.* 6: 112451–112454.

131 Bétard, A. and Fischer, R.A. (2012). *Chem. Rev.* 112: 1055–1083.

132 Yan, R., Zhao, Y., Yang, H. et al. (2018). *Adv. Funct. Mater.* 28: 1802021.

133 Jiang, M., Li, J., Cai, X. et al. (2018). *Nanoscale* 10: 19774–19780.

134 Li, Y., Gao, Z., Chen, F. et al. (2018). *ACS Appl. Mater. Interfaces* 10: 30930–30935.

135 Huang, Y., Zhao, M., Han, S. et al. (2017). *Adv. Mater.* 29: 1700102.

136 Zhu, H., Zhang, Q., and Zhu, S. (2016). *Chem. Eur. J.* 22: 8751–8755.

137 Thakkar, H., Eastman, S., Al-Naddaf, Q. et al. (2017). *ACS Appl. Mater. Interfaces* 9: 35908–35916.

138 Zacharia, R., Cossement, D., Lafi, L., and Chahine, R. (2010). *J. Mater. Chem.* 20: 2145–2151.

139 Lv, Y., Tan, X., and Svec, F. (2017). *J. Sep. Sci.* 40: 272–287.

140 Hu, P., Morabito, J.V., and Tsung, C.-K. (2014). *ACS Catal.* 4: 4409–4419.

141 Lu, G., Li, S., Guo, Z. et al. (2012). *Nat. Chem.* 4: 310–316.

142 He, H., Xue, Y.-Q., Wang, S.-Q. et al. (2018). *Inorg. Chem.* 57: 15062–15068.

143 Zeng, J.-Y., Zhang, M.-K., Peng, M.-Y. et al. (2018). *Adv. Funct. Mater.* 28: 1705451.

144 Ma, X., Qi, K., Wei, S. et al. (2019). *J. Alloys Compd.* 770: 236–242.

145 Li, B., Ma, J.-G., and Cheng, P. (2018). *Angew. Chem. Int. Ed.* 57: 6834–6837.

146 Zheng, Z., Xu, H., Xu, Z., and Ge, J. (2018). *Small* 14: 1702812.

147 Cargnello, M., Doan-Nguyen, V.V.T., Gordon, T.R. et al. (2013). *Science* 341: 771–773.

148 Zhao, M., Yuan, K., Wang, Y. et al. (2016). *Nature* 539: 76.

149 Liu, H., Chang, L., Chen, L., and Li, Y. (2016). *ChemCatChem* 8: 946–951.

150 Wang, S., Fan, Y., Teng, J. et al. (2016). *Small* 12: 5702–5709.

151 Müller, M., Hermes, S., Kähler, K. et al. (2008). *Chem. Mater.* 20: 4576–4587.

152 Guo, C., Su, F., Song, Y. et al. (2017). *ACS Appl. Mater. Interfaces* 9: 41188–41199.

10

Silica-Supported Nanoparticles as Heterogeneous Catalysts

Mahak Dhiman, Baljeet Singh, and Vivek Polshettiwar

Tata Institute of Fundamental Research (TIFR), Homi Bhabha Road, Colaba, Mumbai, Maharashtra 400005, India

10.1 Introduction

Silica has attracted a great deal of attention owing to its unique features, which make it particularly an interesting support for the immobilization of catalysts such as metal nanoparticles (NPs). Some of the key features of silica include high chemical, thermal, and mechanical stability; high surface area; tunable pore size; high adsorption capacity; and a porous network for efficient dispersion of metal precursors or metal NPs and diffusion of substrates. Depending on the pore sizes, silica has been classified into three different categories: microporous (pore size up to 2 nm), mesoporous (pore size 2–50 nm), and macroporous (pores larger than 50 nm).

In the field of heterogeneous catalysis, mesoporous silica finds a great utilization as it possesses an extremely large surface area within a relatively small volume of material. One of the first reports on the synthesis of high surface area mesoporous silica appeared in 1992 by Mobil Research and Development Corporation [1]. They synthesized mesoporous silica (named MCM-41) using liquid crystal template mechanism, which gave rise to materials with hexagonal pores tunable in the range from 3 to 10 nm depending on the choice of cationic surfactants. MCM-41 became one of the most widely explored materials, and further work led to the synthesis of MCM-48 with a cubic arrangement and MCM-50 with a lamellar arrangement [2]. With a completely different strategy, Stucky and coworkers used nonionic triblock copolymers such as alkyl poly(ethylene oxide) (PEO) oligomeric surfactants and poly(alkylene oxide) block copolymers to synthesize SBA-11 (cubic), SBA-12 (3-D hexagonal), SBA-15 (hexagonal), and SBA-16 (cubic cage-structured) class of mesoporous silica [3]. These silica are different from the MCM family as they display larger pores (5–30 nm) and thicker silica walls. FSM-16, the silica synthesized using quaternary ammonium surfactant as a template and layered polysilicate kanemite, presents another type of structure being made of folded sheets of silica [4]. Later in 2010, Polshettiwar et al. prepared dendritic fibrous nanosilica (DFNS, previously known as KCC-1) based on a cyclohexane–water emulsion system [5]. DFNS possessed a unique dendrimer-like structure with centrally radial fibrous channels that

Nanoparticles in Catalysis: Advances in Synthesis and Applications, First Edition.
Edited by Karine Philippot and Alain Roucoux.

Table 10.1 Different types of silica and their characteristics.

Name	Surface area ($m^2\ g^{-1}$)	Pore size (nm)	Pore volume ($cm^3\ g^{-1}$)	References
COK-12	400–550	5–9	0.45–0.88	[6]
DFNS	450–1200	3–25	0.54–2.18	[5]
FSM-16	200–1100	1–4	0.25–0.76	[7]
KIT-6	300–750	4–18	0.57–1.44	[8]
MCM-41	≥1000	3–10	0.7–1.2	[2]
SBA-15	≥1050	5–50	0.75–1.15	[3]
Stober	10–350	1–6	0.01–0.21	[9]
TUD-1	500–1000	5–20	0.6–1.7	[10]

increased gradually from the particle's interior to the surface. Table 10.1 presents some structural characteristics of the cited silica together with those of other types of silica.

10.2 Deposition Methods of Metal NPs

Various methods can be utilized to immobilize metal NPs over different silica supports. This section will briefly discuss synthetic protocols that are employed to support metal NPs over silica. For better understanding and clarity, platinum (Pt) NPs have been chosen as an example to illustrate the different strategies applied focusing on unique support, namely, SBA-15. This section will thus describe the deposition of Pt NPs over SBA-15 silica support by wet impregnation, deposition–precipitation, colloidal immobilization, solid-state grinding, and postsynthetic grafting methods, which can be ideally generalized for various silica supports and diverse metal NPs.

10.2.1 Wet Impregnation Method

The wet impregnation technique requires the mixing of silica with an excess of metal salt solution. Once the maximum adsorption of metal ions over silica occurs, the solid is separated from the solution either by centrifugation or evaporation of solvent. In this method, the amount of metal loading on silica significantly depends on the concentration of the impregnating solution, the porosity of the support, and the precursor affinity onto the silica surface. When strong precursor–support interactions occur through chemical or physical forces (such as surface hydrolysis, ligand substitution, ion exchange, and electrostatic attraction), the quantity of metal NPs immobilized onto the support is generally higher. For example, Chen and coworkers prepared SBA-15-supported Pt NPs by mixing SBA-15 with an aqueous solution of H_2PtCl_6, which led to adsorption of Pt^{4+} ions on the silica support [11].

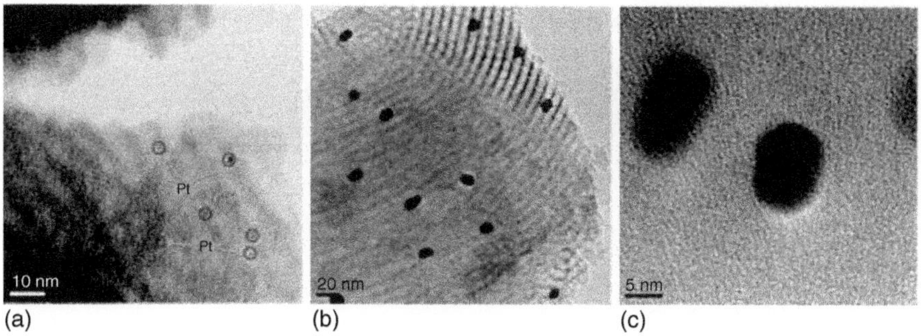

Figure 10.1 TEM images of (a) 0.2 wt% Pt/SBA-15 and (b), (c) 0.2 wt% Pt/SBA-15 catalysts prepared by wet impregnation technique. Source: Reproduced with permission from Wu et al. [11]. Copyright 2015, Royal Society of Chemistry.

The final Pt/SBA-15 catalyst was obtained by vacuum–calcination of the impregnated Pt^{4+}/SBA-15 at 673 K for five hours and subsequent reduction under H_2 at 673 K for five hours. Interestingly, the size and shape of Pt NPs on SBA-15 could be controlled during the calcination step by either heating in vacuum or air. The vacuum–calcination/H_2-reduction-treated materials formed small 2D clusters and single atoms of Pt on SBA-15, while thermal treatment involving air–calcination/H_2 reduction materials resulted in ~4.6 nm rod-like Pt particles (Figure 10.1).

10.2.2 Deposition–Precipitation Method

The principle of the deposition–precipitation method is primarily based on the solubility of metal salts at certain conditions. As long as metal ions can be precipitated, the respective metal-based nanocatalysts can be prepared using this method, which includes three steps: supersaturation, nucleation, and growth of metal NPs on the support. The supersaturation stage in the aqueous solution is achieved either by increasing the concentration of metal ions by evaporating solvent or by lowering the temperature or increasing the pH of the reacting media. Once the concentration of metal ions reaches the supersaturation, nucleation starts and these small nuclei are exclusively deposited onto the silica surface. Generally, precipitation is in hydroxide form, which is later easily converted to the oxide by thermal treatment in air and then reduced under hydrogen to obtain metal particles. It is important to note that the driving force for nucleation and growth of the precipitated metal species is the force of attraction between the metal and the support. The major advantage of this method is that it results in high loading, high dispersion, and uniform NP size. For example, Blekkan and coworkers used this method for the synthesis of SBA-15-supported Pt NPs [12]. Firstly, SBA-15 was dispersed in water, and pH was adjusted to 3.5 using HNO_3 acid. Then, an aqueous solution of urea and an aqueous solution of $Pt(NH_3)_4(OH)_2$ to SBA-15 suspension were added. The mixture was heated to 90 °C, and the pH was monitored and adjusted to 3.5 throughout the duration of the reaction. After 20 hours of heating at 90 °C, the reaction medium was cooled down to room temperature. The silica material was then filtered and washed several times with water before drying at 80 °C for 70 hours. The

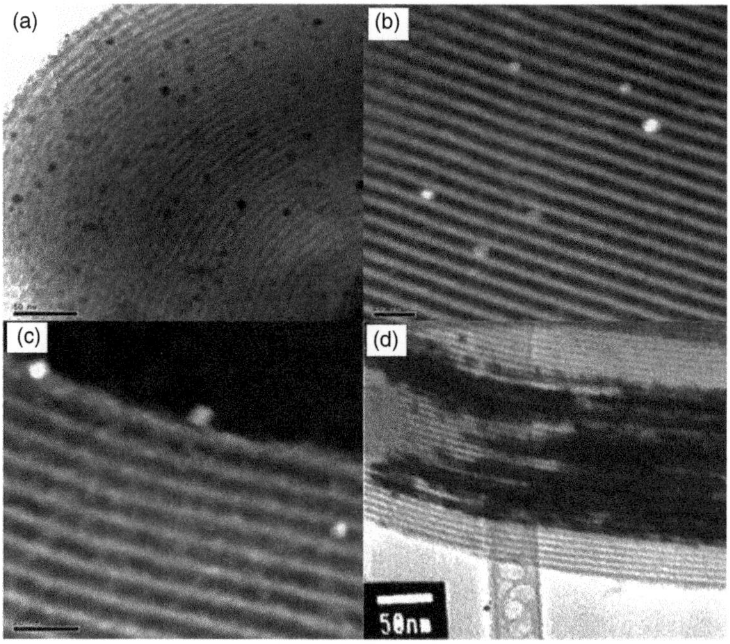

Figure 10.2 TEM images of Pt/SBA-15 prepared by the deposition–precipitation method. (b), (c) Images taken in annular dark-field (ADF) scanning transmission electron microscopy (STEM) imaging mode. The bars in micrographs (b), (c) and (a), (d) are 20 and 50 nm, respectively. Source: Adapted with permission from Chytil et al. [12]. Copyright 2007, Springer.

as-synthesized material was later calcined at 300 °C in air for two hours to obtain Pt NPs immobilized on SBA-15 (Figure 10.2). A significant fraction of the Pt NPs ranging in size from 2 to 4 nm was present within the channels of SBA-15. However, particles of size greater than the pore width of SBA-15, estimated to be 8 nm by the Barrett–Joyner–Halenda (BJH) method, could not be accommodated within the channels of SBA-15 and were present outside the channels.

10.2.3 Colloidal Immobilization Method

In contrast to the aforementioned methods, colloidal immobilization method allows to get supported metal NPs of highly controlled sizes. Indeed, in this method, ligand-capped metal NPs are first prepared separately and then immobilized onto the silica support. Thus, metal NPs are weakly adsorbed onto the silica, and there is no strong bonding, which may often lead to leaching of active sites during catalytic reactions. For example, Somorjai and coworkers synthesized silica-supported monodispersed rhodium (Rh) and Pt NPs of about 1 nm in size using as coating agent dendrimers made of repetitive branched subunits of amine functionalities (Figure 10.3) [13]. Their protocol involved mixing of dendrimer solution with an aqueous solution of $RhCl_3$ or K_2PtCl_4. The complexation reaction between amines

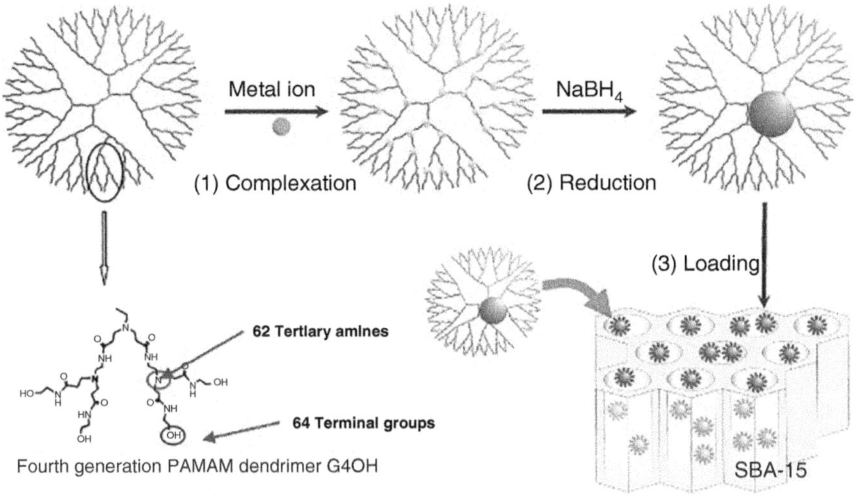

Figure 10.3 Metal NP encapsulation using dendrimeric approach, followed by loading onto a high surface area silica support, SBA-15. Source: Reproduced with permission from Huang et al. [13]. Copyright 2008, American Chemical Society.

and metal ions was studied by absorption spectroscopy, wherein a decrease in the intensities of bands at 200, 226, and 300 nm within the first two hours was attributed to ligand exchange reactions between chloride, water, and the internal amines within the dendrimer. Thereafter, freshly prepared $NaBH_4$ was injected to reduce metal ions to form metal NPs. The obtained colloidal solution was then mixed with SBA-15, and the mixture was sonicated before separation via centrifugation. Interestingly, upon centrifugation, the initial dark brown solution became colorless, while the solid turned brown, which indicated the deposition of the dendrimer-encapsulated Rh and Pt NPs onto the SBA-15 support. This approach represents an efficient way of synthesizing supported NPs without performing any post-treatment step such as calcination or H_2 reduction of the catalyst.

10.2.4 Solid-State Grinding Method

The solid-state grinding method relies on the direct incorporation of a metal salt to silica using the physical grinding principle. Once the metal salt is physically mixed with silica, it is calcined and exposed to a reductive gas environment such as H_2 at high temperatures in order to obtain the MNPs. Because no isolation or washing steps are involved, this method allows metal loading to the highest possible level. For example, Zhou and coworkers used $H_2PtCl_6{\cdot}6H_2O$ as the Pt precursor, which was introduced into SBA-15 [14]. The thoroughly mixed powder was calcined in flowing air at 300 °C for two hours, followed by reduction in H_2/Ar mixture gases at 200 °C to obtain SBA-15-supported Pt NPs. The transmission electron microscopy (TEM) images in Figure 10.4 show 0.5 wt% Pt on the as-synthesized SBA-15 denoted as 0.5PtAS. Similarly, 1.0PtAS (b), 3.0PtAS(c), 5.0PtAS (d), and

Figure 10.4 TEM images for the samples of (a) 0.5PtAS, (b) 1.0PtAS, (c) 3.0PtAS, (d) 5.0PtAS, (e) 10.0PtAS, and (f) 5.0PtCS (template free SBA-15). Scale bars represent 100 nm. Source: Reproduced with permission from Yin et al. [14]. Copyright 2017, Springer.

10.0PtAS (e) are shown. Also shown is 5 wt% Pt on template-free SBA-15, i.e. 5.0PtCS in Figure 10.4f.

10.2.5 Postsynthetic Grafting Method

In addition to previously described features of silica (see introduction part), one property that distinct silica from other support materials is its ability to get functionalized by various organic molecules. Organic functionalities grafted on silica can efficiently be used as anchoring groups to stabilize metal NPs on the surface. In this method, calcined silica supports are often functionalized by amines (e.g. (3-aminopropyl)triethoxysilane) and thiols (e.g. (3-mercaptopropyl)trimethoxysilane) as organic functional groups. The so-obtained functionalized silica can then be used to complex with metal ions, which on reduction will yield monodispersed metal NPs. In this method, the end group of a covalently attached functional motif on silica binds with metal NPs, providing a leach-proof and stable nature to metal NPs. For example, Yang et al. loaded Pt NPs over SBA-15 using an SBA-15 material functionalized with *N*-trimethoxysilylpropyl-*N*,*N*,*N*-trimethylammonium chloride and ethanolic solution of H_2PtCl_6 [15]. After impregnation time, the solid was separated from the reaction mixture by filtration and then washed with ethanol before drying in ambient air. Then, it was reduced in hydrogen flow at 373–573 K to yield uniform and highly dispersed Pt NPs in the hexagonal channels of SBA-15. In these synthesis conditions, the Pt particles in a size range of 1.2–3 nm were almost exclusively formed inside the channels of SBA-15 (Figure 10.5).

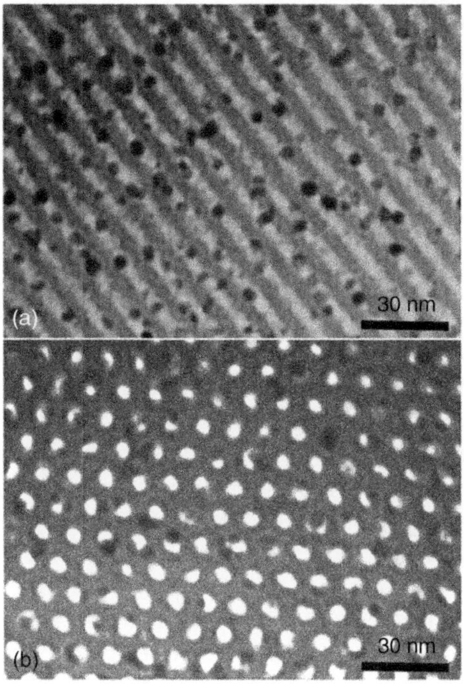

Figure 10.5 TEM images of Pt/SBA-15 prepared by postsynthetic grafting method viewing normal to the pore axis (a) and along the pore axis (b) of support SBA-15. Source: Reproduced with permission from Yang et al. [15]. Copyright 2003, American Chemical Society.

10.3 Application of Silica-Supported NPs in Catalysis

Silica-supported NPs are extensively explored in catalysis for a wide range of reactions with excellent yields, selectivity, stability, and rates. Because there is tremendous activity in the field of catalysis by silica-supported metal NPs, this section will focus on a selection of important catalytic processes involving nanomaterials made of noble metal NPs supported onto silica, i.e. oxidation, hydrogenation, and carbon–carbon (C–C) coupling reactions. These catalytic reactions are extensively used for the production of fine chemicals, fuels, chemical feedstocks, medicines, etc. [16].

10.3.1 Oxidation Reactions

10.3.1.1 CO Oxidation

Carbon monoxide (CO) oxidation has attracted tremendous interest because of its toxic nature. CO is often called the silent killer as even a small exposure to CO can be fatal because it has a high affinity to replace oxygen and binds to hemoglobin in blood cells [17]. It has also become a benchmark reaction for examining the activity of heterogeneous catalysts for oxidation reactions [18]. In the 1980s, silica-supported Au NPs between 2 and 5 nm have been demonstrated to be extremely active catalysts for CO oxidation in particular, thanks to Haruta's group work [19]. Schuth and coworkers also reported an active gold catalyst with well-dispersed Au NPs of size about 4.5 nm in the channels of SBA-15 [20]. Their nanocatalyst showed a reaction rate of 2.7×10^{-4} mmol g^{-1} $cats^{-1}$ for CO oxidation. However, sintering was observed around 100 °C as particles with 10–50 nm in size formed on the outer surface of the silica. This was attributed to the relatively weak interaction between Au and SBA-15. This observation led to further investigation on the role of pore size and structure of SBA-15, as carried out by Datye and coworkers, who could achieve control over the thermal sintering of Au NPs on mesoporous silica [21]. They deposited Au NPs onto different mesoporous silica supports having a range of pore sizes and pore arrangements (2-D hexagonal, 3-D hexagonal, and cubic). Before evaluation for CO oxidation at temperatures ranging from 25 to 400 °C, their materials were reduced at 200 °C in the flow of H_2. Interestingly, among all samples studied, SBA-15 with the thickest pore walls was found to be the most active at regulating the growth of the Au particles. Thus, it was proven that the walls of mesoporous materials play a considerable role in stabilizing Au NPs. Rombi and Ferino et al. used mercaptopropyltrimethoxysilane (MPS)-functionalized silica to support Au NPs and found that the presence of organic residues as well as the size of the as-prepared Au NPs has a significant effect on the catalytic activity for CO oxidation [22]. Other than Au, silver (Ag) NPs have also been reported to show the size-dependent enhancement for CO oxidation. Therefore, Bao and coworkers reported that the silica-supported Ag particles (c. 6–8 nm) had relatively high activity for CO catalytic oxidation [23]. Wang and coworkers then developed a novel and facile method for the synthesis of ultrasmall Ag NPs supported onto silica using the salicylaldimine Schiff base [24]. The obtained Ag NPs were finely dispersed on the silica support with average particle sizes in the range of 1.5–4.4 nm and narrow size distributions, exhibiting a high catalytic activity for complete CO oxidation. This was one of the first reports where direct observation of the Ag particle size effect was correlated with CO oxidation and which allowed to establish that the size in a range of 3–5 nm was favorable for CO oxidation by Ag NPs.

10.3.1.2 Alcohol Oxidation

Selective oxidation of alcohols to carbonyl compounds is another important reaction in modern organic chemistry. Oxidation of alcohols is used to be carried out using stoichiometric amounts of toxic and corrosive reagents [25]. In the past decade, active research was focused on the catalytic oxidation of alcohols using green reagents such as hydrogen peroxide, which leads to water as the sole by-product [26]. One major challenge in alcohol oxidation reactions is to achieve their partial oxidation into aldehydes, but as these latter are more active than alcohols, it is not easy to control their overoxidation into acids. In this regard, silica-supported noble metal NPs have been reported to carry out this transformation much efficiently. Wang et al. prepared gold NPs supported onto fluorous silica gel (FSG), which showed excellent activity as well as selectivity during the partial oxidation of substituted benzyl alcohols to the corresponding aldehydes [27]. They successfully reported that oxidation reactions occurred in an aqueous medium under base-free conditions with only water as the by-product. To further study the effect of surface area and establish the relation between pore structure and size on catalytic activity, Wilson and coworkers synthesized Pd-supported catalysts on four different silica

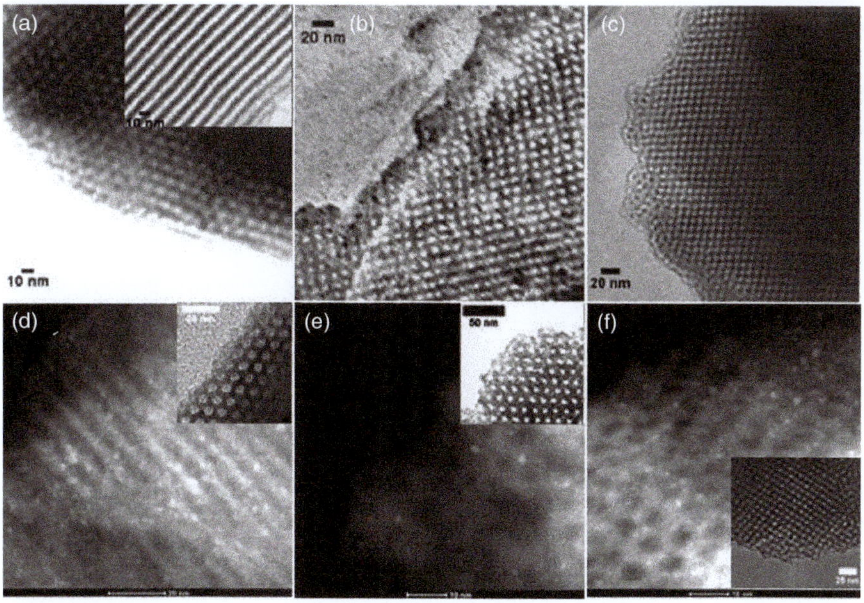

Figure 10.6 High-resolution transmission electron microscopy (HRTEM) bright-field images of (a) SBA-15, (b) SBA-16, and (c) KIT-6 supports and high angle annular dark field (HAADF)-STEM images of (d) 1 wt% Pd/SBA-15, (e) 1 wt% Pd/SBA-16, and (f) 1 wt%/KIT-6 catalysts, Pd NPs dispersed throughout mesopore networks (bright spots); insets show bright-field HRTEM images of SBA-15 and Pd-supported mesoporous silicas. Source: Reproduced with permission from Parlett et al. [28]. Copyright 2013, Wiley.

supports (Figure 10.6): a commercial, low surface area SiO_2 ($200\,m^2\,g^{-1}$); SBA-15, a high surface area support ($950\,m^2\,g^{-1}$); SBA-16 ($820\,m^2\,g^{-1}$); and KIT-6 ($940\,m^2\,g^{-1}$) [28]. Almost three to four times higher activity was observed for all the mesoporous synthesized supports over commercial silica because of the higher Pd dispersions and mean particle sizes of about 2.2, 1.6, and 1.5 nm for SBA-15, SBA-16, and KIT-6, respectively. Moreover, despite similar surface areas and pore diameters in SBA-15, SBA-16, and KIT-6, an additional enhancement factor was observed upon switching from the parallel and noninterconnected channels within SBA-15 to the interconnected pore networks of SBA-16 and KIT-6. The interconnected pores were shown to stabilize dispersed Pd NPs.

The catalytic activity for selective oxidation of crotyl alcohol and cinnamyl alcohol to corresponding aldehydes decreased in the order Pd/KIT-6 ≈ Pd/SBA-16 > Pd/SBA-15 > Pd/SiO_2. Pd/SBA-16 and Pd/KIT-6 resulted in 2–3 times more reactive than Pd/SBA-15, which in turn demonstrated a similar magnitude enhancement over the amorphous Pd/SiO_2. They also proved that instead of zero-valent palladium, highly dispersed palladium oxide NPs (as confirmed by X-ray photoemission spectroscopy [XPS]) were catalytically active species by normalizing the initial catalytic rates to the concentration of either surface palladium oxide or the total surface palladium metal. The constant turnover frequencies (TOFs) for crotyl alcohol and cinnamyl alcohol oxidation were observed after normalization by the

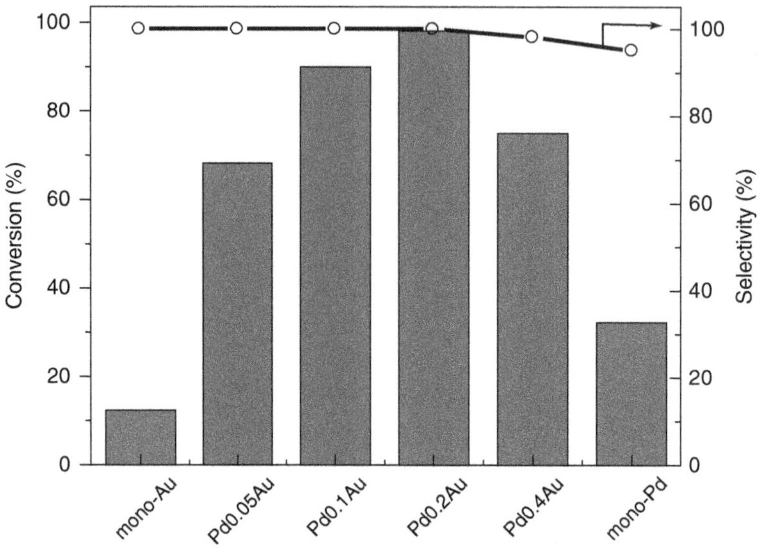

Figure 10.7 Performance of various Pd–Au composition catalysts for benzyl alcohol oxidation. Source: From Yang et al. [30]. Copyright 2013, Elsevier. Reprinted with permission of Elsevier.

palladium oxide surface concentration, whereas a continuously decaying TOF was obtained from normalization by palladium metal. This provided definitive evidence that surface palladium oxide and not metal Pd was the active species responsible for catalyzing allylic alcohol oxidation. In further attempts to increase the catalytic activities, doped mesoporous silica with various elements (Fe, Ti, Ce, etc.) or mesoporous silica grafted with functional groups (SH or NH_2), which limit the growth of gold NPs, has also been reported [29]. Yang et al. synthesized a series of Pd-promoted gold supported onto mesoporous silica using a "green" impregnation-H_2 reduction method [30]. Interestingly, they found that the addition of Pd to gold (Pd_nAu/mesoporous silica nanosphere [MSN], n – the molar ratio of Pd to Au) greatly improved the metal dispersion by inhibiting the sintering of NPs as a result of Pd–Au synergy, which further led to enhanced catalytic activity toward benzyl alcohol oxidation (Figure 10.7). Notably, only 12% and 25% conversions were obtained for monometallic Au and Pd supported on MSN silica, respectively, whereas for Pd-promoted Au catalysts, a maximum conversion of 98% could be achieved with TOF up to 1320 h^{-1}. Similarly, Rostamnia and Kholdi synthesized Pd NPs supported onto polyaniline functionalized hollow silica [31]. Their designed catalyst exhibited an excellent oxidative activity toward benzyl alcohol in aerobic balloon oxygen.

10.3.1.3 Hydrolysis of Silane

Oxidation/hydrolysis of silanes to silanols, which are building blocks for silicon-based polymers and nucleophilic coupling bricks in organic synthesis, is another important chemical reaction [32]. Various metal NPs such as rhodium

(Rh), copper (Cu), rhenium (Re), palladium (Pd), iridium (Ir), tungsten (W), and silver (Ag) NPs supported onto different supports have been employed to carry out silane oxidation. Interestingly, silica-supported Au NPs have performed exceedingly well. Zhang and coworkers prepared the Au/SiO_2 nanomaterial with high dispersion of gold NPs onto the silica support for its evaluation as a catalyst for the selective oxidation of various silanes [33]. Their work established that the Au/SiO_2 nanocatalyst outperforms other reducible oxide-supported gold catalysts owing to the affinity of the silica support for the silane substrates. Taking this work forward, Duan and coworkers embedded Au NPs within poly(1-(4-vinylbenzyl)-3-methylimidazolium chloride that acted as poly(ionic liquid) ligands in the SBA-15 support [34]. In addition to resulting in high catalytic activity and excellent selectivity, this protocol also favored reusability of the catalysts for the oxidation of a variety of silanes into their corresponding silanols under mild conditions.

Camargo and coworkers synthesized an Au catalyst that was very active for dimethylphenylsilane oxidation reaction with resulting TOF and turnover number (TON) of about 14 850 h^{-1} and 98 333, respectively [35]. Although the TON reported was high, it was not sufficient for making a catalyst of industrial interest because for that purpose, TON should exceed or reach around a million mark. Polshettiwar and coworkers thus reported the synthesis of DFNS-supported Au catalyst that resulted in a very high TON value of about 591 000 for the oxidation of dimethylphenylsilane [36]. Moreover, the figure of merit (FOM) defined as $\mathrm{FOM} = \frac{[(\mathrm{TOF}*N*S)]}{[(298+dT)*C]}$, where N is the number of cycles, S is the scale of the reaction, dT is the deviation in the reaction temperature, was calculated to be 633 mmol h^{-1} K^{-1}, which takes into account other key reaction parameters such as temperature, time, scale, and recyclability. Mechanistic study of gold-catalyzed silane oxidation was also carried out, leading to propose a two-step reaction mechanism. Silane was shown to first oxidatively add to the Au NPs, which further reacts with water molecules to yield silanol and hydrides onto the surface (Figure 10.8). Then, hydrides are released from the Au NP surface in the form of dihydrogen as the by-product.

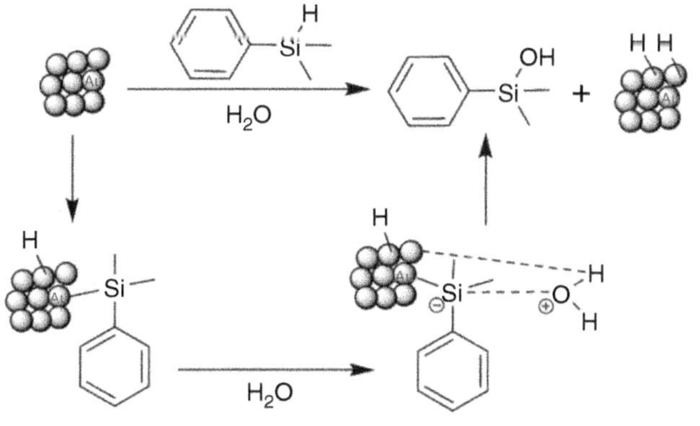

Figure 10.8 Mechanism of dimethylphenylsilane oxidation using Au NPs. Source: Reproduced with permission from Dhiman et al. [36]. Copyright 2017, Royal Society of Chemistry.

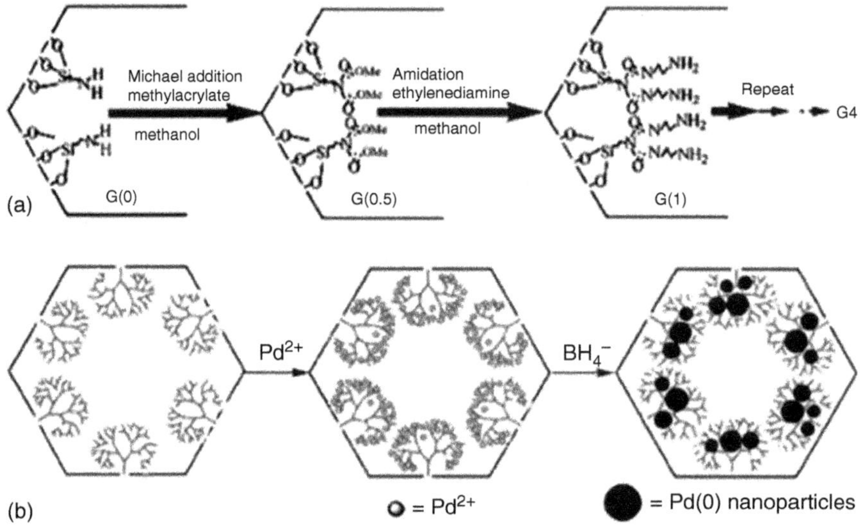

Figure 10.9 Preparation of dendrimer-functionalized SBA-15 and supported Pd NPs. Preparation of SBA-15 supported dendrimers. (b) Formation of Palladium nanoparticles. Source: Reproduced with permission from Jiang and Gao [37]). Copyright 2006, American Chemical Society.

10.3.2 Hydrogenation Reactions

Selective hydrogenation is a very important reaction to access chemicals of pharmaceutical interests. In order to obtain highly active and selective catalysts for such reactions, Gao and coworker reported a very unique and breakthrough work in 2005, where they used a dendrimer-functionalized silica to load Pd NPs for the reduction of double bonds in allyl alcohols [37]. The use of dendrimers as a coating agent posed a considerable problem during the isolation of the nanocatalysts after the catalytic reaction. Therefore, Gao and coworkers used amine-functionalized SBA-15 as a starting material in order to carry out a Michael-type addition reaction with methyl acrylate and form amino propionate esters groups in the pores of the silica. Subsequent amidation of the ester moieties with ethylenediamine completed one cycle of the reaction forming the first generation of dendrimer over silica. Repetition of these two reactions produced dendrimers of fourth generation (Figure 10.9a). After that, Pd(II) ions were introduced into the dendrimers inside the pores of SBA-15 by an impregnation method. Then, a reduction step with $NaBH_4$ resulted in the formation of dark brown powders of Pd(0)-Gn-PAMAM-SBA-15 (Figure 10.9b). These nanomaterials not only resulted in high catalytic activity for the hydrogenation of prop-2-en-1-ol but importantly the hydrogenation rate and selectivity could be controlled using different generations of dendrimers. Based on the different generations of dendrimers utilized, the extent of passivation of the NPs surface as well as accessibility of the precursor to the metal NPs can be tuned. Additionally, the catalysts were found to be stable enough to be recycled five times and preserved for one month under ambient condition, while maintaining their initial catalytic activities.

This protocol was further improved by Asefa and coworkers who also synthesized dendrimer-supported silica for the loading of Pd NPs in order to hydrogenate various alkenes and nitro groups, thus achieving ~100% conversion and high TONs under 10 bar H_2 pressure at r.t. [38]. In order to study the effect of spacer groups on the performance of MCM-41-supported Pt NPs, Lahiri and coworkers functionalized MCM-41 with $(EtO)_3SiCH_2Cl$, $(MeO)_3SiCH_2CH_2CH_2Cl$, and $(CH_3)Cl_2SiCH_2Cl$ [39]. Comparative catalytic hydrogenation studies of various alkenes were carried out wherein the best activities were obtained for the catalyst with chloropropyl ($(MeO)_3SiCH_2CH_2CH_2Cl$) as the spacer group. This observation proved that steric interaction between the spacer group and the incoming substrate played a critical role to provide higher activity. Mesoporous alumino-silicate (Na–Al–SBA-15) was also used by Boutros et al. to prepare supported Rh NPs using quaternary ammonium chloride (*N*,*N*-dimethyl-*N*-cetyl-*N*-(2-hydroxyethyl)ammonium chloride salt) as a stabilizing agent, rhodium chloride hydrate ($RhCl_3{\cdot}xH_2O$) as a metal precursor, and sodium borohydride ($NaBH_4$) as a reducing agent [40]. The use of this quaternary ammonium chloride stabilizer facilitated the adsorption of rhodium(III). Synthesized catalysts were screened for r.t. hydrogenation of styrene at 1 bar of H_2 showing exclusively 100% yield in ethylcyclohexane within six hours.

Understanding the structure-activity relationship of different metal NPs supported on silica is critical to realize efficient catalysts [41]. Polshettiwar and coworkers proposed a new approach for functionalizing silica by replacing expensive dendrimers with inexpensive polyethylenimine on DFNS, which was then loaded with Pd, Rh, Ru, and Pt NPs [42, 43]. The so-obtained nanocatalysts showed good catalytic activities for hydrogenation of styrene under 7 bar H_2 at r.t. and notably in less than two hours. Another approach of loading Pd NPs on silica ($Pd@SiO_2$) that consisted of using supercritical carbon dioxide ($scCO_2$)–acetone mixture was put forward by Bai and coworkers [44]. Porous SiO_2 and palladium(II) acetate ($Pd(OAc)_2$), as a Pd NPs precursor, were placed in a high-pressure cell, and CO_2 was pumped at 20 MPa and 333 K. The mixture was magnetically stirred for 24 hours before isolation and reduction treatment under H_2 at 473 K for five hours. The so-obtained $Pd@SiO_2$ catalysts exhibited higher turnover activity for the hydrogenation of styrene (619 h^{-1}) than conventional silica (347 h^{-1}) and $Pd@TiO_2$ (33 h^{-1}).

Because phenylacetylene is an unwanted feedstock in the styrene polymerization, semihydrogenation of phenylacetylene into styrene is another reaction of industrial importance [45]. Previously, Pd NPs supported onto TiO_2, carbon nanotubes, hexagonal mesoporous silica (HMS), and hydrotalcite have been shown to exhibit styrene selectivity up to 80% during hydrogenation of phenylacetylene under mild reaction conditions [46]. Because the phenylacetylene concentration must be less than 10 ppm, better catalysts are thus needed to improve this reaction [47]. In this regard, Papp et al. synthesized MCM-41-supported Pd NPs [48]. These Pd-MCM-41 catalysts were tested for the liquid-phase hydrogenation of both terminal and internal alkynes under mild reaction conditions. For the semihydrogenation of terminal alkynes, Pd-MCM-14 exhibited high activities and selectivities. Furthermore, these catalysts led to 100% selectivity of alkene formation for the hydrogenation of

3-butyn-1-ol, showing the complete absence of hydrogenolysis and over hydrogenation. These results revealed that the structural order of the mesoporous silica has a positive effect on the catalytic performance. Using the sol–gel technique to support Pd on SiO_2, Panpranot et al. observed the formation of palladium silicide (Pd_4Si) [49]. The Pd/SiO_2 catalysts containing Pd_4Si were found to be superior to commercial SiO_2-supported catalysts, in liquid-phase semihydrogenation of phenylacetylene (styrene selectivity of c. ~60% at 100% conversion of phenylacetylene compared to no styrene selectivity for commercial SiO_2 supported). The same group also reported the role of geometrical confinement of Pd NPs in silica pores during phenylacetylene hydrogenation and unraveled the role of Pd NP dispersion on the styrene selectivity [50]. Very recently, findings by Copéret and coworkers showed that the choice of ligand to functionalize silica can drastically affect the selectivity of the catalyst while maintaining the intrinsic activity [51]. An in-depth kinetic study was carried out to investigate the semihydrogenation of 1-phenyl-1-propyne on Cu/SiO_2 in the absence or in the presence of ligands (tricyclohexylphosphine and N-heterocyclic carbene). It was found that in the presence of ligands, the rate of hydrogenation of the alkyne is unaffected, while over hydrogenation to the alkane is greatly hindered. The observed behavior was attributed to the competitive adsorption of the ligand and the alkynes on a Cu NP surface. This study provided a guideline to improve the selectivity of an inexpensive, earth-abundant Cu catalyst.

Amines are important chemical intermediates in the production of various pharmaceuticals, agrochemicals, cosmetics, herbicides, dyes, and polymers [52]. Homogeneous catalysts based on Ru and Ir have gained a lot of acceptance because of their excellent yield, but their use was limited because of recovery and reuse issues. In order to overcome these issues, Chen et al. supported Au NPs onto silica modified with the 3-glycidoxypropyl group [53]. Au NPs (c. 7–9 nm) were synthesized by mixing 3-glycidoxypropyl-modified silica with aqueous gold chloride ($HAuCl_4$) solution and stirring for two hours at 100 °C before isolating and air drying for two hours at 120 °C. The so-obtained catalyst was evaluated for the hydrogenation of nitro function under H_2 pressure (Table 10.2). Their results highlighted the great potential of silica-supported Au NPs to act as effective catalysts for the selective hydrogenation of aromatic nitro compounds.

Tan et al. utilized 3-aminopropyltriethoxysilane (3-APTS)-functionalized silica, dispersed in aqueous solution to support Au NPs of size around 2 nm in order to hydrogenate a variety of functionalized aromatic nitro groups [54]. The uniformly and highly dispersed Au NPs on the silica resulted in high catalytic reactivity, albeit at relatively harsh conditions, i.e. 140 °C at 40 bar H_2. Synergistic effect in bimetallic NPs is also known to affect conversion as well as yield during nanocatalysis. Kustov and coworkers prepared granulated krupnoporisty silica supported monometallic and bimetallic catalysts of copper (Cu) and zinc (Zn) [55]. Among the different catalysts tested, the bimetallic Cu–Zn system was found to be the most active (99%) and selective (100%) during the hydrogenation of *p*-dinitrobenzene to *p*-phenylenediamine. A further challenge in the field was to specifically convert nitro groups in the presence of other reducible functional groups, which was taken on by Polshettiwar's group using DFNS-supported Pt NPs [43]. Concentration-dependent

Table 10.2 Catalytic hydrogenation of various aromatic nitro compounds over Au/SiO_2.

Entry	Substrate	Product	Time (h)	Conversion (%)	Selectivity (%)
1	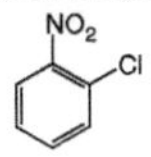	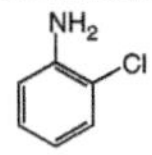	3	100	100
2	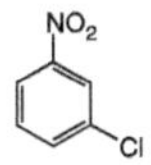	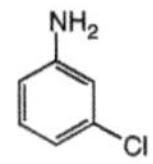	3	100	100
3	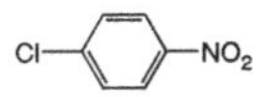		3	100	100
4	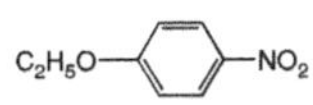	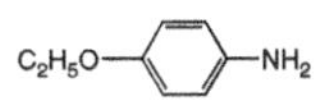	2.5	100	100
5	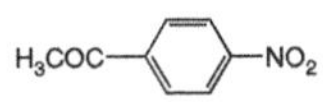	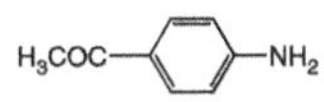	3	100	95.0
6	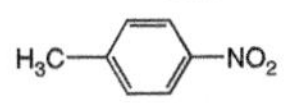	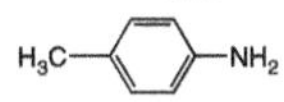	2.5	100	100
7	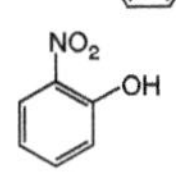	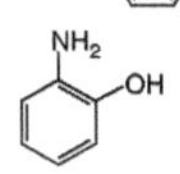	2.5	100	100
8	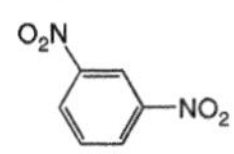	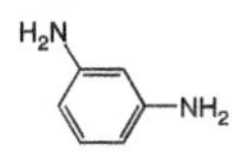	5	100	65.4
9	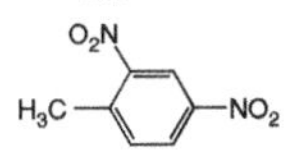	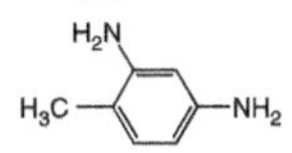	4.5	100	70.5

Reaction conditions: 0.1 g catalyst, 0.5 g substrate, 50 ml ethanol, 140 °C, and 4.0 MPa.
Source: Chen et al. [53].

size effect of Pt NPs was observed on DFNS. At a relatively low concentration (1 wt% of Pt), pseudo-single atoms of Pt were found in addition to NPs. DFNS/Pt (1 wt%) displayed high activity for the hydrogenation of 3-nitrostyrene to 3-vinylaniline with extraordinary selectivity (70%). Subnanometer-sized Pt NPs and high accessibility due to DFNS support were proposed to be responsible for such high catalytic performance and selectivity of the catalyst.

Wei et al. reported an interesting work dedicated to hydrogenation of nitroarenes via transfer hydrogenation reaction using Ru NPs as the catalyst supported on MCM-41-type mesoporous silica (OMS-IL) prepared with an ionic liquid (1-hexadecyl-3-methylimidazolium bromide) as the template [56]. Ru/OMS-IL showed excellent catalytic activity and chemoselectivity for the reduction of various nitroarenes to the corresponding aromatic amines in ethanol with hydrazine hydrate as a hydrogen donor. Verma et al. uniquely utilized the plasmonic effect of Ag deposited onto CeO_2 coated with silica (SBA-15) for the hydrogenation of *p*-nitrostyrene [57]. A comparison of results obtained in dark and under light conditions showed considerable enhancement of the conversion during the hydrogenation reaction under light irradiation conditions (Figure 10.10). The highest conversion (100%) with almost 90% selectivity was achieved on the catalyst with 1.0 wt% loading of Ag NPs and 2 wt% of CeO_2 supported on silica.

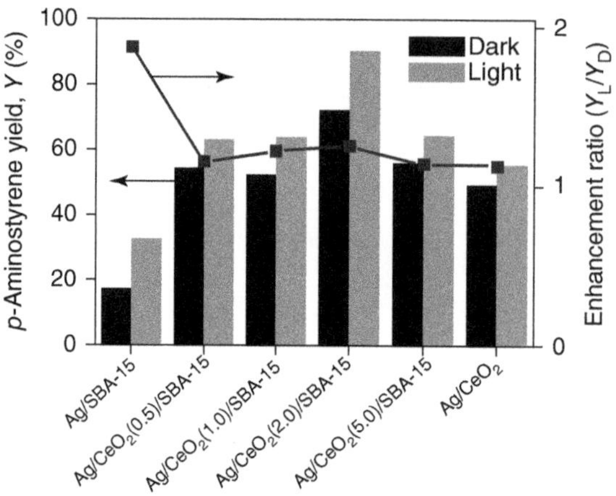

Figure 10.10 Plasmonic effect of Ag deposited onto CeO_2 coated with silica (SBA-15) and comparative yields of *p*-aminostyrene obtained with different catalysts in dark (black) and under light (gray) conditions. Source: From Verma et al. [57]. Copyright 2019, Elsevier. Reprinted with permission of Elsevier.

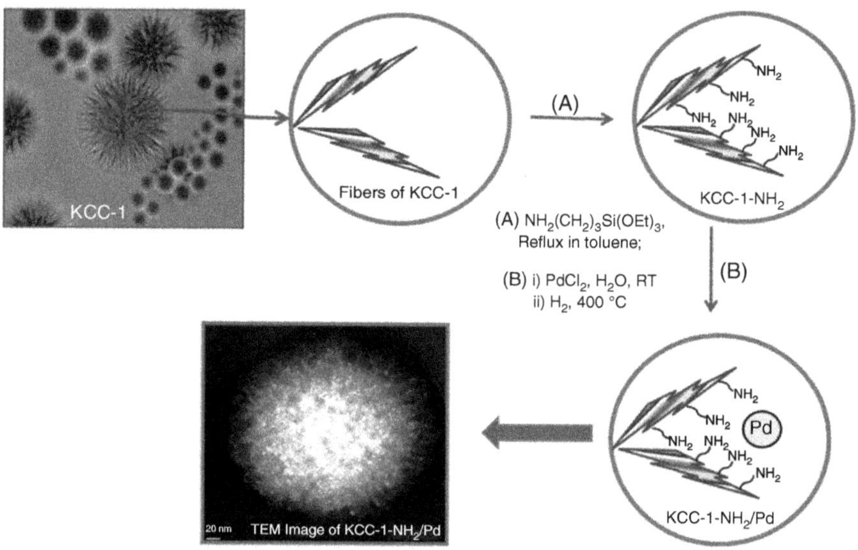

Figure 10.11 Synthesis of DFNS-NH_2/Pd nanocatalysts. Source: Reproduced with permission from Fihri et al. [60]. Copyright 2012, Wiley.

10.3.3 Carbon–Carbon (C–C) Coupling Reactions

The C–C coupling reactions are probably the most important transformations used in the industry to produce drugs, pharmaceuticals, conducting polymers, herbicides, and industrially important starting material compounds [58]. Many reviews have been published concerning the use of supported silica catalysts for C–C coupling reactions [59]. Here, we will highlight a few pertinent works describing the interest of metal NPs supported on silica for this catalytic transformation.

Polshettiwar and coworkers reported the use of DFNS functionalized with amino groups and loaded with Pd NPs for the Suzuki coupling reaction between aromatic halides and aryl boronic acids (Figure 10.11) [60]. DFNS/Pd catalysts were found to be highly active, resulting in yields over 90% toward coupling reaction between a range of aryl bromides and iodides with aryl boronic acids. Interestingly, the catalyst was also active for aryl chlorides (more "sustainable substrates"), with yields as high as 97%. The catalyst was easily recoverable and stable up to 7 cycles, while maintaining the catalytic performance over 90%.

Noori et al. supported Pd NPs on *S*-methylisothiourea-functionalized MCM-41. The so-obtained catalysts were found to be highly active and selective for the corresponding products when evaluated for (i) Suzuki reaction using a variety of aryl halides with phenylboronic acid (up to TOF of $161\,h^{-1}$ with iodobenzene as the substrate), (ii) Stille reaction with triphenyltin chloride (up to TOF of $242\,h^{-1}$ with iodobenzene as the substrate), and (iii) Heck reaction with styrene or *n*-butyl acrylate (up to TOF of $72\,h^{-1}$ with iodobenzene as the substrate) [61]. On the other hand, Yu et al. synthesized Pd NPs supported on diamine-functionalized SBA-15 [62]. Diimine functionalization was found to be specifically efficient for better dispersing Pd NPs on SBA-15. Even at very low loadings of Pd (0.05 mol%), the catalyst exhibited excellent catalytic performance for the Suzuki coupling reaction of electronically diverse aryl halides with phenylboronic acid. The better catalytic performance of Pd-diimine@SBA-15 was attributed to the high dispersion of Pd in the pore channels and high surface area of SBA-15. When compared to Pd/SBA-15 and Pd-diimine@SiO_2 catalysts, the Pd-diimine@SBA-15 showed improved hydrothermal stability and could be used four times without a significant decrease in its catalytic activity.

Also, Polshettiwar and coworkers reported PEI functionalized DFNS as a support to load Pd NPs for carbonylative Suzuki–Miyaura cross-coupling reaction [63]. DFNS at very low loading (0.1 wt% Pd) resulted in excellent catalytic performance for this reaction starting from 4-iodoanisole and phenylboronic acid as substrates. A comparison study revealed that DFNS-PEI/Pd displayed a TON 28 times higher and a TOF 51 times higher than those observed with the best-supported Pd catalysts previously reported in the literature (Table 10.3).

Additionally, DFNS-PEI/Pd was shown to be a versatile catalyst and could be utilized to synthesize different unsymmetrical biaryl ketones. Another work for Suzuki–Miyaura coupling reaction was reported by Cai and coworkers, wherein a fluorescent material FSM was synthesized via grafting fluoroionophore as a fluorescent probe on mesoporous silica [64]. Their results revealed that palladium ions were tightly bound with FSM, impacting leach-proof nature to Pd NPs. The Pd@FSM catalyst displayed exceptional catalytic activity toward the Suzuki–Miyaura coupling reaction in water with only 0.05 mol% of Pd. Interestingly, the catalyst was highly stable and could be stored in air for at least three months without loss of its catalytic activity.

To further improve the recovery of the catalysts, Khodaei and Dehghan prepared a magnetically recoverable nanocatalyst by first functionalizing SBA-15 with a Schiff base ligand and then supporting palladium NPs [65]. SBA-15-supported Pd NPs were further mixed in ethanolic solution of as-prepared Fe_3O_4 NPs under ambient conditions, resulting in a magnetically separable Pd-SBA-15 catalyst (Figure 10.12). These heterogeneous catalysts were evaluated for Heck reaction using iodo-, bromo-, and chloro-benzene derivatives and styrene. After optimization, 3.34 mmol% of Pd and

Table 10.3 Comparison of DFNS-PEI/Pd with other catalysts for the carbonylative Suzuki–Miyaura cross-coupling between 4-iodoanisole and phenylboronic acid.

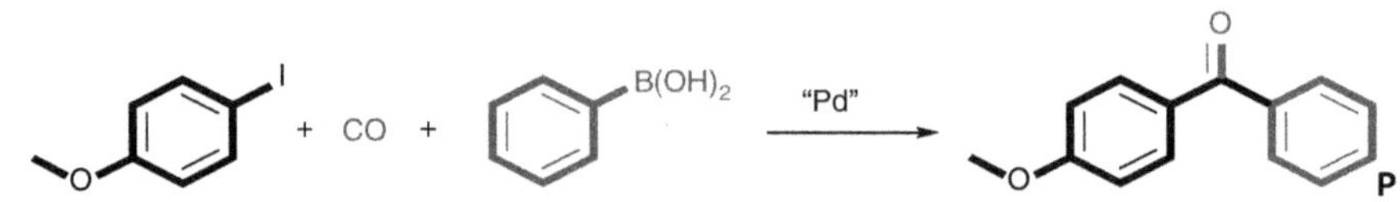

Entry	Catalyst	Experimental conditions	TON	TOF (h^{-1})	References
1	DFNS-PEI/Pd (0.1%)	100 °C, 6 h, 2 bar CO	2456.1	409	[63a]
2	MCM-41-2P-Pd	80 °C, 5 h, 1 bar CO	43.0	8	[63b]
3	Pd/C	100 °C, 8 h, 14 bar CO	45.0	5	[63c]
4	MCM-41-2N-Pd	80 °C, 8 h, 1 bar CO	45.5	5	[63d]
5	Fe_3O_4@SiO_2-SH-Pd	80 °C, 8 h, 1 bar CO	92.0	11	[63e]
6	HMMS-SH-Pd	80 °C, 9 h, 1 bar CO	62.0	6	[63f]
7	P(DVB-NDIIL)-Pd	120 °C, 12 h, 30 bar CO	85.0	7	[63g]
8	Pd supported on aminopropyl polymethylsiloxane microspheres	100 °C, 5 h, 5 bar CO	39.6	7	[63h]
9	Fe_3O_4@SiO_2-2N-Pd	80 °C, 8 h, 1 bar CO	92.0	11	[63i]

Source: Gautam et al. [63].

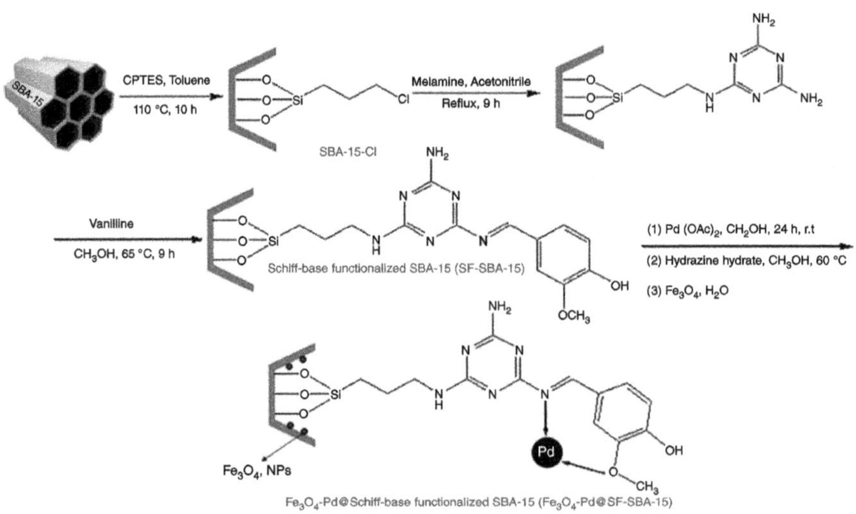

Figure 10.12 Pd NPs immobilized on Schiff base-functionalized mesoporous silica. Source: From Khodaei and Dehghan [65]. Copyright 2019, Wiley. Reprinted with permission of John Wiley and Sons.

Et_3N as a base resulted in best corresponding coupling products. All products were obtained in high yields (up to 97%). The role of magnetic NPs onto the catalysis performance was also investigated. The results obtained indicated that no stilbene product was obtained with only Fe_3O_4 NPs, confirming their unique role in the easy separation of the catalyst from the reaction mixture without Pd NPs leaching and its reusability.

Hitherto, most of the C–C coupling reactions reported in the literature involved development of the reaction scope, nonactivated halides, selection of boron reagents for the reactions, effect of ligand and ligand-free conditions, ability to carry out the reaction at ambient temperature, in an aqueous medium, under solvent-free conditions, in green solvents, with very low catalyst loadings, etc. However, no systematic study was carried out to analyze the effect of the pore size of the mesoporous silica for any coupling reaction. In this regard, Das et al. compared Pd NPs supported onto 3-APTS-grafted MCM-41 and SBA-15 [66]. Interestingly, the pore size was respectively reduced from 3.8 and 7.6 to 2.9 and 6.7 nm because of the grafting. The difference in the catalytic activity of these materials was correlated with their textural properties (surface area and pore sizes) for Suzuki reaction under exactly the same conditions. The best performance of SBA-15-Pd (TOF = 82 h^{-1}) was correlated with the bigger pore sizes as compared to MCM-Pd (TOF = 60 h^{-1}) for coupling reaction between bromobenzene and boronic acid.

Another unique silica material in the form of nanotubes was also reported to support Pd NPs, leading to highly active catalyst for the Suzuki coupling reaction. Bai and coworkers reported a facile *in situ* reduction approach to prepare amine-functionalized silica nanotube (ASNT)-supported Pd NPs [67]. The synthesis procedure for ASNTs@Pd catalyst is shown in Figure 10.13. ASNTs were first synthesized by using C_{14}-D-AlaS as the template, 3-APTS as the grafting agent,

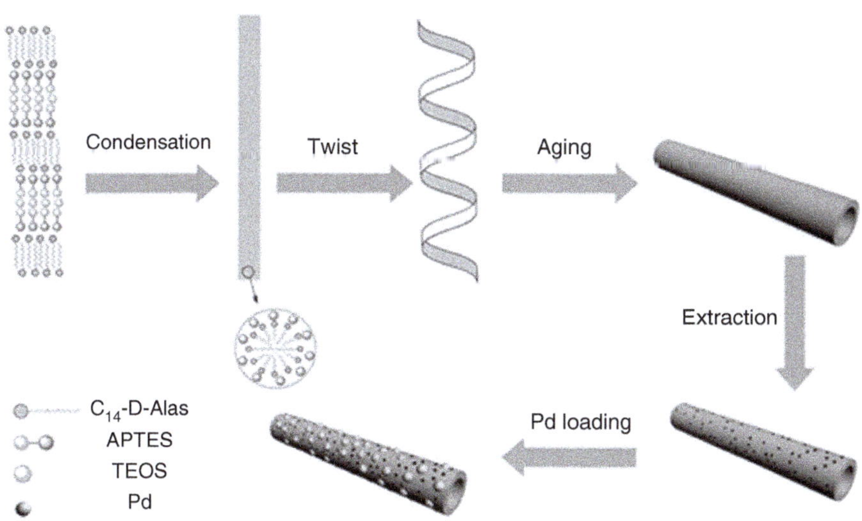

Figure 10.13 Synthetic route to the ASNTs@Pd composite. Source: Reproduced with permission from Liu et al. [67]. Copyright 2018, American Chemical Society.

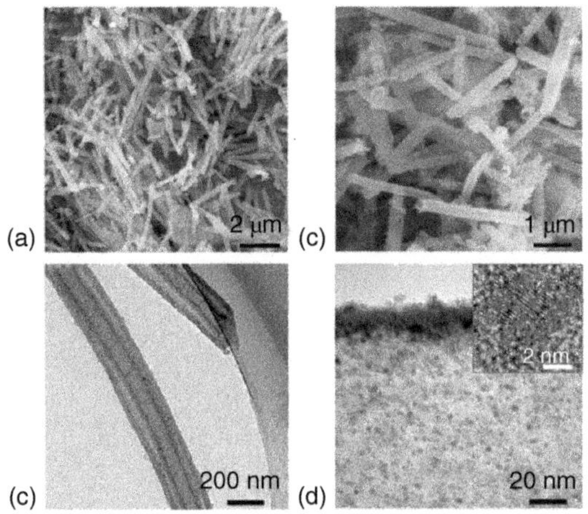

Figure 10.14 Scanning electron microscopy (SEM) and TEM of ASNT support and ASNTs@Pd catalyst. (a) SEM images of ASNTs; (b) ASNTs@Pd composite; (c) TEM image of ASNTs@Pd; (d) high-magnification TEM image of ASNTs@Pd composite (inset is the high-magnification TEM image of a single Pd NP). Source: Reproduced with permission from Liu et al. [67]. Copyright 2018, American Chemical Society.

and tetraethyl orthosilicate (TEOS) as the silicon source. Afterward, Pd NPs were supported on the surface of ASNTs with K_2PdCl_4 as the Pd precursor and $NaBH_4$ as the reducing agent to afford the ASNTs@Pd catalyst. The TEM images revealed well-dispersed Pd NPs on the surface of ASNTs with an average size of 3 nm (Figure 10.14). The catalytic activity of ASNTs@Pd catalyst was evaluated for Suzuki coupling with iodobenzene and phenylboronic acid as test molecules. Results indicated conversion within 30 minutes with a TOF up to 57.4 min^{-1}. In addition, an excellent catalytic yield of ASNTs@Pd was observed for reactants, including 3-methylphenylboronic acid (96.1%), 4-ethylphenylboronic acid (97.2%), 4-ethoxyphenylboronic acid (95.9%), 4-*tert*-butylphenylboronic acid (96.3%), and 3,4,5-trimethoxyphenylboronic acid (98.7%) with phenylboronic acids.

10.4 Conclusion

The use of silica-supported metal NPs is an important field in nanocatalysis and was shown to increase exponentially in recent years. This chapter provides an insight into the design and application of silica-supported metal NPs in catalysis for a selection of important reactions. We presented general synthetic techniques followed to synthesize silica-supported metal NPs and discussed their advantages. We then reviewed the application of different metal NPs supported on silica as nanocatalysts for a selection of reactions that are highly important at the industrial level, namely, oxidation, hydrogenation, and C–C coupling reactions.

On the basis of discussed works, it can be clearly noticed that catalytic performances of the nanocatalysts can be largely controlled by the proper choice of support. The use of silica as the support offers an interesting handle over tuning the pore sizes, their interconnection, surface area, and ease of functionalization. In addition, the stability of NPs was found to increase multifold, and they could be recovered and reused for several cycles. Further advancements in the field of catalyst design and synthesis will lay a better foundation to take the field of nanocatalysis one step closer to realize practical catalysts. More detailed studies on the kinetics of particle formation alongside ease of preparation, as well as systematic studies to control the collective properties of the supported NPs, is greatly desired. Conversely, one major challenge ahead for the improvement of silica-supported metal nanocatalysis is to design procedures to support earth-abundant metal NPs. To achieve this, the development of *in situ* spectroscopic and microscopic techniques will be particularly useful. At the same time, a better understanding of the relationship between different functionalized SiO_2-supported NPs and their catalytic performances will lead to the progress in catalyst synthesis.

References

1 Kresge, C.T., Leonowicz, M.E., Roth, W.J. et al. (1992). *Nature* 359: 710–712.

2 Oye, G., Sjöblom, J., and Stöcker, M. (2001). *Adv. Colloid Interface Sci.* 89–90: 439–466.

3 Zhao, D., Huo, Q., Feng, J. et al. (1998). *J. Am. Chem. Soc.* 120: 6024–6036.

4 Tozuka, Y., Wongmekiat, A., Kimura, K. et al. (2005). *Chem. Pharm. Bull.* 53: 974–977.

5 Polshettiwar, V., Cha, D., Zhang, X., and Basset, J.M. (2010). *Angew. Chem. Int. Ed.* 49: 9652–9656.

6 Jammaer, J., Aerts, A., D, Haen, J. et al. (2009). *J. Mater. Chem.* 19: 8290–8293.

7 Yamamoto, T., Mori, S., Shishido, T. et al. (2009). *Top. Catal.* 52: 657–663.

8 Zhou, B., Li, C.Y., Qi, N. et al. (2018). *Appl. Surf. Sci.* 450: 31–37.

9 Wang, X.D., Shen, Z.X., Sang, T. et al. (2010). *J. Colloid Interface Sci.* 341: 23–29.

10 Telalovi, S., Ramanathan, A., Mula, G., and Hanefeld, U. (2010). *J. Mater. Chem.* 20: 642–658.

11 Wu, H.C., Chen, T.C., Lai, N.C. et al. (2015). *Nanoscale* 7: 16848–16859.

12 Chytil, S., Glomm, W.R., Kvande, I. et al. (2007). *Top. Catal.* 45 (1–4): 93–99.

13 Huang, W., Kuhn, J.N., Tsung, C.K. et al. (2008). *Nano Lett.* 8: 2027–2034.

14 Yin, Y., Yang, Z.F., Wen, Z.H. et al. (2017). *Sci. Rep.* 7: 4509.

15 Yang, C.M., Liu, P.H., Ho, Y.F. et al. (2003). *Chem. Mater.* 15: 275–280.

16 (a) Valange, S. and Védrine, J.C. (2018). *Catalysts* 8: 483–504. (b) Mizuno, N. and Misono, M. (1998). *Chem. Rev.* 98: 199–218. (c) Ali, M.I., Rahman, M.M., Sarkar, S.M., and Hamid, S.B.A. (2014). *J. Nanomat.* 192038: 1–23. (d) Liu, L. and Corma, A. (2018). *Chem. Rev.* 118: 4981–5079. (e) Zaera, F. (2017). *ACS Catal.* 7: 4947–4967. (f) Pagliaro, M., Pandarus, V., Ciriminna, R. et al. (2012). *ChemCatChem* 4: 432–445. (g) Yin, L. and Liebsche, J. (2007). *Chem. Rev.* 107: 133–173.

17 Xu, J., Deng, Y.-Q., Luo, Y. et al. (2013). *J. Catal.* 300: 225–234.

18 Freund, H.J., Meijer, G., Scheffler, M. et al. (2011). *Angew. Chem. Int. Ed.* 50: 10064–10094.

19 (a) Haruta, M., Kobayashi, T., Sano, H., and Yamada, N. (1987). *Chem. Lett.* 16: 405. (b) Haruta, M., Yamada, N., Kobayashi, T., and Iijima, S. (1989). *J. Catal.* 115: 301.

20 Yang, C., Kalwei, M., Schuth, F., and Chao, K. (2003). *Appl. Catal., A* 254: 289–296.
21 Bore, M.T., Pham, H.N., Switzer, E.E. et al. (2005). *J. Phys. Chem. B* 109: 2873–2880.
22 Rombi, E., Cutrufello, M.G., Cannas, C. et al. (2009). *Phys. Chem. Chem. Phys.* 11: 593–602.
23 Qu, Z.P., Huang, W.X., Cheng, M.J., and Bao, X.H. (2005). *J. Phys. Chem. B* 109: 15842.
24 Yu, L., Shi, Y., Zhao, Z. et al. (2011). *Catal. Commun.* 12: 616–620.
25 Mijs, W. J., de Jonge, C. R. H. (1986). Organic Syntheses by Oxidation with Metal Compounds. Plenum Press, NY: Springer US.
26 (a) Ma, B.C., Zhang, Y.S., Ding, Y., and Zhao, W. (2010). *Catal. Commun.* 11: 853–857. (b) Zhao, W., Zhang, Y.S., Ma, B.C. et al. (2010). *Catal. Commun.* 11: 527–531.
27 Wang, L., Yi, W.B., and Cai, C. (2010). *ChemSusChem* 3: 1280–1284.
28 Parlett, C.M.A., Bruce, D.W., Hondow, N.S. et al. (2013). *ChemCatChem* 5: 939–950.
29 (a) Castano, P., Zepeda, T., Pawelec, B. et al. (2009). *J. Catal.* 267: 30–39. (b) Li, X.M., Paraschiv, V., Huskens, J., and Reinhoudt, D.N. (2003). *J. Am. Chem. Soc.* 125: 4279–4284. (c) Liu, W., Yang, X., and Xie, L. (2007). *J. Colloid Interface Sci.* 313: 494–502.
30 Yang, X., Huang, C., Fu, Z. et al. (2013). *Appl. Catal., B* 140: 419–425.
31 Rostamnia, S. and Kholdi, S. (2018). *Adv. Powder Technol.* 29: 1167–1174.
32 (a) Chandrasekhar, V., Boomishankar, R., and Nagendran, S. (2004). *Chem. Rev.* 104: 5847. (b) Denmark, S.E. and Wehrli, D. (2000). *Org. Lett.* 2: 565.
33 Li, W., Wang, A., Yang, X. et al. (2012). *Chem. Commun.* 48: 9183–9185.
34 Ma, L., Leng, W., Zhao, Y. et al. (2014). *RSC Adv.* 4: 6807–6810.
35 da Silva, A.G.M., Kisukuri, C.M., Rodrigues, T.S. et al. (2016). *Appl. Catal., B* 184: 35–43.
36 Dhiman, M., Chalke, B., and Polshettiwar, V. (2017). *J. Mater. Chem. A* 5: 1935–1940.
37 Jiang, Y. and Gao, Q. (2006). *J. Am. Chem. Soc.* 128: 716–717.
38 Biradar, A.V., Biradar, A.A., and Asefa, T. (2011). *Langmuir* 27: 14408–14418.
39 Maity, N., Basu, S., Mapa, M. et al. (2006). *J. Catal.* 242: 332–339.
40 Boutros, M., Shirley, G., Onfroy, T., and Launay, F. (2011). *Appl. Catal., A* 394: 158–165.
41 (a) Schuit, G.C.A. and Van Reijen, L.L. (1958). *Adv. Catal.* 10: 242. (b) Beeck, O. (1945). *Rev. Mod. Phys.* 17: 61.
42 Dhiman, M., Chalke, B., and Polshettiwar, V. (2015). *ACS Sustainable Chem. Eng.* 3: 3224–3230.
43 Dhiman, M. and Polshettiwar, V. (2016). *J. Mater. Chem. A* 4: 12416–12424.
44 Matsuyama, K., Tanaka, S., Kato, T. et al. (2017). *J. Supercrit. Fluids* 130: 140–146.
45 Arena, F., Cum, G., Gallo, R., and Parmaliana, A. (1996). *J. Mol. Catal. A: Chem.* 110: 235–242.

46 (a) Weerachawanasak, P., Mekasuwandumrong, O., Praserthdam, P. et al. (2009). *J. Catal.*: 199–205. (b) Domnguez-Domnguez, S., Berenguer-Muricia, A., Pradhan, B.K. et al. (2008). *J. Phys. Chem. C* 112: 3827–3834. (c) Marin-Astorga, N., Pecchi, G., Pinnavaia, T.J. et al. (2006). *J. Mol. Catal. A: Chem* 247: 145–152.

47 Dominguez-Dominguez, S., Berenguer-Muricia, A., Linares-Solano, A., and CazorlaAmoros, D. (2008). *J. Catal.* 257: 87–95.

48 Papp, A., Molnár, A., and Mastalir, A. (2005). *Appl. Catal., A* 289: 256–266.

49 Panpranot, J., Phandinthong, K., Sirikajorn, T. et al. (2007). *J. Mol. Catal. A: Chem.* 261: 29–35.

50 (a) Tiengchad, N., Mekasuwandumrong, O., Na-Chiangmai, C. et al. (2011). *Catal. Commun.* 12: 910–916. (b) Na-Chiangmai, C., Tiengchad, N., Kittisakmontree, P. et al. (2011). *Catal. Lett.* 141: 1149–1155.

51 Kaeffer, N., Larmier, K., Fedorov, A., and Copéret, C. (2018). *J. Catal.* 364: 437–445.

52 (a) Wang, D. and Astruc, D. (2015). *Chem. Rev.* 115: 6621–6686. (b) Lara, P. and Philippot, K. (2014). *Catal. Sci. Technol.* 4: 2445–2465. (c) Serna, P. and Corma, A. (2015). *ACS Catal.* 5: 7114–7121.

53 Chen, Y., Qiu, J., Wang, X., and Xiu, J. (2006). *J. Catal.* 242: 227–230.

54 Tan, X., Zhang, Z., Xiao, Z. et al. (2012). *Catal. Lett.* 142: 788–793.

55 Shuvalova, E.V., Kirichenko, O.A., Kapustin, G.I., and Kustov, L.M. (2016). *Russ. Chem. Bull. Int. Ed.* 65: 2850–2854.

56 Wei, N., Zou, X., Huang, H. et al. (2018). *Eur. J. Org. Chem.* 2018: 209–214.

57 Verma, P., Kuwahara, Y., Mori, K., and Yamashita, H. (2019). *Catal. Today* 324: 83–89.

58 (a) Pu, L. (1998). *Chem. Rev.* 98: 2405–2494. (b) Hassan, J., Sevignon, M., Gozzi, C. et al. (2002). *Chem. Rev.* 102: 1359–1470.

59 (a) Karimi, B., Mansouri, F., and Khorasani, M. (2016). *Curr. Org. Chem.* 20: 349–380. (b) Sadjadi, S. and Heravi, M. (2017). *RSC Adv.* 7: 30815–30838. (c) Ziarani, G.M., Rohani, S., Ziarati, A., and Badiei, A. (2018). *RSC Adv.* 8: 41048–41100.

60 Fihri, A., Cha, D., Bouhrara, M. et al. (2012). *ChemSusChem* 5: 85–89.

61 Noori, N., Nikoorazm, M., and Ghorbani-Choghamarani, A. (2016). *J. Porous Mater.* 23: 1467–1481.

62 Yu, J., Shen, A., Cao, Y., and Lu, G. (2016). *Catalysts* 6: 181.

63 (a) Gautam, P., Dhiman, M., Polshettiwar, V., and Bhanage, B.M. (2016). *Green Chem.* 18: 5890–5899. (b) Cai, M., Zheng, G., and Zha, L. (2009). Peng, J. *Eur. J. Org. Chem.*: 1585–1591. (c) Khedkar, M.V., Tambade, P.J., Qureshi, Z.S., and Bhanage, B.M. (2010). *Eur. J. Org. Chem.*: 6981–6986. (d) Cai, M., Peng, J., Hao, W., and Ding, G. (2011). *Green Chem.* 13: 190–196. (e) Niu, J.-R., Huo, X., Zhang, F.-W. et al. (2013). *ChemCatChem* 5: 349–354. (f) Niu, N., Liu, M., Wang, P. et al. (2014). *New J. Chem.* 38: 1471–1476. (g) Jiao, N., Li, Z., Wang, Y. et al. (2015). *RSC Adv.* 5: 26913–26922. (h) Zawartka, W., Pośpiech, P., Cypryk, M., and Trzeciak, A.M. (2016). *J. Mol. Catal. A: Chem.* 417: 46–80. (i) Sun, X., Zheng, Y., Sun, L. et al. (2016). *Nano-Struct. Nano-Objects* 5: 7–14.

64 Liang, G., Cai, Q., Zhu, W. et al. (2015). *Anal. Methods* 7: 4877–4880.
65 Khodaei, M.M. and Dehghan, M. (2019). *Appl. Organomet. Chem.* 33: 4618–4627.
66 Das, T., Uyama, H., and Nandi, M. (2018). *New J. Chem.* 42: 6416–6426.
67 Liu, J., Hao, J., Hu, C. et al. (2018). *J. Phys. Chem. C* 122: 2696–2703.

Part III

Application

11

CO_2 Hydrogenation to Oxygenated Chemicals Over Supported Nanoparticle Catalysts: Opportunities and Challenges

Qiming Sun, Zhenhua Zhang, and Ning Yan

Department of Chemical and Biomolecular Engineering, National University of Singapore, 4 Engineering Drive 4, 117585, Singapore

11.1 Introduction

Over the past decades, the continuing worldwide consumption of carbon-rich fossil fuels (for example coal and petroleum) and natural gas has resulted in a sharp increase of carbon dioxide (CO_2) concentration in the atmosphere potentially causing the global climate change and ocean acidification [1]. At present, there are three primary ways to reduce the CO_2 concentration: (i) decrease CO_2 emissions, (ii) capture and store CO_2, and (iii) convert CO_2 into fuels and chemicals [2]. CO_2 storage is effective especially in energy generation and industrial sectors (accounting for more than 40% of the global CO_2 emissions), but it usually suffers from potential leakage of CO_2 [3]. As such, the capture and subsequent reduction of CO_2 to value-added products via electrochemical, photocatalytic, and thermal processes are considered as attractive and promising routes to try to solve or at least limit the CO_2 problem [4]. Although extensive research studies regarding electrochemical [5] and photocatalytic reductions [6] of CO_2 have been reported, scalability is a considerable challenge for these approaches. In comparison, the thermal reduction of CO_2 into chemicals and fuels by using H_2 as a reductant is compatible with existing and well-developed infrastructures in the current refinery.

Thermal reduction of CO_2 over heterogeneous catalytic systems leads to many useful chemicals, such as hydrocarbons, alcohols, and ethers. CO_2 is no longer considered as a waste but as an alternative carbon feedstock. Thus, CO_2 has aroused intense research activities among the community to develop high-performance catalysts for its transformation into valuable chemicals and energy carriers. CO_2 being one of the most stable molecules on earth, its transformation is not an easy task, but promising results have already been obtained. In view of the considerable progresses achieved with supported metal catalysts for CO_2 hydrogenation reactions, this chapter will focus on recent advances using supported nanoparticle (NP) catalysts to convert CO_2 into oxygenated chemicals, including formic acid (FA), methanol, and dimethyl

Nanoparticles in Catalysis: Advances in Synthesis and Applications, First Edition.
Edited by Karine Philippot and Alain Roucoux.

ether (DME). In the last section of the chapter, opportunities and challenges for further advancing this subject are outlined.

11.2 CO_2 Hydrogenation into Formic Acid

Formic acid (FA; HCOOH), possessing a relatively high H_2 content (53 g l^{-1} and 4.4 wt%) and being nontoxic, nonflammable, and easily recharged, has been regarded as a platform for chemical energy carrier for portable H_2 storage devices [7]. Compared with gaseous H_2, the transportation and storage of FA (liquid form at ambient conditions) are straightforward and convenient. Moreover, H_2 can be totally recovered from FA via a dehydrogenation pathway ($HCOOH \rightarrow H_2 + CO_2$) over suitable catalysts [8]. Currently, there are two main industrial methods to prepare FA, including hydrolysis of methyl formate/formamide and oxidation of biomass [9]. Recently, the direct hydrogenation of CO_2 to produce FA has been considered as a promising alternative strategy [3, 10]. This is not only because the target product is valuable but also because this synthesis route is green and cost-effective and ensures the effective utilization of CO_2 compared with traditional synthesis methods. This new route also realizes an economically viable CO_2-mediated H_2 energy cycle, as shown in Figure 11.1.

In principle, the conversion of CO_2 and H_2 into FA can be carried out under a gas phase or an aqueous solution condition. However, the gas-phase hydrogenation of CO_2 into FA requires a high and positive free energy change ($\Delta G = +33\ \text{kJ mol}^{-1}$) because the phase change from gaseous reactants to a liquid product brings about a disfavored entropy. By comparison, the hydrogenation of CO_2 to FA in aqueous solution alters the thermodynamics of the reaction and proceeds more readily, with a lower and negative free energy ($\Delta G = -4\ \text{kJ mol}^{-1}$) [11]. The hydrogenation of CO_2 to FA can be realized by both homogeneous and heterogeneous catalytic systems. In 1976, the catalytic fixation of CO_2 into FA was first reported by Inoue et al. using several transition metal complexes ($Pd(Ph_2PCH_2CH_2PPh_2)_2$

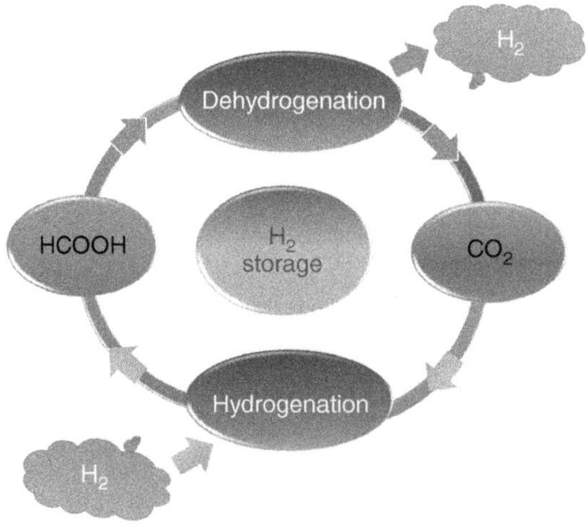

Figure 11.1 Schematic representation of CO_2-mediated H_2 energy cycle.

$Ni(Ph_2PCH_2CH_2PPh_2)_2$, $Pd(PPh_3)_4$, $H_2Ru(PPh_3)_4$, and $H_3Ir(PPh_3)_3$) as homogeneous catalysts under basic conditions (triethylamine (NEt_3), CO_2/H_2 = 25/25 bar, room temperature, and 20 hours) [12]. Among all catalysts, $H_2Ru(PPh_3)_4$ showed the highest FA yield of 87 mol mol^{-1} complex. Since then, a large number of Ir- and Ru-based homogeneous catalysts were developed for the synthesis of FA and formates [13]. The activity for the production of FA via CO_2 hydrogenation over these homogeneous catalysts is truly fascinating. In fact, maximum turnover number (TON) and turnover frequency (TOF) up to 3 500 000 and 1 100 000 h^{-1}, respectively, were reached by using a pyridine-based Ru PNP-Pincer complex [14]. The limitations include unsatisfied production rates due to the low catalyst concentrations in homogeneously catalyzed processes as well as the difficulty to separate and recycle the catalyst from the reaction mixture, hindering large-scale industrial applications. To overcome these limitations, supported metal catalysts have attracted increasing interests [15]. In the following, we mainly focus on the state-of-the-art heterogeneous catalysts reported for the preparation of FA or formates by using CO_2 or bicarbonate as a feedstock.

The initial research on the hydrogenation of CO_2 to formates with heterogeneous catalysts was employing pure metals as catalysts. The synthesis of formates by hydrogenation process was started in 1914 by Bredig and Carter using Pd black as a catalyst and employing alkali/alkaline earth (bi)carbonates as the carbon source [16]. In 1935, Raney nickel-catalyzed formamide synthesis was reported by Farlow and Adkins by using CO_2 and H_2 as raw materials at 20–40 MPa of pressure in the presence of a mixture of ethanol/phenol as the solvents [17]. Although the activities of CO_2 hydrogenation catalyzed by the aforementioned catalysts were very low, they paved the way of using heterogeneous catalysts for CO_2 transformation into value-added chemicals. In 1983, Stalder et al. employed various supported Pd-based catalysts for the reduction of $NaHCO_3$ to HCOONa [18]. An initial activity of 9.8 $mol_{formate}$ $mol_{Pd}{}^{-1}$ h^{-1} was achieved over a 5 wt% Pd/C catalyst starting from 1.0 M $NaHCO_3$ under 1.7 atm H_2. Preti et al. employed a series of transition metal blacks (Fe, Ni, Co, Ru, Rh, Ir, Pd, Pt, and Au blacks) for the decomposition of formate salts to CO_2 and H_2, and by this way, the best candidate for the hydrogenation of CO_2 to formate could be identified [19]. Among the cited metals, the Au black exhibited preferable catalytic activity for the production of FA. However, fast deactivation of the Au black catalyst was observed because of a severe aggregation of metal particles and decrease of active sites. To overcome these drawbacks, dispersing catalytically active metals onto solid supports is widely described in recent studies. Filonenko et al. employed a series of supported and unsupported Au NPs as catalysts for the hydrogenation of CO_2 at 4 MPa of H_2 and CO_2 mixture in EtOH solution and basic conditions [20]. The authors found that the catalytic performance of supported Au NPs was much better than that of unsupported Au NPs and that the support nature strongly affects the catalytic efficiencies. Among TiO_2, Al_2O_3, ZnO, CeO_2, MgAl(Cr)-hydrotalcite, and $CuCr_2O_4$ as materials investigated, the supported Al_2O_3–Au catalyst showed the optimal catalytic performance with TON value reaching up to 215, which is about two times higher than that obtained with Au/TiO_2. Following this, Pd NP-based heterogeneous catalysts were found to exhibit outstanding catalytic activities for the hydrogenation of CO_2 into FA compared with other metallic counterparts. Su et al. systematically studied the hydrogenation of a series of bicarbonates or carbonates to formates over several Pd-based catalysts under various temperatures and pressures [21]. The main finding was that the hydrogenation of bicarbonates was more efficient than that of the carbonates because the protonation of carbonate ions was regarded as the rate-limiting step in aqueous solutions. Moreover, among all the carbon feedstocks tested, NH_4HCO_3 exhibited the highest catalytic activity under the same catalytic condition, as the NH_4HCO_3 possessed the highest concentration of HCO_3^- in the aqueous solution than $NaHCO_3$ and $KHCO_3$, resulting in the highest formate yield during the hydrogenation reaction. They also discovered that by using the activated carbon (AC) as the support, only Pd NP catalysts showed an observable catalytic activity, whereas other transition metals such as Ru, Rh, Pt, and Ni were inactive under the applied conditions. In addition, the nature of the supports significantly affected the catalytic performance. Thus, Pd/AC showed more than a sevenfold improvement of the TON value compared with those of Pd/Al_2O_3, Pd/$CaCO_3$, and Pd/$BaSO_4$, which can be attributed to the higher dispersion of Pd NPs on the AC than that on other supports. Similar support effects were also reported in other works, such as supported Au catalysts [20] and supported Ru catalysts [22].

Novel carbon materials, such as reduced graphite oxide (r-GO) [23], nitrogen-containing mesoporous graphitic carbon nitride (mpg-C_3N_4) [24], and carbon nanotube-graphene (CNT-GR) [25] have recently been employed as supports to stabilize metal NPs for CO_2 hydrogenation to FA, which led to superior catalytic activity as compared with traditional AC supports. In view of the significant cost of the noble metal Pd, cheap metal species were introduced in order to improve Pd-based catalysts. This strategy toward bimetallic catalysts allowed to reduce the use of Pd and to further enhance the catalytic performance in CO_2 hydrogenation. Nguyen et al. reported the synthesis of Pd–Ni alloys onto CNT-GR and performed the hydrogenation of CO_2 in water without any base additive [25]. The CNT-GR supports improved the dispersion of Pd–Ni alloy NPs, and thanks to a synergetic effect of the two associated metals in catalyst as well as good stability, an FA yield of 1.92 mmol with a TON of 6.4 was achieved under mild reaction conditions (40 °C and 50 bar, $H_2/CO_2 = 1$), providing a higher yield than that of the monometallic Pd catalyst. Recently, Mori et al. prepared a series of Pd–Ag NPs (including Pd@Ag (a $Pd_{core}Ag_{shell}$ structure), Pd–Ag alloy, and Ag@Pd (a $Ag_{core}Pd_{shell}$ structure)) supported on TiO_2, which exhibited improved selectivity toward FA under 2.0 MPa of pressure ($CO_2/H_2 = 1$) and 100 °C, which is regarded as a relative soft reaction condition for CO_2 hydrogenation [26]. The catalytic activities of the Pd–Ag catalysts strongly depend on the surface compositions (Figure 11.2).Thanks to the existence of neighboring Ag atoms, the Pd@Ag/TiO_2 catalyst with isolated and electron-rich Pd atoms exhibits a maximum TON of 14839 based on Pd surface dispersion, increasing more than 10 times relative to a monometallic Pd/TiO_2. The authors also found that the isolated and electron-rich Pd atoms in Pd@Ag NPs boosted the electronegativity of the dissociated hydride species, which was beneficial for

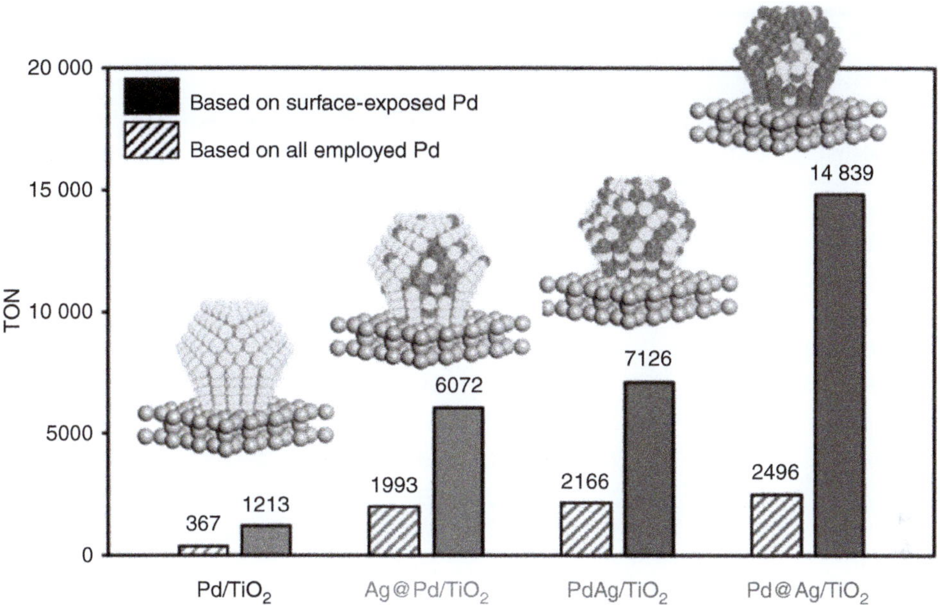

Figure 11.2 Comparison of the catalytic activities of various supported Pd–Ag catalysts with different surface compositions and Pd/TiO_2 in CO_2 hydrogenation. Conditions: catalyst (10 mg), 1.0 M aqueous $NaHCO_3$ solution (10 ml), $H_2/CO_2 = 1:1$ (total 2.0 MPa), and 100 °C. Source: Reproduced with permission from Mori et al. [26]. Copyright 2018, American Chemical Society.

the hydrogenation of the adsorbed HCO_3^- species, which is the rate-determining step. This was further evidenced by first-principles calculations and kinetic measurements. Very recently, Sun et al. developed a facile, one-pot, ligand-protected synthesis strategy to encapsulate ultrafine bimetallic Pd–CoO NPs with a good dispersion inside the channels of mesoporous silica nanospheres (MSNs). Thanks to the small size of metal NPs (~1.8 nm) and electron-enriched Pd species due to the electron transfer from CoO to Pd, the optimized $Pd_{0.8}Co_{0.2}$@MSN catalyst exhibits a high TOF value of 1824 h^{-1} under a pressure of 2.0 MPa ($H_2/CO_2 = 1:1$) in a basic solution at 373 K [27]. Table 11.1 summarizes the reaction conditions and the catalysis performance (expressed in TON and TOF) obtained with various metal NP-based catalysts previously reported for the CO_2 hydrogenation to FA or formates.

Besides the bulk metals and supported metal catalysts, there are also reports on heterogenized molecular catalysts for FA synthesis via CO_2 hydrogenation. These catalytic systems obtained by anchoring homogeneous complexes onto heterogeneous supports such as mesoporous silica, covalent organic frameworks, and polymers combine the advantages of high activity and selectivity of homogeneous species with the easy separation of heterogeneous ones [10, 31]. Because this chapter mainly focuses on the supported metal NP-based catalysts for catalytic CO_2 conversions, the grafted homogeneous molecular complexes for the CO_2 hydrogenation will not be further introduced.

Table 11.1 Reaction conditions and catalytic performance of various metal NP-based catalysts for the hydrogenation of CO_2 into FA or formates.

Catalyst	Temperature (°C)	$P_{(H_2)}$ (MPa)	$P_{(CO_2)}$ (MPa)	Solvent/ additive	$C_{(FA)}$ (mol l^{-1})	Time (h) /TON	TOF (h^{-1})	References
Pd black	70	6	0	$H_2O/KHCO_3$	0.37	23/5	0.22	[16]
Pd black	70	5	2	H_2O/none	0.02	4.5/0.28	0.06	[16]
5% Pd/C	25	0.17	0	$H_2O/NaHCO_3$	0.54	46/115	25	[18]
5% Pd/γ-Al_2O_3	25	—	0	$H_2O/NaHCO_3$	0.235	53/50	1.5	[18]
5% Pd/$BaSO_4$	25	—	0	$H_2O/NaHCO_3$	0.09	50/19	2.1	[18]
Au/TiO_2	70	2	2	EtOH/NEt_3	0.09	20/111	5.5	[20]
Au/Al_2O_3	70	2	2	EtOH/NEt_3	0.20	20/215	11	[20]
Au/ZnO	70	2	2	EtOH/NEt_3	<0.01	20/2	<1	[20]
Au/CeO_2	70	2	2	EtOH/NEt_3	<0.01	20/8	<1	[20]
Pd/C	20	2.75	0	$H_2O/NaHCO_3$	0.28	1/527	527	[21]
Pd/C	20	2.75	0	$H_2O/KHCO_3$	0.30	1/567	567	[21]
Pd/C	20	2.75	0	H_2O/NH_4HCO_3	0.24	1/728	728	[21]
Pd/C	20	2.75	0	H_2O/Na_2CO_3	<0.01	1/<1	<1	[21]
Pd/C	20	2.75	0	H_2O/NH_4CO_3	0.15	1/278	278	[21]
Pd/C	20	2.75	0	H_2O/NH_4HCO_3	0.95	15/1 769	118	[21]
Pd/C	20	5.52	0	H_2O/NH_4HCO_3	0.90	2/1 672	836	[21]
Pd/Al_2O_3	20	2.75	0	H_2O/NH_4HCO_3	0.08	1/278	278	[21]
Pd/$CaCO_3$	20	2.75	0	H_2O/NH_4HCO_3	<0.01	1/20	20	[21]
Pd/$BaSO_4$	20	2.75	0	H_2O/NH_4HCO_3	0.02	1/212	212	[21]
Pd/r-GO (1 wt%)	100	4	0	$H_2O/KHCO_3$	4.53	32/7 088	221	[23]
PdNi/CNT-GR	40	2.5	2.5	H_2O/none	0.02	15/6.4	<1	[25]
Pd/TiO_2	100	1	1	$H_2O/NaHCO_3$	—	24/1 213	—	[26]
Ag@Pd/TiO_2	100	1	1	$H_2O/NaHCO_3$	—	24/6 072	—	[26]
AgPd/TiO_2	100	1	1	$H_2O/NaHCO_3$	—	24/7 126	—	[26]
Pd@Ag/TiO_2	100	1	1	$H_2O/NaHCO_3$	—	24/14 829	—	[26]
$Pd_{0.8}Co_{0.2}$ @MSN	100	1	1	$H_2O/NaHCO_3$	—	1/1 824	—	[27]
Ru/γ-Al_2O_3	80	5	8.5	EtOH/NEt_3	0.68	1/731	731	[28]
PdAg/ SBA-15	100	1	1	$H_2O/NaHCO_3$	—	24/874	36	[29]
Ru/LDH	100	1	1	$H_2O/NaHCO_3$	—	24/461	19	[30]
Ru/LDH	100	1	1	H_2O/NaOH	—	24/698	29	[30]

11.3 CO_2 Hydrogenation to Methanol

Methanol (CH_3OH) is one of the most important and versatile platform chemicals in the petrochemical industry [32]. This molecule is largely used as a key intermediate to produce a variety of chemicals, such as formaldehyde, acetic acid, and methyl *tert*-butyl ether [33]. Moreover, methanol has excellent combustion characteristics, high energy density, and high energy efficiency, making it a suitable fuel for internal combustion engine-driven vehicles and direct fuel cell [34]. With the successful development of a commercially applicable methanol-to-olefins (hydrocarbons) technology in recent years, the demand for methanol has risen sharply [32]. Before 1923, methanol was a by-product of the preparation of charcoal by destructive distillation of wood. At that time, methanol was just applied for lighting, cooking, and heating purposes [35]. Beginning in the 1920s, scientists from BASF in Germany successfully developed and commercialized a syngas-to-methanol process by using $ZnO–Cr_2O_3$ catalysts under typical operating conditions of 320–450 °C and 25–35 MPa of CO and H_2 mixture [36]. Subsequently, this technique is widely used in the large-scale production of methanol. Until now, methanol is still exclusively produced from fossil fuel-derived syngas. In coal-rich countries, for example, South Africa and China, syngas derived from coal is the major source for methanol production, but these processes are associated with the generation of significant amounts of CO_2 placing a serious burden on the environment.

Recently, the research and development trend in methanol synthesis is shifting toward a more sustainable process via CO_2 hydrogenation, provided hydrogen is able to be produced in a renewable manner. Nevertheless, the hydrogenation of CO_2 to methanol is not a new concept. Indeed, this reaction can be dated back for more than 90 years, initially developed almost at the same time with the syngas-to-methanol process. As early as 1920s–1930s, several methanol plants in the United States started to use CO_2 as raw material for methanol production. Later, more efficient catalysts based on Cu and other metal/metal oxide components have been developed for CO_2 hydrogenation conversion. In the 1960s, Imperial Chemical Industries (ICI) for the first time developed a highly selective $Cu–ZnO–Al_2O_3$ catalyst for the methanol synthesis via CO_2 hydrogenation, and until now, Cu and Zn species remain indispensable in any industrial methanol synthesis catalysts because of their excellent performances as well as economic competitiveness.

Figure 11.3 shows the chemical equation (A) of CO_2 hydrogenation to methanol. The value of ΔH°_{298K} is $-49.5\ kJ\ mol^{-1}$ indicates that this reaction is exothermic. Often, reverse water gas shift (RWGS) occurs as a major side reaction in the process of methanol formation from CO_2 (Figure 11.3, Eq. B). As such, there is the possibility that the CO formed via RWGS undergoes further reduction into

$$CO_2 + 3H_2 \rightarrow CH_3OH + H_2O \quad \Delta H^{\circ}_{298K} = -49.5\ kJ/mol \quad (A)$$

$$CO_2 + H_2 \rightarrow CO + H_2O \quad \Delta H^{\circ}_{298K} = +41.2\ kJ/mol \quad (B)$$

$$CO + 2H_2 \rightarrow CH_3OH \quad \Delta H^{\circ}_{298K} = -90.7\ kJ/mol \quad (C)$$

Figure 11.3 Reaction equations of CO_2 hydrogenation and the formation of methanol.

methanol (Figure 11.3, Eq. C). Based on the thermodynamic properties of the reactions and Le Chatelier's principle, the increase of pressures and decrease of temperatures are favorable for methanol formation. While for the RWGS reaction, increasing the temperature is more advantageous, and the pressure has essentially no effect on the equilibrium. Screening optimized reaction conditions, fabricating high-performance catalysts, and achieving outstanding selectivity for methanol synthesis are of great significance and challenge where catalyst development is the key.

In the past decade, more than 70% of the published reports concerning CO_2 hydrogenation to methanol are still associated with Cu–ZnO composites. In general, the metallic Cu and ZnO serve as an active phase and promoter, respectively. For further improving the catalytic performance, additives such as Al_2O_3, ZrO_2, and CeO_2 were usually introduced into the catalysts. Typically, a commercial Cu–ZnO–Al_2O_3 catalyst for methanol synthesis is prepared by a coprecipitation method where the molar percentages of Cu, ZnO, and Al_2O_3 are in the range of 50–70%, 20–50%, and 5–20%, respectively. These catalysts are usually calcined in air and subsequently reduced in H_2 under atmospheric pressure before the reaction to generate zero-valent Cu NPs with high metal stability and dispersion as well as large surface areas [37]. So far, there are several synthesis methods for the preparation of Cu–ZnO-based catalysts, including coprecipitation, sol–gel, citrate, and impregnation methods. Among these synthesis methods, the coprecipitation method is the most popularly and frequently used one, which accounts for about 70% of all reported methods for Cu–ZnO-based catalysts in the past decades. A typical coprecipitation process can be described as the precipitation of $M(OH)_x$ or/and $M(OH)_x(CO_3)_y$ (M = Cu, Zn, Al, Zr, etc.) by mixing with the metal precursor solution and basic precipitating agents (usually carbonates, bicarbonates, or hydroxides). The different addition orders of reactants and experimental parameters (e.g. temperature, pH value, aging period, and calcination or reduction treatment) can significantly affect the microcosmic structures and physicochemical properties that determine catalytic activity. In 2008, Baltes et al. systematically investigated the preparation of ternary Cu–ZnO–Al_2O_3 catalysts through a coprecipitation route under various experimental parameters. The best methanol synthesis activity was achieved over the catalyst obtained from the precursors precipitated in the pH range from 6 to 8 at 70 °C, an aging time of 20–60 minutes, and optimum calcination at 250–300 °C [38]. Moreover, the authors observed that some malachite-type residues resulting from the initial hydroxy carbonate precipitate could be present, which most likely favored the formation of highly active sites (CuO crystallites and Cu^0 NPs) during the reduction. Fan et al. reported the synthesis of Cu–ZnO–Al_2O_3 catalyst via the coprecipitation method with the assistance of microwave (MW) irradiation. The authors found that the MW irradiation during the aging process was beneficial for the activity and stability of the catalyst, and the optimized catalyst (aging at 80 °C for one hour) possessed the highest methanol yield among all obtained catalysts [39]. For Cu-based catalysts, the catalytic activity is structure sensitive and usually depends on Cu NP structures. Yoshihara and Campbell reported Cu (110) to be more active for CO_2 hydrogenation to FA than Cu (111) and Cu (100) [40]. A monolayer of adsorbed formates without other species such as carbon or oxygen impurity was observed on the surface of Cu (110) after reactions. However, according to ultra-high-vacuum experiments, Nakamura et al. found that Cu (110) did not show an expected high activity for CO_2 dissociation [41]. In disparate catalytic systems, probably, the active structures of catalysts have different effects on the reactivity. Moreover, some ions such as Na^+ and NO_3^- ions from precipitation agents caused severe Cu agglomeration, low metal dispersion, and unsatisfactory catalytic performance. To solve these problems, modified synthesis methods, such as using NH_4HCO_3 as a precipitation agent [42] and decomposition of metal ammonia complexes [43], were developed to improve the Cu dispersion and catalytic activity.

Introduction of metal oxides, such as ZrO_2, CeO_2, and Ga_2O_3, can also promote the catalytic activities of Cu-based catalysts. Xiao et al. employed an oxalate coprecipitation method to synthesize a quaternary Cu–ZnO–ZrO_2–TiO_2 composite catalyst [44]. The presence of ZrO_2 and TiO_2 in the materials enhanced the methanol productivity, which can be attributed to the decreased size of CuO and ZnO and therefore increased interactions between Cu and Zn species, as a result of Zr and Ti additions. Recently, Graciani et al. reported that the Cu–CeO_x interface is highly efficient for the synthesis of methanol [45]. Using ambient pressure X-ray Photoelectron Spectroscopy (XPS) and infrared reflection absorption spectroscopies, authors found that CO_2 is prone to adsorb onto a CeO_2(111) surface producing strongly bound unstable carbonate species, which fast transform over CeO_x/Cu(111) and Cu/CeO_x/TiO_2(110) to generate methanol. A more rapid methanol generation rate than that of traditional Cu–ZnO catalysts was observed. Recently, Fornero et al. synthesized a ternary Cu–Ga_2O_3–ZrO_2 catalyst (2 wt% Cu) by depositing Ga_2O_3 onto a dried and calcined Cu–ZrO_2 precursor via an incipient wetness impregnation method [42]. This ternary Cu–Ga_2O_3–ZrO_2 catalyst exhibited progressively higher yield to methanol than that of the binary Cu–ZrO_2 and Cu–Ga_2O_3 counterparts. These works above promise that fabricating multicomponent catalysts with various active metal oxides is a powerful strategy for improving the catalytic performance of the methanol production.

Besides Cu-based catalysts, some Pd NP-based catalysts were utilized for methanol synthesis from CO_2 hydrogenation. In 1995, Fujitani et al. for the first time reported the synthesis of Pd/Ga_2O_3 catalysts by a coprecipitation method for the methanol production, which was more active than the Cu/ZnO catalyst [46]. The authors also investigated the support effect on various Pd-based catalysts and found that the methanol yield highly depended on different metal oxide supports and decreased gradually in the following order: $Ga_2O_3 > ZnO > A1_2O_3 > TiO_2 = Cr_2O_3 > SiO_2 = ZrO_2$. Subsequently, other Pd/Ga_2O_3-based catalysts were developed [47]. The atomic hydrogen readily formed on Pd^0 or/and Pd–Ga bimetallic NPs and then spills over on Ga_2O_3 sites to reach the reactive carbon-containing species and finished the catalytic cycle. Moreover, the authors also inferred that the bimetallic phase may inhibit the methanol decomposition and/or reverse water gas shift reaction, resulting in low CO production. Besides Ga_2O_3, several other species also allowed to improve the catalytic performance of Pd-based catalysts. Koizumi et al. reported mesoporous material-encaged Pd-base catalysts, showing excellent Pd dispersion, while introducing alkali/alkaline earth metal additives, which further enhanced the catalytic performance of the catalysts [48]. Jiang et al. synthesized a SiO_2-supported Pd–Cu bimetallic catalyst (Pd/(Pd + Cu) = 0.25–0.34) [49]. The Pd(0.25)–Cu/SiO_2 exhibited two folds higher than the physical mixture of those over monometallic Cu and Pd catalysts, and the Pd(0.34)–Cu/SiO_2 showed the maximum CH_3OH generation rate among all bimetallic catalysts. Pd–Cu alloy formation is regarded as a crucial factor for the generation rate of methanol because of the bimetallic promotion and strong synergistic effect. Similarly, some recent works revealed that Pd–Zn alloy NPs were conducive to the methanol production from CO_2 hydrogenation [50].

In recent years, developing a facile strategy to produce active heterogeneous catalysts for low-pressure CO_2 hydrogenation into CH_3OH has attracted great interests. Studt et al. reported a series of Ni–Ga intermetallic compounds with different ratios of Ni/Ga [51]. By employing a descriptor-based analysis of the process and computational methods, the authors identified that Ni–Ga intermetallic compounds were stable candidates for highly efficient reduction of CO_2 to methanol at atmospheric pressure and with a CO_2-to-H_2 ratio of 1:3. Among all Ni–Ga NP catalysts, the Ni_5Ga_3 NPs exhibited the highest activity and selectivity for the methanol production, a performance superior to that of the existing Cu–ZnO–Al_2O_3 catalyst with respect to their ability to limit the amount of CO in favor of methanol production. Li et al. reported a hybrid catalyst consisting of MnO_x NPs supported onto mesoporous spinel Co_3O_4, which exhibited 30% methanol selectivity and $0.18\,s^{-1}$ yield under low pressures (<6 bar of CO_2/H_2/He mixture) [52]. The superior catalytic activity was attributed to the existence of an active interface between Co_3O_4 surface layers and MnO_x NPs as confirmed by X-ray absorption and electron energy loss spectroscopies. More recently, the CO_2 hydrogenation to methanol under low-pressure condition was also achieved over Au NPs activated on a CeO_x/TiO_2 interface [53] and Cu-decorated ZnO nanorod array [54].

11.4 CO_2 Hydrogenation to Dimethyl Ether

DME, as the simplest ether, is a colorless, nontoxic, noncorrosive, and environmentally friendly chemical. The cetane number of DME is between 55 and 60, higher than that of the conventional diesel fuel, rendering it a promising fuel substitute for the diesel engine [55]. DME can also be conveniently used as a fuel for cooking, domestic heating, and small-scale electric power generation applications. Similar to methanol production, DME is currently exclusively produced from fossil fuel-based syngas. In recent years, direct CO_2 hydrogenation to DME has drawn great interests among the scientific community. As shown in Figure 11.4, in the single-step conversion, methanol synthesis and methanol dehydration simultaneously occur over a bifunctional catalyst.

Indeed, this reaction can be regarded as a cascade process, where the bifunctional catalysts contain an active component for methanol synthesis (often in the form of metal NPs) [45, 56], as well as an active component for methanol dehydration (often

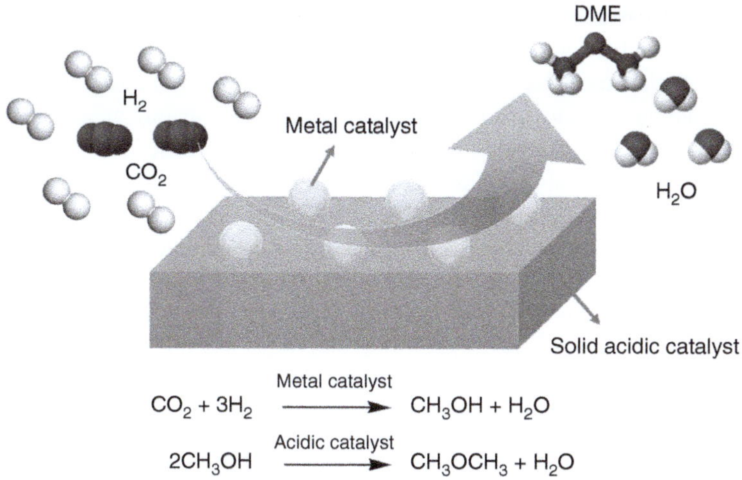

Figure 11.4 Schematic illustration of the CO_2 hydrogenation into DME via a bifunctional catalyst.

conducted by solid acid catalysts) [57]. There are two main protocols to prepare the hybrid bifunctional catalysts, including the physical or integrated mixing methods. γ-Al_2O_3 is a commonly used catalyst for methanol dehydration reaction because of its acid functions and relatively high surface area. However, investigations revealed that the presence of water decreased the activity of γ-Al_2O_3. Water was prone to absorb on the catalyst surface, thus inhibiting the adsorption of methanol and blocking the active sites responsible for dehydration [58]. Many efforts were devoted to developing higher performance catalysts with enhanced tolerance toward water. The acidic zeolites, such as H-ZSM-5, H-USY, and SAPO-34 zeolites, possessing both Lewis and relatively stronger Brønsted acid sites relative to γ-alumina, exhibited many advantages including high activity, superior water resistance, and long-term stability during the reaction. Bonura et al. combined a Cu–ZnO–ZrO_2 methanol synthesis catalyst and a commercial H-ZSM-5 zeolite by a physical mixing method to realize a one-step CO_2 hydrogenation reaction to DME [59]. They also found that the synergetic effect of metal/oxide(s)–acidic zeolite interface was the main driving force for the enhanced CO_2 conversion and product yield. Moreover, high DME yield was achieved over bifunctional catalyst systems by introducing several cations, such as La and Zr, which further strengthen the acid sites of zeolites. Apart from the H-ZSM-5 zeolite, other zeolites were also successfully employed for methanol dehydration reaction. Dubois et al. reported several physically mixed catalysts combining a Cu–ZnO–Al_2O_3 catalyst and different solid acid catalysts [60]. Among them, the zeolites, especially Y-zeolite (Si/Al = 3) and MOR-zeolite (Si/Al = 5), exhibited satisfactory selectivity and productivity toward DME. Compared with NaY with weak acid sites, the CuNaY zeolite was the most efficient catalyst as a result of their moderate acidity, further indicating that the acid parameter of zeolite is the most important factor toward DME production.

Apart from the physical mixing method, the catalysts can be synthesized by an integrated mixing method, often achieved by using a one-pot synthesis strategy to manipulate the catalytic functions of the two cascade reactions by incorporating spatially adjacent active sites to promote the DME synthesis. Naito et al. first reported such an integrated catalyst for the direct DME synthesis [61]. They combined many transition metal chlorides with graphite and sodium as a catalyst and found that graphite–$PdCl_3$–Na was the most promising catalyst. The integrated mixing method was also employed for the catalyst preparation combining ZSM-5-type zeolite catalyst and Cu-based methanol synthesis catalyst. Sun et al. prepared a series of Pd NP-modified CuO–ZnO–Al_2O_3–ZrO_2/HZSM-5 catalysts by coprecipitating sedimentation method and found that the incorporation of Pd significantly improved the reaction attributed to the enhanced hydrogen spillover from zero-valent Pd NPs to the adjacent active sites [62]. Afterward, many researchers used this integrated mixing method to manufacture several hybrid catalysts combining Cu-based and zeolite-based components, which not only simplified synthesis steps but also exhibited better selectivity toward DME compared with the conventional hybrid-type catalysts [63].

11.5 Perspectives and Conclusion

Thanks to the continuous efforts from both academia and industry in recent decades, important advancements in heterogeneous catalysis for the hydrogenation of CO_2 to value-added chemicals have been achieved. Nevertheless, there is still a long way to go for practical applications of CO_2 hydrogenation on a large scale. For these processes to become economically feasible, several key issues should be addressed in future investigations. Foremost, developing more efficient technologies for the capture of pure CO_2 and production of green H_2 is the basis for CO_2 hydrogenation. So far, the CO_2 capture in massive amounts from the air highly depends on the development of scalable membrane technology. The future advancements of membrane separation may make it possible to utilize CO_2 as an attractive feedstock. Developing more facile and cost-effective methods for the efficient generation of H_2 is also crucial and challenging. Until now, electrochemical catalysis powered by photovoltaics and wind power, as well as steam reforming of natural gas, exhibited some advantages to produce the H_2, but the efficiency of catalysts and production costs should be further enhanced and decreased, respectively.

Besides the production of oxygenated chemicals mentioned in this chapter, plenty of hydrocarbon chemicals, such as methane, light olefins, and aromatic hydrocarbons, can also be produced via CO_2 hydrogenation process. These chemicals are widely used as raw materials for the manufacture of various high value-added fine chemicals. Developing more efficient hybrid heterogeneous catalysts with multiple active components and novel catalytic processes should facilitate the development of CO_2 hydrogenation reactions in mild conditions. In addition, for the fundamental research on the heterogeneous catalysis of CO_2 hydrogenation, more efforts should be directed to the understanding of reaction mechanisms, identification of active sites and their structural dynamics, and determination of rate-limiting steps. High-throughput synthetic platforms and testing systems with the aid of artificial intelligence will be another driving force for catalyst development. These, as we envisage, will lead to the discovery of the next-generation catalysts and subsequently offer new opportunities for more environmentally benign and economically more competitive CO_2 hydrogenation processes featured by decreased temperature, decreased pressure, and increased productivity. We hope this chapter provides an overview of the history and current status of CO_2 hydrogenation using supported NP catalysts, and we believe that this is an area of significant further potential to be exploited in the next decade and beyond.

Acknowledgment

We thank the NUS Flagship Green Energy Program for the Financial Support.

References

1 (a) Knutson, T.R. and Tuleya, R.E. (2004). *J. Clim.* 17: 3477–3495. (b) Hansen, J., Sato, M., Ruedy, R. et al. (2006). *Proc. Natl. Acad. Sci. U.S.A.* 103: 14288–14293. (c) Kahn, B. (2016). *Earth's CO_2 Passes the 400 PPM Threshold – Maybe Permanently.* Climate Central. (d) Princiotta, F. (2011). *Global Climate Change – The Technology Challenge*, 1–50. Springer.

2 (a) Sakakura, T., Choi, J.-C., and Yasuda, H. (2007). *Chem. Rev.* 107: 2365–2387. (b) Otto, A., Grube, T., Schiebahn, S., and Stolten, D. (2015). *Energy Environ. Sci.* 8: 3283–3297.

3 Álvarez, A., Bansode, A., Urakawa, A. et al. (2017). *Chem. Rev.* 117: 9804–9838.

4 (a) Wang, W.-H., Himeda, Y., Muckerman, J.T. et al. (2015). *Chem. Rev.* 115: 12936–12973. (b) Porosoff, M.D., Yan, B., and Chen, J.G. (2016). *Energy Environ. Sci.* 9: 62–73. (c) Li, W., Wang, H., Jiang, X. et al. (2018). *RSC Adv.* 8: 7651–7669.

5 (a) Gattrell, M., Gupta, N., and Co, A. (2006). *J. Electroanal. Chem.* 594: 1–19. (b) Whipple, D.T. and Kenis, P.J.A. (2010). *J. Phys. Chem. Lett.* 1: 3451–3458. (c) Xie, H., Wang, T., Liang, J. et al. (2018). *Nano Today* 21: 41–54.

6 (a) Ran, J., Jaroniec, M., and Qiao, S.-Z. (2018). *Adv. Mater.* 30: 1704649. (b) Habisreutinger, S.N., Schmidt-Mende, L., and Stolarczyk, J.K. (2013). *Angew. Chem. Int. Ed.* 52: 7372–7408. (c) Low, J., Cheng, B., and Yu, J. (2017). *Appl. Surf. Sci.* 392: 658–686.

7 (a) Grasemann, M. and Laurenczy, G. (2012). *Energy Environ. Sci.* 5: 8171–8181. (b) Boddien, A. and Junge, H. (2011). *Nat. Nanotechnol.* 6: 265–266.

8 (a) Sun, Q., Wang, N., Bing, Q. et al. (2017). *Chem* 3: 477–493. (b) Wang, N., Sun, Q., Bai, R. et al. (2016). *J. Am. Chem. Soc.* 138: 7484–7487. (c) Li, Z. and Xu, Q. (2017). *Acc. Chem. Res.* 50: 1449–1458. (d) Zhu, Q.-L. and Xu, Q. (2015). *Energy Environ. Sci.* 8: 478–512.

9 (a) Wang, W., Niu, M., Hou, Y. et al. (2014). *Green Chem.* 16: 2614–2618. (b) Reutemann, W. and Kieczka, H. (2012). Formic Acid. In: *Ullmann's Encyclopedia of Industrial Chemistry*, 14–33. Weinheim: Wiley-VCH.

10 (a) Shao, X., Yang, X., Xu, J. et al. (2019). *Chem* 5: 693–705. (b) Su, X., Yang, X.-F., Huang, Y. et al. (2018). *Acc. Chem. Res.* 52: 656–664. (c) Bulushev, D.A. and Ross, J.R.H. (2018). *Catal. Rev.* 60: 566–593.

11 Jessop, P.G., Ikariya, T., and Noyori, R. (1994). *Nature* 368: 231.

12 Inoue, Y., Izumida, H., Sasaki, Y., and Hashimoto, H. (1976). *Chem. Lett.* 5: 863–864.

13 (a) Klankermayer, J., Wesselbaum, S., Beydoun, K., and Leitner, W. (2016). *Angew. Chem. Int. Ed.* 55: 7296–7343. (b) Jessop, P.G., Ikariya, T., and Noyori, R. (1995). *Chem. Rev.* 95: 259–272. (c) Jessop, P.G., Joó, F., and Tai, C.-C. (2004). *Coord. Chem. Rev.* 248: 2425–2442.

14 Filonenko, G.A., van Putten, R., Schulpen, E.N. et al. (2014). *ChemCatChem* 6: 1526–1530.

15 Gunasekar, G.H., Park, K., Jung, K.-D., and Yoon, S. (2016). *Inorg. Chem. Front.* 3: 882–895.

16 Bredig, G. and Carter, S.R. (1914). *Ber. Dtsch. Chem. Ges.* 47: 541–545.

17 Farlow, M.W. and Adkins, H. (1935). *J. Am. Chem. Soc.* 57: 2222–2223.

18 Stalder, C.J., Chao, S., Summers, D.P., and Wrighton, M.S. (1983). *J. Am. Chem. Soc.* 105: 6318–6320.

19 Preti, D., Resta, C., Squarcialupi, S., and Fachinetti, G. (2011). *Angew. Chem.* 123: 12759–12762.

20 Filonenko, G.A., Vrijburg, W.L., Hensen, E.J.M., and Pidko, E.A. (2016). *J. Catal.* 343: 97–105.

21 Su, J., Yang, L., Lu, M., and Lin, H. (2015). *ChemSusChem* 8: 813–816.

22 Hao, C., Wang, S., Li, M. et al. (2011). *Catal. Today* 160: 184–190.

23 Bi, Q.-Y., Lin, J.-D., Liu, Y.-M. et al. (2014). *Angew. Chem. Int. Ed.* 53: 13583–13587.

24 Lee, J.H., Ryu, J., Kim, J.Y. et al. (2014). *J. Mater. Chem. A* 2: 9490–9495.

25 Nguyen, L.T.M., Park, H., Banu, M. et al. (2015). *RSC Adv.* 5: 105560–105566.

26 Mori, K., Sano, T., Kobayashi, H., and Yamashita, H. (2018). *J. Am. Chem. Soc.* 140: 8902–8909.

27 Sun, Q., Fu, X., Si, R. et al. (2019). *ChemCatChem* 11: 5093–5097. https://doi.org/10.1002/cctc.201901167.

28 Liu, N., Du, R.J., and Li, W. (2013). *Adv. Mater. Res.* 821–822: 1330–1335.

29 Mori, K., Masuda, S., Tanaka, H. et al. (2017). *Chem. Commun.* 53: 4677–4680.

30 Mori, K., Taga, T., and Yamashita, H. (2017). *ACS Catal.* 7: 3147–3151.

31 (a) Zhang, Y., Fei, J., Yu, Y., and Zheng, X. (2004). *Catal. Commun.* 5: 643–646. (b) Xu, Z., McNamara, N.D., Neumann, G.T. et al. (2013). *ChemCatChem* 5: 1769–1771. (c) Lo, H.-K. and Copéret, C. (2019). *ChemCatChem* 11: 430–434.

32 (a) Tian, P., Wei, Y., Ye, M., and Liu, Z. (2015). *ACS Catal.* 5: 1922–1938. (b) Sun, Q., Xie, Z., and Yu, J. (2017). *Natl. Sci. Rev.* 5: 542–558. (c) Chinchen, G.C., Denny, P.J., Jennings, J.R. et al. (1988). *Appl. Catal.* 36: 1–65. (d)

Pérez-Fortes, M., Schöneberger, J.C., Boulamanti, A., and Tzimas, E. (2016). *Appl. Energy* 161: 718–732.

33 Jadhav, S.G., Vaidya, P.D., Bhanage, B.M., and Joshi, J.B. (2014). *Chem. Eng. Res. Des.* 92: 2557–2567.

34 Wasmus, S. and Küver, A. (1999). *J. Electroanal. Chem.* 461: 14–31.

35 Olah, G.A., Goeppert, A., and Prakash, G.K.S. (2009). *J. Org. Chem.* 74: 487–498.

36 Mittasch, A., Pier, M., and Winkler, K. (1923). BASF. German Patents 1923, 415,686, 441,433, 462,837; US Patents 1923, 1,558,559, 1,569,775.

37 Bansode, A. and Urakawa, A. (2014). *J. Catal.* 309: 66–70.

38 Baltes, C., Vukojević, S., and Schüth, F. (2008). *J. Catal.* 258: 334–344.

39 Fan, H., Zheng, H., and Li, Z. (2010). *Front. Chem. Eng. China* 4: 445–451.

40 Yoshihara, J. and Campbell, C.T. (1996). *J. Catal.* 161: 776–782.

41 Nakamura, J., Rodriguez, J.A., and Campbell, C.T. (1989). *J. Phys.: Condens. Matter* 1: SB149.

42 Fornero, E.L., Sanguineti, P.B., Chiavassa, D.L. et al. (2013). *Catal. Today* 213: 163–170.

43 Wang, D., Zhao, J., Song, H., and Chou, L. (2011). *J. Nat. Gas Chem.* 20: 629–634.

44 Xiao, J., Mao, D., Guo, X., and Yu, J. (2014). *Energy Technol.* 3: 32–39.

45 Graciani, J., Mudiyanselage, K., Xu, F. et al. (2014). *Science* 345: 546–550.

46 Fujitani, T., Saito, M., Kanai, Y. et al. (1995). *Appl. Catal., A* 125: L199–L202.

47 (a) Collins, S.E., Chiavassa, D.L., Bonivardi, A.L., and Baltanás, M.A. (2005). *Catal. Lett.* 103: 83–88. (b) Collins, S.E., Delgado, J.J., Mira, C. et al. (2012). *J. Catal.* 292: 90–98.

48 Koizumi, N., Jiang, X., Kugai, J., and Song, C. (2012). *Catal. Today* 194: 16–24.

49 Jiang, X., Koizumi, N., Guo, X., and Song, C. (2015). *Appl. Catal., B* 170–171: 173–185.

50 (a) Xu, J., Su, X., Liu, X. et al. (2016). *Appl. Catal., A* 514: 51–59. (b) Díez-Ramírez, J., Valverde, J.L., Sánchez, P., and Dorado, F. (2016). *Catal. Lett.* 146: 373–382.

51 Studt, F., Sharafutdinov, I., Abild-Pedersen, F. et al. (2014). *Nat. Chem.* 6: 320.

52 Li, C.-S., Melaet, G., Ralston, W.T. et al. (2015). *Nat. Commun.* 6: 6538.

53 Yang, X., Kattel, S., Senanayake, S.D. et al. (2015). *J. Am. Chem. Soc.* 137: 10104–10107.

54 Du, S., Tang, W., Lu, X. et al. (2018). *Adv. Mater. Interfaces* 5: 1700730.

55 (a) Arcoumanis, C., Bae, C., Crookes, R., and Kinoshita, E. (2008). *Fuel* 87: 1014–1030. (b) Semelsberger, T.A., Borup, R.L., and Greene, H.L. (2006). *J. Power Sources* 156: 497–511.

56 (a) Behrens, M., Studt, F., Kasatkin, I. et al. (2012). *Science* 336: 893–897. (b) Wang, J., Li, G., Li, Z. et al. (2017). *Sci. Adv.*: 3, e1701290.

57 Xu, M., Lunsford, J.H., Goodman, D.W., and Bhattacharyya, A. (1997). *Appl. Catal., A* 149: 289–301.

58 Aguayo, A.T., Ereña, J., Sierra, I. et al. (2005). *Catal. Today* 106: 265–270.

59 Bonura, G., Cordaro, M., Spadaro, L. et al. (2013). *Appl. Catal., B* 140–141: 16–24.

60 Dubois, J., Sayama, K., and Arakawa, H. (1992). *Chem. Lett.* 21: 1115–1118.

61 Naito, S., Ogawa, O., Ichikawa, M., and Tamaru, K. (1972). *J. Chem. Soc., Chem. Commun.* 3: 1266–1266.

62 Sun, K., Lu, W., Wang, M., and Xu, X. (2004). *Catal. Commun.* 5: 367–370.

63 (a) Frusteri, F., Bonura, G., Cannilla, C. et al. (2015). *Appl. Catal., B* 176–177: 522–531. (b) Zhang, Y., Li, D., Zhang, Y. et al. (2014). *Catal. Commun.* 55: 49–52.

12

Rebirth of Ruthenium-Based Nanomaterials for the Hydrogen Evolution Reaction

Nuria Romero, Jordi Creus, Jordi García-Antón, Roger Bofill, and Xavier Sala

Universitat Autònoma de Barcelona, Facultat de Ciències, Departament de Química, Bellaterra, Catalonia 08193, Spain

12.1 Introduction

Global warming due to elevated fossil fuel consumption has put forward the imperative need for a clean and carbon-neutral energy source. One attractive way to face this challenge is the production of H_2 by the splitting of water (Eq. (12.1)) either photochemically or photoactivated by sunlight. By analogy with what green plants, algae, and cyanobacteria do to store sunlight energy in the form of chemical bonds, the production of H_2 from water is one of the processes in the so-called artificial photosynthesis [1]. Thus, most frequently, water is oxidized to O_2 at the anode (Eq. (12.2)), and the released electrons and protons generate H_2 at the cathode (Eq. (12.3)).

$$2H_2O \rightarrow O_2 + 2H_2 \tag{12.1}$$

$$2H_2O \rightarrow O_2 + 4H^+ + 4e^- \tag{12.2}$$

$$4H^+ + 4e^- \rightarrow 2H_2 \tag{12.3}$$

However, oxidation of water (12.2) and generation of H_2 (12.3) are kinetically slow processes. Therefore, the use of a catalyst is required to speed up their rates and make the whole reaction competitive. Nanoparticles (NPs) constitute a very interesting option to catalyze these processes in view of their elevated specific surface area and the possibility of depositing/embedding them onto various supports as well as functionalizing their surface with different molecules [2]. Consequently, the use of NPs as the active species onto solid electrodes in electrocatalytic systems has become widespread in recent years.

For the hydrogen evolution reaction (HER), platinum has been traditionally the most used metal in the solid phase because it is the most active metal in reducing protons, especially at acidic pH. The key parameter that determines the efficiency of a heterogeneous HER catalyst is its M—H bond energy. Thus, a high M–H binding energy facilitates the adsorption of protons at acidic pH but hinders the ulterior

Nanoparticles in Catalysis: Advances in Synthesis and Applications, First Edition.
Edited by Karine Philippot and Alain Roucoux.

desorption of H_2. On the contrary, a low M–H binding energy has the exact opposite effect. Equilibrium between both extreme situations is thus desirable. Platinum fulfills exactly this condition because it is located at the center of the volcano plot for proton reduction catalysts because its M–H binding energy is neither too low nor too high [3, 4]. In comparison to Pt, ruthenium, presents a M—H bond only slightly weaker, thus hardly decreasing the HER catalytic efficiency [5]. Besides, metallic Ru accommodates versatile working conditions because it is very stable under both acidic and basic pH, in clear contrast to Pt, whose long-term stability at basic pH is not optimal [6]. In the case of neutral or basic conditions, though, it is the Ru–O binding strength (either Ru–OH_2 at neutral pH or Ru–OH at basic pH) that initially rules the capacity of the NPs to perform the reaction. A strong bonding energy promotes water/OH^- adsorption at the surface of the NPs, which at the same time favors water/OH^- dissociation to eventually generate H_2. Finally, and very importantly and judiciously, the cost of Ru is significantly less than that of Pt (at least six times). These characteristics altogether explain the rebirth of Ru metal in nanoparticulate form as a HER catalyst as it is seen for the past five years [7].

Given a recent exhaustive review, [7] this chapter is not intended to provide a comprehensive summary of the most active Ru-based NP systems in HER catalysis but to present the key factors that govern the HER electrocatalytic and photocatalytic performance of Ru-containing nanocatalysts. For this aim, a set of remarkable examples will be described to illustrate these factors. Given the variability of Ru nanoparticulate systems published, the first section will focus on the most common accepted ways to objectively assess (benchmark) the HER activity of NPs for comparative purposes. Then, the general trends observed among the most promising electrochemical and photochemical systems will be critically discussed in order to finally point to possible future research directions in the field.

12.2 Relevant Figures of Merit

The most common way to test the HER catalytic activity of Ru-based NP systems is through electrochemical measurements after their deposition onto conducting electrodes, typically glassy carbon (GC). For that purpose, the use of rotating disk electrodes (RDEs) is highly recommended because H_2 is continuously evolving from the solution during HER. The constant rotation of the disk induces a laminar flow of solution toward and across the electrode, which controls the steady-state current, while H_2 gas accumulation at the electrode surface is minimized. Therefore, by the use of RDE, a good accessibility of the substrate to the catalyst active sites during catalysis as well as a more trustable determination of the amount of evolved H_2 with specific probes can be ensured. However, only five years ago, there was no objective way to compare the HER catalytic activity of NPs in front of the widespread set of conditions (types of support, catalyst loadings, size, shape, and chemical nature of NPs and their stabilizers, working potentials, pH, etc.) employed throughout the literature. Thus, the electrochemical activity of heterogeneous catalysts was traditionally normalized by the total amount of catalyst present, yielding the so-called mass activity parameter (expressed as A mg^{-1} or A g^{-1}). [8] However, mass activity represents an average activity and does not take into account the existence of different sites (e.g. exposed/active sites vs. unexposed/inactive sites) in the catalytic sample. Fortunately, from then on, a new set of benchmarking parameters were proposed and widely accepted/employed by the researchers in the field in order to meaningfully assess the intrinsic activity of heterogeneous HER catalytic systems in a relatively fast and affordable manner [6, 8].

The first benchmarking parameter is the overpotential at the beginning of the electrocatalytic process (η_0), that is, the difference between the thermodynamic equilibrium potential of the desired reaction (in this case, the reduction of protons) and the actual potential where the catalytic reaction starts for each catalyst. η_0 can be determined by the appearance of a clear change on the current intensity (i) during a voltammetric measurement, supposing that no other reduction processes involving nanoparticle stabilizers, metal centers, and so on take place in the same potential region. However, the accurate determination of the real point where the slope of a voltammetric experiment starts to change is not always trivial, and thus, η_0 is not absent of subjectivity. In order to overcome this difficulty, the overpotential required to achieve a current density ($|j|$) of 10 mA cm^{-2} (η_{10}), which is the current density for a solar water-splitting device exposed to one sun illumination and operating at a standard solar-to-fuel efficiency of 10%, has appeared as one of the most commonly used benchmarking parameters. Furthermore, with the aim of assessing the stability of the studied systems, η_{10} is frequently compared at the beginning and after a certain period of time after uninterrupted catalysis, generally two hours but also even more than 12 hours for long-term stability tests [6].

The precise determination of the current density ($|j|$, in mA cm^{-2}) is also crucial for correctly finding a sensible value of η_{10}. From this point of view, it is very important to calculate the real electrochemically active (or specific) surface area of the tested electrocatalysts. In many works, η_{10} has been calculated by dividing the experimental current intensity by the geometric area of the electrode instead of that of the electrocatalyst, thus providing no information about the intrinsic activity of the catalyst. To overcome this hurdle, two alternative methods have been proposed. The first one is the Brunauer–Emmett–Teller (BET) method, in which the specific surface area is determined by gas adsorption measurement. BET yields relatively accurate values of the *total* specific surface of the analyzed electrocatalyst. However, not all surface sites are necessarily active, and thus, this method leads to an overestimation of the catalytically active surface area and thus to an underestimation of the intrinsic activity of the real active sites [6]. The second method is based on electrochemical measurements in order to assess the *electrochemically active* surface area (ECSA). To do so, the double-layer capacitance (C_{DL}) of the material needs to be measured previously, which can be assessed in two different ways: (i) the measurement of cyclic voltammograms (CVs) at different scan rates in nonfaradaic regions [6] and (ii) electrochemical impedance spectroscopy (EIS) measurements also in nonfaradaic regions. The C_{DL} measured by both ways tend to agree within $\pm$15%, although some authors claim that the latter may be preferable. [8] After having in hand the ECSA value, the roughness factor (RF) is calculated as RF = ECSA/geometrical electrode area. Then, the current density $|j|$ is normalized by the RF, yielding the specific activity or specific current density ($|j_s| = |j|/RF$), which is a very useful parameter to meaningfully compare the performance of very diverse heterogeneous systems with different RFs and even to normalize the electroactivity of differently loaded electrodes of the same catalyst.

Also, the turnover number (TON) and turnover frequency (TOF) per real catalytically active site are two other very important benchmarking parameters. Normally, for the particular case of Ru, the determination of the total number of metallic active sites is achieved by the Cu underpotential deposition method (UPD), in which all metallic Ru active sites are initially saturated with Cu atoms by selective Cu^{2+} electroreduction. Then, a controlled oxidation leads to the release into solution of the previously deposited Cu atoms. These liberated Cu atoms can be quantified by the integration of the area of the Cu oxidation wave, thus informing on the number of accessible metallic Ru active sites [9]. Then, the TON and TOF per real active site can be calculated by dividing the total amount of H_2 gas molecules detected by a selective probe or the amount of evolved gas per time lapse by the total amount of active sites, respectively. Additionally, these values can also be deduced from voltammetric measurements if the total number of active sites is known [8].

Another very useful benchmarking parameter is the Tafel plot, the representation of log $|j|$ vs. η. When taking $\log(|j|)$, a linear relationship appears between η and $\log(|j|)$ in the form $\eta = a + b\log(|j|)$, whose slope b, expressed in mV dec^{-1}, gives important information on the kinetics and thus on the possible mechanistic pathway followed during the electrochemical process [8].

HER involves the formation of one H—H bond by the reduction of two H^+ groups. In heterogeneous systems, this process has been described to occur through two different reaction pathways, namely, the Volmer–Heyrovsky or the Volmer–Tafel pathways (Eqs. (12.4, 12.5) and Eqs. (12.4, 12.6), respectively) [10].

$$\text{Volmer step}: \text{Cat}_{(s)} + \text{H}^+_{(aq)} + e^- \rightarrow \text{Cat} - \text{H}_{(ad)} \tag{12.4}$$

$$\text{Heyrovsky step}: \text{Cat} - \text{H}_{(ad)} + \text{H}^+_{(aq)} + e^- \rightarrow \text{H}_{2(s)} + \text{Cat}_{(s)} \tag{12.5}$$

$$\text{Tafel step}: 2\,\text{Cat} - \text{H}_{(ad)} \rightarrow \text{H}_{2(s)} + 2\,\text{Cat}_{(s)} \tag{12.6}$$

The Volmer step (Eq. (12.4)) corresponds to the adsorption of one proton onto the catalyst surface, a common process for all HER electrocatalysts. Then, H_2 evolution can either take place through the electrodesorption of the adsorbed hydrogen atom with H^+ in solution (Heyrovsky step, Eq. (12.5)) or, in the case of colloidal catalysts, through the recombination of two Cat-H units from a unique NP or from two different ones (Tafel step, Eq. (12.6)). In HER electrocatalysis, the rate-determining step (rds) of the whole process depends on the binding energy of the M—H bond. Thus, when the rds is the Volmer step, a Tafel slope of $\approx$120 mV dec^{-1} is obtained, whereas when the rds is the Heyrovsky or the Tafel step, slopes of $\approx$40 mV dec^{-1} or $\approx$30 mV dec^{-1}, respectively, are obtained (Figure 12.1). However, sometimes, Tafel plots need to be analyzed with caution because of the inherent experimental inaccuracy in their assessment and the reported dependence of slope b with the applied

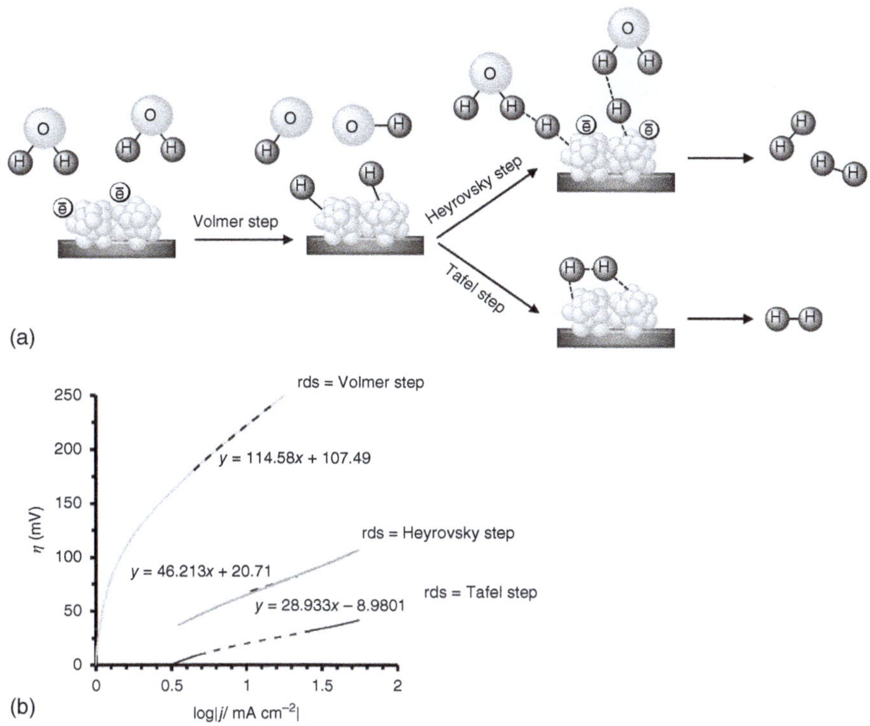

Figure 12.1 Schematic representation of the HER pathways on nanoparticulate electrocatalysts (a) and their correlation with the Tafel slope (b).

potential [10]. Moreover, strict steady-state conditions are required for the calculation of Tafel plots, which are not always accomplished [10].

Finally, the selectivity of an electrocatalyst is measured by the faradaic efficiency of a bulk electrolysis experiment. The faradaic efficiency is the percentage of electrons that has been effectively used for evolving gas according to Faraday's law. The evolved gas is typically analyzed by means of Clark-type electrodes either in solution or in the gas phase. Interestingly, the use of a rotating ring disk electrode (RRDE) as a working electrode allows determining the faradaic efficiency of the electrocatalysts deposited on its surface with no need of additional probes and at the same time avoids H_2 accumulation at its surface [8].

12.3 Factors Ruling the Performance of Ru-Based NPs in HER Electrocatalysis

In the following subsections, the main factors that govern the HER catalytic activity of Ru-based NPs after a critical comparison of the results of a set of noticeable systems will be discussed. The number of publications centered on the use of these nanomaterials for HER electrocatalysis has significantly risen since 2016, with most of the relevant literature published very recently. Within those, the great majority mainly consists in Ru-based NPs deposited or supported/embedded onto conductive C-based materials that are electrochemically triggered, although some works are also based on metallic and semimetallic supports.

12.3.1 Surface Composition

The first important parameter to consider when synthesizing metal-based NPs is the targeted composition of their surface (oxidation state/s, stabilizing agent/s, coordination/interaction modes, etc.) and whether or not it evolves during electrocatalysis. Thus, the systematic use of surface composition analysis techniques, such as X-ray photoelectron spectroscopy (XPS), among others, is extremely recommended. Among the different methods to synthesize metal-based NPs, the organometallic approach has been proven as an appropriate methodology to obtain NPs with controlled size, shape, oxidation state, and surface chemical composition (Figure 12.2) [11]. This method consists in the decomposition of an organometallic precursor under mild conditions (e.g. 3 bar of H_2, r.t.) in the presence of a stabilizing agent (also called as the capping agent, usually a ligand, but also the solvent itself – that can be an ionic liquid – or even supports of different natures, including polymers).

By using the organometallic method, surface clean and homogeneous in size NPs are obtained. For example, homogeneous 21.4 nm porous Ru NPs in a MeOH/THF mixture have been synthesized and successfully deposited onto GC electrodes [12]. These NPs show good activity in HER catalysis both at pH 7 and pH 0 (η_0 close to 0 mV, η_{10} of 83 mV, a Tafel slope of 46 mV dec^{-1}, a TOF$_{100\,mV}$ of 0.87 s^{-1}, and a specific current density $|j_s|$ of 0.36 mA cm^{-2} at 100 mV overpotential in 0.5 M H_2SO_4) and excellent durability during 12 hours at acidic pH thanks to their high elevated ECSA value (173.07 m^2/g) and the absence of strongly coordinated stabilizers onto their

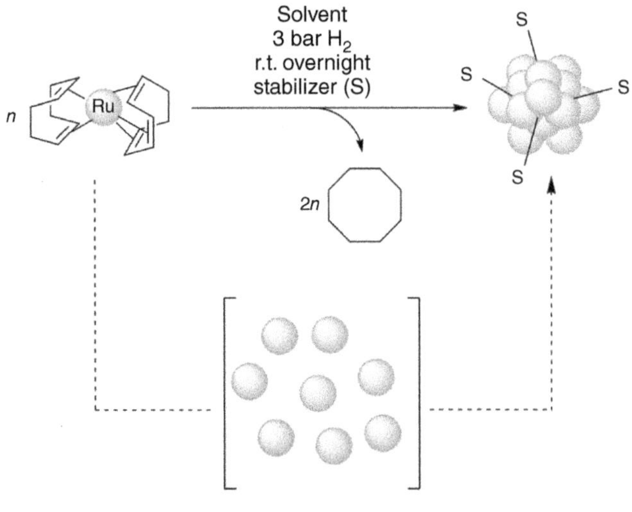

Figure 12.2 Schematic representation of the organometallic method for the synthesis of Ru-based NPs.

surface. Alternatively, by the addition of a coordinating ligand, metal NPs stabilized by the added capping ligand instead of the solvent can be obtained. Thus, we have synthesized 1.5 nm 4-phenylpyridine(PP)-capped Ru NPs in THF, [13] which were afterward drop-cast onto a GC electrode for electrochemical analysis. This system, called PP–Ru–GC, spontaneously gets partially oxidized at the surface to RuO_2 upon exposure to air, generating a mixed Ru/RuO_2 system where the PP ligand is still present (Figure 12.3).

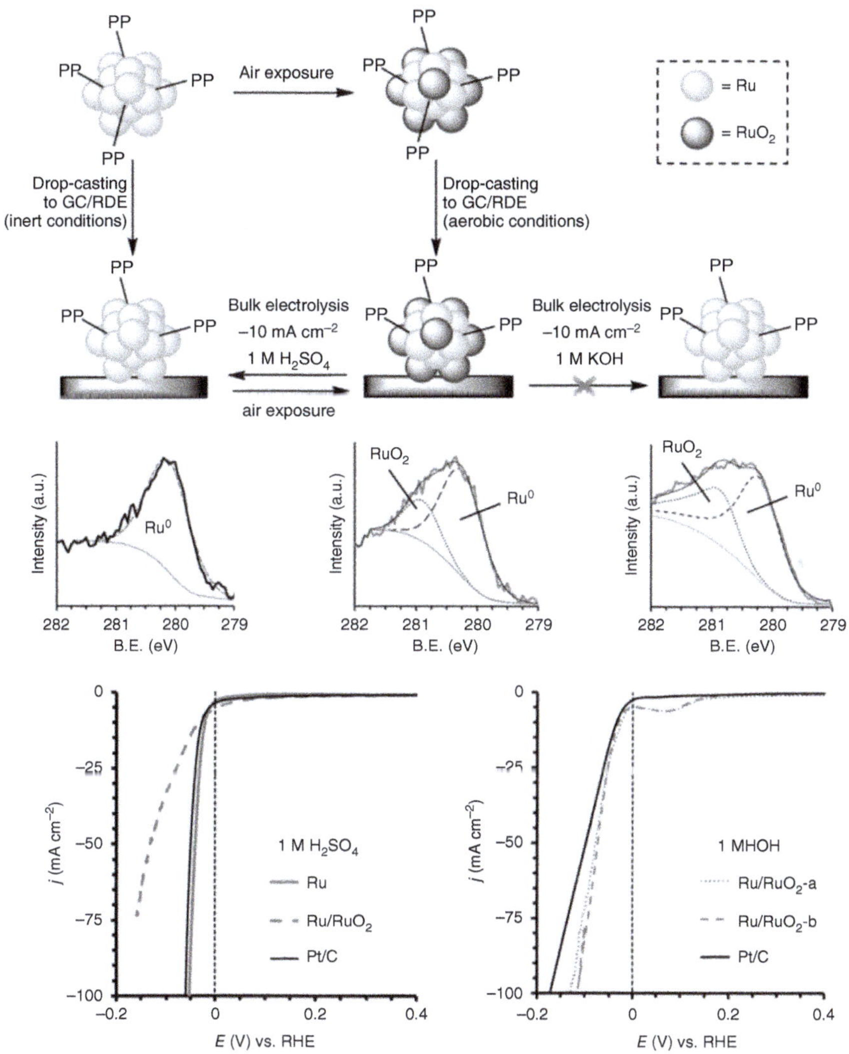

Figure 12.3 Schematic representation of the conversion processes between the Ru and RuO_2 states in the PP–Ru–GC system at acidic and basic pH (top), their corresponding XPS spectra (middle) and their LSV polarization curves (bottom) at acidic pH (lower-left side) and basic pH (lower-right side) before (a) and after (b) 20 minutes of bulk electrolysis at $-10\,mA\,cm^{-2}$. Source: Modified from Creus et al. [13].

Interestingly, the HER activity of the PP–Ru–GC system is clearly dependent on the oxidation state of the NPs. Thus, pure Ru NPs are very stable and more active in 1 M H_2SO_4 than the mixed Ru/RuO_2 system [13], the formers yielding η_0 close to 0 mV, η_{10} of 20 mV, a Tafel slope of 29 mV dec^{-1}, a very high $TOF_{100\,mV}$ of 17.4 s^{-1}, and a $|j_s|_{100\,mV}$ of 0.55 mA cm^{-2}, thus improving the results of the Ru–MeOH/THF system in the absence of capping ligands and also those of commercial Pt/C [12]. Also, at acidic pH, the Ru/RuO_2 state can be converted back into pure Ru form by applying a reductive potential of −10 mA cm^{-2} during 20 minutes (Figure 12.3) [13]. Very interestingly, other authors have also observed that RuO_2 NPs can catalyze the HER [14, 15]. However, in these works, no analysis of the oxidation state of the catalyst has been reported after HER catalytic experiments, and therefore, the possibility that the real active species may be metallic Ru cannot be disregarded. On the other hand, in our PP–Ru–GC system, the pure Ru form is not stable in 1 M NaOH; instead, the only stable form of the PP–Ru–GC system is the mixed Ru/RuO_2 form. This system, as stated before, is less active and slightly less stable than the pure Ru system at acidic pH (η_0 close to 0 mV, η_{10} close to 25 mV, and Tafel slope of 65 mV dec^{-1}), although still improves the benchmarks of commercial Pt/C (Figure 12.3). In this work, we conclude that the presence of PP molecules onto the surface of Ru NPs induces good mechanical stability, which allows the maintenance of their structure even after a long catalytic run. DFT calculations support well this hypothesis by showing the coordination of 11 PP capping agents onto the surface of the model $Ru_{55}H_{53}$ NPs (1 nm in size) through N-σ and π-coordination modes (Figure 12.4). Moreover, the presence of hydride species modifies the *d*-band energy levels of the surface Ru atoms, thus having a stabilizing effect, while these energy levels are basically not modified by the PP ligands. This demonstrates that the PP capping ligands only show a moderate adsorption strength to the NP surface, making more hydride species present onto the NP surface than PP molecules (53 compared to 11), which explains the excellent H_2 evolution behavior of the PP–Ru NPs.

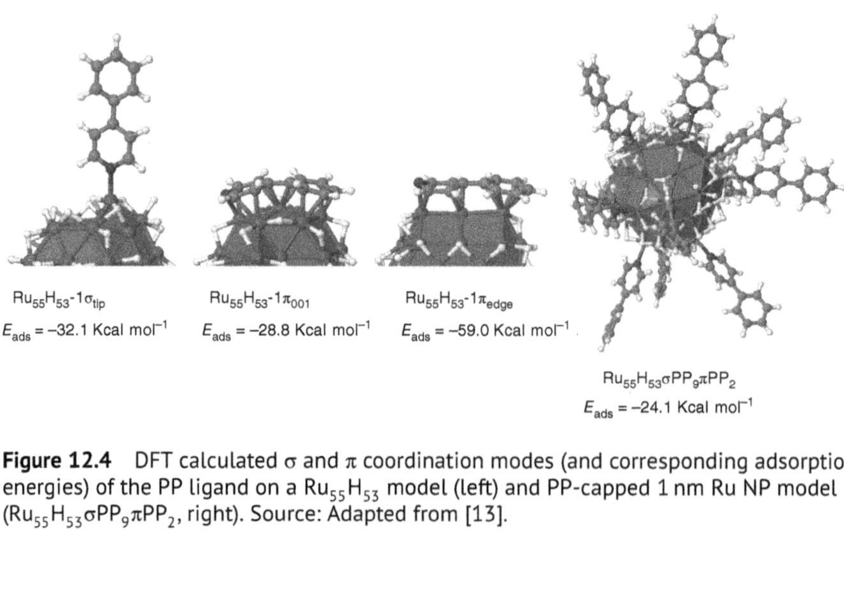

Figure 12.4 DFT calculated σ and π coordination modes (and corresponding adsorption energies) of the PP ligand on a $Ru_{55}H_{53}$ model (left) and PP-capped 1 nm Ru NP model ($Ru_{55}H_{53}\sigma PP_9\pi PP_2$, right). Source: Adapted from [13].

Also, the PP–Ru NPs have been shown to be active in HER catalysis under acidic conditions when being deposited onto the surface of carbon microfibers (CFs), thus directly providing cathodes for HER [16]. In this case, the PP-stabilized Ru NPs have been deposited on top of bare and COOH-functionalized CFs by impregnation of these supports with a colloidal solution of preformed NPs (ex situ systems) and their performance compared to that of systems obtained by the direct organometallic synthesis of Ru NPs in the presence of bare and COOH-functionalized CFs (in situ systems) without any added ligand. The ex situ systems (with PP ligands as stabilizers) show clearly lower η_0 and η_{10} values and higher stabilities than their in situ counterparts (only stabilized by the supports), thus evidencing once more the decisive role of the capping ligand. Furthermore, the high mechanical stability of the cathodes made of bare CFs and PP–Ru NPs is attributed to the π–π interactions between PP and the graphitic regions of the support.

With the previous examples, the potentiality of the organometallic approach for tuning the surface of Ru-based NPs and, consequently, the catalytic performance of the generated systems has been evidenced. The main objective of this approach is to unravel the effect that the inexhaustible pool of stabilizing ligands (or supports) can exert on the composition, redox properties, and reaction mechanisms and thus on the activity and stability of Ru-based NPs (as well as first-row transition-metal-based NPs) [17] in energy-related transformations such as the HER and the OER. Thus, with the aim of rationally tuning the surface active sites by the effect of such ligands, a thorough work is required by the systematic combination of experimental electrochemical, spectroscopic, and structural data (i.e. surface oxidation states and their evolution under turnover conditions, real active sites, surface hydrides count, etc.) with DFT theoretical modeling of the prepared systems.

12.3.2 Phase Structure and Degree of Crystallinity

It has been reported that metallic Ru NPs usually adopt an hcp structure [18]. However, several groups have observed the formation of an anomalous Ru-fcc structure or a mixture of Ru-fcc and Ru-hcp structures after annealing a Ru molecular precursor on C-based supports. This has been observed in the following three cases: (i) the solvothermal reduction of Ru^{3+} mixed with Cu^{2+}, followed by selective etching of Cu [18], (ii) the thermolysis of Ru-modified Cu-1,3,5-benzenetricarboxylic acid [19], or (iii) the copyrolysis of $RuCl_3$ and dicyandiamide [20]. Thus, researchers should take caution and not take for granted the formation of Ru-hcp NPs regardless of the synthetic methodology used for their preparation. It is thus advisable to systematically check the real crystalline nature of the obtained Ru NPs with either X-ray techniques (WAXS and XRD) or microscopy experiments (HRTEM).

In the three previous examples, the helpful effect of the fcc-Ru phase in the HER catalytic performance is observed, especially under alkaline conditions. A particularly interesting system is the third of the above examples where Ru-fcc is observed [20], in which 2 nm fcc-Ru NPs were deposited onto graphitic carbon nitride supported on C (Ru/g-C_3N_4/C), showing good activity (η_0 of 15–20 mV and a $TOF_{100\,mV}$ of 4.2 s^{-1}) and stability for 50 hours at pH 13, even outperforming commercial Pt/C.

The atypical Ru-fcc structure, induced by the g-C_3N_4 support, is responsible for the enhanced HER activity on the basis of DFT calculations of H adsorption energies (ΔG_{H*}) and water dissociation barriers (ΔG_B). Thus, on the one hand, the values of ΔG_{H*}, following the trend Pt < Ru_{fcc} < Ru_{hcp}, would indicate that Pt should be the most active HER electrocatalyst. However, on the other hand, fcc-Ru shows the lowest value of ΔG_B among the three systems, whereas Pt shows the highest. In this sense, the kinetics of water dissociation is clearly favored for fcc-Ru at basic pH, thus explaining the better performance of fcc-Ru over Pt. This work shows the importance of considering not only the H adsorption energy but also the water dissociation kinetics when comparing the HER performance of Ru-based nanocatalysts in alkaline media.

Another important parameter that has been described to affect the HER catalytic activity and stability of Ru-based NPs is their degree of crystallinity. In a recent work, polydopamine (PDA)-coated Ru NPs were thermally annealed on top of N-doped graphene, generating 4 nm hcp-Ru NPs coated with a thin layer of N-doped C [21]. As the annealing temperature is increased from 400 °C up to 700 °C, the degree of crystallinity increases in parallel with the HER electrocatalytic activity. Thus, the system synthesized at 700 °C (hcp-Ru@NC-700) shows high activity (η_{10} of 27.5 mV, a Tafel slope of 37 mV dec^{-1}, and a $TOF_{25\,mV}$ of 1.6 s^{-1}) and durability (up to six hours or 20 000 CV cycles) at acidic pH. The more efficient (100) and (002) planes at the NP surface are progressively more exposed during the temperature-induced crystallization process according to DFT calculations. This is the main explanation for the improved activity of the highly crystalline hcp-Ru NPs.

Finally, as stated before, RuO_2 NPs have also been described to catalyze the HER in two different works [14, 15]. In both works, a clear decrease of the activity when increasing the degree of crystallinity of the initially hydrous RuO_2 NPs is observed after annealing.

In conclusion, with the available data so far, the most active HER electronanocatalysts would be either highly crystalline Ru NPs (ideally showing either fcc structure or (100) and (002) hcp planes) or amorphous hydrous RuO_2 NPs.

12.3.3 Influence of the C Matrix or the C-Based Support

In order to be able to measure the HER electrocatalytic performance of Ru-based nanocatalysts, the NPs are normally deposited on or embedded in C-based supports, which can also be tuned/modified by the addition of specific dopants (generally nitrogen). Very frequently, these supports dramatically affect the behavior of the catalyst due to a reciprocal influence between the NPs and the supports. In the literature, two main procedures for obtaining such supported systems have been described: (i) the direct deposition of Ru-based NPs onto a C-based support (graphene, N-doped graphene, hollow C spheres, C nanotubes, N-doped C nanotubes, carbon nitride, etc.) by thermal annealing of a Ru precursor or (ii) the copyrolysis of C, N, and Ru sources at the same time, generally forming complex multielement composite materials.

Concerning the former, the thermal annealing of a Ru precursor, very stable and active systems, generally outperforming the HER activity of commercial Pt/C, have been obtained. For example, a very active system at pH 14 composed of 1.5 nm Ru NPs onto a C-matrix (Ru/C-300) was obtained after adsorption and pyrolysis at 300 °C of $[Ru_3(CO)_{12}]$, achieving a η_0 close to 0 mV, a η_{10} of only 14 mV, and a Tafel slope of 32.5 mV dec^{-1} [22]. A dependence of the size and the dispersion degree of the NPs and thus of the HER catalytic activity with the pyrolysis temperature is observed. Further, the authors conclude that the main reason for the superior activity of Ru/C-300 at basic pH comes from the favorable H_2O binding and dissociation energies in comparison to Pt/C.

One of the supports used in the literature has been graphene and its derivatives. Thus, a close performance to that of Pt/C has been reported for Ru NPs deposited onto graphene-layered carbon (Ru-GLC), with η_0 = 3 mV, η_{10} = 35 mV, and a Tafel slope of 46 mV dec^{-1} at acidic pH [23]. Better results, outperforming Pt/C, have been even obtained after coordination of Ru^{3+} onto edge-carboxylic acid-functionalized graphene nanoplatelets, followed by reduction with $NaBH_4$ and annealing at 600–800 °C, yielding 2 nm Ru NPs onto graphene nanoplatelets (Ru@GnP) [24]. This is one of the most active Ru nanoparticulate HER catalysts at acidic pH, achieving η_0 close to 0 mV, η_{10} = 22 mV and a Tafel slope of 28 mV dec^{-1}. This nanomaterial also shows significant activity at pH 14 (η_0 of practically 0 mV, η_{10} = 13 mV, and a Tafel slope of 30 mV dec^{-1}).

Another interesting point to be discussed is the beneficial effect of N-doping in C-supported Ru nanoparticulate HER electrocatalysts obtained through thermal annealing of a Ru precursor. Thus, high Ru–H_2O binding energy and increased rate of water capture at the NP surface at pH 14 promoted by the presence of N has been reported on the basis of DFT calculations for 5.8 nm Ru nanoclusters deposited onto conductive N-doped graphene (Ru/NG-750) [25]. This system was obtained after nucleation of $[Ru(NH_3)_6]^{3+}$ onto graphene oxide (GO) at 160 °C and reduction with NH_3/Ar at 750 °C and constitutes to our knowledge the second most active Ru-based nanocatalyst at basic pH, with η_0 = 0 mV, η_{10} = 8 mV, and a Tafel slope of 30 mV dec^{-1}. However, it is slightly unstable at acidic pH, probably because of decomposition and further dissolution.

The presence of N has also been related to an increase in resistance against NP aggregation and an increase in the electrocatalytic activity. Thus, the N-doping of C-black by pyrolysis at 800 °C of $[Ru(bpy)_3]^{2+}$ yielded 1.1 nm Ru NPs on top of N-doped C (Ru/CN-800) [26], and the N-doping of 3D carbon supports by pyrolysis at 500 °C of $RuCl_3$ and urea as N source generated 1.3 nm Ru nanoclusters deposited onto a three-dimensional N-doped C scaffold (Ru/3DNPC-500) [27]. Both systems are very active, showing low η_{10} (14–15 mV) and Tafel slopes (30–31 mV dec^{-1}) values and good stability at pH 14, outperforming commercial Pt/C. Another work in the field reports on the pyrolysis of C foam in the presence of $RuCl_3$ at 800 °C in the air, which yielded ca. 5 nm Ru NPs deposited on N-doped graphite (Ru-NGC) [28]. This catalyst presented good HER catalytic activity at both acidic and basic pH (η_{10} = 25 mV and a Tafel slope of 31 mV dec^{-1} at pH 0, and η_{10} = 37 mV and a Tafel slope of 40 mV dec^{-1} at pH 14).

A positive effect of a carbon nitride coating to Ru NPs deposited onto MWCNTs (CN_x@Ru/MWCNT) has also been described, showing η_{10} = 39 mV and a Tafel slope of 28 mV dec^{-1} at pH 14 [29]. However, very curiously, when the graphene nanoplatelets of the Ru@GnP system reported above [24] were modified by N-doping, the authors observed a decrease of their activity, which was assigned to the blocking of Ru active sites by N atoms. To our knowledge, this is the only one example in the literature in which the doping with N is detrimental for the HER catalytic activity.

With respect to the copyrolysis methods, the key role of the N-doping into the C-based support/matrix was also evidenced. Thus, copyrolysis of glucosamine, melamine, and $RuCl_3$ led to the formation of 2.4 nm Ru NPs on the surface of N-doped graphene-like nanosheets (Ru@CN). This nanomaterial showed good stability (it maintains 90% of the initial activity after 10 hours of operation at 60 °C) at basic pH, η_0 very close to 0 mV, and η_{10} of 32 mV [30]. The authors explained this behavior from the strong interaction between the Ru atoms and the N sites of the support, thus preventing the NPs to agglomerate or leach. Nevertheless, the activity is moderate at acidic and neutral pH.

A significantly more active system both at acidic and basic pH has been obtained after polycondensation of hexaketocyclohexane, hexaaminobenzene, and $RuCl_3$, followed by reduction with $NaBH_4$ at 175 °C [5]. By this way, 1.6 nm Ru NPs embedded into a nitrogenated, holey, two-dimensional carbon structure (Ru@C_2N) were obtained, with excellent activity both under acidic (η_{10} = 22 mV, Tafel slope = 30 mV dec^{-1}, exchange current density ($|j_0|$) of 1.9 mA cm^{-2}, and $TOF_{50\,mV}$ = 1.95 s^{-1}) and basic (with η_{10} = 17 mV, Tafel slope = 38 mV dec^{-1}, and $TOF_{50\,mV}$ = 1.66 s^{-1}) conditions, improving the values for Pt/C for the latter. DFT calculations indicate the prominent role of the C_2N framework in the global output of the system. Under basic conditions, the C_2N framework provides an increase of the Ru–OH_2 binding energy – thus facilitating water adsorption – and of the water dissociation rate – thus increasing the proton supply rate – compensating the unfavorable elevated Ru–OH binding energy in comparison to that of Pt–OH. This explains the increased efficiency of the Volmer step and in the end of the whole HER process.

The favorable effects of the C_2N supports in HER electrocatalysis of Ru_x@C_2N systems have theoretically been modeled, showing both an increased electroactivity and stability against aggregation [31]. However, caution should be taken when comparing data from different authors because single-site catalysis from individual Ru atoms has been described in Ru–C_xN_y composite systems formed by copyrolysis methods at basic pH, with superior catalytic activity than the Ru NPs themselves [32, 33]. These results clearly indicate the need of thoroughly characterizing the cathodic materials at an atomic scale in order to discern whether the catalytic active species are Ru atoms, Ru NPs, or both.

Another key parameter to control in order to maximize the HER activity of Ru-derived NPs onto C-based supports is the degree of exposure of the NPs. Thus, if the NPs are completely deposited onto the surface of the C support instead of being placed inside the C matrix, a maximized reactivity can be expected.

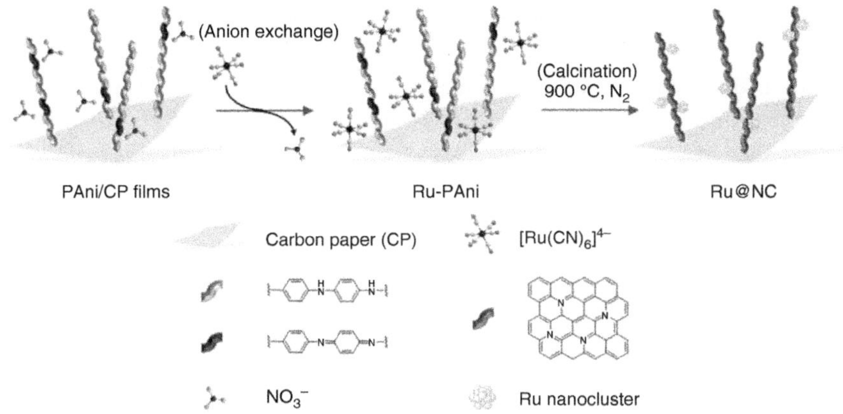

Figure 12.5 Schematic representation of the synthesis of the hierarchically ordered Ru@CN system. Source: Modified from Wang et al. [34].

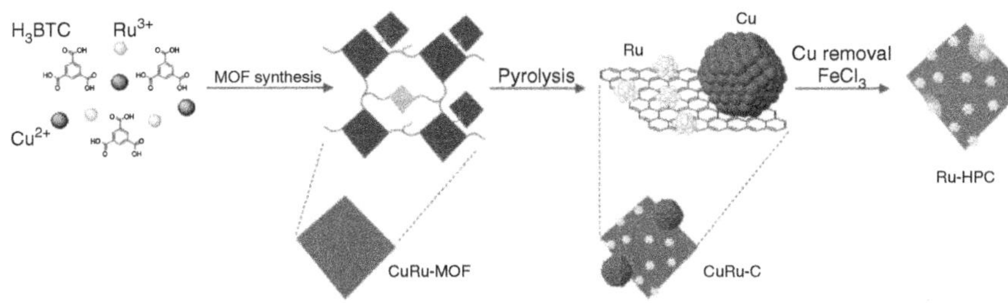

Figure 12.6 Schematic representation of the synthesis of the hierarchically porous Ru–HPC system. Source: Modified from Qiu et al. [35].

This behavior has been recently achieved by two different methods: first, by the potential-controlled assembly of nitrate onto polyaniline fibers (PAni) deposited onto carbon paper as a support (CP), followed by anion exchange with a Ru molecular precursor (NO_3^- substituted by $[Ru(CN)_6]^{4-}$) and a subsequent calcination at 900 °C under N_2 (Figure 12.5) [34] and second, by the annealing at high temperature of a CuRu–MOF precursor based on the 1,3,5-benzenetricarboxylic acid (H_3BC) motif, followed by Cu removal under treatment with Fe^{3+} (Figure 12.6) [35]. Hierarchically ordered C-based supports containing highly exposed monodisperse Ru NPs were obtained in both cases, showing high HER activity and stability at pH 14 at very low Ru loadings. Thus, in the case of the polyaniline fiber method, 1.6 nm Ru NPs were deposited onto a N-doped C support (Ru@NC), achieving at pH 14 a η_{10} as low as 26 mV, a Tafel slope of 36 mV dec^{-1}, and an impressive TOF$_{100\,mV}$ of 10.8 s^{-1}, the highest TOF ever reported so far under basic conditions [34].

Even better results have been obtained by the CuRu–MOF synthetic method, in which 2.9 nm Ru NPs were deposited onto a hierarchically porous C support (Ru–HPC), yielding at pH 14 and with only 5.6 wt% Ru a η_{10} of ~5 mV (the lowest one ever reported), a Tafel slope of $33.9\,mV\,dec^{-1}$, and the highest reported $|j_0|$, $6.59\,mA\,cm^{-2}$, as well as a $TOF_{100\,mV}$ as high as $9.2\,s^{-1}$ [35]. This synthesis method takes advantages from the high surface area and porosity of the bimetallic CuRu–MOF precursor, which are somehow transferred into the final carbonized product and from the presence of isolated and uniformly distributed Ru sites that are converted into ultrafine and well-dispersed Ru NPs. Furthermore, the insulation of the Ru sites by the Cu sites and the organic ligands in the CuRu–MOF precursor probably prevents Ru aggregation during the pyrolysis step, while the ulterior Cu etching process generates a large amount of meso- and macropores in Ru–HPC, thus increasing the accessibility of the final Ru sites. All these factors together explain the extraordinary performance and stability of the Ru–HPC system.

In conclusion, Ru NPs deposited/embedded on/in C-based supports have been prepared by two main methods: either by directly attaching a Ru molecular precursor onto the C support, followed by an annealing process at high temperature to form the Ru NPs or by the condensation of organic precursors that control the growth of the NPs with a subsequent carbonization process at high temperature in order to produce the C matrix around the preformed NPs. Additionally, the N-doping of the C-supports has been proven as a valid tool to increase both the stability and activity of the generated cathodes. However, given the complexity of processes occurring during the copyrolysis (and even the simple pyrolysis) methods, a proper control over the final C-supported Ru NP systems is not easy to achieve. In consequence, the fine-tuning of the NPs active sites is compromised, which challenges the possibility of improving their catalytic performance through the rational analysis of structure–activity relationships.

12.3.4 Influence of Heteroatoms

Apart from nitrogen, the introduction of other elements in the coordination sphere of Ru can also be used as a tool to increase the HER catalytic activity. Therefore, in the past years, a very common strategy for transition-metal-based HER nanoparticulate electrocatalysts – including Ru – has been the introduction of phosphorous. This strategy is based on the adequate H binding energy of P and the increased resistance against corrosion at acidic pH of the so-obtained nanomaterials induced by the presence of this element [36]. In the same way, it has also been reported that the synergistic interaction of Ru with other metal/semimetal-based nanostructures yields improved HER nanocatalytic systems compared to the respective separated metals, thanks to superior electron conductivity and optimized H adsorption energy.

12.3.4.1 Phosphorous

The introduction of P has been proven to tune the electronic properties of the NPs, yielding very active systems specially under acidic conditions [37], although also under basic conditions [38–41]. This is basically due to an improvement in electronic conductivity and in H adsorption energy, as will be exposed below.

One of the most prominent examples deals with Ru NPs dispersed onto a carbonized collagen scaffold (Ru–CCS), in which the binding of P and S atoms to

Ru promotes the separation of holes and electrons and consequently increases the number of localized electrons on the Ru surface, thus helping the adsorption of protons by decreasing $\Delta G_H{}^*$ close to zero [37]. This system achieves $\eta_{10} = 27.3$ mV, a Tafel slope = 33 mV dec^{-1}, and a TOF$_{50\,mV}$ of 3.70 s^{-1} at acidic pH. Other examples are the RuP_x NPs encapsulated inside N,P-doped C shells [39] (RuP_2@NPC) and nanospheres [38] (RuP_x@NPC), showing good activity and stability at pH 0 (e.g. $\eta_0 \approx 0$ mV, η_{10} = 38 mV, Tafel slope = 38 mV dec^{-1}, and $|j_0|$ = 1.99 mA cm^{-2} for RuP_2@NPC).

Even more active systems based on RuP_x NPs in C-based supports have been described, where a rise in electron mobility and charge density redistribution between the C support and the NPs is observed as the result of their synergistic interaction, yielding to H adsorption energy very close to 0. Thus, Ru_2P NPs deposited on reduced graphene oxide (Ru_2P/RGO) are very stable and active both at pH 0 ($\eta_0 \approx 0$ mV, $\eta_{10} = 22$ mV, Tafel slope = 29 mV dec^{-1}, and $|j_0| = 2.2$ mA cm^{-2}) and at pH 14 ($\eta_0 \approx 0$ mV and $\eta_{10} = 13$ mV) [40], representing one of the most promising Ru-based nanoparticulate HER catalysts so far. Very interestingly, Ru_2P/RGO shows a lower η_{10} value at pH 14 than pH 0. This may derive from a partial transfer of electrons from Ru_2P to the C surface atoms of RGO, which favors the splitting of water and thus the adsorption and recombination of hydrogen. This catalytic system contrasts with the vast majority of noble metal-based HER nanocatalysts, which present a decreased activity at basic pH because of an increase of the hydrophobicity of their surface, hampering the adsorption of water [42]. In addition, even slightly more active Ru_2P NPs at acidic pH have been described recently [41]. These NPs, partially encapsulated by a C layer, have been deposited on top of P/N-doped C nanofibers on a C cloth (Ru_2P@PNC/CC-900) by a 900 °C calcination process of polyaniline fibers in the presence of $RuCl_3 \cdot 3H_2O$ (Figure 12.7). Based on DFT calculations, the authors proposed that the P/N-doped C surface modifies the ΔG of H adsorption of the Ru_2P NPs, thereby enhancing the adsorption and reduction of protons [41].

Finally, exceptionally stable and active agglomerates of 32 nm RuP NPs dispersed onto C (RuP/C; $\eta_0 \approx 0$ mV, η_{10} = 19 mV, Tafel slope = 37 mV dec^{-1}, $|j_0|$ = 3.22 mA cm^{-2}, and TOF$_{20\,mV}$ = 3.35 s^{-1} at pH 0; $\eta_0 \approx 0$ mV, η_{10} = 18 mV, Tafel slope = 34 mV dec^{-1}, $|j_0|$ = 3.07 mA cm^{-2}, and TOF$_{20\,mV}$ = 3.62 s^{-1} at pH 14) have been described [43]. In this work, the superior stability and performance of the

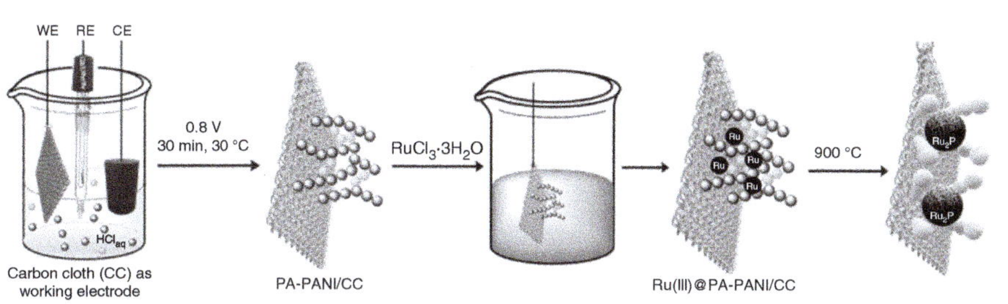

Figure 12.7 Schematic representation of the synthesis of the Ru_2P@PNC/CC-900 system. Source: Modified from Liu et al. [41].

agglomerates compared to smaller isolated 3 nm RuP NPs has been ascribed to a better stabilization of the P species in the agglomerates because of their lower surface energy.

12.3.4.2 Metals and Semimetals

The use of other metals or semimetals in combination with Ru-based NPs has also been assayed lately. Thus, Ru was combined with Co [44, 45], Ni [46], NiCo [47], and NiCoMo [48]. For example, the Ru-doping of Co nanocubes and their subsequent 600 °C annealing generated very stable and active N-doped graphene shell-encapsulated RuCo nanoalloys (RuCo@NC, with $\eta_0 \approx 0$ mV, $\eta_{10} = 28$ mV, Tafel slope = 31 mV dec^{-1}, and $|j_0| = 3.31$ mA cm^{-2}) at pH 14 [44]. DFT calculations show that Ru eases the transfer of electrons from the core to the C shell, thereby decreasing the H adsorption energy and increasing the HER kinetics. Also, the combination of Ru NPs with Ni@Ni_2P nanorods (Ru–Ni@Ni_2P–HNRs) has been reported to accelerate H_2 desorption, thanks to an optimal H adsorption energy and a fast electron transfer in comparison to the Ru-devoid Ni_2P system, showing good HER performance at pH 0 and 14 (Tafel slope = 35 mV dec^{-1} at pH 0, and $\eta_0 \approx 5$ mV, $\eta_{10} = 31$ mV, and Tafel slope = 41 mV dec^{-1} at pH 14) [46].

On the other hand, Ru NPs have also been deposited on metal or semimetal-based supports, such as MoS_2 [49], MoO_2 [50], TiO_2 [51], W [52], Si [53], Te [54], $Y(OH)_3$ [55], or CeO_2 [56], showing synergistic effects in HER similar to those above-mentioned for bi- or multimetallic electrocatalysts. Thus, Ru NPs deposited on top of MoS_2 nanosheets on C paper (Ru/MoS_2/CP) yield $\eta_0 \approx 0$ mV and $\eta_{10} = 13$ mV at pH 14 for 12 hours [49]. The higher activity observed in alkaline conditions than under acidic pH is attributed to the fact that at pH 14, the splitting of water over Ru is coupled with the H adsorption on MoS_2, thus demonstrating the existence of synergy between the NPs and the support. Another interesting example concerns the pyrolysis of $RuCl_3$ onto a suspension of a Mo-based MOF, yielding a Ru–MoO_2 composite made of Ru NPs anchored onto bigger MoO_2 particles (Ru–MoO_2) [50]. This system is very active at pH 14 ($\eta_0 = 0$ mV, $\eta_{10} = 29$ mV, and Tafel slope = 31 mV dec^{-1}). DFT calculations together with experimental XPS measurements showed that the particular HER activity comes from the sum of two factors, namely, the weak H binding force of MoO_2 and the fast electron transfer between MoO_2 and Ru. Analogously, DFT calculations demonstrated that the electronic interactions between W and Ru in a nanocomposite consisting in Ru NPs onto C dispersed into carbon black and deposited onto bigger W NPs (W + Ru/C) decrease the Ru–H binding energy, thus leading to improved HER activity at pH 0 as compared to Ru NPs in the absence of W [52].

12.4 Factors Ruling the Performance of Ru-Based NPs in HER Photocatalysis

Ru NPs have also been introduced in photocatalytic systems. However, this task is challenged by the observed difficulty in transferring electrons from a photosensitizer

(PS) to the Ru nanocatalyst and circumventing, at the same time, the nondesired back-electron transfer processes. In fact, the molecular PS that has been generally used in water splitting photocatalysis, $[Ru(bpy)_3]^{2+}$, is not adequate in the case of Ru NPs to achieve proper electron transfer between the PS and the NPs. [57] Thus, together with a sacrificial electron donor (SED), which supplies the electrons needed for the reduction process when half-cell systems are assayed, the use of an electron mediator such as methyl viologen is frequently required. Only when the PS can achieve long charge-separated states after light exposure, no electron mediator will be needed, thus yielding more efficient and chemically less complex systems.

Very interestingly, a bifunctional molecule that acts both as a PS and an electron mediator (the 2-phenyl-4-(1-naphthyl)-quinolinium ion, $QuPh^+$-NA) has been successfully reported to be active in photocatalytic HER when employing Ru NPs stabilized by the polyvinylpyrrolidone (PVP) as a catalyst under alkaline conditions [58–60]. The authors have observed that the photocatalytic activity does not increase after a determined ruthenium concentration because of the increase in light dispersion and opacity. The dependency of the activity with the size of the Ru NPs is also reported in this study. Small NPs show elevated negative charge density on their surface, thus facilitating the proton reduction step but hindering the H atom association step because of low H atom density on the surface. On the contrary, big NPs present more H atoms at their surface, thus facilitating their association process, but because their negative charge density is lower, the previous proton reduction step is hindered. Therefore, intermediate size NPs display the best results [58]. Apart from $QuPh^+$–NA, Eosin Y dye [61] and the combined use of $[Ru(bpy)_3]^{2+}$ with 9-phenyl-10-methyl-acridinium derivatives as electron mediators [62] have also led to successful photocatalytic HER when employing Ru-based NPs.

Finally, oxide-derived materials (SiO_2, TiO_2, CeO_2, etc.) have also been used as supports in order to avoid Ru NP agglomeration during photocatalytic HER when using $QuPh^+$–NA as PS [57]. Also, the molecular PS has been replaced by an organic bulk heterojunction layer that can absorb light and push photoinduced charge separation when using RuO_2 NPs as a catalyst, obtaining high voltage and current density [63]. Interestingly, in the case of using solid heterostructures as light absorbers instead of molecular compounds, 1D RuO_2 (110) nanowires deposited onto 1D rutile TiO_2 (110) nanorods have been proven to be active HER photocatalysts under visible light radiation. This happens thanks to their rod-like RuO_2/TiO_2 heterostructure, as opposed to the very poor visible light absorption of RuO_2-loaded bulk TiO_2 NPs [64]. Thus, although not in the form of NPs, RuO_2 deposited onto TiO_2 nanorods narrows the band gap of the semiconductor, making it possible to absorb visible light. Additionally, tandem particle overall water splitting photocatalysts have been reported recently, in which Ru NPs act as HER cocatalysts (Ru–$SrTiO_3$:Ru/$BiVO_4$ [65] and doped derivatives [66]). Although the development of these tandem photocatalytic systems is still at the beginning, they seem to be very promising candidates in Ru NP-based HER photocatalysis owing to their theoretical elevated efficiency (close to 28%) and low cost of fabrication.

12.5 Summary and Conclusions

The interest in the use of Ru nanomaterials as HER catalysts has been continuously increasing during the past five years, thus promising new discoveries in the years to come. Luckily, the vast majority of the systems have been analyzed in terms of the already reported benchmarking parameters described in Section 12.2, and additionally, their mechanistic behavior has been correlated with their structural and electronic properties deduced both from experimental and theoretical (DFT) simulations. A graphical comparison of the η_{10} and the Tafel slope (b) values both at acidic and at basic pH for the electrocatalytic systems reported within this chapter is presented in Figure 12.8. This figure clearly highlights the very promising results

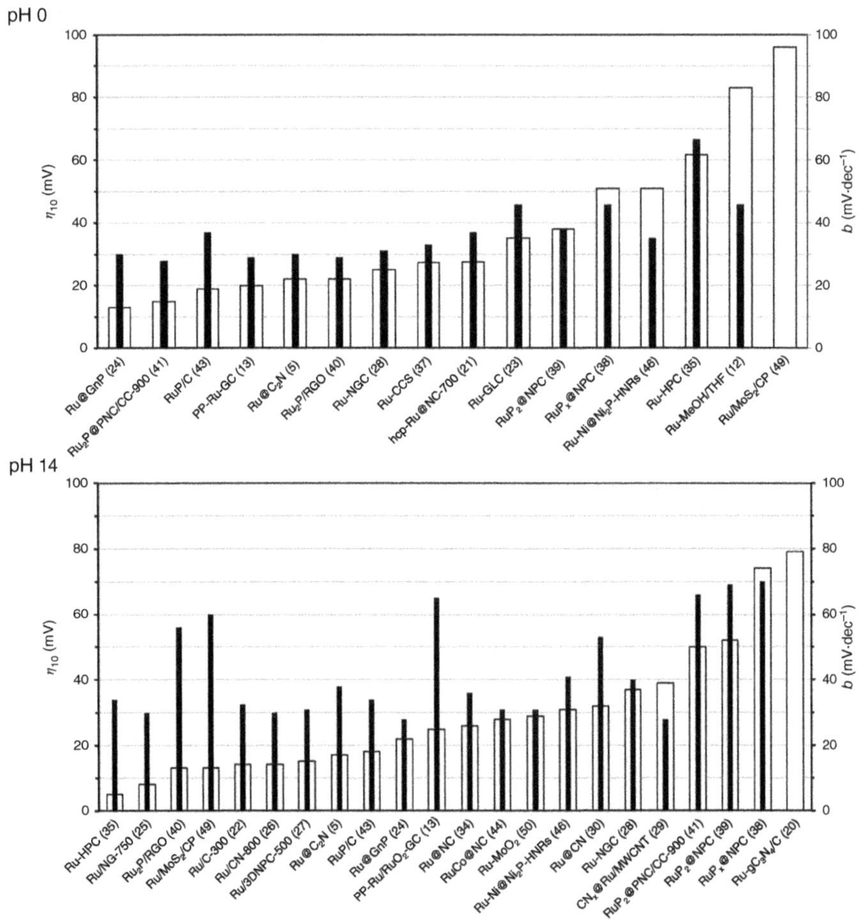

Figure 12.8 Schematic representation of the η_{10} (white bars) and the Tafel slope (b, black bars) values for the Ru-based HER nanoelectrocatalysts described within this. The name of the catalyst as appearing in the text and its corresponding citation reference are indicated for each case.

recently obtained with Ru-based nanomaterials, thus indicating the real potentiality of Ru for replacing Pt in cathodes for the overall water splitting process.

Two main families of systems can be discerned depending on the role of the C-based support used: those in which the support does not affect the activity of the Ru-based NPs and those in which the support has a relevant effect. For the formers, the Ru-based NP active sites can be easily tuned because they are physically simpler systems in which the chemistry at the NP surface is similar to a certain extent to that of molecular systems. From this viewpoint, the use of the organometallic method brings countless possibilities if considering the large panel of ligands that can be used to stabilize NPs. Furthermore, with the help of thorough electrochemical analysis, surface characterization, and DFT calculations, structure–activity and structure–stability relationships can be deduced, leaving space for further rational design of practical electrocatalytic systems.

Contrastingly, the tunability of the second family of systems is far more difficult given the more complex nature of these hybrid/composite systems. In these cases, the aggregation of the Ru NPs is limited, thanks to their interaction with the C-based support/matrix, thereby increasing the stability of the whole systems. Furthermore, in some cases, these intricate interactions end up with very promising catalytic properties. For example, very suitable H/H_2O adsorption energies and water dissociation rates were achieved at alkaline pH. These phenomena, together with their long-term stability at basic pH, allow to seriously considering the Ru-based nanoparticulate systems as candidates to replace Pt in cathodes in the near future. However, serious drawbacks that need to be overcome in the near future are their scalability and their applicability in light-driven processes. In this sense, the main challenges are the development of simple, cheap, and scalable syntheses of Ru-based NPs and their efficient triggering by sunlight by using molecular or semiconducting photoabsorbers.

Acknowledgments

Sustained support from MINECO, FEDER, AGAUR, and the GDRI HC3A Franco-Catalan action are gratefully acknowledged. Recent research projects and grants include CTQ2011-26440, CTQ2015-64261-R, and CTP2013-0016. J.C. thanks the UAB and the "Euroregió Pirineus Mediterrània" for predoctoral grants and J.G.-A. acknowledges the Serra Húnter Program.

References

1 Lewis, L.N. (2016). *Science* 351: 6271.
2 Amiens, C., Chaudret, B., Ciuculescu-Pradines, D. et al. (2013). *New J. Chem.* 37: 3374–3401.
3 Nørskov, J.K., Bligaard, T., Logadottir, A. et al. (2005). *J. Electrochem. Soc.* 152: J23–J26.

4 Zheng, Y., Jiao, Y., Jaroniec, M., and Qiao, S.Z. (2015). *Angew. Chem. Int. Ed.* 54: 52–65.
5 Mahmood, J., Li, F., Jung, S. et al. (2017). *Nature Nanotech.* 12: 441–446.
6 McCrory, C.C.L., Jung, S., Ferrer, I.M. et al. (2015). *J. Am. Chem. Soc.* 137: 4347–4357.
7 Creus, J., De Tovar, J., Romero, N. et al. (2019). *ChemSusChem.* 12: 2493–2514. https://doi.org/10.1002/cssc.201900393
8 Anantharaj, S., Ede, S.R., Karthick, K. et al. (2018). *Energy Environ. Sci.* 11: 744–771.
9 Gree, C.L. and Kucernak, A. (2002). *J. Phys. Chem. B* 106: 1036–1047.
10 Shinagawa, T., Garcia-Esparza, A.T., and Takanabe, K. (2015). *Sci. Rep.* 5: 13801.
11 Amiens, C., Ciuculescu-Pradines, D., and Philippot, K. (2016). *Coord. Chem. Rev.* 308: 409–432.
12 Drouet, S., Creus, J., Collière, V. et al. (2017). *Chem. Commun.* 53: 11713–11716.
13 Creus, J., Drouet, S., Suriñach, S. et al. (2018). *ACS Catal.* 8: 11094–11102.
14 Lee, J., Shah, S.A.S., Yoo, P.J., and Lim, B. (2017). *Chem. Phys. Lett.* 673: 89–92.
15 Park, H.-S., Yang, J., Cho, M.K. et al. (2019). *Nano Energy* 55: 49–58.
16 Creus, J., Mallón, L., Romero, N. et al. (2019). *Eur. J. Inorg. Chem.*: 2071–2077.
17 De Tovar, J., Romero, N., Denisov, S.A. et al. (2018). *Mater. Today Energy* 9: 506–515.
18 Gao, K., Wang, Y., Wang, Z. et al. (2018). *Chem. Commun.* 54: 4613–4616.
19 Yang, K., Xu, P., Lin, Z. et al. (2018). *Small* 14: 1803009.
20 Zheng, Y., Jiao, Y., Zhu, Y. et al. (2016). *J. Am. Chem. Soc.* 138: 16174–16181.
21 Li, Y., Zhang, L.A., Qin, Y. et al. (2018). *ACS Catal.* 8: 5714–5720.
22 Xu, C., Ming, M., Wang, W. et al. (2018). *J. Mater. Chem. A* 6: 14380–14386.
23 Chen, Z., Lu, J., Ai, Y. et al. (2016). *ACS Appl. Mater. Interfaces* 8: 35132–35137.
24 Li, F., Han, G.-F., Noh, H.-J. et al. (2018). *Adv. Mater.* 30: 1803676.
25 Ye, R., Liu, Y., Peng, Z. et al. (2017). *ACS Appl. Mater. Interfaces* 9: 3785–3791.
26 Cheng, X., Wang, H., Ming, M. et al. (2018). *ACS Sustainable Chem. Eng.* 6: 11487–11492.
27 Wang, H., Xu, C., Chen, Q. et al. (2019). *ACS Sustainable Chem. Eng.* 7: 1178–1184.
28 Song, Q., Qiao, X., Liu, L. et al. (2019). *ChemComm* 55: 965–968.
29 Gou, W., Li, J., Gao, W. et al. (2019). *ChemCatChem* 11: 1970–1976.
30 Wang, J., Wei, Z., Mao, S. et al. (2018). *Energy Environ. Sci.* 11: 800–806.
31 Zhang, X., Chen, A., Zhang, Z. et al. (2018). *J. Mater. Chem. A* 6: 11446–11452.
32 Zhang, J., Liu, P., Wang, G. et al. (2017). *J. Mater. Chem. A* 5: 25314–25318.
33 Lu, B., Guo, L., Wu, F. et al. (2019). *Nat. Commun.* 10: 631.
34 Wang, Z.L., Sun, K., Henzie, J. et al. (2018). *Angew. Chem. Int. Ed.* 57: 5848–5852.
35 Qiu, T., Liang, Z., Guo, W. et al. (2019). *Nano Energy* 58: 1–10.
36 Shi, Y. and Zhang, B. (2016). *Chem. Soc. Rev.* 45: 1529–1541.
37 Luo, D., Zhou, B., Li, Z. et al. (2018). *J. Mater. Chem. A* 6: 2311–2317.
38 Chi, J.Q., Gao, W.K., Lin, J.H. et al. (2018). *ChemSusChem* 11: 743–752.

39 Pu, Z., Amiinu, I.S., Kou, Z. et al. (2017). *Angew. Chem. Int. Ed.* 56: 11559–11564.

40 Liu, T., Wang, S., Zhang, Q. et al. (2018). *Chem. Commun.* 54: 3343–3346.

41 Liu, T., Feng, B., Wu, X. et al. (2018). *ACS Appl. Energy Mater.* 1: 3143–3150.

42 Cheng, T., Wang, L., Merinov, B.V., and Godard, W.A. III, (2018). *J. Am. Chem. Soc.* 140: 7787–7790.

43 Yu, J., Guo, Y., She, S. et al. (2018). *Adv. Mater.* 30: 1800047.

44 Su, J., Yang, Y., Xia, G. et al. (2017). *Nature Comm.* 8: 14969.

45 Xu, Y., Li, Y., Yin, S. et al. (2018). *Nanotechnology* 29: 225403.

46 Liu, Y., Liu, S., Wang, Y. et al. (2018). *J. Am. Chem. Soc.* 140: 2731–2734.

47 Liu, S., Liu, Q., Lv, Y. et al. (2017). *Chem. Commun.* 53: 13153–13156.

48 Shervedani, R.K., Torabi, M., and Foroushani, M.S. (2018). *J. Phys. Chem. C* 122: 17621–17631.

49 Liu, J., Zheng, Y., Zhu, D. et al. (2017). *Nanoscale* 9: 16616–16621.

50 Jiang, P., Yang, Y., Shi, R. et al. (2017). *J. Mater. Chem. A* 5: 5475–5485.

51 Demir, E., Akbayrak, S., Önal, A.M., and Özkar, S. (2018). *J. Colloid Interface Sci.* 531: 570–577.

52 Joshi, U., Malkhandi, S., Ren, Y. et al. (2018). *ACS Appl. Mater. Interfaces* 10: 6354–6360.

53 Zhu, L., Cai, Q., Liao, F. et al. (2015). *Electrochem. Commun.* 52: 29–33.

54 Yang, X., Zhao, Z., Yu, X., and Feng, L. (2019). *Chem. Commun.* 55: 1490–1493.

55 Liu, Y., Lu, X., Che, Z. et al. (2018). *Chem. Commun.* 54: 12202–12205.

56 Demir, E., Akbayrak, S., Önal, A.M., and Özkar, S. (2018). *ACS Appl. Mater. Interfaces* 10: 6299–6308.

57 Yamada, Y., Shikano, S., and Fukuzumi, S. (2013). *J. Phys. Chem. C* 117: 13143–13152.

58 Yamada, Y., Miyahigashi, T., Kotani, H. et al. (2011). *J. Am. Chem. Soc.* 133: 16136–16145.

59 Fukuzumi, S. and Yamada, Y. (2012). *J. Mater. Chem.* 22: 24284–24296.

60 Yamada, Y., Miyahigashi, T., Ohkubo, K., and Fukuzumi, S. (2012). *Phys. Chem. Chem. Phys.* 14: 10564–10571.

61 Kong, C., Li, Z., and Lu, G. (2015). *Int. J. Hydrogen Energy* 40: 5824–5830.

62 Yamada, Y., Yano, K., and Fukuzumi, S. (2012). *Aust. J. Chem.* 65: 1573–1581.

63 Francàs, L., Burns, E., Steier, L. et al. (2018). *Chem. Commun.* 54: 5732–5735.

64 Nguyen-Phan, T.-D., Si, L., Vovchok, D. et al. (2016). *ACS Catal.* 6: 407–417.

65 Melo, M.A. Jr.,, Wu, Z., Nail, B.A. et al. (2018). *Nano Lett.* 18: 805–810.

66 Wang, Q., Hisatomi, T., Jia, Q. et al. (2016). *Nat. Mater.* 15: 611–615.

13

Nanocatalytic Architecture for the Selective Dehydrogenation of Formic Acid

Ismail B. Baguc[1], Gulsah S. Kanberoglu[1], Mehmet Yurderi[1], Ahmet Bulut[1], Metin Celebi[1], Murat Kaya[2], and Mehmet Zahmakiran[1]

[1] *Van Yuzuncu Yil University, Nanomaterials and Catalysis Research Group, Department of Chemistry, Campus, Van 65080, Turkey*
[2] *Atilim University, Department of Chemical Engineering and Applied Chemistry, Ankara 06836, Turkey*

13.1 Introduction

Today, the advancement of secure and effectual hydrogen (H_2) storage is a crucial concern in the hydrogen-based energy economy, which would aid the changeover from fossil fuel-based energy to renewable energy [1, 2]. In this respect, an increasing number of studies are still reported for testing various solid materials such as carbon nanotubes [3], zeolites [4], metal–organic framework (MOF) materials [5], organic polymers [6], boron hydrides ($LiBH_4$, $NaBH_4$, etc.) [7], and boron–nitrogen compounds (NH_3BH_3, $N_2H_4BH_3$, $NH_3B_3H_7$, etc.) [8] in the chemical hydrogen storage [9]. Additionally, recent studies have shown that the liquid organic hydrogen carrier (LOHC) strategy is also an effective way for storing hydrogen by hydrogenation/dehydrogenation cycle storage system [10]. Among various LOHC materials, formic acid (FA; HCOOH), which is one of the major stable products formed in biomass processing and contains 4.4 wt% of H_2, has recently attracted significant attention as a potential hydrogen carrier for fuel cells designed toward portable use because of its inherent following advantages [11, 12]:

(i) FA is in liquid form at ambient temperature, which simplifies its handling, storing, and transportation;
(ii) FA has less toxic nature so that its dilute solutions can be used as food additives (US Food and Drug Administration) [13];
(iii) FA has a simple structure and can be decomposed into small molecules, which have low molecular weights such as H_2 and CO_2, providing simple reaction control;
(iv) FA can also be used directly in formic acid fuel cells (FAFC), which has significant advantages over methanol fuel cells (MFC) in crossover flux and electromotive force parameters [14]; and

Nanoparticles in Catalysis: Advances in Synthesis and Applications, First Edition.
Edited by Karine Philippot and Alain Roucoux.

(v) Carbon dioxide (CO_2) as a by-product of FA dehydrogenation (vide infra) can be recycled through a carbon-neutral fuel cycle to (Scheme 13.1) [11].

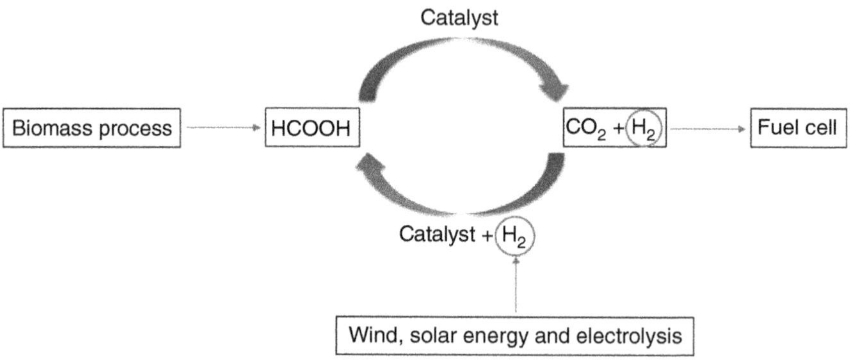

Scheme 13.1 Carbon-neutral fuel cycle diagram. Source: Modified from Yadav and Xu [11].

FA can be catalytically decomposed via dehydrogenation (Eq. (13.1)) and dehydration (or decarbonylation) (Eq. (13.2)) pathways depending on the selectivity of the catalyst, pH of the reaction solution, and temperature [12].

$$HCOOH \rightarrow H_2 + CO_2 \quad \Delta G_{298K} = -35.0\ \text{kJ mol}^{-1} \tag{13.1}$$

$$HCOOH \rightarrow H_2O + CO \quad \Delta G_{298K} = -14.9\ \text{kJ mol}^{-1} \tag{13.2}$$

The selective decomposition of FA through the dehydrogenation pathway (Eq. (13.1)) is vital for the production of ultrapure H_2, as toxic carbon monoxide (CO) produced by the dehydration route (Eq. (13.2)) significantly reduces the activity of platinum (Pt)-based metal catalysts used in proton exchange membrane fuel cells (PEMFC) because of poisoning effect [15]. Recently, serious efforts have focused on the development of homogeneous catalysts for the selective dehydrogenation of FA [16–23]. Although remarkable catalytic activities have been achieved in some of these molecular systems [21–23], the substantial difficulties associated with recovery–isolation–purification procedures considerably hamper the practical use of such homogeneous catalysts in onboard applications. In this connection, because of the advantages of heterogeneous catalysis [24], the recent research studies have been directed toward the design of heterogeneous catalysts that can offer high activity, selectivity, and stability in the FA decomposition.

The first attempt for the decomposition of FA in the presence of heterogeneous catalysts was done by Sabatier and Mailhe in 1911, 152, 1212 [25]. Thenceforth, catalytic FA decomposition has been employed as a model test reaction to investigate the surfaces and reactivities of different heterogeneous catalysts [26–28]. For example, Rienacker performed decomposition of FA vapor over Cu-, Ag-, and Au-based heterogeneous catalysts under ambient pressure and found that bimetallic CuAu and AgAu alloys provide higher activities than monometallic counterparts [29]. Next, the same group carried out more systematic studies in which the surface reactivities of transition metals such as Fe, Ni, Cu, Ag, Pd, Au, and Pt in both monometallic and bimetallic alloyed forms were surveyed in FA decomposition [30–32].

In 1950s, heterogeneous catalytic systems based on Ni, Cu, and NiCu alloys were tested in FA decomposition at elevated temperatures. A monometallic Ni-based heterogeneous catalyst showed highest activity at ~250 °C with a H_2 generation rate of 20 ml min^{-1}, whereas the purity of H_2 was not reported [33]. The surface reactivity of heterogeneous catalysts in FA decomposition was also examined by using spectroscopic methods. For instance, Sexton carried out FA decomposition on Cu(111) surface and examined this catalytic transformation by electron energy loss spectroscopy (EELS) and found that formate ($HCOO^-$) species formed on the surface of copper and transformed into H_2 and CO_2 at temperature >150 °C [34]. In a subsequent similar study [35], the decomposition of FA on both Pd(100) and K-promoted Pd(100) surfaces was studied. FA decomposition on Pd(100) was shown to start below −73 °C, whereas for K-promoted Pd(100), the preadsorbed potassium led to HCOOK formation on Pd surface above 127 °C. The analysis of the reaction products revealed the formation of equal amounts of CO, CO_2, and H_2. As aforementioned, the majority of the earlier studies focused on the decomposition of FA in the gas phase at elevated temperatures (>100 °C) that usually yields CO as a by-product. In addition to the low selectivity, the integration of gas-phase reactor H_2 generation system into PEMFC is usually not effective and safe [12]. Therefore, the development of heterogeneous catalysts that provide high activity and selectivity toward FA dehydrogenation in aqueous solution under mild conditions has great importance [11, 12].

Of particular interest, metal nanoparticles (NPs) have been shown to afford better catalytic performances than their bulk counterparts not only in FA dehydrogenation but also in various organic/inorganic catalytic reactions [36, 37]. Metal nanoparticles have higher surface-to-volume ratio matched to their bulk counterparts. Hence, in metal nanoparticles, majority of catalytically active atoms are on the surface, and these surface atoms do not order themselves in the same way as those in bulk metals [38]. Besides, the electrons in nanoparticles are restricted to spaces that can be as small as a few atoms widths across giving rise to quantum size effects [38, 39]. Moreover, recent progress in the synthesis of metal nanoparticles provides the mastering of their size and surface composition, while these parameters cannot be controlled throughout the synthesis of classical heterogeneous catalysts. As well as their superior catalytic performances, metal nanoparticles are usually acting as isolable and reusable catalysts and, thus, also satisfy some necessities of the modern concept of "green catalysis" [40].

At this concern, huge numbers of studies about the evaluation of various metal nanoparticles catalysts in the decomposition of FA have been reported in recent years [41–45]. Although platinum (Pt) [46], gold (Au) [46, 47], cobalt (Co) [48], and ruthenium (Ru) [49] nanoparticles catalysts have been employed for FA dehydrogenation, the majority of the nanocatalysts used in the dehydrogenation of FA were composed of palladium (Pd), which was known as the most reactive metal for this reaction. Unfortunately, monometallic Pd nanoparticles cannot maintain their initial activity in the long term because of strong adsorption of poisonous CO formed as an intermediate during FA decomposition [50]. There are three possible routes for

FA dehydrogenation (Scheme 13.2) [50]. The first one is the formate route, in which FA dissociates into formate, which then forms CO_2. In the second direct route, FA decomposes into CO_2 and 2H. The third possible pathway is an indirect one in which FA dehydrates into CO and then it goes into water gas shift reaction.

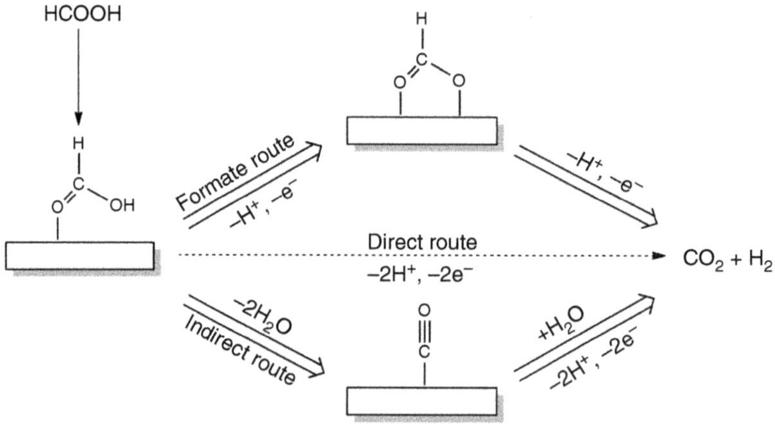

Scheme 13.2 Proposed FA dehydrogenation pathways. Source: Modified from Luo et al. [50].

In this context, the survey of the literature shows that the presence of other transition metals/metal oxides (in alloy, core@shell, or physical mixture form) can help to prevent CO poisoning of active Pd centers in Pd-based nanocatalysts for FA decomposition. These strategies are discussed below and form the framework of this chapter, where monometallic and multimetallic (alloy, core@shell, and physical mixture) Pd-based nanocatalytic architectures will be reviewed.

13.2 Monometallic Palladium-Based Nanocatalysts

The inspection of monometallic Pd-based nanocatalysts reported until now reveals two common issues in all of these catalytic systems: (i) all catalysts have been used in solid supported form; there is no report on the employment of colloidal Pd nanoparticles as a catalyst in FA decomposition possibly because of the low stability in acidic aqueous FA solution and (ii) wet impregnation (or precipitation), followed by chemical reduction technique has been employed in all of the synthesis protocols [51].

Zhu et al. [52] prepared Pd nanoparticles supported on nanoporous carbon MSC-30, Vulcan XC-72R, and Al_2O_3 and tested them as catalysts in the dehydrogenation of aqueous FA-SF (sodium formate) mixture. It should be noted that SF (HCOONa) has mostly been used as a promoter in the catalytic FA decomposition to enhance the dehydrogenation selectivity of the catalytic systems by forcing metal–formate formation in the first step of the catalytic reaction. The synthesis protocol to prepare Pd/MSC-30, Pd/Vulcan XC-72R, and Pd/Al_2O_3 comprised precipitation of Pd precatalyst salt with sodium hydroxide (NaOH), followed by sodium borohydride ($NaBH_4$) reduction of impregnated Pd^{2+} cations to metallic Pd at room temperature (RT). Among supported Pd materials, the highest catalytic performance relative to H_2 generation for the dehydrogenation of FA/SF = 1/1 was achieved with Pd/MSC-30. A turnover frequency TOF = 750 h^{-1} defined as mol product/mol catalyst × time at 100% H_2 selectivity ([moles of dehydrogenation products/mol of dehydration products] × 100%) and 25 °C was obtained because of the large surface area (3000 m^2 g^{-1}) of the support compared to the two others (Vulcan XC-72R and Al_2O_3), which possibly enhances the interaction between Pd and carbon.

In ensuing a separate study, the commercial activated carbon was used as a support material for guest Pd nanoparticles with a mean diameter of 1.8 nm (Pd/C), which were made by impregnation–reduction technique [53]. The authors defined the activity of these supported Pd nanoparticles in terms of both TOF (400 h^{-1} at 60 °C in 4.0 M aqueous FA) and Arrhenius plot estimated activation energy (57.8 kJ mol^{-1}) values. The most important drawback of this catalytic system is its short lifetime, which was explained by the blockage of active Pd sites with adsorbed protons throughout the catalytic reaction.

In addition to the surface area parameter, the functionalization of support materials is crucial because it can affect the catalytic reactivity of guest Pd nanoparticles in FA dehydrogenation. In this line, Xu and coworkers [54] reported that various amine (PDA = 1,4-phenylenediamine, AMA = 4-(aminomethyl)aniline, DAH = 1,6-diaminohexane, DMPDA = *N*,*N*-dimethyl-1,4-phenylenediamine, and A = aniline)-functionalized reduced graphene oxide (rGO)-supported Pd nanoparticles (Figure 13.1a,b) show significant better activities than that of Pd supported on neat reduced graphene oxide (Pd/rGO). The authors claimed that the amine groups present onto the support surface play several crucial roles such as (i) they achieve high dispersion of Pd nanoparticles onto the support material, (ii) they act as a stabilizing agent during the growth of Pd nanoparticles onto the support leading to the formation of small-sized Pd nanoparticles, and (iii) according to mechanistic insights, they behave as proton scavengers to facilitate metal–formate formation and enhance the rate of FA dehydrogenation (Figure 13.1c).

The effect of particle size on the catalytic reactivity of Pd nanoparticles was investigated by Lia et al. [55], who prepared Pd/C catalysts with Pd particle sizes ranging from 4.5 ± 0.5 to 2.1 ± 0.3 nm. Expectedly, they found that the highest catalytic activity (TOF = 835 h^{-1} at 25 °C) can be reached with Pd/C with smallest Pd particle size (2.1 ± 0.3 nm). Additionally, the electronic effect in Pd nanocatalysis was also examined, and it was concluded that the decrease in Pd particle size leads to formation of larger fraction of positively charged Pd species that has crucial influence on activity enhancement.

Akbayrak et al. investigated the effect of various metal oxide support materials on the activity of Pd nanoparticles in the catalytic dehydrogenation of aqueous FA + SF mixture [56]. They used a series of metal oxides including SiO_2, Al_2O_3, TiO_2, CeO_2, ZrO_2, and HfO_2 to stabilize Pd nanoparticles. The best activity was achieved with CeO_2. This is explained by the production of cationic Pd species on its

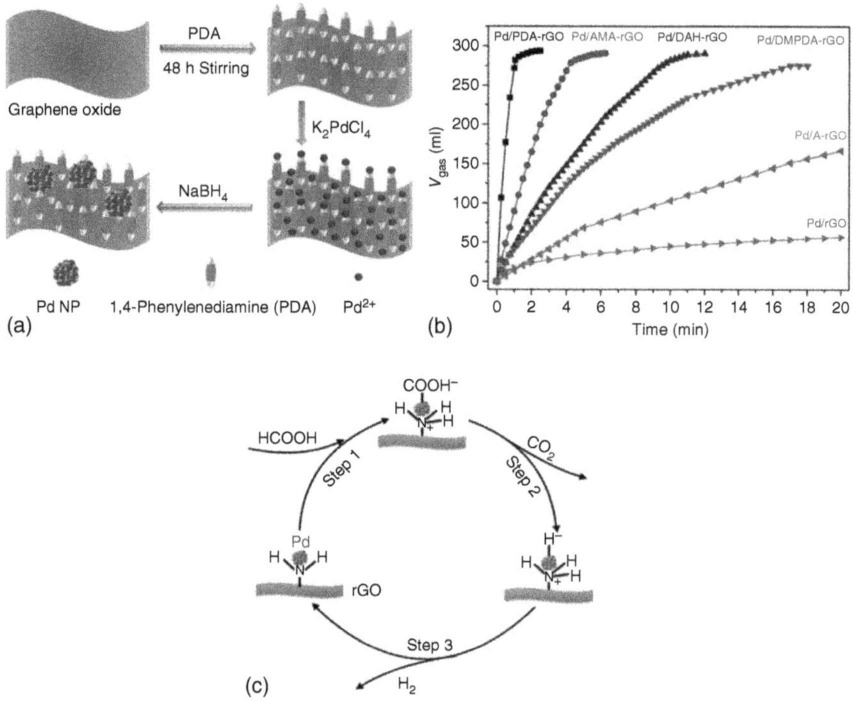

Figure 13.1 (a) Schematic illustration for the fabrication of Pd/PDA-rGO, (b) volume of the generated gas ($CO_2 + H_2$) vs. time graph for the dehydrogenation of FA over Pd/PDA-rGO, Pd/AMA-rGO, Pd/DAH-rGO, Pd/DMPDA-rGO, and Pd/rGO (nPd/nFA = 0.015, FA/SF = 1 : 1, 50 °C), (c) possible reaction pathway of amine-associated decomposition of FA over Pd/PDA-rGO. Source: Reproduced with permission from Ref. [54].

surface defect sites, resulting from oxygen vacancies [51]. Pd/CeO_2 provides a TOF value of 1400 h^{-1} at 25 °C with high dehydrogenation selectivity (~100%). Along with metal oxide and carbon materials, microporous materials such as zeolites were also used as host material to catalytically active Pd nanoparticles. Wang et al. prepared silicate-1 (MFI) zeolite-encapsulated ultrafine Pd nanoparticles under direct hydrothermal conditions by using $[Pd(NH_2CH_2CH_2NH_2)_2]Cl_2$ as a Pd precursor and tetrapropylammonium hydroxide (TPAOH) as a template [57]. Small Pd nanoparticles (~1.5 nm) in situ encapsulated within MFI zeolite were fabricated by starting with a gel at hydrothermal conditions (170 °C for four days), followed by H_2 reduction. The resulting Pd nanoparticles afforded TOF values of 856 and 3027 h^{-1} at 25 and 50 °C, respectively. Furthermore, attributable to a confinement effect on the catalytic durability of Pd nanoparticles, Pd/silicate-1 catalyst displays great stability throughout the performed recyclability experiments, which prevents almost its inherent activity up to fifth catalytic recycle.

The effect of the presence of amine groups on support material has also been recently studied by Li et al., who modified MSC-30 carbonaceous support by tandem nitrogen functionalization (N-MSC-30) [58]. Pd nanoparticles generated on N-MSC-30 by coprecipitation–reduction technique (Scheme 13.3) afforded a TOF

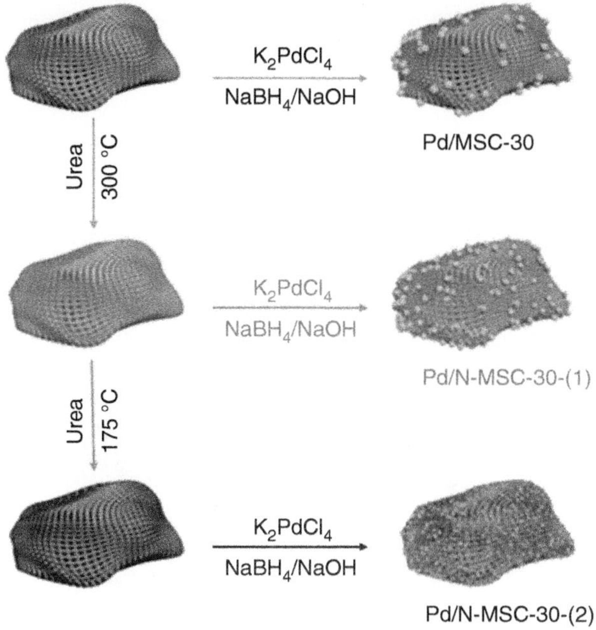

Scheme 13.3 Schematic illustration for the synthesis of Pd/MSC-30 and Pd/N-MSC-30 nanocatalysts [58].

value of 8414 h^{-1} at 60 °C in the dehydrogenation of aqueous FA + SF mixture. This remarkable activity was assigned to the existence of nitrogen-containing groups onto the support, thus providing effective anchoring of Pd nanoparticles, which prevents agglomeration as well as basic sites that expedite FA deprotonation and favor Pd formate form that direct reaction pathway toward dehydrogenation.

Dimitratos and coworkers have recently reported on the development of Pd nanoparticles supported on carbon nanofibers (CNFs) via sol immobilization and impregnation techniques [59]. They found that Pd/CNFs prepared by the sol immobilization method display higher catalytic activity in FA decomposition than those prepared by the impregnation method. This result was assigned to (i) the higher surface atomic Pd percentage (defined as surface Pd atoms/total Pd atoms × 100% and (ii) smaller Pd particle size in Pd/CNFs. A TOF value of 979 h^{-1} was achieved in the catalytic decomposition of aqueous FA solution with high selectivity for H_2 (>99%) at 30 °C. More recently, Wan and coworkers prepared ultrafine Pd particles uniformly embedded in N-enriched mesoporous carbon (Pd/NMC) by hard template, followed by deposition–precipitation method [60]. It was found that the carbonization temperature (600–1000 °C) of NMC has great influence on the type and content of N species and thus size, distribution, and dispersion of guest Pd nanoparticles through the interaction between Pd- and N-containing functionalities. Under optimum conditions, the resulting Pd nanoparticles in Pd/NMC demonstrated not only over 12 times higher H_2 production amount than Pd/MC (amine-free mesoporous carbon) but also very promising stability (>95% retention efficiency within four cycles eight hours of total reaction time) without any generation of a CO by-product.

13.3 Bimetallic Palladium-Based Nanocatalysts

Palladium-based bimetallic nanocatalysts (Pd-M, M = transition metal) will be examined in three separate categories depending on their architectures such as (i) physical mixture (Pd-M) vide infra, (ii) alloy (PdM), and (iii) core@shell (M@Pd), which have unique advantages in catalysis for FA decomposition (Scheme 13.4).

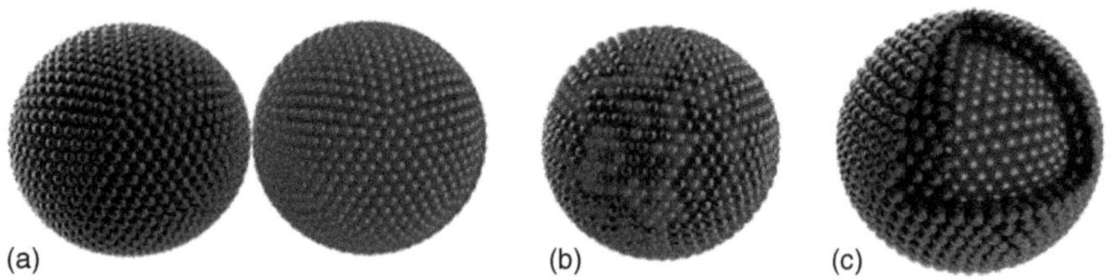

Scheme 13.4 Different forms of Pd-containing bimetallic catalysts: (a) physical mixture (Pd-M), (b) alloy (PdM), and (c) core@shell (M@Pd) forms.

13.3.1 Bimetallic Pd-Containing Nanocatalysts in the Physical Mixture Form

For the first time, we have developed supported bimetallic Pd–MnO_x nanoparticles in the physical mixture form (existing of monometallic Pd and MnO_x nanoparticles on the support surface) on the surface of 3-aminopropyltriethoxysilane (APTS)-functionalized SiO_2 (Pd–MnO_x/SiO_2–NH_2) by the deposition–reduction technique [61]. This bimetallic supported nanocatalyst catalyzes the additive-free (in the absence of SF or other bases) dehydrogenation of FA with notable activity (TOF = 1300 h^{-1}) at high selectivity (>99%) and conversion (>99%) at 50 °C. To shed some light on the effect of MnO_x nanoparticles, CO-stripping voltammetry measurements were performed on Pd–MnO_x/SiO_2–NH_2 and Pd/SiO_2–NH_2 catalysts. The onset potential for CO oxidation on Pd–MnO_x/SiO_2–NH_2 (0.45 V) was found to be lower than that on Pd/SiO_2–NH_2 (0.68 V). Additionally, CO oxidation peak current decreased severely at the end of first scan for Pd–MnO_x/SiO_2–NH_2 catalyst, which reveals a weaker CO adsorption on Pd–MnO_x/SiO_2–NH_2 than on Pd/SiO_2–NH_2. These two findings clearly disclose that CO poisoning resistance of Pd nanoparticles is significantly enhanced by the presence of separately nucleated MnO_x NPs, which acts as a CO sponge around the catalytically active Pd nanoparticles.

Next, Lee and coworkers investigated the catalytic activities of Pd–MnO_x nanoparticles supported on various NH_2-functionalized mesoporous silica materials (KIE-6, SBA-15, and MCM-41), which were prepared by wet impregnation–borohydride reduction method, in the FA decomposition [62]. Among these catalysts, Pd (2.0 wt%)–MnO_x (Mn basis 4.0 wt%)/NH_2–KIE catalyst provides the highest catalytic activity (TOF = 593 h^{-1}). More recently, Alotaibi et al. have reported that palladium (Pd)–iridium (Ir) nanoparticles supported on activated carbon black (CB) and titania (TiO_2) can be prepared by deposition–reduction method. The authors claimed that these catalysts display high activity (activity of Pd/C > Pd/TiO_2) and stability in FA dehydrogenation, albeit they have not defined the activity, selectivity, and stability of these catalytic systems [63].

13.3.2 Bimetallic Pd-Containing Nanocatalysts in the Alloy Structure

Liu and coworkers prepared PdAg alloy nanostructures in the absence of any support material and stabilizing agents by a borohydride reduction of an aqueous mixture of $Pd(NO_3)_2$ and $AgNO_3$. They obtained agglomerated PdAg alloy nanostructures, which showed good activity in the dehydrogenation of FA + SF mixture (TOF = 31.5 h^{-1}) at RT [64]. Zhou et al. prepared bimetallic PdAu, PdAg, and PdCu alloy nanoparticles on activated carbon by a two-step procedure composed of wet impregnation, followed by chemical reduction, and used them as catalysts in the catalytic decomposition of aqueous FA + SF mixture at 82 °C [65]. Size analysis of these catalysts showed a mean particle size trend as follows: Pd/C > PdCu/C > PdAg/C > PdAu/C, thus revealing that addition of a second metal decreases the size of the bimetallic alloy nanoparticles. The catalytic activity of these nanomaterials was found to have the inverse tendency, namely, Pd/C < PdCu/C < PdAg/C < PdAu/C under identical conditions, thus leading to the conclusion that smaller NPs were more active. Interestingly, it was also noticed that the catalytic activities of PdAu/C and PdAg/C can be significantly enhanced by codeposition with CeO_2 (addition of CeO_2 at the initial stage of the preparation protocol). The resulting PdAg/C–CeO_2 and PdAu/C–CeO_2 catalysts provided turnover frequencies of 227 and 76 h^{-1}, respectively, in the catalytic decomposition of aqueous FA + SF mixture at 82 °C with maximum 80 ppm CO generation.

Xu and coworkers prepared MIL-101 (MIL: MOF material from Material Institut Lavoisier) and ethylenediamine (ED)-functionalized MIL-101 (ED-MIL-101) as support materials to fabricate PdAu and PdRu bimetallic alloy nanoparticles by using simple impregnation–reduction method [66]. Among them, PdAu/ED-MIL-101 catalyst offered the best activity in the decomposition of aqueous FA + SF mixture at 90 °C. They found that the presence of ED in the MOF support contributes to the formation of highly dispersed and smaller sized PdAu alloy nanoparticles as the result of strong interaction with metal precursors and nanoparticles. Additionally, they also pointed out that when Pd is alloyed with Au in PdAu/ED-MIL-101, the electronic structure is modified, which effectively inhibits the adsorption of CO on the surface of the catalyst since the adsorption strength of CO on Au is much smaller than that of CO on Pd. Moreover, Au does not form stable complexes with CO, which results in persistent activity. Similarly, Gao et al. fabricated PdAg alloy nanoparticles on amine-functionalized UiO-66 (UiO: Universitetet i Oslo), which is a zirconium (IV)-based MOF material with unique properties such as light harvesting, high thermal and chemical stability, tunable pore size, and large surface area [67]. They found that amino groups present in NH_2–UiO-66 play a crucial role in both nucleation and surface growth steps of guest PdAg alloy nanoparticles. The comparison of the catalytic activity of Ag/NH_2–UiO-66, Pd/NH_2–UiO-66, and PdAg/NH_2–UiO-66 nanomaterials in the dehydrogenation of FA showed that PdAg/NH_2–UiO-66 is the most performing (TOF = 103 h^{-1}) even in the absence of SF and at RT. This result is attributed to a synergic effect between Pd and Ag.

Other research groups also probed employment of MOFs as a suitable support material of nanocatalyst for the catalytic decomposition of FA. For example, Dai et al. prepared PdAg alloy nanoparticles supported on zeolitic imidazole framework (ZIF-8) by using impregnation–reduction method [68]. Different metal compositions (Pd, Ag, and PdAg alloy) on ZIF-8 (ZIF: zeolitic imidazolate framework) as well as different support materials (C, Al_2O_3, and SiO_2) were used for comparison purpose. PdAg alloy nanoparticles supported onto ZIF-8 provided not only the best activity (TOF = 580 h^{-1}) but also the best catalytic stability (~90% activity preserved at fifth run) in the dehydrogenation of aqueous FA + SF mixture at 80 °C. In another interesting study, Feng et al. used MOF-derived porous carbon as the support material to prepare PdAg/MOF-5 nanocatalysts, whose preparation route is illustrated in Scheme 13.5 [69]. PdAg/MOF-5-C-900 (900 denotes carbonization temperature) prepared in this manner provides ultrapure H_2 generation (no CO detected) with an initial TOF value of 854 h^{-1} at RT.

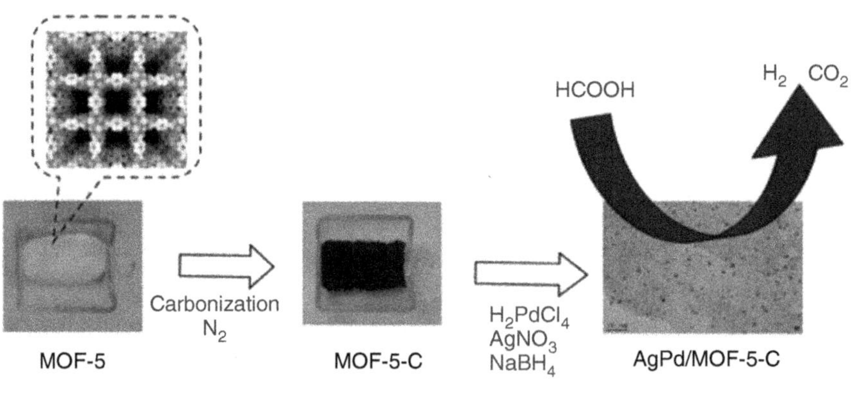

Scheme 13.5 Schematic illustration of AgPd/MOF-5-C-900 preparation for catalytic formic acid dehydrogenation. Source: From Feng et al. [69]. © Elsevier.

The influence of amine functionalities on the support material was also investigated by Yamashita and coworkers [70], who developed PdAg alloy nanoparticles on resin (Amberlite-type) bearing -$N(CH_3)_2$ functional groups. The resulting resin-supported PdAg alloy nanoparticles produced high-quality H_2 from the decomposition of FA + SF mixture at 65 °C with a TOF value of 1900 h^{-1}. This

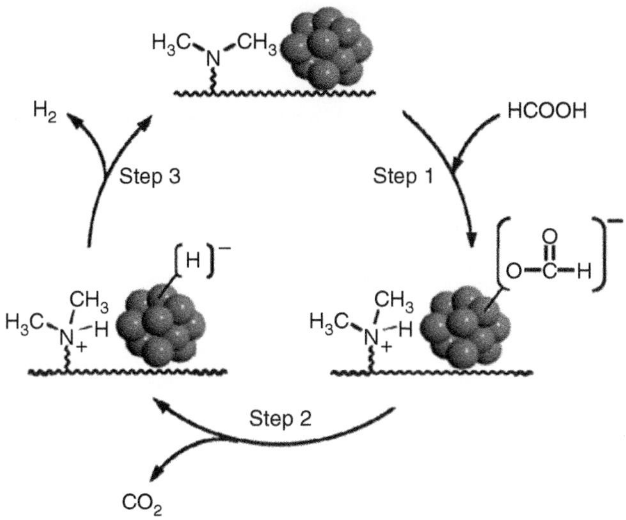

Scheme 13.6 Possible reaction pathway for the dehydrogenation of FA over PdAg/resin-N(CH_3)$_2$ catalyst. Source: Reproduced with permission from Feng et al. [69].

remarkable TOF value was assigned to (i) small size of PdAg alloy nanoparticles, (ii) synergic effect between Pd and Ag nanoparticles, and (iii) the existence of amine functionalities on the support material. Detailed mechanistic studies were conducted in order to understand the effect of –N(CH_3)$_2$ functional groups on the catalytic decomposition of FA. As presented in Scheme 13.6, the obtained results revealed that: (i) O—H bond cleavage is assisted by weakly basic –N(CH_3)$_2$ groups that act as a proton scavenger, affording formation of Pd-formate species along with a $-^{+}$HN–(CH_3)$_2$ group; (ii) Pd-formate species undergo β-hydride elimination, leading to the formation of CO_2 and a Pd-hydride species; and (iii) the reaction of Pd-hydride species with –(NH(CH_3)$_2$)$^+$ produces H_2, along with regeneration of active Pd(0) species.

rGO was also examined as a suitable host material to stabilize PdAg alloy nanoparticles [71]. Thus, rGO-dispersed PdAg alloy nanoparticles were easily obtained by performing coreduction of metal precursors in the presence of rGO as a dispersing support material. PdAg/rGO so-prepared displayed 100% H_2 selectivity during FA dehydrogenation at RT with an initial TOF value of 105 h^{-1}. Metin and coworkers prepared PdAg alloy nanoparticles by a coreduction method in which oleic acid and oleylamine (OAm) were used as a surfactant and both a cosurfactant and mild reducing agent, respectively [72]. Nearly monodispersed PdAg alloy nanoparticles were formed and then supported onto activated carbon (PdAg/C) for evaluation in the catalytic neat FA dehydrogenation. It was found that PdAg alloy nanoparticles provide better activity than Pd or Ag and physically mixed Pd–Ag nanoparticles in the dehydrogenation of FA under identical conditions because of the synergic effect in PdAg alloy nanoparticles. The same authors also developed active carbon-supported PdAu alloy nanoparticles (PdAu/C) for the catalytic decomposition of FA [73], which provided better activity (TOF = 230 h^{-1}) than the counterpart PdAg/C catalyst (TOF = 105 h^{-1}).

Huang and coworkers prepared Pd and PdPt and PdAg bimetallic alloy nanoparticles on graphitic carbon nitride (g-C_3N_4) by impregnation and photoreduction techniques [74]. Amid Pd/g-C_3N_4, PdAg/g-C_3N_4, and PdPt/g-C_3N_4 tested catalysts, the best catalytic performance was achieved by PdPt/g-C_3N_4, which provided a TON (turnover number) value of 478 and a TOF value of 239 h^{-1} in the dehydrogenation of aqueous FA.

Aside from classical impregnation–reduction or coprecipitation–reduction techniques, Navlanigarcia et al. [75] synthesized a PdAg/C nanocatalyst through a two-steps procedure, namely, the preparation of colloidal polyvinylpyrrolidone (PVP)-stabilized PdAg alloy nanoparticles, followed by their deposition onto the carbon support. PVP was found to have impact on particle size, electronic properties, and surface composition of PdAg/C nanocatalysts. Under optimum [PVP]/[Pd + Ag] (~1.0), the catalytic system achieves an initial TOF value of 855 h^{-1} in the catalytic decomposition of aqueous FA + SF mixture at 30 °C [75]. Recently, Yang et al. reported on a non-noble metal sacrificial method as an alternative toward small-sized, highly dispersed PdAu alloy nanoparticles onto the rGO support [76]. Thus, they first used cobalt (Co(acetate)$_2$) as a sacrificial agent to form CoPdAu$_0$/rGO and then introduced phosphoric acid (H_3PO_4) as a cobalt scavenging agent to remove Co particles to get the $Pd_{0.4}Au_{0.6}$/rGO nanocatalyst, which provided a TOF value of 4880 h^{-1} in the hydrogen generation from aqueous FA + SF solution. The authors emphasized that the sacrifice of the cobalt compound can prevent AuPd NPs from aggregation, which benefits their catalytic performance. Nozaki et al. reported that nanoporous CeO_2 prepared by chemical dealloying of Ce–Al amorphous alloy acts as a better support material than other metal oxide supports for guest PdAu alloy nanoparticles (formed on support material by impregnation–reduction method). As reported, the PdAu/CeO_2 catalyst displayed a TOF value of 72 h^{-1} at 50 °C in the catalytic decomposition of aqueous FA solution. [77].

Aforementioned catalytic systems usually involve the use of platinum (Pt), silver (Ag), or gold (Au) as a secondary metal to form catalytically active PdM (M = Pt, Ag, and Au) alloy nanoparticles for the dehydrogenation of FA. In order to change the electronic properties of catalytically active Pd nanoparticles and limits the consumption of precious noble metals, the incorporation of a first-row transition metal as a secondary metal has great importance. Literature survey revealed that nickel (Ni) [78], copper (Cu) [79], and cobalt (Co) [80] have been already employed as a secondary metal to get catalytically active PdM (M = Ni, Co, and Cu) alloy nanoparticles for the dehydrogenation of FA. Zhang et al. used a composite composed of graphene nanosheets (GNs) and CB as a host material for PdNi alloy nanoparticles. This catalyst was prepared by sodium borohydride reduction of $Pd(NO_3)_2 \cdot 2H_2O$ and $NiCl_2 \cdot 6H_2O$ salts deposited onto the support surface in aqueous solution. The resultant PdNi/GNs-CB nanocatalyst provides an initial TOF value of 529 h^{-1} in the catalytic decomposition of aqueous FA + SF mixture at RT [78]. Although PdNi/GNs–CB nanocatalyst shows notable activity, its catalytic stability was found to be very low because of agglomeration of GNs–CB surface-decorated PdNi nanoparticles. Resins (Amberlite-type) have been commonly exploited in nanocatalysis as a support material because of their high stability and suitability to surface functionalization. Encouraged by these advantages, Mori et al. generated PdCu alloy nanoparticles on Amberlite-type basic resin (bearing $-NH_2$ functionalities) named IRA96SB [79] for their evaluation in the catalytic decomposition of aqueous FA + SF mixture. PdCu/IRA96SB at [Pd]/[Cu] = 1/1 exhibits 100% selectivity over FA dehydrogenation with an initial TOF value of $810\,h^{-1}$.

In a very recent study, CoPd alloy nanoparticles were generated onto a covalent triazine-based framework (CTF; prepared by the trimerization of terephthalonitriles under ionothermal conditions) as a support through an impregnation–reduction method [80]. The as-obtained CoPd/CTF nanocatalyst achieved a TOF value of $882\,h^{-1}$ at 30 °C in the dehydrogenation of an aqueous FA + SF solution. Additionally, this CoPd/CTF nanocatalyst was found to be highly stable for reusability. This property has been assigned to synergetic effect between CoPd nanoparticles and nitrogen-rich CTF support.

13.3.3 Bimetallic Pd-Containing Nanocatalysts in the Core@Shell Structure

The first well-defined palladium-based core@shell nanocatalyst has been developed by Huang et al. [81]. This catalyst was synthesized by a simultaneous reduction method without using any stabilizing agent on activated carbon (PdAu@Au/C). It was found to catalyze dehydrogenation of aqueous FA + SF solutin at a H_2 generation rate of 7.1 ml min^{-1}), which is nine times higher than the activity achieved with a monometallic Au/C nanomaterial prepared under identical conditions. Also, a low amount of CO releasing (30 ppm) was observed.

In a subsequent study, Tsang and coworkers prepared a series of colloidal suspensions of PVP-stabilized M@Pd core@shell (M = Au, Ag, Pt, Rh, and Ru) nanoparticles by a successive reduction method and immobilized them onto activated carbon (Au@Pd/C, Ag@Pd/C, Pt@Pd/C, Rh@Pd/C, and Ru@Pd/C) [82]. Among these catalysts, the best activity was attained with carbon-supported Ag@Pd (1 : 1) nanoparticles (TOF = $252\,h^{-1}$ at 50 °C). The most important insight into this study relies on the characterization of core@shell structures. Direct imaging by high-resolution transmission electron microscopy (HRTEM) failed, as there was a similar contrast between the core and shell elements (image contrast primarily depends on the difference in atomic number, but Ag and Pd are next to each other in the periodic table). Atom probe tomography (APT) revealed the internal structure of the core@shell nanoparticles, and APT technique relies on generating high electric fields (3–5 V $Å^{-1}$) at the apex of a needle-shaped specimen. On top of this, an additional short (ns) pulse removes ions in a controlled manner into a high-resolution time-of-flight mass spectrometer equipped with a single-ion-sensitive detector. APT thus combines atomic-scale position mapping with single-atom chemical

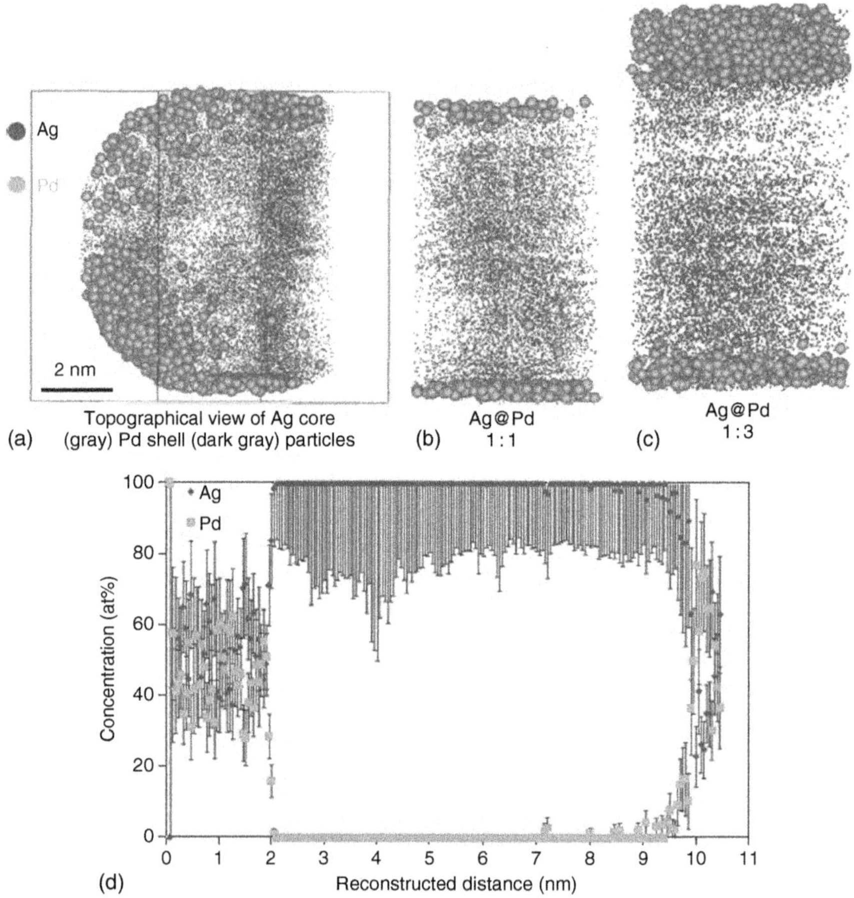

Figure 13.2 Atom map (a) of an individual Ag@Pd nanoparticle, clearly showing the core@shell structure, and column sections of atom maps, (b) and (c) taken through the Ag@Pd nanoparticles with as-synthesized compositions as indicated, showing the differences in shell thickness. (d) Composition profile of Ag and Pd through a (1 : 3) Ag@Pd core@shell nanoparticle, showing a pure Ag core (light gray) with a 5–10 atomic layer thickness of Pd shell (dark gray) atoms. Source: From Tedsree et al. [82]. © Springer Nature.

identification (see Figure 13.2). The interfaces between core and shell are very sharp, and clear differences are seen in the thicknesses of Pd shells. For Ag@Pd (1 : 1) samples, the shells were no more than 1–2 atomic layers thick, whereas in Ag@Pd (1 : 3) samples, Pd shells were all significantly thicker of the order of 5–10 atomic layers. The higher reactivity of Ag@Pd with (1 : 1) ratio than that composed of (1 : 3) ratio was attributed by the authors to electronic promotion by the underlying Ag, which has a rather short range of just a few atomic distances.

rGO, which has high surface area and suitability for functionalization, was also used to support catalytically active M@Pd nanoparticles for FA dehydrogenation. For instance, Wang et al. [83] reported a green and facile strategy (Scheme 13.7) for the direct nucleation and growth of ultrafine (1.8 nm) and well-dispersed

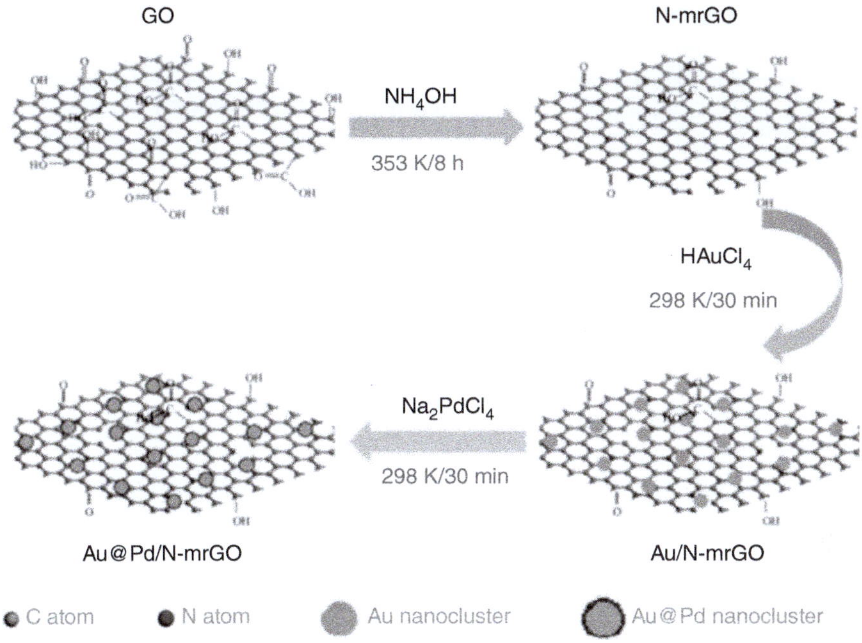

Scheme 13.7 Schematic illustration for the preparation of Au@Pd/N-rGO nanocatalyst. Source: Adapted from Ref. [83].

Au@Pd (core@shell) nanoparticles on nitrogen-doped mildly reduced graphene oxide (Au@Pd/N-rGO) without additional protective and reducing agents. In this synthesis, the N-doped N-rGO support acted as both the reducing agent because of surface functionalities and the support by taking advantage of its moderate reducing ability and high dispersing capacities. Because of the strong synergic effect in core@shell nanocatalysts, the catalytic activity of this Au@Pd/N-rGO nanomaterial (TOF = 89 h^{-1}) was found to be higher than its monometallic (Au/N-rGO, Pd/N-rGO) and alloy (AuPd/N-rGO) counterparts in the dehydrogenation of FA in the absence of SF at RT. Next, it was shown that not only N-doped rGO but also N-doped graphene–carbon nanotube aerogel (synthesized via hydrothermal reaction) is an appropriate support material for immobilizing Ag@Pd nanoparticles for FA dehydrogenation catalysis [84]. Ag@Pd core–shell nanoparticles were generated in the presence of nitrogen-doped graphene–carbon nanotube aerogel as a support by subsequent reduction method; first, $AgNO_3$ was reduced with ethylene glycol to lead Ag seeds and then reduction of $PdCl_2$ on support surface allowed growth of Pd shell on Ag seeds. The resulting Ag@Pd nanoparticles achieved to catalyze additive-free decomposition of FA at high dehydrogenation selectivity (>99%) and activity (TOF = 413 h^{-1}) even at RT.

For comparison purpose, the turnover activities (TOF value) of various Pd-based bimetallic alloy and core@shell nanocatalysts in FA dehydrogenation are summarized in Table 13.1.

Table 13.1 The catalytic activity comparison of Pd-based bimetallic alloy and core@shell nanocatalysts in the catalytic decomposition of aqueous FA or FA + SF solution.

Catalyst	Substrate	Temperature (°C)	Turnover frequency (TOF, h^{-1})	References
PdAu/C	FA + SF	82	—	[65]
PdAg/C	FA + SF	82	—	[65]
PdCu/C	FA + SF	82	—	[65]
PdAu/ED-MIL-101	FA + SF	90	—	[66]
Pd_4Ag_1/NH_2-UiO-66	FA	25	103	[67]
Pd_4Ag_1/NH_2-UiO-66	FA	50	525	[67]
Pd_4Ag_1/NH_2-UiO-66	FA	80	893	[67]
PdAg/resin	FA + SF	75	1900	[70]
PdAg/rGO	FA	25	105	[71]
$Pd_{58}Ag_{42}/C$	FA	50	382	[72]
PdAu	FA	50	230	[73]
PdAg	FA + SF	25	32	[64]
PdPt/g-C_3N_4	FA	25	239	[74]
$Pd_{0.82}Ag_{0.18}$/ZIF-8	FA + SF	80	580	[68]
$Pd_{12}Ag_3$/MOF-5-C-900	FA	25	854	[69]
PdAg/C PVP stabilized	FA + SF	30	855	[75]
$Pd_{0.4}Au_{0.6}$/rGO	FA + SF	50	497	[76]
PdNi/GNs-CB	FA + SF	—	529	[78]
PdCu/IRA96SB	FA + SF	75	810	[79]
CoPd/CTF	FA + SF	30	882	[80]
Ag@Pd/C	FA	50	252	[82]
Au@Pd/N-rGO	FA + SF	25	89	[83]

13.3.4 Trimetallic Pd-Containing Nanocatalysts

With the aim to enhance the catalytic activity of PdAu/C nanocatalysts [84] in FA dehydrogenation, in a first attempt to develop trimetallic Pd-containing nanocatalysts for FA dehydrogenation, Zhou et al. added rare-earth elements (REs) to carbon-supported PdAu alloy nanoparticles (PdAu/C) [65]. They prepared physical mixtures of REs-PdAu/C (REs = dysprosium [Dy], europium [Eu], and holmium [Ho]) through consecutive impregnation–reduction method and found that the activities of PdAu alloy nanoparticles increased by the addition of REs in the physical mixture form. The activity order was found to be Dy–PdAu/C > Eu–PdAu/C > Ho–PdAu/C > PdAu/C and Dy–PdAu/C. The highest activity observed with Dy–PdAu/C catalyst provided a TOF value of 269 h^{-1} at 92 °C with <100 ppm CO generation. Particle size, electronic, and synergic effects were indicated to have key influences on the promotion effect that leads to enhanced activity.

Wang and coworkers used wet impregnation reduction method to prepare activated carbon-supported trimetallic PdAuNi alloy nanoparticles. Because of the high metal loading (~10 wt%), the sizes of the resulting alloy nanoparticles were estimated in the range of 16–35 nm. The optimum molar metal composition regarding reactivity was found to be $Pd_{0.45}Au_{0.15}Ni_{0.40}/C$. This nanomaterial exhibits the best catalytic performance in terms of both selectivity (100%) and activity (TOF = $12.4\,h^{-1}$) at 25 °C without using any additive or promoter [85]. By the same methodology, we prepared activated carbon-supported trimetallic PdNiAg alloy nanoparticles (PdNiAg/C) [86], whose mean diameter (5.6 ± 2.2 nm) is smaller than that of the previous PdAuNi/C system owing to low metal loading (~2.0 wt%) onto the support. PdNiAg/C nanomaterial catalyzes the catalytic decomposition of aqueous FA + SF mixture with moderate activity (TOF = $85\,h^{-1}$) high selectivity (~100%) toward dehydrogenation. More importantly, these carbon-supported PdNiAg alloy nanoparticles display high stability against agglomeration, leaching, and CO poisoning. PdNiAg/C was shown to be a highly reusable catalyst retaining almost inherent activity (>94%) even at fifth reuse in the dehydrogenation of aqueous FA + SF solution with high selectivity (~100%) at complete conversion. We used the similar wet impregnation reduction method to fabricate APTS-functionalized SiO_2-supported trimetallic PdAuCr alloy nanoparticles in different molar compositions [87]. For the molar composition of $Cr_{0.15}Au_{0.25}Pd_{0.60}/N\text{–}SiO_2$ (~2.6 nm), we achieved excellent selectivity (100%) and activity (TOF = $730\,h^{-1}$) because of the sum of the effect of Cr alloying and $-NH_2$ functionalization in the complete dehydrogenation of FA even at RT.

We also investigated the effect of adding MnO_x (mixtures of Mn(II)–Mn(III)–Mn(IV) oxides) nanoparticles as a physical mixture form on the activity of amine-functionalized SiO_2-supported monometallic Pd [61] and bimetallic PdAg [88] and PdAu [89] alloy nanoparticles in FA dehydrogenation under mild reaction conditions. Besides catalytic performance test experiments, detailed mechanistic studies were performed by using electrochemical methods and in situ spectroscopic techniques. The obtained results revealed that incorporation of Ag or Au sites into Pd nanoparticles yields PdAg or PdAu alloys by decreasing CO adsorption strength at the active Pd surface, thus increasing the CO poisoning tolerance. Also, the addition of superoxophilic MnO_x nanoparticles offers sacrificial CO anchoring sites forming carbonates. As a result of these, the active Pd sites in $PdAg\text{–}MnO_x/N\text{–}SiO_2$ and $PdAu\text{–}MnO_x/N\text{–}SiO_2$ catalysts demonstrate high selectivity (100%) and activity ($TOF_{(PdAg)} = 330\,h^{-1}$ and $TOF_{(PdAu)} = 785\,h^{-1}$) in FA dehydrogenation in the absence of additives even at RT. Yan et al. [90] also prepared $PdAu\text{–}MnO_x$ nanoparticles but using ZIF-8/graphene composite as a support. This catalyst was found to be active (TOF = $382\,h^{-1}$) in FA dehydrogenation. The lower activity of $PdAu\text{–}MnO_x$/ZIF-8-graphene ($382\,h^{-1}$) with respect to $PdAu\text{–}MnO_x/N\text{–}SiO_2$ ($785\,h^{-1}$) can be attributed to the effect of $-NH_2$ functionalization plus employing of different support materials.

The aforementioned promotion effect of amine functionalities on the catalytic activity of Pd-based trimetallic nanoparticles in FA dehydrogenation was also used in basic resin (Amberlite type)-supported PdCuCr [91], AuPdCeO_2-decorated rGO [92], and GN-supported PdAuNi [93] alloy nanoparticles. Mori et al. used impregnation–borohydride reduction method to prepare PdCuCr alloy nanoparticles within a macroreticular basic resin, which possesses $-N(CH_3)_2$ functional groups [91]. In the catalytic decomposition of aqueous FA + SF solution, this nanocatalyst offers a TOF value of 830 h^{-1} at RT. The functionality impact in the resin support was also examined, and it was found that the resin with weakly basic $-N(CH_3)_2$ groups displays more capability of O–H cleavage (Scheme 13.6) compared to resins functionalized with $-SO_3H$, –COOH, and $-N(CH_2Cl)_2$ groups. Wang et al. fabricated AuPdCeO_2 onto rGO [92] by sodium borohydride reduction of aqueous solution, $HAuCl_4$, Na_2PdCl_4, and $Ce(NH_4)_2(NO_3)_6$ metal salts in the presence of rGO. In the absence of any stabilizing agent, the resultant clean surface of AuPdCeO_2 explains their high activity (TOF = 52.9 h^{-1}) in the dehydrogenation of aqueous FA solution at 25 °C without CO generation. Recently, we have developed trimetallic PdAuNi alloy nanoparticles on the surface of APTS-functionalized graphene nanosheets (PdAuNi/f-GNS) [93] by using double-solvent technique [94, 95] combined with the liquid-phase chemical reduction technique. By this way, small-sized (3.2 nm) PdAuNi alloy nanoparticles in high dispersion throughout the support surface were obtained. The experiments performed in the catalytic decomposition of FA showed that the PdAuNi/f-GNS material selectively catalyzes the target reaction through the dehydrogenation pathway (100% H_2 selectivity) with a TOF value of 1090 h^{-1} at almost complete conversion (>92%) and 25 °C. The so-obtained catalytic activity was found to be higher than those described with heterogeneous catalysts reported to date for the additive-free dehydrogenation of FA.

As for all catalysis, achieving high catalytic stability with nanocatalysts while avoiding sintering and leaching during catalytic reaction is one of the most important goals to consider in the design of novel catalytic architectures for FA dehydrogenation. For this purpose, our recent efforts focused on the development of oxide layer-protected metal nanoparticles supported on various metal oxides through atomic layer deposition (ALD) method. The ALD technique is a thin-film growth method similar to chemical vapor deposition (CVD), except that the deposition occurs from cycles each of which is a sequence of two surface reactions [96]. Contrary to previous surface coating techniques, ALD offers the option for atomically controlled postmodification of supported metal nanoparticles by applying protective overcoats [97]. Moreover, the self-limiting, layer-by-layer deposition feature of ALD allows exact control over the thickness of the protective layer, which protects solid-supported metal nanoparticles against sintering and leaching. Thus, ALD is also the most encouraging technique for overcoming the mass transfer resistance problem [96].

Inspired by these technical advantages, we prepared ALD–SiO_2 layer-protected PdCoNi alloy nanoparticles supported onto TiO_2 nanopowders [98]. Our synthesis protocol involves successive steps such as (i) the preparation of TiO_2-supported

PdCoNi alloys nanoparticles by the conventional wet impregnation, followed by simultaneous reduction method, (ii) the protecting of PdCoNi nanoparticle surface with OAm, which acts as a blocking agent to prevent growing of SiO_2 layers onto PdCoNi nanoparticle surface, (iii) the growing of SiO_2 layer around surface-protected PdCoNi alloy nanoparticles to optimum thickness for catalysis, and then (iv) removing OAm blocking agent through acetic acid washing, followed by calcination and H_2 reduction steps (Scheme 13.8a). SiO_2 layers between TiO_2 nanopowder-supported PdCoNi alloy nanoparticles were generated using sequential exposures to APTS and water (H_2O) vapors (Scheme 13.8b). Starting from the optimized $Pd_{0.60}Co_{0.18}Ni_{0.22}/TiO_2$ nanocatalyst, various numbers (1–20) of SiO_2–ALD cycles were performed to generate protective layers of progressively increasing thickness. Under optimized SiO_2 layer thickness (Figure 13.3a–c) conditions, the resulting $Pd_{0.60}Co_{0.18}Ni_{0.22}/TiO_2$–ALD–$SiO_2$ (six cycles) nanocatalyst exhibited activity two times higher than that obtained with the non-ALD–SiO_2-protected $Pd_{0.60}Co_{0.18}Ni_{0.22}/TiO_2$ catalyst in the catalytic decomposition of aqueous FA + SF solution at 25 °C. This novel nanocatalyst provided an *initial* TOF value of 207 h^{-1} with >99% dehydrogenation selectivity at almost complete conversion. Moreover, these SiO_2 layer-protected PdCoNi alloy nanoparticles displayed the exceptional catalytic stability throughout the reusability experiments against sintering and leaching. Therefore, they preserved ≥83% of their catalytic activity and selectivity at >90% catalytic conversion even at the 20th catalytic reuse, whereas $Pd_{0.60}Co_{0.18}Ni_{0.22}/TiO_2$ totally lost its catalytic reactivity under identical conditions even at seventh catalytic reuse (Figure 13.3d). The activity comparison of different trimetallic catalytic systems in FA dehydrogenation is given in Table 13.2.

13.3.5 Other Pd-Free Nanocatalysts

Sadeghzadeh reported the synthesis of a novel nanocatalyst based on PbS nanoparticles immobilized into the nanospaces of ionic liquid (IL)-based fibrous nanosilica KCC-1 (KCC-1: fibrous nanosilica) (KCC-1/IL/PbS) (Scheme 13.9) [99]. The catalytic performances of various KCC-1/IL/X (X = Au, Cu, Pd, Ag, Pt, Mn, Ni, Zn, Co, HgS, and PbS) nanoparticle systems were examined in the catalytic decomposition of FA + SF mixture under identical conditions (40 °C and FA/SF = 9/1). The activity was found to be in the order of PbS > ZnS > HgS > Au > Cu > Pd > Ag > Pt > Mn > Ni > Zn > Co. The catalytic activity of KCC-1/IL/PbS led to a TOF value of 604 h^{-1} for the catalytic decomposition of aqueous FA + SF mixture at 40 °C. KCC-1/IL/PbS catalyst could be recovered and reused more than 10 times with no decrease in its activity. Based on these results, it was concluded by the authors that even if the KCC-1/IL nanostructure acts as cobweb for the soluble PbS species, it can also operate as a nanoscaffold to maintain the PbS nanoparticles into the mesofibers.

Ting et al. prepared small-sized $PtRuBiO_x$ nanoparticles supported on carbon ($PtRuBiO_x/C$) by using citrate method [100]. Although the authors did not give clear explanation for the structural nature of their trimetallic $PtRuBiO_x$ nanoparticles, findings in transmission electron microscopy (TEM) and catalytic studies pointed

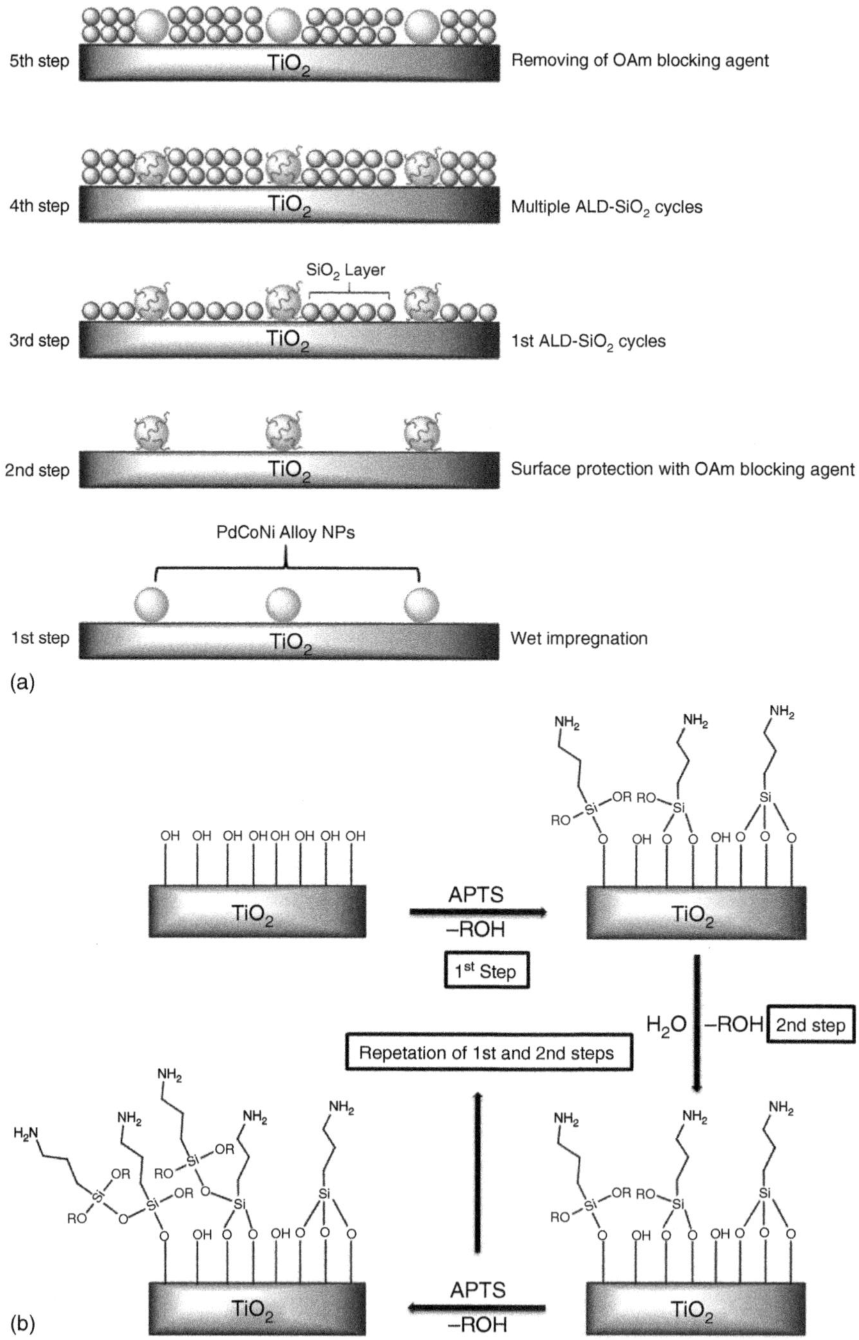

Scheme 13.8 (a) The schematic illustration of the synthesis protocol followed in the preparation of ALD–SiO_2 layer-protected PdCoNi alloy nanoparticles supported on TiO_2 nanopowders. (b) A schematic description of the ALD-sequential deposition of APTS and water vapor on the TiO_2 nanopowders' surface. Source: From Caner et al. [98]. © Elsevier.

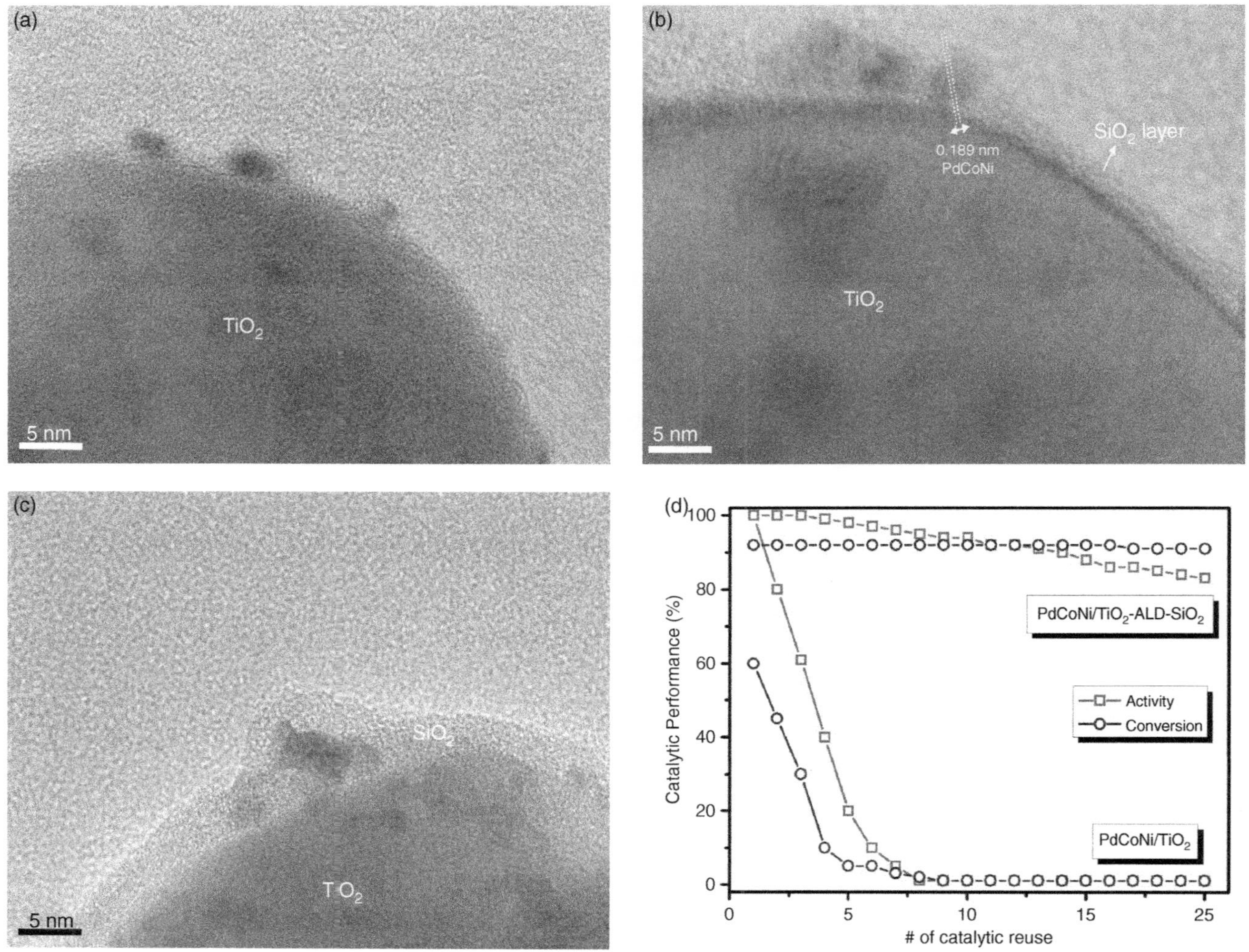

Figure 13.3 HRTEM images of (a) $Pd_{0.60}Co_{0.18}Ni_{0.22}/TiO_2$–ALD–$SiO_2$ (6-cycles), (b) $Pd_{0.60}Co_{0.18}Ni_{0.22}/TiO_2$–ALD–$SiO_2$ (10-cycles), (c) $Pd_{0.60}Co_{0.18}Ni_{0.22}/TiO_2$–ALD–$SiO_2$ (20-cycles), and (d) the percentage of retained catalytic performance vs. number of catalytic reuse for $Pd_{0.60}Co_{0.18}Ni_{0.22}/TiO_2$ and $Pd_{0.60}Co_{0.18}Ni_{0.22}/TiO_2$–ALD–$SiO_2$ (six cycles) catalyzed dehydrogenation of aqueous FA + SF (0.175 M FA + 0.175 M SF) at RT. Source: From Caner et al. [98]. © Elsevier.

Table 13.2 The catalytic activity comparison of Pd-based trimetallic nanocatalysts in the catalytic decomposition of aqueous FA or FA + SF solution.

Catalyst	Substrate	Temperature (°C)	Turnover frequency (TOF, h^{-1})	References
Pd–Au–Dy/C	FA + SF	92	269	[84]
$Pd_{0.45}Au_{0.15}Ni_{0.40}/C$	FA	25	12.4	[85]
PdNiAg/C	FA + SF	50	85	[86]
$Cr_{0.15}Au_{0.25}Pd_{0.60}$/N–$SiO_2$	FA	25	730	[87]
PdAg–MnO_x/N–SiO_2	FA	25	330	[88]
PdAu–MnO_x/N–SiO_2	FA	25	785	[89]
AuPd–MnO_x/ZIF-8–rGO	FA	25	382	[90]
PdCuCr/resin	FA + SF	75	830	[91]
AuPd–CeO_2/N-rGO	FA	25	52.9	[92]
PdAuNi/f-GNS	FA	25	1090	[93]
PdCoNi/TiO_2–ALD–SiO_2	FA + SF	25	207	[98]

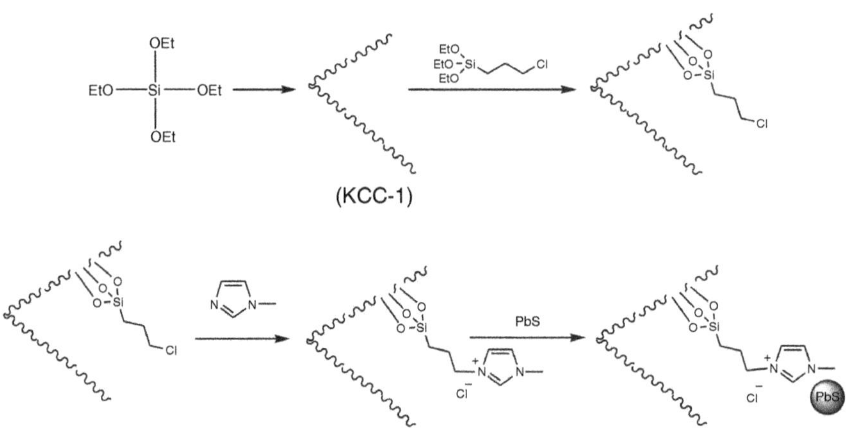

Scheme 13.9 Schematic illustration of the synthesis for KCC-1/IL/PbS nanoparticles. Source: Sadeghzadeh [99]. © Elsevier.

out that the metals exist in the physical mixture of PtRu alloy and BiO_x nanoparticles. They reported that bismuth exists under the form of different carbonates including $(BiO)_4(OH)CO_3$ and bismuth subcarbonate $(BiO)_2(CO_3)$ that can be formed from CO_2, water, and Bi_2O_3 [101] and that these bismuth carbonates play a possible role in removing the product or forming the intermediate as adsorbed CO_2, which enhances the dehydrogenation selectivity. PtRuBiO_x/C catalyst provided a TOF value of 312 h^{-1} based on only Pt and Ru atoms at 80 °C in the dehydrogenation of FA.

Bide et al. [102] synthesized selenium (Se)-doped graphene/$CoFe_2O_4$ via a one-step polyol method that afforded a TOF value of $306\,h^{-1}$ in the dehydrogenation of aqueous FA solution in the absence of additives at 60 °C [103]. The authors indicated that the Se dopant increases electron transfer and induces strain to carbon material and high polarizability and that these facts enhance the catalytic activity of $CoFe_2O_4$ nanoparticles.

13.4 Summary and Conclusions

FA has recently been recognized as one of the most promising hydrogen storage materials not only owing to its simple storage, nontoxicity, and low-cost production but also because of the generation of only gaseous CO_2 along with H_2, which can be recycled to FA through catalytic hydrogenation. The development of active, selective, and reusable nanocatalysts has thus great importance for the widespread usage of FA in the chemical hydrogen storage. Until now, many different nanocatalyst systems have been prepared and employed for the liquid-phase dehydrogenation of FA and some of them exhibit notable catalytic performances. There are many features that impact on the catalytic dehydrogenation efficiency of FA, but there is still room for research in the improvement of the existing nanocatalysts. Following aspects and requirements are expected to be included:

(1) *Active metals.* The common active metals in FA dehydrogenation are Pt, Pd, and Au, and among them, Pd-based nanocatalysts provide best catalytic performances in terms of both activity and selectivity. Firstly, checking the activity of other prestigious transition-metal-based nanocatalytic systems especially composed of iridium (Ir), rhodium (Rh), and ruthenium (Ru), which are known as highly active metals in various hydrogenation [104–107] and dehydrogenation [108–111] reactions, needs to be done in FA dehydrogenation. Additionally, from the view of cost of metals, tungsten (W) and molybdenum (Mo) carbide (WC, MoC) nanoparticles, which showed platinum-like activity in different catalytic transformations [112–115], should also be tested as low cost nanocatalysts.
(2) *Structure of catalyst.* Aforementioned studies have revealed that the electronic structure of active metal (Pd and Pt) should be modified by incorporating second or third metal in the forms of alloy and core@shell nanostructures. In this context, (i) low-cost earth-abundant metals should be considered as second and/or third metal to minimize the total cost of the catalytic system and (ii) it is better to exploit firstly core@shell structure rather than alloy structure for Pd- or Pt-based nanocatalyst system to achieve atom economy and minimize the cost of the catalytic system.
(3) *Support material.* The expectations from ideal support material should (i) contribute to the control of particle size of guest active metal nanoparticles, (ii) assist the dispersion of active metal nanoparticles, (iii) show synergic effect with different active metals, and (iv) have available sites for amine functionalization (or nitrogen-doping), which provide a basic environment around active metals and

facilitates O—H bond cleavage that yields metal–formate formation in FA dehydrogenation.

(4) *Dopant material.* When the catalytic system comprises an active metal (monometallic or bimetallic), dopant materials (such as Se, Mn, and Bi oxides), which usually exist in the physical mixture form, can act as a CO sponge around the catalytically active metal nanoparticles and thus increase the CO poisoning resistance of active centers. In this context, to achieve homogeneous distribution of dopants in the catalytic system, their direct use as a support material might be interesting. For example, solid MnO_x and BiO_x deserve to be tested as a support material for active metal nanoparticles in FA dehydrogenation. Another interesting approach at this point would be employment of ALD technique to prepare thin thickness layer between active metal nanoparticles and support to get homogeneous distribution of MnO_x or BiO_x throughout the catalytic material.

(5) *Large-scale FA dehydrogenation.* Until now, all of the aforementioned nanocatalysts have been employed in small-scale FA dehydrogenation. More research efforts are thus required in the future in order to perform large-scale FA dehydrogenation if one wants to make progress in the usage of this approach as hydrogen source specifically for mobile fuel cell applications.

Acknowledgments

Authors acknowledge Van Yuzuncu Yil University Scientific Research Project Coordination Unit (FDP-2018-6944) for financial support to their laboratory.

References

1 Turner, J.A. (2004). *Science* 305: 972–974.
2 Hetland, J. and Mulder, G. (2007). *Int. J. Hydrogen Energy* 32: 736–747.
3 Laurent, C., Morishima, R., Suzuki, H., and Lelay, M. (2004). *J. Phys. Chem. B* 108: 12718–12723.
4 Dong, J., Wang, X., Xu, H. et al. (2007). *Int. J. Hydrogen Energy* 32: 4998–5004.
5 Tranchemontagne, D.J., Park, K.S., Furukawa, H. et al. (2012). *J. Phys. Chem. C* 116: 13143–13151.
6 McKeown, N.B. and Budd, P.M. (2006). *Chem. Soc. Rev.* 35: 675–683.
7 Sakintuna, B., Darkrim, F.L., and Hirscher, M. (2007). *Int. J. Hydrogen Energy* 32: 1121–1140.
8 Hamilton, C.W., Baker, T., Staubitz, A., and Manners, I. (2009). *Chem. Soc. Rev.* 38: 279–293.
9 Turner, J., Sverdrup, G., Mann, K. et al. (2008). *Int. J. Energy Res.* 32: 379–407.
10 Saksaa, P.V.T., Cook, C., Kiviaho, J., and Rapo, T. (2018). *J. Power Sources* 396: 803–823.
11 Yadav, M. and Xu, Q. (2012). *Energy Environ. Sci.* 5: 9698–9713.

12 Enthaler, S., Langermann, J.V., and Schmidt, T. (2010). *Energy Environ. Sci.* 3: 1207–1217.
13 Ha, S., Larsen, R., Zhu, Y., and Masel, R.I. (2004). *Fuel Cells* 4: 337–343.
14 Demirci, U.B. (2007). *J. Power Sources* 169: 239–246.
15 Kordesch, K.V. and Simader, G.R. (1995). *Chem. Rev.* 95: 191–207.
16 Myers, T.W. and Berben, L.A. (2014). *Chem. Sci.* 5: 2771–2777.
17 Zell, T., Butschke, B., Ben-David, Y., and Milstein, D. (2013). *Chem. Eur. J.* 19: 8068–8072.
18 Wang, W.H., Xu, S., Manaka, Y. et al. (2014). *ChemSusChem* 7: 1976–1983.
19 Johnson, T.C., Morris, D.J., and Wills, M. (2010). *Chem. Soc. Rev.* 39: 81–88.
20 Grasemann, M. and Laurenzcy, G. (2012). *Energy Environ. Sci.* 5: 8171–8181.
21 Fukuzumi, S., Kobayashi, T., and Suenobu, T. (2010). *J. Am. Chem. Soc.* 132: 1496–1497.
22 Boddien, A., Mellmann, D., Gartner, F. et al. (2011). *Science* 333: 1733–1736.
23 Scholten, J.D., Prechtl, M.H.G., and Dupont, J. (2010). *ChemCatChem* 2: 1265–1270.
24 White, R.J., Luque, R., Budarin, V.L. et al. (2009). *Chem. Soc. Rev.* 38: 481–501.
25 Sabatier, P. and Mailhe, A. (1912). *Compt. Rend.* 12: 1212–1230.
26 Trillo, J.M., Munuera, G., and Criado, J.M. (1972). *Catal. Rev.* 7: 51–58.
27 Columbia, M.R. and Thiel, P.A. (1994). *J. Electroanal. Chem.* 369: 1–6.
28 Tamaru, K. (1997). *Appl. Catal., A* 151: 167.
29 Rienacker, G. (1936). *Z. Anorg. Allg. Chem.* 227: 353–359.
30 Rienacker, G. and Hansen, N. (1956). *Z. Anorg. Allg. Chem.* 285: 283–289.
31 Rienacker, G. and Volter, J. (1959). *Z. Anorg. Allg. Chem.* 302: 299–309.
32 Rienacker, G. and Mueller, H. (1968). *Z. Anorg. Allg. Chem.* 357: 255–268.
33 Dowden, D.A. and Reynold, P.W. (1950). *Discuss. Faraday Soc.* 8: 184–190.
34 Sexton, B.A. (1979). *Surf. Sci.* 88: 319–328.
35 Solymosi, F. and Kovacs, I. (1991). *Surf. Sci.* 259: 95–102.
36 Zahmakiran, M. and Özkar, S. (2011). *Nanoscale* 3: 3462–3481.
37 Zhou, B., Han, S., Raja, R., and Somorjai, G.A. (2007). *Nanotechnology in Catalysis*, vol. 3, New York: Springer.
38 Pool, R. (1990). *Science* 248: 1186–1192.
39 Schmid, G. (2004). *Nanoparticles*. Weinheim: Wiley-VCH.
40 Poliakoff, M., Fitzpatrick, J.M., Farren, T.R., and Anastas, P.T. (2002). *Science* 297: 807–812.
41 Joo, F. (2008). *ChemSusChem* 1: 805–808.
42 Singh, A.K., Singh, S., and Kuma, A. (2016). *Catal. Sci. Technol.* 6: 12–40.
43 Mellmann, D., Sponholz, P., Junge, H., and Beller, M. (2016). *Chem. Soc. Rev.* 45: 3954–3988.
44 Li, J., Zhu, Q.L., and Xu, Q. (2015). *Chimia* 69: 348–352.
45 Zhang, L., Wu, W., Jian, Z., and Fang, T. (2018). *Chem. Pap.* 72: 2121–2135.
46 Ojeda, M. and Iglesia, E. (2009). *Angew. Chem. Int. Ed.* 48: 4800–4803.
47 Yi, N., Saltsburg, H., and Stephanopoulos, M.F. (2013). *ChemSusChem* 6: 816–819.

48 Tang, C., Surkus, A.-E., Chen, F. et al. (2017). *Angew. Chem. Int. Ed.* 56: 16616–16620.

49 Liu, H., Mei, Q., Wang, Y. et al. (2016). *Sci. China, Ser. B Chem.* 59: 1342–1347.

50 Lu, Q., Beller, M., and Jiao, H. (2013). *J. Theor. Comput. Chem.* 12: 1330001–1330029.

51 White, R.J., Luque, R., Budarin, V.L. et al. (2009). *Chem. Soc. Rev.* 38: 481–502.

52 Zhu, Q.L., Tsumori, N., and Xu, Q. (2013). *Chem. Sci.* 5: 195–199.

53 Hu, C., Pulleri, J.K., Ting, S.W., and Chan, K.Y. (2014). *Int. J. Hydrogen Energy* 39: 381–390.

54 Song, F.Z., Zhu, W.L., Tsumori, N., and Xu, Q. (2015). *ACS Catal.* 5: 5141–5144.

55 Lia, J., Chen, W., Zhao, H. et al. (2017). *J. Catal.* 352: 371–381.

56 Akbayrak, S., Tonbul, Y., and Özkar, S. (2017). *Appl. Catal., B* 206: 384–392.

57 Wang, N., Sun, Q., Bai, R. et al. (2016). *J. Am. Chem. Soc.* 138: 7484–7492.

58 Li, Z., Tsumori, N., Liu, Z. et al. (2017). *ACS Catal.* 7: 2720–2724.

59 Sanchez, F., Alotaibi, M.H., Motta, D. et al. (2018). *Sustainable Energy Fuels* 2: 2705–2716.

60 Sun, J., Qiu, H., Cao, W. et al. (2019). *ACS Sustainable Chem. Eng.* 7: 1963–1972.

61 Bulut, A., Yurderi, M., Karatas, Y. et al. (2015). *Appl. Catal., B* 164: 324–333.

62 Jin, M.H., Ohi, D., Park, J.H. et al. (2016). *Sci. Rep.* 6: 33502–33510.

63 Alotaibi, M.H., Alotaibi, R.J., and Aldosar, O.F. (2018). *Org. Chem. Curr. Res.* 7: 43–44.

64 Liu, J., Lan, L., Li, R. et al. (2015). *Int. J. Hydrogen Energy* 41: 951–958.

65 Zhou, X., Huang, Y., Xing, W. et al. (2008). *Chem. Commun.*: 3540–3542.

66 Gu, X., Lu, Z.H., Jiang, H.L. et al. (2011). *J. Am. Chem. Soc.* 133: 11822–11825.

67 Gao, S.T., Liu, W., Feng, C. et al. (2016). *Catal. Sci. Technol.* 6: 869–874.

68 Dai, H., Xia, B., Lan, W. et al. (2015). *Appl. Catal., B* 165: 67–72.

69 Feng, C., Wang, Y., Gao, S. et al. (2016). *Catal. Commun.* 78: 17–21.

70 Mori, K., Dojo, M., and Yamashita, H. (2013). *ACS Catal.* 3: 1114–1119.

71 Ping, Y., Yan, J.M., Wang, Z.L. et al. (2013). *J. Mater. Chem. A* 1: 12188–12195.

72 Zhang, S., Metin, O., Su, D., and Sun, S. (2013). *Angew. Chem. Int. Ed.* 52: 3681–3684.

73 Metin, O., Sun, X., and Sun, S. (2013). *Nanoscale* 5: 910–912.

74 Liu, X., Su, P., Chen, Y. et al. (2018). *New J. Chem.* 42: 9449–9454.

75 Navlanigarcia, M., Mori, K., Ai, N. et al. (2015). *Ind. Eng. Chem. Res.* 88: 5244–5250.

76 Yang, X., Pachfule, P., Chen, Y. et al. (2016). *Chem. Commun.* 52: 4171–4174.

77 Nozaki, A., Yamashita, A., Fujiwara, R. et al. (2018). *Bull. Chem. Soc. Jpn.* 91: 1710–1714.

78 Qin, Y.L., Wang, J.W., Wu, Y.M., and Wang, L.M. (2013). *Chem. Commun.* 49: 10028–10030.

79 Mori, K., Tanaka, H., Dojo, M. et al. (2015). *Chem. Eur. J.* 3: 1114–1119.

80 Feng, T., Wang, J.-M., Gao, S.T. et al. (2019). *Appl. Surf. Sci.* 469: 431–436.

81 Huang, Y., Zhou, X., Yin, M. et al. (2010). *Chem. Mater.* 22: 5122–5128.

82 Tedsree, K., Li, T., Jones, S. et al. (2011). *Nat. Nanotechnol.* 6: 302–307.

83 Wang, Z.-L., Yan, J.-M., Wang, H.-L. et al. (2013). *J. Mater. Chem. A* 1: 12721–12725.
84 Zhou, X., Huan, Y., Liu, C. et al. (2010). *ChemSusChem* 3: 1379–1382.
85 Wang, Z.-L., Ping, Y., Yan, J.-M. et al. (2014). *Int. J. Hydrogen Energy* 39: 4850–4856.
86 Yurderi, M., Bulut, A., Zahmakiran, M., and Kaya, M. (2014). *Appl. Catal., B* 160: 514–524.
87 Yurderi, M., Bulut, A., Caner, N. et al. (2015). *Chem. Commun.* 51: 11417–11420.
88 Bulut, A., Yurderi, M., Karatas, Y. et al. (2015). *ACS Catal.* 5: 6099–6110.
89 Karatas, Y., Bulut, A., Yurderi, M. et al. (2016). *Appl. Catal., B* 180: 586–595.
90 Yan, J.-M., Wang, Z.-L., Gu, L. et al. (2015). *Adv. Energy Mater.* 5: 1500107–1500113.
91 Mori, K., Naka, K., Masuda, S. et al. (2017). *ChemCatChem* 9: 3456–3462.
92 Wang, Z.L., Yan, J.M., Zhang, Y.F. et al. (2014). *Nanoscale* 6: 3073–3077.
93 Bulut, A., Yurderi, M., Kaya, M. et al. (2018). *New J. Chem.* 42: 16103–16114.
94 Aijaz, A., Karkamkar, A., Choi, A.J. et al. (2012). *J. Am. Chem. Soc.* 134: 13926–13939.
95 Zhu, Q.L., Li, J., and Xu, Q. (2013). *J. Am. Chem. Soc.* 135: 10210–10213.
96 George, S.M. (2010). *Chem. Rev.* 110: 111–131.
97 Feng, H., Elam, J.W., Libera, J.A. et al. (2010). *Chem. Mater.* 22: 3133–3142.
98 Caner, N., Bulut, A., Yurderi, M. et al. (2017). *Appl. Catal., B* 210: 470–483.
99 Sadeghzadeh, S.M. (2016). *Microporous Mesoporous Mater.* 234: 310–316.
100 Ting, S.W., Cheng, S., Tsang, K.Y. et al. (2009). *Chem. Commun.* 47: 7333–7335.
101 Taylor, P., Sunder, S., and Lopata, V.J. (1984). *Can. J. Chem.* 62: 2863–2868.
102 Bide, Y., Nabid, M.R., and Etemadi, B. (2016). *Int. J. Hydrogen Energy* 41: 20147–20155.
103 Li, X., Feng, J., Zhu, H. et al. (2014). *RSC Adv.* 4: 33619–33625.
104 Zahmakiran, M. and Özkar, S. (2008). *Langmuir* 24: 7065–7067.
105 Zahmakiran, M., Ayvali, T., and Philippot, K. (2012). *Langmuir* 28: 4908–4914.
106 Bayram, E., Zahmakiran, M., Ozkar, S., and Finke, R.G. (2010). *Langmuir* 26: 12455–12464.
107 Zahmakiran, M., Leshkov, Y.R., and Zhang, Y. (2012). *Langmuir* 28: 60–64.
108 Zahmakiran, M. and Ozkar, S. (2009). *Inorg. Chem.* 48: 8955–8964.
109 Zahmakiran, M., Tristany, M., Philippot, K. et al. (2010). *Chem. Commun.* 46: 2938–2940.
110 Clark, T.J., Whittel, R.G., and Manners, I. (2007). *Inorg. Chem.* 46: 7522–7527.
111 Jaska, C.A. and Manners, I. (2004). *J. Am. Chem. Soc.* 126: 2698–2699.
112 Levy, R.B. and Boudart, M. (1973). *Science* 181: 547–549.
113 Scanlon, M.D., Bian, X., Vrubel, H. et al. (2013). *Phys. Chem. Chem. Phys.* 15: 2847–2857.
114 Oyama, S.T. (1992). *Catal. Today* 15: 179–200.
115 Colton, R.J., Huang, J.T.J., and Rabalais, J.W. (1975). *Chem. Phys. Lett.* 34: 337–339.

Part IV

Activation and Theory

14

Magnetically Induced Nanocatalysis for Intermittent Energy Storage: Review of the Current Status and Prospects

Julien Marbaix, Nicolas Mille, Julian Carrey, Katerina Soulantica, and Bruno Chaudret

Université de Toulouse, CNRS, INSA, UPS, LPCNO, 135 avenue de Rangueil, Toulouse 31077, France

14.1 Introduction

Facing the depletion of fossil fuels and the increase of greenhouse gases leading toglobal warming, innovative technologies based on renewable intermittent energies attract a lot of interest. In this context, magnetic heating has emerged as an alternative way to activate catalysis in both liquid and gas phases. Focusing on carbon feedstock valorization including CO_2 and CO, it has been shown that it is possible to convert these reactants to fuels using magnetically induced catalysis. Gas-phase catalysis has been carried out using both magnetic heating of micro- (metallic pellets, disks, and rods) and macrostructured materials (metallic reactors), while magnetic nanoparticles (MNPs) have been more recently applied to activate catalytic reactions. Thus, the development of magnetic monodomain nanoalloys mostly based on iron enables to reach a broad temperature range (300–900 °C) in a very short time. This review is focused on catalysts and catalytic reactions achieved so far, both in liquid and gas phases. A special focus on nanostructured magnetic materials is justified by the increased interest they received in the past decade, as far as their synthesis, characterization, and heating performances are concerned. Future perspectives toward magnetic, catalytic, and heating property optimization will also be addressed. This will highlight the interest of using magnetic heating to perform catalysis with possible local hot spots, which in some cases achieves better yields than the ones anticipated from thermodynamic considerations. Furthermore, the potential of magnetically induced catalysis in the frame of renewable energy production and especially CO_2 valorization will be illustrated. This technology exhibiting extremely high heating rates paves the way to an efficient coupling between intermittent energy production and catalytic processes using magnetic heating. Finally, the perspectives of improvement of catalysis performed using magnetic heating will be emphasized with respect to the energy efficiency of the technology and its environmental impact assessment in comparison to traditional external heating. As shown, the design of optimized MNP can contribute to

Nanoparticles in Catalysis: Advances in Synthesis and Applications, First Edition.
Edited by Karine Philippot and Alain Roucoux.

increasing the global energy efficiency of the system by reducing the frequency and the amplitude required to perform catalysis. Every step and resources that are involved in the catalysis process have to be integrated in a multicriteria life cycle assessment in order to (i) compare the technology to traditional analogs, (ii) establish perspectives of decreasing the environmental impact, and (iii) assess the sustainability potential of catalysis using magnetic heating.

Recently, a comprehensive review describes the remarkable advantages of the induction heating technology in catalysis, which today has a renewed interest because of energy concerns [1]. The present chapter will therefore concentrate on some historical aspects of the technique and on recent developments in our group or related to our objectives.

14.2 General Context and Historical Aspects

At the end of the nineteenth century, scientists discovered and investigated induction heating, i.e. the fact that a material can release heat when excited by a time-varying electromagnetic field [2]. The early applications of this heating technique were foundry and cooking. Nowadays, it is also applied to thermal treatments in metallurgy and even medicine [2]. In the industry, the main advantages of this technique are rapidity, repeatability, reliability, and easy automation [2]. The arguments for using this technology are high heating rates and energy efficiency, but this last point is still controversial because it depends on cooking conditions (e.g. volume to be heated) [3, 4].

This technique to provide heat to a chemical reaction is applied at three different scales. The first and most simple case is heating the macroscopic reactor using electromagnetic induction. This has very few advantages and is nearly never studied by current research groups. One example using macroscale induction heating comes from Leclercq et al. [5]. They supplied heat to a steel tube on which a thin film of catalyst had been deposited. The fast heating constitutes the first and main advantage of induction technology. The target application is the removal of harmful gases from plants, and fast commissioning is very valuable in the case of accidental contamination.

The second advantage of electromagnetic induction is its contactless heating ability. Indeed, it ispossible to heat a substance directly inside the reactor without having to supply heat externally. This has been experimented with millimeter- and micrometer-scale objects inside reactors for decades. This field of study started in the 1960s with a first patent from Shell Company [6]. A piece of rotating metal was heated by induction inside a reactor and this heat was transferred to a fluid nearby. The advantages were the very high temperature which could be reached (continuously 2500 °C) and the large temperature sweeping rate for heating and cooling. This fast temperature kinetics allowed to avoid side reactions during the heating and cooling processes or resulting from temperature inhomogeneity and thus increased the product yield. This increased reaction efficiency is still one of the most significant advantages outlined by people using such “inside heating.” It is also proposed to overcome the heat transfer limitation of a traditional heating reactor [7]. Fast and localized heating can also be achieved using microwave heating, which has been developed in parallel [8]. Electromagnetic heating has the advantage of being simpler to design and operate, as advanced by Ceylan et al. [9].

Electromagnetic induction can also be used for contactless selective heating. The first publication on the subject is from Bartley [10] in 1975. He performed dual functional catalysis within one reactor, thanks to iron spheres of 20 μm that were electromagnetically heated and alumina-silica particles of millimeter size that were not heated, where both acted as catalysts. This enabled to simultaneously perform different reactions in one reactor. He even used the different reactions occurring on the different catalytic components with a calibration procedure to have an insight into the temperature difference between the heated iron spheres and the alumina-silica particles. This possibility to heat only parts of the reactor is still used at the millimeter scale, for example, by Chatterjee et al. [11]. They heated steal beads and a catalyst separated by 10 of millimeters of glass beads. They pointed out the possibility "to perform consecutive reactions" within one reactor thanks to this design.

Selective heating can be pushed a step further with the possibility of heating electromagnetically directly the catalytic bed. This is achieved at the millimeter scale by Koch et al. in a patent dated from 1999 [12]. Several groups have adopted this approach as detailed below.

This selective heating of the catalyst has been widely used, especially using nanoparticles (NPs). The first advantage of NPs is their high surface area-to-volume ratio, thus affording original surface reactivities. Moreover, because NPs can be tuned in composition and size, they can be designed to heat electromagnetically and act at the same time as a catalyst. The first work describing catalytic NP heating has been performed by Ovenston and Walls [13]. They heated nickel NPs, which acted as a catalyst for the reforming of hydrocarbons.

Direct heating of the catalyst and of its close environment can lead to a strong temperature gradient between the heated catalyst (or catalyst vicinity) and the rest of the reactor [14, 15]. This idea of hot spots provided by NPs at the nanoscale arises from the field of magnetic hyperthermia applied to medicine (heating NPs inside cancer cells to induce their death) when cell death is observed without any global warming of the cell itself [16–18]. For chemical reaction catalysis, this concept has been proposed since the 1980s to enhance microwaved heated catalytic reactions, but without strong evidence, as explained by Thomas [19]. Temperature measurements at the nanoscale are indeed very challenging. The first paper providing such measurements was from Pellegrino's group [20]. They measured a temperature gradient of 70 K within 5 nm. Some other measurements by Dong and Zink [21] and Piñol et al. [22] reported temperatures higher in the vicinity of the NPs than in their surroundings. Nevertheless, these results are very specific to the systems studied and, probably, depend strongly on the nature of the interface between the NPs and the surrounding medium. To our knowledge, no evidence of nanoscale temperature gradient in a magnetically activated chemical reaction has been provided yet.

14.3 Characteristics of the Nanocatalysts Used in Magnetic Hyperthermia

14.3.1 Metal Oxide Nanomaterials

Since 1980, magnetic NPs have been widely investigated for biomedical applications such as cancer treatment by magnetic hyperthermia or drug delivery [22]. Thanks to their biocompatibility and easy synthesis, mostly iron oxide-based materials have been employed. The saturation magnetization (M_s) of these materials can reach $70\,\text{A}\,\text{m}^2\,\text{kg}^{-1}$ and their specific absorption rate (SAR) up to a few hundred $\text{W}\,\text{g}^{-1}$, normalized at 100 kHz (SAR: quantifies the heating capacity of the material). The development of magnetic oxides has been so far essentially limited to a role of support in solution by Kirschning [9, 23, 24]. However, recent developments of magnetic core–shell nanostructures or cobalt ferrites ($Co_xFe_{3-x}O_4$) could be of interest in catalysis.

14.3.2 Iron (0) Nanoparticles

Well-controlled Fe(0) NPs were obtained by reduction under H_2 of the complex $\{Fe\{N[Si(CH_3)_3]_2\}_2\}_2$ in the presence of long-chain carboxylic acids and amines [25, 26]. These particles were originally shown to display high SAR [27] compared to oxides and thus have been used as a starting material for the synthesis of several iron-based heating agents. The Fe(0) particles' heating capability has been assessed in solution at 100 kHz, 50 mT to be up to $560\,\text{W}\,\text{g}^{-1}$ [15]. Upon deposition of a cobalt precursor at the surface of these NPs, a FeCo shell can be formed at the surface. The resulting core–shell Fe@FeCo NPs display the same SAR value as the Fe(0) NPs but an increased saturation magnetization up to $M_s = 206\,\text{A}\,\text{m}^2\,\text{kg}^{-1}$ at 300 K and 3 T. As expected, given the nonmagnetic nature of Ru, the formation of a Ru shell decreases the coercive field and the M_s down to $165\,\text{A}\,\text{m}^2\,\text{kg}^{-1}$, leading to a degraded SAR down to $380\,\text{W}\,\text{g}^{-1}$. Moreover, it was shown that once impregnated on a support, typically alumina, the dipolar interactions and relative mobility of the particles are reduced, and the SAR values are reduced to $128\,\text{W}\,\text{g}^{-1}$ and $58\,\text{W}\,\text{g}^{-1}$ for Fe@FeCo and Fe@Ru, respectively. However, while the Fe@Ru NPs start heating at 8 mT, Fe(0) and Fe@FeCo are activated at amplitudes above 20 mT.

14.3.3 Iron Carbide Fe(C) Nanomaterials

Iron carbides are attracting materials because they behave as soft magnets displaying a saturation magnetization close to that of Fe(0) but are more air resistant than iron metal and, in principle, biocompatible. The first NPs presenting a FeC surface were synthesized by Meffre et al. A core–shell Fe@FeC configuration was achieved by decomposing $Fe(CO)_5$ in the presence of H_2 [28]. Two FeC phases have been identified as $Fe_{22}C$ and Fe_5C_2, and the magnetic properties are different from those of pure Fe(0) as a result of the low anisotropy of the carbide phase. More recently, inspired by the Fischer–Tropsch reaction and using syngas as the carbidizing agent,

Bordet et al. showed the possibility to gradually incorporate carbon up to the full carbidization of preformed Fe(0) NPs after 140 hours. The FeC NPs thus produced were monodisperse (15 nm mean size diameter) and composed of 82.3% $Fe_{22}C$, identified as the most efficient heating phase, and 17.7% Fe_5C_2 [29]. Excellent magnetic properties are displayed by these NPs, namely, $M_s = 170\,A\,m^2\,kg^{-1}$ at 300 K, which start heating at 15 mT when suspended in solution, leading to an outstanding SAR of 3200 $W\,g^{-1}$ at 100 kHz, 47 mT. Moreover, the time needed for a full transformation of the Fe(0) to the FeC NPs can be drastically reduced by removing the water formed during the carbidization [30]. The evolution of the heating performances as a function of the agglomeration state on the FeC NPs has been studied. It was shown that the relative mobility of the particles allows chain formation, which is crucial for their increased heating capability [31].

In a recent work, FeC NPs were used as a soft magnetic material to trigger the magnetic activation of hard magnetic materials such as Co nanorods under moderate field amplitude at 300 kHz [32]. It has to be noted that Co nanorods are not well adapted as a heating agent because of their high magnetic anisotropy, which does not allow activation by the commonly used magnetic fields. However, the heat released from FeC NPs enables to reduce the coercivity of the Co nanorods, which in turn can function as heating agents at lower magnetic fields. This strategy allows heating the catalytic bed up to the Co Curie temperature, which is much higher than that of FeC NPs (>1000 °C and 400–450 °C, respectively) [33, 34]. In the same perspective, the use of nanoalloys composed of different magnetic metals offers the possibility of tuning the magnetic properties and therefore the power of the heating agent.

14.3.4 Bimetallic FeNi Nanoparticles

Bimetallic $Fe_{30}Ni_{70}$ NPs were synthesized by codecomposition of $\{Fe[N(SiMe_3)_2]_2\}_2$ and $Ni[^iPrNC(CH_3)N^iPr]_2$ complexes under 3 bar of H_2 using palmitic acid (PA) (1.8 eq. per Fe complex) as a stabilizer. The resulting $FeNi_3$ NPs display the alloy fcc structure but with a surface enrichment in Fe, which explains the relatively poor reactivity of these particles for CO_2 hydrogenation. However, further deposition of a Ni layer through decomposition of $Ni[^iPrNC(CH_3)N^iPr]_2$ under 3 bar of H_2 on preformed $FeNi_3$ NPs leads to $FeNi_3$@Ni NPs, which display a depleted SAR of 350 $W\,g^{-1}$, but with excellent catalytic properties [35].

14.3.5 Bimetallic FeCo Nanoparticles

FeCo NPs initially synthesized by using $Co(\eta^3\text{-}C_8H_{13})(\eta^4\text{-}C_8H_{12})$ or $Co(N(SiMe_3)_2)_2$ and $Fe(CO)_5$ as precursors of cobalt and iron species, respectively, in the presence of hexadecylamine (HDA), dodecylamine, palmitic acid, stearic acid, and oleic acid [36] exhibited a saturation magnetization value of 170 vs. 245 $A\,m^2\,kg^{-1}$ at 300 K for the bulk material. In this case, a thermal treatment above 400 °C was necessary in order to restore 90% of the bulk value of M_s of FeCo NPs [37]. More recently, Garnero et al. showed that FeCo NPs with a M_s value of 226 $A\,m^2\,kg^{-1}$ could be obtained without thermal treatment by decomposing $\{Fe\{N[Si(CH_3)_3]_2\}_2\}_2$ and $Co(N(SiMe_3)_2)_2$

metallic precursors in the presence of HDA and either palmitic acid (PA) or hexadecylammonium chloride (HDA·HCl) under H_2 atmosphere [38]. The resulting spherical FeCo NPs with a size diameter of 7.5 or 11 nm, respectively, are monodisperse and adopt a bulk bcc structure. Detailed structural analyses by high-resolution transmission electron microscopy (HRTEM), Mössbauer spectroscopy, and Co Nuclear Magentic Resonance (NMR) evidenced the formation of NPs displaying a gradient of concentration of Co into the Fe lattice (PA) or the formation of an ordered structure (HDA·HCl).

14.3.6 CoNi Nanoparticles

Mortensen's group used CoNi and CoNi@Cu NPs of 40 and 50 nm mean size, respectively, for methane steam reforming. A SAR between 3 and 20 W g^{-1} under the operating conditions is claimed, which is comparable to the values published by Meffre et al. in the case of functionalized Fe-based NPs. Interestingly, the authors studied the evolution of saturation magnetization and hysteresis area of their sample as a function of temperature between 200 and 950 °C. Therefore, they could determine accurately the ferro to paramagnetic transition of CoNi and CoNi@Cu, which occurs at the Curie temperature (892 and 875 °C, respectively). This information is of high interest with respect to the targeted working temperature for catalysis. Thus, compared to pure Fe (T_c = 750 °C) or FeC (T_c = 400–450 °C), alloying enables to activate reactions that operate at higher temperatures [6, 39].

Varsano et al. produced alloys of micrometric size starting from Co chunks and Ni wire inside an arc-melting furnace. These microalloys were found active for methane dry reforming under magnetic induction [40] (Table 14.1).

14.4 Catalytic Applications in Liquid Solution and Gas Phase

14.4.1 Gas-Phase Catalysis

Heterogeneous catalysis using magnetic induction has not only been based on magnetic metals. Indeed, Eddy currents arising either from the reactor itself (bulk metallic materials) or from large metallic particles (microstructured materials) have also proved to be efficient. More recently, the use of nanoscaled materials as both magnetic heating agents and catalysts emerged in the literature.

14.4.1.1 Catalysis Activated by Magnetically Heated Micro- and Macroscaled Materials

In 2014, Pérez-Camacho et al. reported the Ni-catalyzed dry reforming of methane activated by magnetic heating of the stainless steel reactor [41]. The Curie temperature of stainless steel (up to 900 °C) was attained with a very fast temperature increase (>100 °C min^{-1}), resulting in 90% CH_4 conversion to CO. The authors highlighted the difficulty of the system temperature measurement, showing that the

Table 14.1 Main features of selected magnetic nanoparticles that are adapted for use as heating agents for catalysis activation.

Nature	Size (nm)	$M_{s, RT}$ ($A\,m^2\,kg^{-1}$)	$SAR_{100\,kHz\ solution,\ RT}$, ($W\,g^{-1}$)	$SAR_{100\,kHz,\ supported,\ RT}$ ($W\,g^{-1}$)	Amplitude (mT)	References	Year
Fe(0)	11	198	560	—	50	[15]	2015
Fe(0)@FeCo	12.2	206	568	128			
Fe(0)@Ru	12.4	165	380	58			
FeC	15.1	170	3200	—	47	[29]	2016
FeC@Ni	15.2	164	2000	—			
FeC	15.1	170	3000	—	47	[29]	2018
CoNi	40	35 (200 °C)	—	3–20	42	39	2018
CoNi@Cu	50	35 (200 °C)	—				

surface of the reactor and the catalytic bed exhibited very different values (550 vs. 860 °C). The compatibility of the system with a highly intermittent source of energy has been demonstrated by coupling the magnetic induction to a dynamic source of electricity generation (e.g. wind turbines). In a different application domain, an automotive industry, namely, the Canadian company Advanced Technology Emission Solution Inc., patented the magnetic heating of catalytic reactors for the reduction of harmful emissions in car exhausts [42]. They use an exhaust system composed of metallic parts, which under magnetic induction attains 300 °C, a temperature at which the catalyst particles reduce the CO and NO_x concentrations in exhaust gasses. The HCN production in magnetically heated cylindrical reactors at a temperature between 950 and 1400 °C and field frequency between 50 Hz and 3 kHz has been patented as well [43]. The same group patented the magnetic heating of microstructured materials in the shape of pellets, rings, or rods, resulting in the activation of a catalytic foam [44].

Other groups performed catalysis by heating microstructured materials rather than heating the catalytic reactor itself. Thus, Rebrov and coworkers reported the citronellol hydrogenation by using a Pd/TiO_2 catalyst heated by nickel ferrite micrometric particles [11]. In 2015, Varsano et al. performed methane dry reforming using both as a heating agent and as catalysts' micrometric powders of $Ni_{60}Co_{40}$ at 190 kHz [40]. At a temperature rate of nearly 200 °C min^{-1}, the measurement by the pyrometer showed that the system could reach 950 °C and a 75% CO yield. The authors measured the heating properties (SAR) of their micrometric heating agents as a function of temperature, showing that the heating performances dramatically drop for temperatures above 700 °C. The Curie temperature of the NiCo alloy being estimated between 700 and 800 °C, the authors evidenced the correlation of the composition of the magnetically heating agent to the maximum temperature reachable under magnetic induction (ferro- to paramagnetic transition). This feature was already reported in several works [45, 46], thus highlighting the importance of this heating agent's property to activate catalysis at the appropriate temperature. This aspect is particularly interesting when considering alloy NPs of Ni, Fe, or Co, for instance, and it will be discussed in the next section.

Finally, Ni nanocatalysts supported on magnetically heated oxidized carbon-felt disks enabled to perform CO_2 methanation through Sabatier reaction with 74% CO_2 conversion and 97% selectivity for CH_4. The authors could report excellent repeatability and stability of the catalyst after 80 hours on stream [47].

Magnetic heating has been used on both micro- and macrostructured materials to perform catalysis. However, the miniaturization of the heating agents offers the advantage of exploiting their high surface area when using them also as a catalyst.

14.4.1.2 Catalysis Activated by Magnetic Heating of Nanoparticles

The first results of catalysis employing nanoscale magnetic heating agents were patented by Chaudret et al. [48]. Following these patented results, they published the first proof of concept of batch heterogeneous catalysis (Fischer–Tropsch reaction) activated by magnetic induction using Fe(0) NPs, pure or surrounded by a Ru or FeCo shell [15]. By working at 54 kHz and 50 mT, the magnetic particles allowed reaching temperatures high enough to activate the reaction. Gas-phase ^{13}C NMR and mass spectrometry monitoring showed the formation of light alkanes (C1–C6) and alkenes in the case of pure Fe(0) NPs. Ru functionalization offered the best activity, up to complete CO conversion into light alkanes.

Bordet et al. developed the synthesis of iron carbide (FeC)-based NPs as magnetic heating agents and/or catalysts for the activation of CO_2 hydrogenation into methane (Sabatier reaction) [29]. This exothermic reaction ($\Delta H_{400\,K} = -165\,kJ\,mol^{-1}$) can be achieved in one or several steps (water gas shift and CO hydrogenation) depending on the catalyst used. Pure FeC NPs supported on aluminosilicate support (Siralox) were first used alone. Although they are highly efficient heating agents, they display a low catalytic activity (5% conversion at 300 kHz and 64 mT field amplitude). However, they were able to activate classical Sabatier reaction catalysts located either on their surface (Ni) or on the support (Ru). Thus, the continuous flow reaction applying the flow rates of 100 ml min^{-1} and H_2/CO_2 gas mixture in a ratio 4/1 enabled to reach a CH_4 yield of 15% in the case of supported core@shell FeC@Ni at 40 mT and up to 93% at 28 mT in the case of FeC supported on Siralox impregnated with Ru NPs (Ru-Siralox). In all cases, IR thermography revealed a very rapid temperature increase, from 25 °C up to 350 °C in only 30 seconds.

In 2017, Mortensen et al. used CoNi bimetallic NPs as a magnetically activated heating agent but also as a catalyst for the steam reforming of methane at high temperature (>780 °C) [6]. Methane steam reforming is a highly endothermic reaction ($\Delta H_{400\,K} = 206\,kJ\,mol^{-1}$), which requires high temperatures and is generally coupled to the water gas shift reaction (WGSR) in order to use the produced CO. Using flow rates higher than 300 ml min^{-1}, they converted 98% of the incoming methane with a selectivity for CO of 70% at 68 kHz and 40 mT. In a subsequent publication, these authors doped the NiCo alloy with Cu, which allowed reducing the field amplitude (24 mT) while maintaining excellent methane conversion (>95%) [39]. They have also demonstrated the importance of the gas residence time in the inducted area and highlighted the dramatic catalytic performances decrease if the flow rate becomes too high. The authors patented their results: first regarding the reactor heating by Eddy currents for nitrile synthesis [49], then for high-temperature magnetic heating of nanocatalysts for endothermic applications occurring between 400 and 950 °C [50]. In this latter patent, the inventors emphasized the importance of the Curie temperature of the magnetic material to be heated.

In 2017, Haldor Topsoe patented a magnetic induction process for ammonia dissociation and local hydrogen production for renewable energy storage. Requiring a temperature of 500 °C, this endothermic reaction has been achieved at frequencies between 100 and 500 kHz using Fe, FeCo, and an appropriate catalyst supported on alumina [51].

Following the work on FeC@Ni NPs developed by Bordet et al. in 2016, Niether et al. employed the same catalyst to optimize water electrolysis at 300 kHz and 48 mT in such a way that the potential necessary for oxygen production (oxygen evolution reaction, OER) could be reduced by 200 mV in an alkaline electrolyzer [52]. The hydrogen evolution reaction (HER) potential could be reduced as well by c. 100 mV. These performances would in principle require a temperature of 200 °C, while using magnetic hyperthermia, only a temperature increase of 5 °C was measured, hence offering promising opportunities relative to the degradation of electrodes and the energy efficiency of the process.

The main characteristics of the heterogeneous catalysis processes performed in the presence of magnetic heating, including the nature of the heating agent, the magnetic field, and the achieved performances, are presented in Table 14.2.

14.4.2 Catalytic Reactions in Solution

Besides heterogeneous catalysis, induction heating of NPs is an interesting way of heating solutions and could lead to novel overheating effects. In 2008, Kirschning's group was the first one to perform chemical reactions including catalytic ones activated by magnetoinduction [9]. They used iron oxide spherical NPs (10–40 nm) coated with a silica shell to activate a large panel of organic reactions in solution by applying an alternating magnetic field at a frequency of 25 kHz. Thanks to an inductor surrounding a continuous line allowing the injection of six reactants, the temperature of the system reached 220 °C under magnetic field. According to the authors, the silica shell did not change the magnetic properties but offered a better mechanical resistance and the possibility to functionalize the surface of the heating iron oxide, as exemplified by Pd NPs, which when immobilized on the silica shell catalyzed cross-coupling reactions. By working in a temperature range between 60 and 170 °C, they successfully performed diverse reactions such as transesterification, condensation, Claisen rearrangement, Suzuki or Heck coupling, and Hartwig–Buchwald amination. In the case of Claisen rearrangement and Hartwig–Buchwald amination, a better reaction yield was achieved by using magnetic heating than with a traditional external heating source owing to the fact that heat is generated directly in the catalytic bead, which leads to a global temperature heterogeneity in the system.

Kirschning's team also used steel beads (0.8–4.8 mm) as magnetic heating agents as their size can confer good microstructural properties for a fixed bed catalytic reactor [23]. They showed that under induction, increasing the size of the beads increases the temperature up to 350 °C. Transfer hydrogenation, heterocyclic condensations, and C–C and C–X couplings have been achieved, enlarging the organic reaction panel already explored. Still working on the Claisen rearrangement of allyl aryl ethers into phenols, they compared the yield of reactions in batch heated by magnetic hyperthermia, traditional oil bath, or microwave heating. The traditional external heating displayed the lowest yield (17%) microwave and magnetic heating enabled to reach yields of 38% and 39%, respectively, thus confirming their previous results. However, they concluded that magnetic heating is more interesting because it is easier to set up than microwave.

Finally, in 2013, the same group increased importantly the working field frequency up to 800 kHz. This showed that 300 °C and even 400 °C could be reached by using silica-encapsulated iron oxide NPs and steel beads, respectively [24]. At this point,

Table 14.2 Heterogeneous catalysis achieved by magnetic heating of nanocatalysts.

Reaction	$\Delta H_{300\,K}$ (kJ mol^{-1})	Nature of the magnetic agent	Magnetic agent loading (wt%)	Support	Field frequency (kHz)	Field amplitude (mT)	Flow rate (ml min^{-1})	Temperature (°C)	Conversion (%)	References	Year
Fischer–Tropsch	−206	Fe(0)	—	—	54	50	Batch	>200	—	[15]	2015
		Fe(0)@FeCo							—		
		Fe(0)@Ru							Total		
Sabatier	−165	FeC	10	SiRAlOx	300	64	100	350	5	[29]	2016
		FeC@Ni				40			50		
		FeC, Ru-Siralox				28			98		
		FeCo, Ni-Siralox			100	47	Batch	—	92.5	Submitted	2020
Steam methane reforming	206	NiCo	21.6	$MgAl_2O_4$	68	40	330	780	98	[18]	2017
		NiCo	15	Al_2O_3		27	850		95	[39]	2018
		NiCo@Cu				24	850		95		
		NiCo@Cu				24	2500		<60		
Ammonia decomposition	46	Fe	—	Al_2O_3	100–500		—	360–550	—	[51]	2017
		FeCo						400–700	—		

Table 14.3 Selected chemical reactions in solution activated by magnetic hyperthermia achieved by Kirschning's group (2008–2013).

Heating source	Functionalization	Field frequency (kHz)	Reaction	T (°C)	Yield (%)	References	Year
Fe_3O_4/Fe_2O_3@Silica	x	25	Transesterification	60	88	[9]	2008
Fe_3O_4/Fe_2O_3@Silica	x	25	Buchwald–Hartwig amination	90	75		
Fe_3O_4/Fe_2O_3@Silica	Pd^0	25	Suzuki–Miyaura/Heck coupling	120	Up to 84		
Fe_3O_4/Fe_2O_3@Silica	x	25	Transfer hydrogenation	70	Up to 92		
Fe_3O_4/Fe_2O_3@Silica	x	25	Pericyclic reactions	120	52	[23]	2011
Steel beads	x				64		
Steel beads	x	25	Mannich	65	88		
Fe_3O_4/Fe_2O_3@Silica	x	25	Claisen rearrangement: allyl aryl ether	200	39	[23]	2013
Fe_3O_4/Fe_2O_3@Silica	x	800		205	89		
Fe_3O_4/Fe_2O_3@Silica	x	800		225	Decomposition		
Oil bath	x	x		200	17		
Microwave	x	x		200	38		

the yield of Claisen rearrangement of allyl aryl ether reached 89%, but complete reaction could not be achieved because of the substrate decomposition at these high temperatures. Magnetic heating has also been implemented in a multistep process for the production of neuroleptic drugs, showing the huge potential of this technology including the benefits from reaction time reduction and reactant quantity decrease. The main results of Kirschning's group are summarized in Table 14.3, highlighting that only one reaction was catalyzed by Pd.

In our group, we developed the use of magnetic heating in solution with NPs displaying both a magnetic core and a catalytic surface. FeCx@Ru NPs were prepared by decomposing $Ru_3(CO)_{12}$ onto preformed iron carbide NPs. These NPs catalyze the selective hydrodeoxygenation (HDO) of acetophenone derivatives, furfural and hydroxymethylfurfural, in mild conditions (3 bar H_2), which are otherwise carried out heterogeneously at high pressures and temperatures. Thus, quantitative HDO of furfural and hydroxymethyl furfural into methyl furan and 2–5 dimethylfuran, respectively, were observed using 1.0 mol% of Ru [53]. These results demonstrate the high potential of this technique (Scheme 14.1).

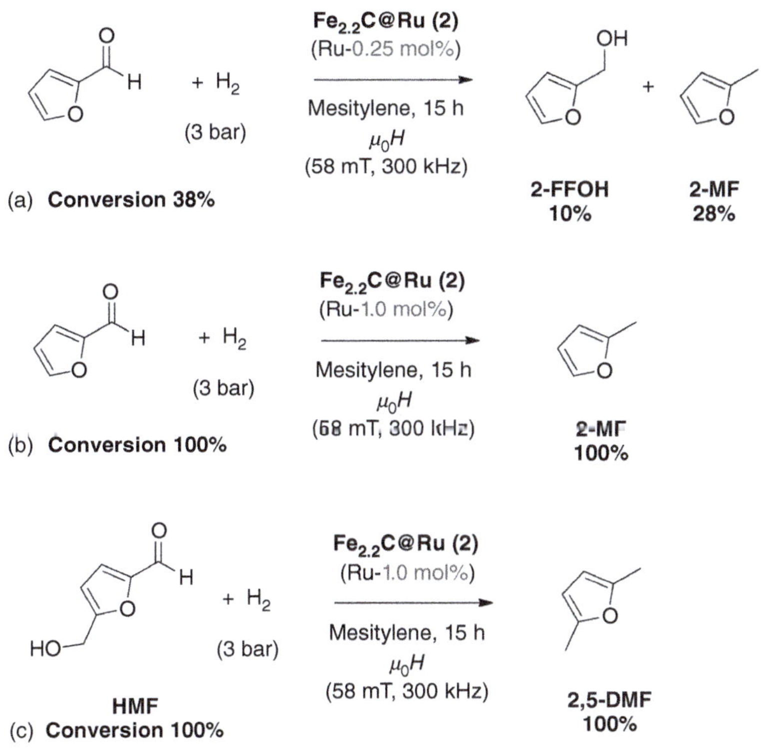

Scheme 14.1 Hydrodeoxygenation (HDO) reactions catalyzed in solution in the presence of magnetic heating. a) HDO of furfural in the presence of 0.25 mol%Ru, b) HDO of furfural in the presence of 1 mol% Ru, c) HDO of hydroxymethylfurfural (HMF) in the presence of 1 mol% Ru.

14.5 Perspectives

Besides the interesting activity it demonstrated that magnetic heating has to prove its efficiency in terms of stability, heat management, and energy consumption to be eventually part of future processes.

14.5.1 Stability of the Catalytic Bed During Catalysis by Magnetic Heating

The literature available on the long-term stability of the catalytic bed under magnetic heating conditions is not yet abundant. However, Bordet et al. reported a good stability of the catalysts over 150 hours on stream at 28 mT (350 °C), and surprisingly, despite a slight sintering of particles, an increase of methane yield was observed as the Sabatier reaction proceeded [29]. Sintering has also been mentioned by Mortensen et al. at 800 °C on CoNi particles, whose size increased from 12 to 32 nm after 24 hours on stream [7]. Finally, Meffre et al. reported sintering together with a thin carbon layer deposition on the surface of Fe-based particles and carbide-phase formation, leading potentially to the catalyst deactivation [15]. Depending on the composition of the NPs, their SAR, and their catalytic activity, they suggested that sintering could be dramatic (Fe(0) NPs) or minor (Fe@Ru).

14.5.2 Thermal Management and Process Chemistry Using Magnetic Heating for Catalytic Applications

The groups who have worked on catalysis by magnetic hyperthermia agree on the fact that the major advantage of this technology compared to traditional external heat sources such as oven, oil bath, or even microwaves is that the heat is produced directly in the catalytic bed. This enables to drastically reduce the energy losses and to operate under conditions of "cold catalysis" by local hot spots. Another point, which remains to be studied in detail, is that the catalytic performances seem to be better when magnetic induction is employed. This is likely due to a reaction rate limited by kinetic factors rather than heat transfer through the catalyst. Indeed, in a traditional reactor of heterogeneous catalysis, the heat is provided by an external source, and the global catalytic bed temperature measured is most likely higher than the surface of the catalyst for an endothermic reaction or lower for an exothermic reaction. In the case of magnetic hyperthermia, local hot spots provide heat in the system locally. Thus, a nanoscale temperature gradient is expected, which may greatly influence the catalytic performances. This has been observed in the case of Sabatier reaction (93.7% yield vs. 86% in traditional conditions) [29, 32]. It is expected that the high-temperature gradient next to the heating agent enables to efficiently evacuate the water out of the catalytic area. This would enhance the yield of reaction.

Such observations should incite the scientific community to design adapted reactors for catalysis by magnetic hyperthermia. Working on a wide range of temperatures (up to 900 °C) on both endothermic and exothermic reactions as well as material constrains associated with the use of a magnetic field imposes specific developments in terms of chemical engineering. In the case of an endothermic reaction, the heat has to be conserved and the losses minimized. In the case of steam methane reforming, Mortensen's group developed an optimized fixed bed reactor for endothermic reactions in which the thermal losses are minimized [54]. It is composed of a narrow thermal transfer isolated zone to maximize the conversion of electrical energy into magnetic energy, claiming more than 90% efficiency. This design enables to minimize the thermal losses and maximize the catalytic performances for an endothermic reaction.

In the case of an exothermic reaction, thermal management is more complex, as it is already for traditional heterogeneous catalysis. As the reaction is activated, it provides heat by itself and has to be cooled down in order to maintain an adequate temperature for optimal catalytic performances and avoid any dramatic increase of temperature, leading to the potential overheating of the system. Strategies that are efficient for traditional heating sources may be more complicated to implement in magnetically activated systems but should constitute an interesting perspective of technology amelioration that has not been studied so far.

14.6 Perspective of the Integration for Renewable Energy Use

14.6.1 Interest of Power to Gas and Catalysis Using Magnetic Heating for Renewable Energy Use

Because of the dramatic increase of greenhouse gases leading to global warming, political agreements have set a target limit of 2 °C of global temperature increase by 2050. A main requirement for sustainable development is a massive and efficient deployment of renewable energies. Because of their intermittency, an efficient way to store them appears to be the use of electricity overproduction for producing energetic gaseous vectors; this is the concept of power-to-gas (P2G). This process could use large amounts of clean energy to produce hydrogen and to enable the use of sustainable reactants, such as vegetal derivate chemicals or even CO_2 as building blocks for the production of useful products. Combined with emerging technologies of carbon capture and storage (CCS), greenhouse gases can be converted through renewable energy into high value-added functionalized (C—O, C—N, and C—C bond formation) and high energetic fuels (lower carbon oxidation state) for which infrastructures are already available [55, 56]. Magnetic heating could play a major and disruptive role in this whole process at different steps, including water electrolysis and ammonia decomposition for local hydrogen production, followed by its use in methanation units (Figure 14.1).

Despite the development required in the reactor design, magnetic hyperthermia has the major advantage of allowing a rapid activation. Indeed, as mentioned by Bordet et al., in the case of Sabatier reaction, a steady state at 350 °C is reached in 30 seconds [29], without mentioning that the temperature at the surface of the magnetic agents is expected to be even higher at time scales corresponding to the

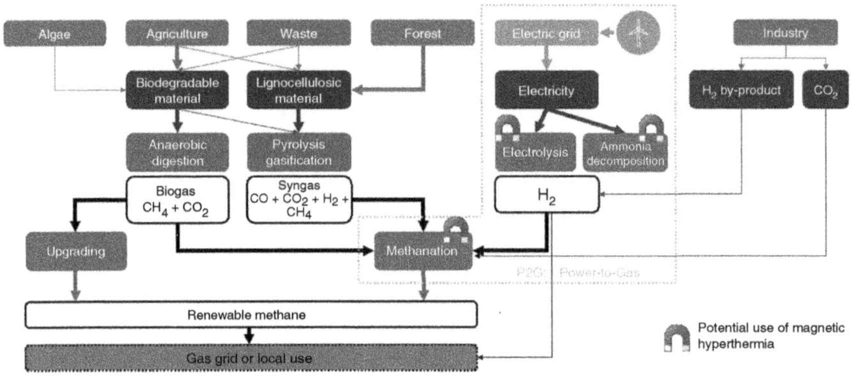

Figure 14.1 Potential use of magnetic heating for low carbon footprint use of renewable energy sources.

magnetic moment relaxation. The system dynamics are thus compatible with the intermittency of renewable energy production. Indeed, using magnetic induction, Pérez-Camacho et al. reported an excellent adaptability of their catalytic system to electricity generation by wind turbines for dry reforming of methane [4], and the Sabatier reaction is expected to be efficient even if coupled to a highly reversible "on/off" energy input [57]. The influence of a catalyst concentration profile could be studied under these conditions, introducing in the scope of technological assessment additional parameters such as transient regime, safety, and durability of the process [58]. These last results show the importance of process chemistry for the integration of CO_2 methanation in a high intermittent energy production frame. Facing huge inertia of heating and cooling in traditional catalysis by external sources, magnetic hyperthermia offers the possibility of an on/off function enabling to respond accurately to the time-scale intermittency of renewable energy production and to bring an appreciable adaptability to the requirements of a reaction such as Sabatier.

14.6.2 Energy Efficiency and Environmental Considerations of Catalysis by Magnetic Heating

The energy efficiency of catalysis by magnetic heating has to be competitive with the traditional heterogeneous catalysis in order to become a potential alternative. The process can be roughly divided into two main consecutive steps: (i) transformation of magnetic energy into thermal energy and (ii) the use of the heat released for energy production or more generally for products of reaction. The efficiency of each step can be expressed as $\eta_{\text{magn - therm}}$ (Eq. (14.1)) and $\eta_{\text{therm - NRJ}}$ (Eq. (14.2)), respectively. The power needed for magnetic field generation, as a function of its frequency and amplitude $P_{\text{field}}(f, \mu_0 H)$, is converted into heat by excitation of the magnetic NPs, whose ability to release heat is a function of their $\text{SAR}(f, \mu_0 H)$ and their mass m_{NPmagn}. As a result, we have

$$\eta_{\text{magn-therm}} = \frac{\text{SAR}(f, \mu_0 H) \cdot m_{\text{NPmagn}}}{P_{\text{field}}(f, \mu_0 H)} \quad (14.1)$$

where $\eta_{magn-therm}$ increases by increasing the SAR, the magnetic NP mass, and by lowering power consumption. Moreover, the more the magnetic sample heats at a low frequency and amplitude, the higher the $\eta_{magn-therm}$ will be. This highlights the importance of the SAR optimization through the synthesis of new magnetic heating agents for magnetic hyperthermia. The course of the reaction is given by Y_{reac}. If the products of the reaction are fuels, the produced energy can be quantified by using their higher heating value ($HHV_{products}$) and the mass flow rates of production $D_{m,\,products}$. As a result, we can compare the energy produced with the energy content of the reactants:

$$\eta_{\text{therm-NRJ}} = \frac{\sum_i Y_{\text{reac},i} \cdot D_{\text{m,products},i} \cdot \text{HHV}_{\text{products},i}}{\sum_j \cdot D_{\text{m,react},j} \cdot \text{HHV}_{\text{react},j}} \tag{14.2}$$

We can further consider the global efficiency $\eta_{hyperthermia}$ of the system operating at a given frequency and amplitude of magnetic field (Eq. (14.3)). This can be calculated by considering the energy content of the products over the total energy needed to achieve the reaction. In our case, the required energy is given by the sum of the incoming energy from the reactants (fuels) and electricity input, which is ideally coming from sustainable sources.

$$\eta_{\text{hyperthermia}} = \frac{\sum_i Y_{\text{react},i} \cdot D_{\text{m,products},i} \cdot \text{LHV}_{\text{products},i}}{\sum_j \cdot D_{\text{m,reac},j} \cdot \text{LHV}_{\text{reac},j} + P_{\text{field}}(f, \mu_0 H)} \tag{14.3}$$

As an example, in the case of Sabatier reaction achieved by Bordet et al. using the FeC NPs supported on Ru-Siralox, 100 mg of FeC (75% iron) can transform up to 19.5% of the electrical energy into heat at 300 kHz, 28 mT [29]. This heat enables to reach a methane yield of 93.7% by using a flow of 100 ml min^{-1} of incoming H_2 and 25 ml min^{-1} of CO_2 (which is not considered as a fuel), giving $\eta_{therm-NRJ} = 78.1\%$. Because of the difference of HHV between H_2 and CH_4, the theoretical maximum of $\eta_{therm-NRJ}$ is 83.1%. Thus, in this case, the chemical conversion is not an issue because the power used to generate the required field is too important. As a result, 16.1 W (H_2 incoming flow) is converted into 12.6 W (CH_4 outgoing flow) by using 1.82 kW power input for field generation, giving rise to $\eta_{hypherthermia} = 0.7\%$. Meffre et al. reported a 0.5% in the case of Fischer–Tropsch [15] reaction while Mortensen et al. did not mention the global energy efficiency when studying steam reforming of methane [39]. The above-mentioned low values do not consider any projected optimization for $\eta_{therm-NRJ}$ nor $\eta_{magn-therm}$. Obviously, there is much room for improvement regarding (i) the development of NPs of optimized heating capability at low frequency and/or amplitude, (ii) the process chemistry related to the energy necessary for field generation, and (iii) thermal management (Figure 14.2).

Moreover, the above-mentioned considerations are restricted to the catalytic system itself but do not consider the energy required for reactant production and preparation (H_2, CH_4, and CO_2) and its potential environmental impact. Indeed, considering fossil methane or hydrogen produced by methane reforming does not allow an accurate assessing of the sustainability potential of magnetic hyperthermia for clean energy storage. A more precise approach would be to compute every single step from raw resources up to the use of the products and their recycling, through

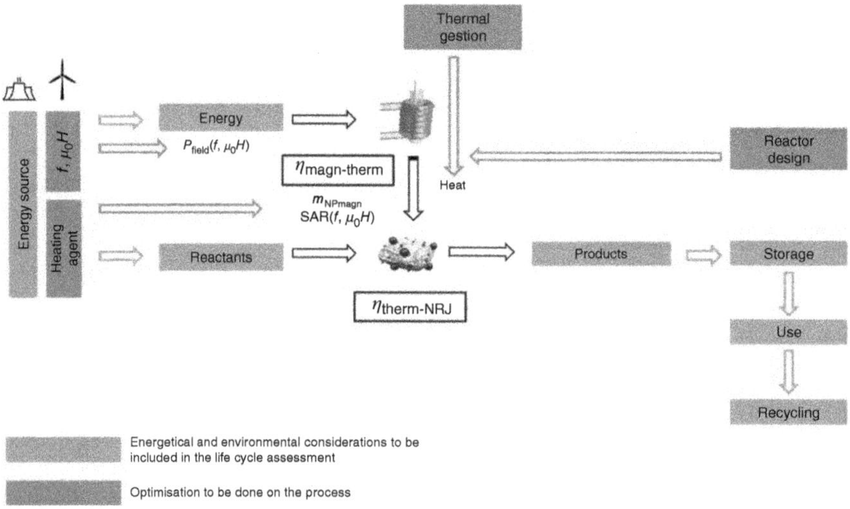

Figure 14.2 Catalysis using magnetic heating integration into a life cycle assessment and major perspectives of optimization of the process.

a multicriteria environmental analysis by performing the life cycle assessment of catalysis by magnetic hyperthermia (Figure 14.2). Such analyses have been performed in the field of traditional heterogeneous catalysis regarding hydrogen production [59], CO2 capture and storage (CCS) [60, 61], methane [62, 63], and synthetic fuels production [55, 56, 64]. We have recently reported such an analysis for the CO_2 methanation activated by magnetic heating. In this work, the life cycle assessment of the process, which integrates water electrolysis and CO_2 capture and separation in the system boundaries, was performed in order to highlight the environmental potential of the technology. The perspectives of technology improvement were considered to assess the ecological impact of the process at both 2020 and 2050 horizons. It is shown that the total flow rate of reactants, the source of CO_2, and the energy mix play a major role on the environmental impact of sustainable CH_4 production, whereas the lifetime of the considered catalyst has no significant influence. This illustrates the high environmental potential of the methanation activated by magnetic heating when coupled with industrial exhausts and renewable electricity production. As a result, the whole process is expected to reduce by 75% its environmental impact toward 2050, consuming 0.81 MJ of nonrenewable primary energy per MJ of produced CH_4. Thus, the technology is expected to be environmentally competitive compared to existing similar processes using external heating sources with the additional interest of being extremely dynamic in response in line with the intermittency of renewable energy production [65].

14.7 Conclusion

In this chapter, we intended to highlight the potential of using magnetic induction to perform catalysis. During the past decade, researchers have been developing systems that allowed reaching a broad range of temperatures and activating a large scope of reactions using both micro- and macroscaled metallic materials. More recently, mono- and bimetallic NPs of Fe, FeC, FeCo, FeNi, and CoNi, exhibiting both heating and catalytic behaviors have been developed. Reactions based on CO, CO_2, and light alkanes such as CO_2 methanation (Sabatier reaction), Fischer–Tropsch, and reforming of methane were successfully activated. The state-of-the-art is promisingly subject to evolution in a near future as new magnetic nanomaterials should enable to adjust their heating performances thanks to a fine control of the magnetic properties. Appropriate thermal management and reactor design are required toward coupling this technology to a highly intermittent energy production in order to store renewable energies. Finally, the perspectives of improvement regarding the global energy efficiency of the system are promising, especially by reducing the frequency and the amplitude of the required magnetic field, while the life cycle assessment of this innovative activation method would enable to shed some light on the sustainability potential of the technology.

References

1 Wang, W., Tuci, G., Duong-Viet, C. et al. (2019). *ACS Catal.* 9: 7921–7935.
2 Lucia, O., Maussion, P., Dede, E.J., and Burdio, J.M. (2014). *IEEE Trans. Ind. Electron.* 61: 2509–2520.
3 Department of Energy. (2013). Energy Conservation Program: Test Procedures for Conventional Cooking Products. Fed. Regist., vol. 79, no. 232, Docket No. EERE-2012-BT-TP-0013.
4 Sweeney, M., Dols, J., Fortenbery, B., and Sharp, F. (2014). Induction Cooking Technology Design and Assessment. Presented at the ACEEE Summer Study on Energy Efficiency in Buildings.
5 Leclercq, J., Giraud, F., Bianchi, D. et al. (2014). *Appl. Catal., B* 146: 131–137.
6 Schaafsma, Y. (1963). GB925025A, 01 May.
7 Mortensen, P.M., Engbæk, J.S., Vendelbo, S.B. et al. (2017). *Ind. Eng. Chem. Res.* 56: 14006–14013.
8 Baxendale, I.R., Griffiths-Jones, C.M., Ley, S.V., and Tranmer, G.K. (2006). *Chem. Eur. J.* 12: 4407–4416.
9 Ceylan, S., Friese, C., Lammel, C. et al. (2008). *Angew. Chem. Int. Ed.* 47: 8950–8953.
10 Bartley, B.H. (1975). *J. Catal.* 40: 86–93.
11 Chatterjee, S., Degirmenci, V., and Rebrov, E.V. (2015). *Chem. Eng. J.* 281: 884–891.
12 Koch, T., Mehdizadeh, M., Ashmead, J. et al. (1999). WO/1999/001212, 15 January.
13 Ovenston, A. and Walls, J.R. (1983). *J. Chem. Soc., Faraday Trans.* 79: 1073–1084.
14 Mehdaoui, B., Meffre, A., Carrey, J. et al. (2011). *Adv. Funct. Mater.* 21: 4573–4581.

15 Meffre, A., Mehdaoui, B., Connord, V. et al. (2015). *Nano Lett.* 15: 3241–3248.

16 Asín, L., Ibarra, M.R., Tres, A., and Goya, G.F. (2012). *Pharm. Res.* 29: 1319–1327.

17 Sanchez, C., El Hajj Diab, D., Connord, V. et al. (2014). *ACS Nano* 8: 1350–1363.

18 Iovino, N., Bohorquez, A.C., and Rinaldi, C. (2014). *Nanomedicine* 9: 937–939.

19 Thomas, J.R. (1997). *Catal. Lett.* 49: 137–141.

20 Riedinger, A., Guardia, P., Curcio, A. et al. (2013). *Nano Lett.* 13: 2399–2406.

21 Dong, J. and Zink, J.I. (2014). *ACS Nano* 8: 5199–5207.

22 Piñol, R., Brites, C.D.S., Bustamante, R. et al. (2015). *ACS Nano* 9: 3134–3142.

23 Ceylan, S., Coutable, L., Wegner, J., and Kirschning, A. (2011). *Chem. Eur. J.* 17: 1884–1893.

24 Hartwig, J., Ceylan, S., Kupracz, L. et al. (2013). *Angew. Chem. Int. Ed.* 52: 9813–9817.

25 Dumestre, F., Chaudret, B., Amiens, C. et al. (2004). *Science* 303: 821–823.

26 Lacroix, L., Lachaize, S., Falqui, A., and Chaudret, B. (2009). *J. Am. Chem. Soc.* 131: 549–557.

27 Mehdaoui, B., Meffre, A., Lacroix, L.-M. et al. (2010). *J. Magn. Magn. Mater.* 322: L49–L52.

28 Meffre, A., Mehdaoui, B., Kelsen, V. et al. (2012). *Nano Lett.* 12: 4722–4728.

29 Bordet, A., Lacroix, L.M., Fazzini, P.F. et al. (2016). *Angew. Chem. Int. Ed.* 55: 15894–15898.

30 Bordet, A., Asensio, J.M., Soulantica, K., and Chaudret, B. (2018). *ChemCatChem* 10: 4047–4051.

31 Asensio, J.M., Marbaix, J., Mille, N. et al. (2019). *Nanoscale* 11: 5402–5411.

32 Kale, S.S., Asensio, J.M., Estrader, M. et al. (2019). *Catal. Sci. Technol.* 9: 2601–2607.

33 Le Caer, G., Dubois, J.M., Pijolat, M. et al. (1982). *J. Phys. Chem.* 86: 4799–4808.

34 Pfeiffer, F. and Radeloff, C. (1980). *J. Magn. Magn. Mater.* 19: 190–207.

35 De Masi, D., Asensio, J.M., Fazzini, P.-F. et al. (2020). *Angew. Chem. Int. Ed.* https://doi.org/10.1002/anie.201913865.

36 Desvaux, C., Dumestre, F., Amiens, C. et al. (2009). *J. Mater. Chem.* 19: 3268–3275.

37 Desvaux, C., Lecante, P., Respaud, M., and Chaudret, B. (2010). *J. Mater. Chem.* 20: 103–109.

38 Garnero, C., Lepesant, M., Garcia-Marcelot, C. et al. (2019). *Nano Lett.* 19: 1379–1386.

39 Vinum, M.G., Almind, M.R., Engbaek, J.S. et al. (2018). *Angew. Chem. Int. Ed.* 57: 10569–10573.

40 Varsano, F., Bellusci, M., La Barbera, A. et al. (2019). *Int. J. Hydrogen Energy* 44: 21037–21044.

41 Pérez-Camacho, M.N., Abu-Dahrieh, J., Rooney, D., and Sun, K. (2015). *Catal. Today* 242: 129–138.

42 Crawford, R. and Douglas, J. (2015). US2015/0075137 A1.

43 Blackwell, B. (2003). WO03/078054 A1.

44 Mehdizadeh, M., and Koch, T.. (2001). USOO6315972B1.

45 Grasset, F., Duguet, E., Mornet, S., and Vasseur, S. (2004). *J. Mater. Chem.* 14: 2161–2175.
46 Pollert, E., Knížek, K., Maryško, M. et al. (2007). *J. Magn. Magn. Mater.* 316: 122–125.
47 Wang, W., Duong-Viet, C., Xu, Z. et al. (2019). *Catal. Today* https://doi.org/10.1016/j.cattod.2019.02.050.
48 Chaudret, B., Carrey, J. P., Fazzini, F. et al. (2013). FR 3003 774.
49 Nielsen Hojlund, P. E.. (2017). WO2017/186615 A1.
50 Mortensen, P. M.. (2017). WO2017/036794 A1.
51 Nielsen Hojlund, P. E.. (2017). WO2017/186614 A1.
52 Niether, C., Faure, S., Bordet, A. et al. (2018). *Nat. Energy* 3: 476–483.
53 Asensio, J.M., Miguel, A.B., Fazzini, P.-F. et al. (2019). *Angew. Chem. Int. Ed.* 58: 11306–11310.
54 Mortensen, P. M. (2017). WO2017/186437 A1.
55 Olah, G.A., Goeppert, A., and Prakash, G.K.S. (2009). *J. Org. Chem.* 74: 487–498.
56 Graves, C., Ebbesen, S.D., Mogensen, M., and Lackner, K.S. (2011). *Renewable Sustainable Energy Rev.* 15: 1–23.
57 Fache, A., Marias, F., Guerré, V., and Palmade, S. (2020). *Waste Biomass Valorization* 11: 447–463.
58 Fache, A., Marias, F., Guerré, V., and Palmade, S. (2018). *Chem. Eng. Sci.* 192: 1124–1137.
59 Holladay, J.D., Hu, J., King, D.L., and Wang, Y. (2009). *Catal. Today* 139: 244–260.
60 Markewitz, P., Kuckshinrichs, W., Leitner, W. et al. (2012). *Energy Environ. Sci.* 5: 7281–7305.
61 Cuéllar-Franca, R.M. and Azapagic, A. (2015). *J. CO2 Util.* 9: 82–102.
62 Meylan, F.D., Piguet, F.P., and Erkman, S. (2017). *J. Energy Storage* 11: 16–24.
63 Collet, P., Flottes, E., Favre, A. et al. (2017). *Appl. Energy* 192: 282–295.
64 Naik, S.N., Goud, V.V., Rout, P.K., and Dalai, A.K. (2010). *Renewable Sustainable Energy Rev.* 14: 578–597.
65 Marbaix, J., Kerroux, P., Montastruc, L. et al. (2020). *Int. J. Life Cycle Assess.* https://doi.org/10.1007/s11367 020 01731 8.

15

Sabatier Principle and Surface Properties of Small Ruthenium Nanoparticles and Clusters: Case Studies

Iker del Rosal and Romuald Poteau

Université de Toulouse, INSA, UPS, CNRS, LPCNO (IRSAMC), 135 avenue de Rangueil, Toulouse F-31077, France

15.1 Introduction

Nanocatalysis has now become a major field that involves the use of nanomaterials for transforming molecules into value-added compounds. Nanoparticles (NPs) of metals, oxides, semiconductors, or other compounds are today used to catalyze major chemical reactions [1, 2]. Among them, metal nanoparticles (MNPs) have emerged as a complement to conventional homogeneous and heterogeneous catalysts [3], potentially showing high activities and selectivities. One main interest in MNPs *inter alia* relies on their high surface area-to-volume ratios [4], with a high concentration of potential and different active sites and therefore an expected high catalytic activity. According to the underlying reaction mechanism, increasing particle size may increase, decrease, or have no effect on the reaction rates of a given catalytic system [5]. These last years have witnessed that quantum chemistry calculations can bring a molecular perspective to experimental investigations. As a matter of fact, recent progresses in computational catalysis and nanochemistry have significantly improved our theoretical understanding of particle-dependent reactivity changes in nanocatalysis.

In this regard, one of the major achievements of computational quantum chemistry is its ability to address multistep chemical reactions by the location of reactants, products, intermediates, and transition states (TSs) on the potential energy surface (PES) of the system. Nowadays, quantitative approaches based on the harmonic approximation are routinely applied in the context of homogeneous catalysis [6, 7], most of the time within density functional theory (DFT) methods [8]. Computational chemistry has also proven its ability to shed light on surface organometallic chemistry [9] or on catalysis on transition metal (TM) surfaces [10–13]. However, exploring the surface chemistry of ligand-stabilized NPs remains challenging. The reason lies in the difficulty to accurately describe such complex systems, which exhibit a nanoscale metallic core with ligands on its surface, often partially mobile and/or labile, and not so easy to characterize experimentally. Yet, setting up a theoretical calculation requires to define a relevant structural model, as close as possible

Nanoparticles in Catalysis: Advances in Synthesis and Applications, First Edition.
Edited by Karine Philippot and Alain Roucoux.

to the compound experimentally investigated. Beyond semiempirical approaches [14], theoretical chemists face a dilemma, either considering scale models, i.e. small organometallic clusters, or taking advantage of periodic-boundary methods, i.e. to consider crystalline MNPs as a superposition of infinite surfaces with low to high crystallographic indexes [15]. Such models may shed light on the properties of NPs. However, given that these metal NPs are often said to have physical and chemical properties intermediate between those of small molecular compounds and bulk materials, resulting from surface or quantum size effects [16], it is possible that both models lead to artifactual conclusions, especially in the case of small MNPs.

In the present chapter, we will focus on ultrasmall to small (i.e. less than ~4 nm) colloidal MNPs, which have shown a remarkable catalytic activity for many reactions [17–19]. Solution approaches for preparing MNPs present key advantages because they allow to precisely control their size, shape, spatial distribution, surface composition, and electronic structure, as well as their thermal and chemical stability. In brief, MNPs are obtained by reducing organometallic precursors or olefinic ligands under mild conditions, usually by dihydrogen, and in the presence of ligands that act as capping agents of the resulting MNPs [20]. Good protective agents prevent the coalescence of MNPs, do not poison the surface, and to some extent may also favor the catalytic activity. The presence of activated H_2 on the surface of such MNPs may also be of utmost importance according to the target chemical transformation. Although the surface composition can dramatically act on yield and selectivity of a given reaction [21, 22], it has never been evaluated in a systematic way and the surface compositionperformance relationships are poorly understood. Is it possible to rationalize trends in nanocatalysis and to develop nanocatalyst design principles? A positive answer relies on joint theoretical and experimental studies. Regarding the theoretical part, it requires an atomic-scale description, usually provided by first-principles methods such as DFT, coupled to microkinetic (MK) models [23]. They are part of a conceptual framework that crosses the microscopic and macroscopic scales and that drive the quest for an optimum nanocatalyst. This powerful framework is the so-called Sabatier principle, which states that the best reaction rate corresponds to an optimal adsorption strength on a surface [24]. In these last years, these ideas have successfully been developed and applied in heterogeneous catalysis to rationalize trends in some major chemical reactions (Haber–Bosch process [25, 26] and Fischer–Tropsch reaction [27],...) and have allowed the theoretical design of efficient transition metal heterogeneous catalysts [28]. A major outcome of this approach was the development of an efficient albeit low-cost bimetallic catalyst for ammonia synthesis [29]. It relied on the correlation between theoretical turnover frequencies (TOFs) obtained through a MK modeling and adsorption energies of N_2, where dissociation is the rate-limiting step (the dependence of the catalytic activity on the nitrogen adsorption energy is a consequence of the Brønsted–Evans–Polanyi (BEP) linear scaling relationship between activation and adsorption energies [30]). This correlation is seen under the form of volcano plots where optimal catalysts lie on top of the volcano, following the seminal proposition of Balandin in the late 1950s [31]. The top of the volcano is sometimes called the Sabatier optimum.

Since these early studies, the catalytic decomposition of formic acid (HCOOH = H_2 + CO_2, i.e. a reverse pathway of CO_2 hydrogenation) has been thoroughly investigated as a test case of the Sabatier principle [32–34], both experimentally and theoretically. Yet, theoretical studies often focus on bare surfaces, either because they deal with heterogeneous catalysts under UHV conditions or either because the impact of surface composition is neglected. The surface state of MNPs is however an important parameter that may dramatically modify adsorption energies and kinetic data, notwithstanding the possible active interaction between reactants and surface species. In the following, two case studies, namely, the H/D isotopic exchange at C atoms and the hydrogen exchange reaction, will illustrate how the surface composition or the nature of capping ligands can influence the reactivity of ruthenium nanoparticles (RuNPs). Previously published results [35–37] have been supplemented here with new calculations and analysis for the purpose of this chapter.

15.2 C–H Activation and H/D Isotopic Exchange in Amino Acids and Derivatives

15.2.1 Reference Activation and Dissociation Energies

In these last years, C–H activation and functionalization has opened the way to the diversification of functional molecules [38]. It also affords rapid access to deuterated labeled compounds, which are important tools in many areas of chemical and biochemical science [39]. They are essential for mechanistic studies in biology and in organic and organometallic chemistry. Yet, C–H bonds are very strong, typical dissociation enthalpy values, $\Delta_d H_{298}$, ranging from 89 to 133 kcal mol^{-1} [40]. The use of transition metal (TM) complexes or surfaces can counterbalance the energetic cost of the cleavage by the formation of strong M–C and M–H bonds. In the case of ruthenium, Ru(0001)–H and Ru(0001)–(CH_3) bond energies are 64.7 and 39.4 kcal mol^{-1}, respectively, at the DFT-Perdew, Burke, and Ernzerhof (PBE) level of calculation [41], to be, for example, compared to the first C–H bond dissociation energy of methane, 109.0 kcal mol^{-1} (104.0 kcal mol^{-1} experimentally [40]). From a kinetic perspective, the trapping-mediated dissociative chemisorption study of ethane on Ru(0001) yielded activation energies for the C–H bond cleavages of 9.3 kcal mol^{-1} [42]. The TSs of the elementary reactions for the dissociation of methane on the Ru(0001) surface were also investigated by Ciobîcă et al. with DFT periodic calculations and the nudged elastic band (NEB) method for 25% coverages [43]. The first C–H activation energy was found to be 20.3 kcal mol^{-1}, whereas it is thermodynamically favored by a small amount (−2.1 kcal mol^{-1}). The reactivity of methane on a stepped Ru(11$\bar{2}$0) was also studied by DFT [44]. The calculated barrier for the first dissociation of methane, 13.4 kcal mol^{-1}, is smaller than that for the Ru(0001) surface. A microcanonical unimolecular rate theory (MURT) was used to characterize methane dissociative chemisorption on Ru(0001) [45]. It indicated that the threshold activation energy for CH_4 dissociative chemisorption on Ru(0001) is 14.1 kcal mol^{-1} only.

15.2.2 H/D Exchange Mechanism

Let us now get back to the isotopic H/D exchange at carbon atoms catalyzed by RuNPs. Experimentally, a selective C(sp^3)–H activation/deuteration process was first achieved, by using RuNPs dispersed in a polymer matrix, namely, polyvinylpyrrolidone, PVP (RuNP@PVP), which can be considered as "naked particles". These NPs are stabilized by this matrix and by hydrides adsorbed at their surface, which originate from the synthesis process [46, 47]. They allowed effective and selective deuteration of a large diversity of nitrogen-containing compounds in the α-position of the nitrogen atom, in mild conditions (1–2 bar of deuterium gas at room temperature or 55 °C) [48]. It must be emphasized at this point that NMR characterization evidenced the mobility of surface hydrides and their easy exchange with hydrogen molecules in the gas phase [49]. The exposure of RuNP@PVP to deuterium gas thus involves the substitution of surface H by surface D, as schematized on Figure 15.1.

RuNPs thus behave both as catalysts with their exposed metal atoms and as reservoirs of deuterides with various coordination modes (edge-bridging, face-capping, and on-top, see Refs. [50–52]). This H/D exchange method was then generalized to the deuteration of amino acids [35]. The H/D reaction occurred enantiospecifically and with very high yield at the $C_\alpha(sp^3)$ position with full retention of configuration. We studied the isopropylamine model, **1**, by DFT-based mechanistic explorations in order to explain the experimental outcome [35].

An Eley–Rideal mechanism has been dismissed, whereas two Langmuir–Hinshelwood-type mechanisms were found. The most favorable is based on oxidative addition, the other one involving σ-bond metathesis. The oxidative addition mechanism of the C–H bond breaking is summarized in Figure 15.2. It first starts by the adsorption of the amine on one Ru atom through the lone pair of the nitrogen atom to give $\mathbf{1}^{N*,CH}$, followed by a one-step C–H activation on a neighboring Ru site that leads to the formation of a four-membered dimetallacycle $\mathbf{1}^{N*,C*}$ together with the coadsorption of the H atom on the Ru surface. This key type of intermediate also permitted the rationalization of the enantiospecificity of this reaction in the case of asymmetric carbon atoms. Indeed, the joint grafting of the N and C atoms onto two

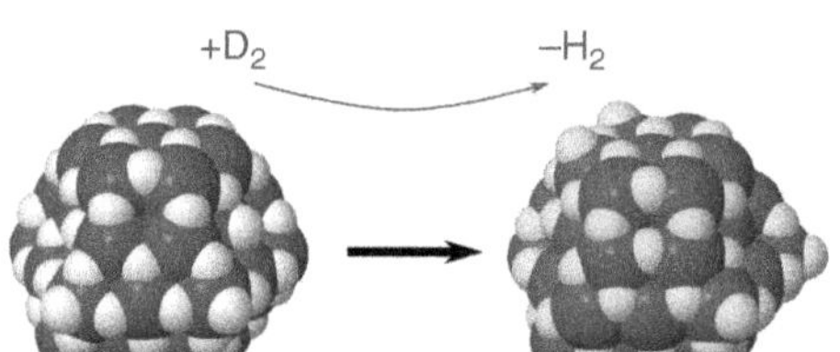

Figure 15.1 Schematized H (white atoms) /D (light gray atoms) substitution at the surface of a 1 nm RuNP (dark gray atoms).

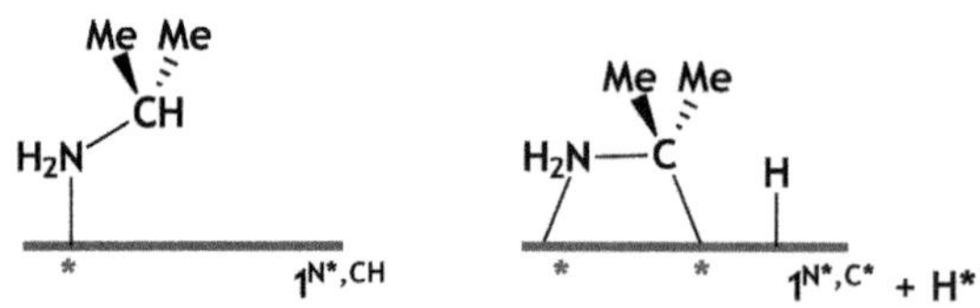

Figure 15.2 Adsorbed isopropylamine model before ($\mathbf{1}^{N*,CH}$) and after ($\mathbf{1}^{N*,C*}$) C–H activation on a metal surface.

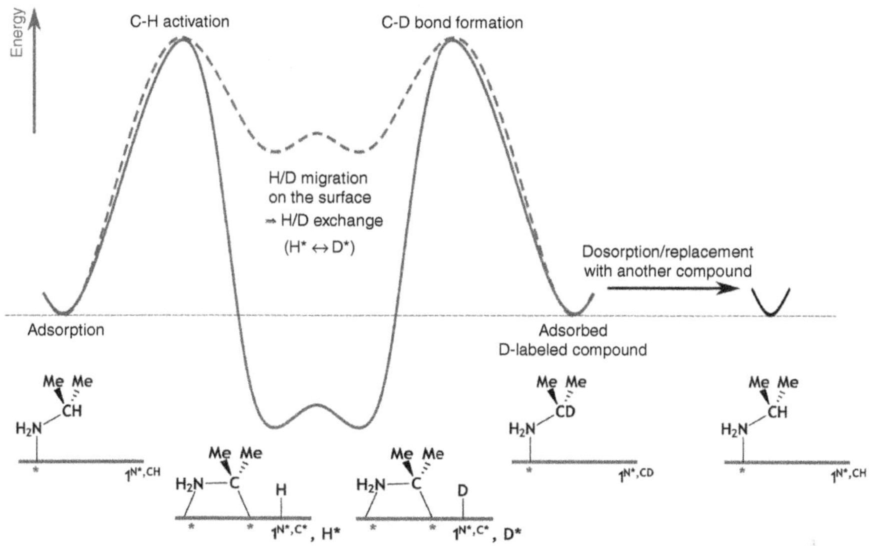

Figure 15.3 Simplified H/D exchange mechanism plotted as energy profiles. Dashed line: optimal pathway; plain line: unfavorable pathway.

vicinal Ru atoms of the surface proceeds with retention of configuration because of the rigidity of the dimetallacycle. This reaction has been studied on a cluster model made of 13 atoms, which provides energies in close agreement with a 1 nm Ru_{55} model [35]. In the following, we shall only consider Ru_{13} clusters with various deuteride coverage values.

After analysis of the mechanism and of several energy profiles, some clues regarding the optimal and worst energy profiles are summarized in Figure 15.3 (dashed and plain lines, respectively). After the C–H activation step, the replacement in the vicinity of the Ru*–C active site of the H resulting from the C–H bond breaking by a deuteride is a mandatory prerequisite for the D-labeling reaction to occur. In both cases, the migration of H and D on the surface (symbolized by H* ↔ D*) is supposed to be easy. This is an acceptable assumption, owing both to experimental evidence [49, 53] and to DFT calculations (barrier height from fcc to hcp sites on a Ru(0001) surface: 5.0 kcal mol^{-1}, in agreement with experimental data [53]). Once a deuteride has moved in the appropriate reaction site, the C–D bond formation can occur. The corresponding barrier height will depend on the relative stability of the $\mathbf{1}^{N^*,C^*}$ dimetallacycle intermediate. Provided that it is low enough, and that thermodynamics is favorable, the outcome of the reaction is the $\mathbf{1}^{N^*,CD}$ D-labeled counterpart of $\mathbf{1}^{N^*,CH}$. It can then desorb and be replaced on the same active site by another candidate to the labeling process. What makes the difference between the plain and dashed profiles is the relative stability of the ($\mathbf{1}^{N^*,C^*}$, **H*/D***) intermediates. Provided that the C–H activation would involve similar barrier heights, the unfavorable reaction coordinate (plain profile) goes through an intermediate significantly more stable than the $\mathbf{1}^{N^*,CH/D}$ reactants and products. This energy well is thus a trap that hampers the H/D exchange reaction. On the other hand, in the reaction profile plotted in dashed lines, the stability of the final $\mathbf{1}^{\mathbf{N^*,CD}}$ with respect to the $\mathbf{1}^{\mathbf{N^*,C^*}}$ **, H*/D*** intermediate is a driving force to the H/D exchange reaction. A MNP catalyst with given characteristics (metal M, crystal packing, available surface crystallographic planes, diameter, nature, and number of surface species) that would favor the dashed profile with a moderate to low $\mathbf{1}^{\mathbf{N^*,CH}} \rightarrow \mathbf{1}^{\mathbf{N^*,C^*}}$ **+ H*** barrier height could be considered as the *Sabatier optimum* for this reaction, i.e. an efficient catalyst that would lie on top of the volcano plot (*vide supra*). We shall see in the following that such optimum can be attained with a given metal (Ru) and shape (small icosahedron) only by changing the H coverage value (see also the short discussion in the SI of Ref. [35]). However, let us first consider some dissociation and adsorption energies obtained from PBE calculations on the bare Ru_{13} icosahedron. In gas phase, the homolytic C–H cleavage in isopropylamine compounds requires 91 kcal mol^{-1}. Thanks to the metal surface, this dissociation can be thermodynamically counterbalanced by the adsorption energy of the 1-aminoisopropyl radical and of the hydrogen atom on different sites of the surface. On Ru_{13}, adsorption energies for the model fragments NH_3, CH_3, and H are found to be −23 kcal mol, −52 kcal mol^{-1}, and −64 kcal mol^{-1}, respectively. This makes the adsorption process of the dimetallacycle thermodynamically stable by ~−140 kcal mol^{-1}, which compensates the C–H bond cleavage by almost 50 kcal mol^{-1}! Given the −23 kcal mol^{-1} adsorption energy of the amine on Ru_{13}, only the unfavorable H/D exchange pathway plotted in Figure 15.3 is expected to be available on bare Ru surfaces. However, we know from experiments that RuNPs@PVP constitute a very efficient catalyst for this reaction [35, 48]. We will see hereafter that it may be related to the deuteride/Ru surface ratio, which changes the reaction profile from unfavorable to favorable.

15.2.3 Bare Cluster

On an almost bare cluster, $Ru_{13}D$, the overall energy profile reported in Figure 15.4 in black is qualitatively similar to the unfavorable profile plotted as a plain line in Figure 15.3. After the σ bonding of the amine through the lone pair of the nitrogen atom (E_{ads}: −24.2 kcal mol^{-1}, in line with the −23 kcal mol^{-1} previously calculated for the model NH_3 fragment), the C–H bond dissociation reaction is exothermic by c. 12 kcal mol^{-1}, with a low activation barrier of ~18 kcal mol^{-1}. The resulting species are the dimetallacycle $\mathbf{1}^{\mathbf{N^*,C^*}}$ and the hydrogen atom (**H***) adsorbed in the closest μ_3 site of the metal surface (ΔE: −36.4 kcal mol^{-1}, slightly weaker than the −50 kcal mol^{-1} estimated value, probably owing to the geometry constraint involved by the dimetallacyle on the contrary to two separated moieties). In the full energy profile, the easy migration of H and D on the surface and the replacement in the vicinity of the active site of H by D (H* ↔ D*) is shown with a horizontal dashed line. Nevertheless, on an hypothetical bare RuNP, the replacement on the same site of the μ_3-H resulting from the C–H bond breaking by a deuteride would probably be a rare event from a dynamical point of view. In addition, as can be seen from the right part of the energy profile, the barrier height for the C–D bond formation is rather

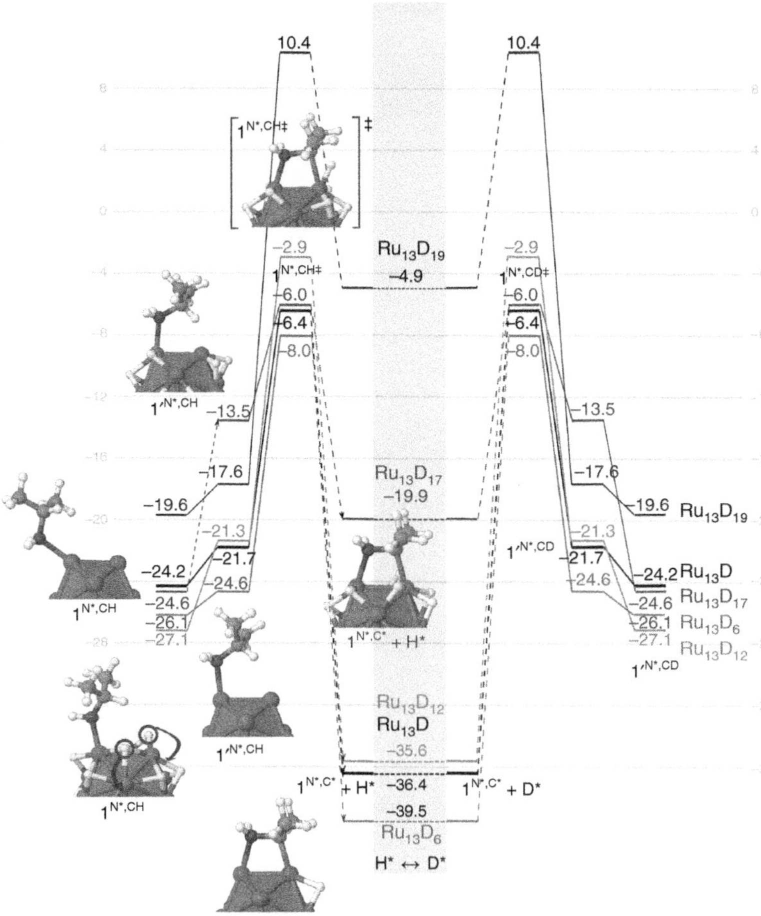

Figure 15.4 Energy diagram for the Langmuir–Hinshelwood-type H/D exchange mechanism investigated on $Ru_{13}D_x$ clusters with various hydrogen coverages (considering that the presence of D atoms on the Ru surface is purely formal because in standard quantum chemistry calculations, no isotopic effects are taken into account in the solution of the Schrödinger equation); difference energies, given in kcal mol^{-1}, are calculated with $Ru_{13}D_x$ (*x*:1, 6, 12, 17, 19) and gas-phase isopropylamine taken as a reference. D atoms are shown in dark gray. $\mathbf{1}^{N^*,C^*}$, **H*** species are not directly connected along the reaction coordinate to the $\mathbf{1}^{N^*,CH‡}$ TS; an optimal coordination site has systematically been searched in line with the easy motion of hydrides on the surface.

high (30 kcal mol^{-1}), and no thermodynamic driving force can pull the reaction toward the $\mathbf{1}^{N^*,CD^*}$ species. In summary, this profile and some considerations regarding the dynamics of H and D on the surface suggest that hypothetical naked RuNPs would be very efficient catalysts for C–H activation but that the resulting $\mathbf{1}^{N^*,C^*}$ intermediate would remain thermodynamically stuck in its energy well.

15.2.4 $Ru_{13}D_{19}$

With ~1.6 D/Ru_{surf}, the D surface coverage in $Ru_{13}D_{19}$ is slightly greater than the usual measured values for RuNPs [54]. The energy profile, plotted in Figure 15.4, is very different from $Ru_{13}D$. The adsorption energy of the amine is found to be −19.6 kcal mol^{-1} (to be compared to −24.2 kcal mol^{-1} in $Ru_{13}D$). It means that neither electronic nor steric effects significantly lower the adsorption strength of the diethylamine reactant on this cluster. The steric hindrance on the surface involved by the 19 deuterides prevents an easy activation of the C–H bond. The $\mathbf{1}^{\mathbf{N^*,CH‡}}$ TS actually lies 30.0 kcal mol^{-1} above $\mathbf{1}^{\mathbf{N^*,CH}}$, a barrier height uneasy to overcome under mild conditions. The dissociated H atom then also feels this steric clash on the crowded surface, which makes the resulting coadsorption compound, $\mathbf{1}^{\mathbf{N^*,C^*}}$, **H***, barely stable on the surface (ΔE: −4.9 kcal mol^{-1}). The resulting energy profile is now qualitatively close to the optimal pathway schematized in Figure 15.3, except that the barrier height makes this kinetically controlled reaction uneasy. In summary, a high number of deuterides on the surface could facilitate the H/D isotopic exchange, except that the kinetic barrier is high.

15.2.5 $Ru_{13}D_n$, $n = 6–17$

The reaction energy profiles for $Ru_{13}D_n$, with $n = 6$ and 12, are very similar to that of $Ru_{13}D$ catalyst. The small differences probably originate from the motion of the deuterides along the reaction coordinate, which involve minor energy differences (for example, the $\mathbf{1}^{\mathbf{N^*,CH‡}}$ TS lies at −2.9 kcal mol^{-1} for $Ru_{13}D_12$ instead of −6.4 and −8.0 kcal mol^{-1} for $Ru_{13}D$ and $Ru_{13}D_6$, respectively). The main conclusion is that a fairly small hydride coverage value (up to 1 H/Ru_{surf}) does not significantly change the reaction profile of the reaction, which falls in the unfavorable case plotted in Figure 15.3. Regarding $Ru_{13}D_{17}$, the resulting energy profile, plotted in gray in Figure 15.4, is now intermediate between the $Ru_{13}D_{19}$ and $Ru_{13}D_x$ ($x \leq 12$) catalysts, thus making this surface composition optimal for the target reaction.

This $Ru_{13}D_{17}$ catalyst has been designed from $Ru_{13}D_{19}$ by removing two deuterides in the vicinity of the active site in $\mathbf{1}^{\mathbf{'N^*,CH}}$ (atoms circled in black in Figure 15.4) in order to lower the steric hindrance that comes with the formation of the dimetallacycle intermediate (the present numbers slightly differ from Ref. [35]). All calculations have been done again but with different H grafting modes for the sake of consistency with the other surface compositions considered in the present study. The resulting energy profile does not qualitatively differ. It costs 11.1 kcal mol^{-1} w.r.t. $\mathbf{1}^{\mathbf{'N^*,CH}}$. However, it turns out to yield an energy profile intermediate between the $Ru_{13}D_n$ ($n \leq 12$) and $Ru_{13}D_{19}$ cases, with a low 18.6 kcal mol^{-1} apparent C–H activation barrier height, a dimetallacycle intermediate ($\mathbf{1}^{\mathbf{N^*,C^*}}$, **H***) less stable than the $\mathbf{1}^{\mathbf{N^*,CD}}$ final product by 4.7 kcal mol^{-1}, and a very low C–D formation barrier height (13.9 kcal mol^{-1}).

15.2.6 Short Discussion

The numbers reported in Figure 15.4 must be taken with caution, in particular because the motion of the deuterides on the surface during the course of a reaction

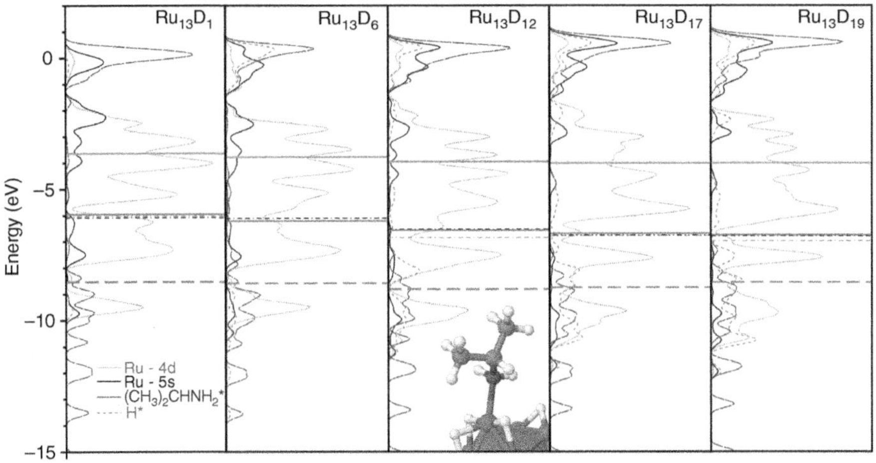

Figure 15.5 DOS of the $[(CH_3)^2CHNH_2]$ $^*Ru_{13}D_x$ compound. Long dashes: DOS projected on all valence AOs of the $[(CH_3)_2CHNH_2]$ substrate. Short dashes: DOS projected on the 1s AOs of the surface hydrides. Plain line: DOS projected on the Ru 5s AOs. Dots: DOS projected on the Ru 4d AOs. Horizontal lines: Fermi energy (plain line around −3.6 eV), *d*-band center of Ru surface atoms (plain line around −6 eV), *d*-band center of the Ru core atom (dashed line), *d*-band center of the N* coordination site (dashed dotted gray line), and *d*-band center of the CH activation site (dashed dotted black line). The 3D optimized structure of $(CH_3)^2CHNH_2$ adsorbed on $Ru_{13}D_{12}$ reactant is also shown. Similar geometry of the adsorbate/Ru_{13} cluster has been considered for the other surface compositions.

can change the energy outcome. A preliminary molecular dynamics study could be appropriate in order to identify the most stable conformations of the reactants and reaction intermediates on the metal surface. The present theoretical study however clearly evidences a four-membered dimetallacycle as the key intermediate (see also Ref. [35]). This investigation also suggests that a collective motion of surface species can optimize the C–H activation step, in conjunction with an optimal deuteride coverage value, in line with the Sabatier principle.

The density of states (DOS) of the $\mathbf{1}^{N*,CH}$ substrate adsorbed on all the $Ru_{13}D_n$ clusters considered in this work is plotted in Figure 15.5, together with the Fermi energy and the overall and atomic *d*-band centers [55]. As already shown in previous studies [56, 57], the presence of surface species stabilize the *d*-band center of surface metal atoms, whereas upon hydrogenation, the *d*-band center of the inner Ru atom tends to get closer to the Fermi energy, a trend that is stronger upon carbonylation [57]. It is worth noting that the electronic properties of the metal part do not differ much between $Ru_{13}D_n$ clusters (n = 12–19) both in terms of energy profile, of global *d*-band center, and of atomic *d*-band center for the two active sites (see Figure 15.5: gray dashed dotted line for the grafting site and black dashed dotted line for the neighboring C–H cleavage site). On the contrary, the DOS profiles are very different for $Ru_{13}D$ and $Ru_{13}D_6$. This suggests that the reaction outcome is ruled both by the modulation of the local electronic structure by surface species and by the steric hindrance around the C–H activation site. In addition, as discussed in the $Ru_{13}D_{17}$ case, the dynamics of surface species may also play a role.

15.3 Hydrogen Evolution Reaction

15.3.1 Introduction

Our modern society is facing the need to replace fossil fuels by clean and renewable energy sources. Hydrogen is an excellent candidate that can be produced from water through the water-splitting process. Under acidic conditions, it involves two successive reactions, namely, the oxygen evolution reaction (OER) and the hydrogen evolution reaction (HER) [58]:

$$\text{OER: } 2H_2O \longrightarrow O_2 + 4H^+ + 4e^-$$

$$\text{HER: } 2H^+ + 2e^- \longrightarrow H_2$$

Regarding HER, although Pt-based catalysts show the highest efficiencies in acidic media with minimum overpotentials and fast kinetics, their cost and stability issues at other pH values require finding cheaper substitutes. In that respect, nanostructured metal catalysts can present attractive properties. As reported by Drouet et al. [59], a porous Ru nanomaterial was proven to be an efficient electrocatalyst for the HER under acidic and neutral conditions. More recently, 4-phenylpyridine (4PP)-protected RuNPs were shown to be highly active for the HER in both acidic and basic media. From a general point of view, this study highlighted the potential of designing and preparing ligand-stabilized NPs as catalysts for the HER and, therefore, "paved the way to the fine-tuning of the catalytic properties of these nanocatalysts through the limitless strategy of ligand capping" [36]. From a theoretical point of view, it is not easy to anticipate which kind of metal or surface ligands would favor this reaction. According to the DFT calculations published reported by Sala and coworkers [36], the contribution of 4PP ligands in the HER process could stem from their rather moderate adsorption strength, which allows a large number of hydrogen atoms to be present on the Ru surface. This behavior is different from that, for example, of strongly coordinated ligands such as CO, which impede the presence of surface H atoms in a large range of experimental conditions.

DFT calculations can even better play a role in the rational design of good catalysts for such application, following, for example, the seminal work of Nørskov et al. [60]. A correlation has been shown between H_2 dissociative adsorption Gibbs free energies $\left(\Delta G_{H^*} = \frac{1}{n}\left[G(nH^*) - G(\text{surf}) - \frac{n}{2}G(H_2)\right]\right)$ and experimental exchange current densities i_0 for the HER. After consideration of several metals, a Balandin–Sabatier volcano curve had been found, with platinum electrodes on top of the volcano and ΔG_{H^*} for Pt that lies between −2.0 and +0.7 kcal mol^{-1} as a function of H coverage. The main conclusion of this study is that the requirement for an optimal HER catalyst is a dissociative adsorption Gibbs free energy of H_2 that lies as close as possible

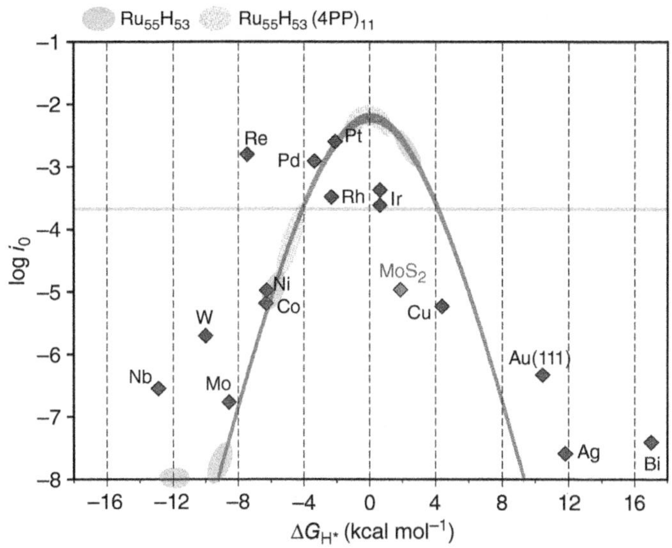

Figure 15.6 Volcano-like behavior of the experimental exchange current density i_0 (in A cm^{-2}) reported as a function of the DFT-calculated dissociative adsorption Gibbs free energy ΔG_{H^*} (in kcal mol^{-1}) for pure metals [60, 61] – dark gray diamonds – and nanoscale MoS_2 – light gray diamond. The gray horizontal line is the i_0 value ($\log i_0 = -3.69$) experimentally measured for very small (1.5 ± 0.3 nm) and homogeneous in size, 4-PP-capped RuNPs in acidic conditions [36]. The ideal volcano, in gray, was not schematized in the original figure. The logarithm of the current density for the RuNP models considered in present study is predicted from our DFT calculations of ΔG_{H^*}: gray shapes for the probed sites in the unprotected $Ru_{55}H_{53}$, and checkerboard pattern shapes for $Ru_{55}H_{53}(\sigma\text{-PP})_9(\pi\text{-PP})_2$. Source: Based on Jaramillo et al. [61].

to 0 kcal mol^{-1}. The resulting experimental exchange current density i_0 is plotted in Figure 15.6 as a function of DFT-PBE ΔG_{H^*} values [11, 60], [61]. The ideal volcano was not schematized in the original figures. It is not a rigorous interpolation of the experimental results, the intention being only to give a guideline to qualitatively predict the activity of new materials (the results obtained for [Ru_{55}] nanoclusters will be described in the next section).

15.3.2 4-Phenylpyridine-Protected RuNPs

Although only the coordination of 4PP ligands at the surface of a 55-atoms RuNP model was considered in Ref. [36], that preliminary study is now going to be supplemented by new computational and theoretical studies.

The ΔG_{H^*} probing index has been calculated on six representative adsorption sites of this model. DFT calculations performed on a 1.2 H/Ru_{surf} hydrogenated Ru_{55} model showed that the coordination of the 4PP ligand at the ruthenium surface can occur through two coordination modes. There is a competition between a vertical adsorption mode, ruled out by the σ-donation of the nitrogen lone pair and an aromatic π-to-metal surface interaction, with a flat-lying configuration of PP where

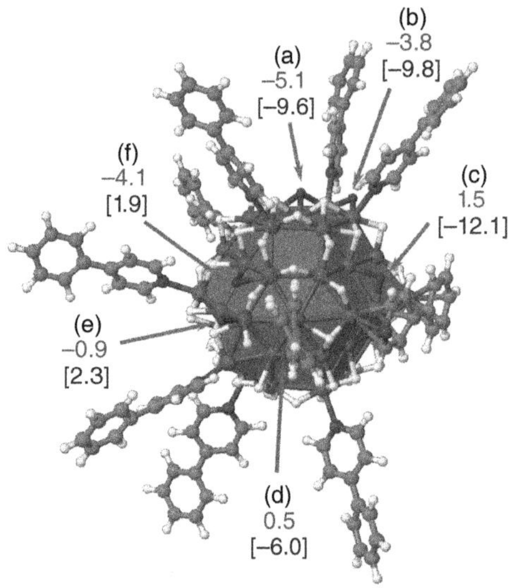

Figure 15.7 Probed adsorption sites (a–f) for HER on a 4PP-protected 1 nm RuNP, $Ru_{55}H_{53}(\sigma\text{-PP})_9(\pi\text{-PP})_2$ (see also the description of the model in Ref. [36]). Dissociative adsorption Gibbs free energy, ΔG_{H^*}, are given in kcal mol^{-1}. ΔG_{H^*} values between square brackets were calculated on the unprotected $Ru_{55}H_{53}$ nanocluster. Source: Based on Creus et al. [36].

each aromatic cycle exhibits the well-known $\mu_3 - \eta^2 : \eta^2 : \eta^2$ face-capping mode of benzene. These previous DFT calculations suggested that the saturation is reached around 11 4PPs on this model, with the coadsorption of 9 σ-4PPs and 2 π-4PPs being more stable than the adsorption of 11 σ-4PPs by c. 20 kcal mol^{-1} [36].

The resulting compound, $Ru_{55}H_{53}(\sigma\text{-PP })_9(\pi\text{-PP})_2$, is shown in Figure 15.7; the six hydrides selected for the calculation of ΔG_{H^*} are colored in dark gray. The μ_3-H^a lies on the Ru(0001) facet, the edge-capping μ-H^b is in between the (001) and (101) facets, μ-H^{c,d,f} lie on the (101) facets, whereas the η-H^e is close to the tip (see Ref. [57] for a description of the Ru_{55} core). On the basis of ΔG_{H^*}, found to be between −5.1 and 1.5 kcal mol^{-1}, some sites (H^{c-e}) in the 4PP-capped RuNPs are expected to be highly active for the HER. The ligth gray shapes on Figure 15.6 help to qualitatively evaluate the possible current densities that could be produced by using such RuNPs. ΔG_{H^*} has also been calculated for the same sites, but in $Ru_{55}H_{53}$, the unprotected counterpart of $Ru_{55}H_{53}(\sigma\text{-PP })_9(\pi\text{-PP}_2)$. The values are reported in Figure 15.7 between square brackets. Without the ligands, the H_2 dissociative adsorption Gibbs free energy lies in the vicinity of the usual value, i.e. ~-10 kcal mol^{-1}, with the exception of hydrides H^{e,f}, which are weakly coordinated on the surface. The ligands thus tend to significantly lower adsorption energies, up to c. 0 kcal mol^{-1}. The only noticeable exception is observed for H^f, owing to a significant reorganization of surface hydrides in $Ru_{55}H_{52}$, i.e. after removal of H^f (see computational details). Regarding the exchange current density i_0 for the HER catalyzed by hypothetical unprotected RuNPs, it is expected to be very low, see also the gray shapes in Figure 15.6.

We have tried to interpret this ligand effect that significantly favors the HER in the 4PP-protected nanocluster in terms of atomic charges of surface Ru atoms or coordination *d*-band center [56]. No linear correlation was found (*vide infra*) between these descriptors and ΔG_{H^*}, even though there is an overall relationship between the stabilization of the coordination *d*-band center upon coordination of the 11 4PP ligands and the decrease of adsorption strengths of hydrides (4PP-protected nanocluster:

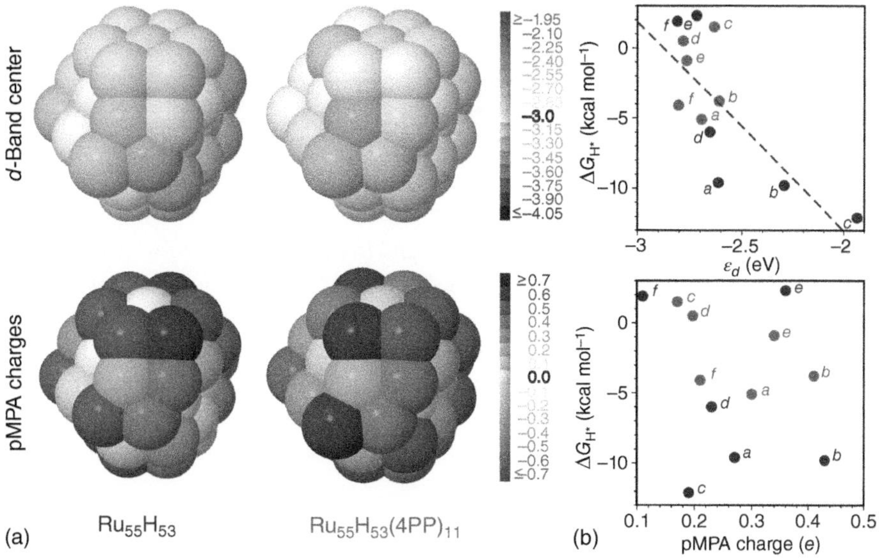

Figure 15.8 (a) *d*-Band center and pMPA charges of surface metal atoms in $Ru_{55}H_{53}$, on the left, and $Ru_{55}H_{53}(\sigma\text{-PP})_9(\pi\text{-PP})_2$, on the right, plotted as 3D color maps (see also Ref. [62] and references therein); the *d*-band center and charge scales are in eV and in elementary positive charge unit, *e*, respectively. Source: Based on Creus et al. [36]. (b) ΔG_{H^*} plotted as a function of the coordination *d*-band center (upper part) or as a function of the average charge of the Ru atoms on which the H-grafting occurs (lower part); dark gray dots: $Ru_{55}H_{47}$, light gray dots: $Ru_{55}H_{47}(\sigma\text{-PP})_9(\pi\text{-PP})_2$; labels a–f refer to the same sites as in Figure 15.7.

$\varepsilon_{d_{surf}} = -2.94$ eV; unprotected nanocluster: $\varepsilon_{d_{surf}} = -2.87$ eV) [36]. The atomic *d*-band centers and charges of the surface metal atoms are reported as 3D color maps in Figure 15.8a.

Regarding $\varepsilon_{d_{surf}}$, even though its decrease upon coordination of the 11 4PP ligands is moderate, some atoms exhibit a stronger stabilization (light gray to white or even black, i.e. a decrease of ε_d by c. 0.3–0.6 eV). On the contrary, and this is not really surprising given that they are weak L ligands, the 11 4PP ligands do not significantly change the oxidation state of the surface metal atoms. The correlation analysis between the ΔG_{H^*} index and ε_d or pMPA charges is reported in Figure 15.8b. pMPA charges of the adsorption sites $H^{a\text{-}f}$ do not significantly change between the hydrogenated Ru_{55} model (light gray dots) and its 4PP-protected counterpart (dark gray dots). This property does not account for the variation of ΔG_{H^*} between the two models, and it does not seem to be a good descriptor of the possible activity of RuNPs. Regarding *d*-band centers, there is a fair linear correlation between ε_d and ΔG_{H^*} in the unprotected model (dark gray dots and dashed line). The deviation from a linear behavior could stem from the observed significant relaxation of the geometry of surface species after removing of an hydride. Upon protection by the 11 4PP ligands, it is interesting to notice that whereas the average $\varepsilon_{d_{surf}}$ index is stabilized by 0.07 eV only, ε_d is stabilized by several tenths of eV for sites b and c, in agreement with weaker adsorption strengths. Nevertheless, the significantly weak

H-adsorption on sites a and d is not explained. Some additional effects should probably be taken into account, such as some steric effects involved by the π electrons of the 4PP ligands. ε_d nevertheless gives interesting trends, with a narrowing of the ε_d domain for the six probed sites, -2.7 ± 0.1 eV instead of the (-2.8 eV, -1.9 eV) range for $Ru_{55}H_{47}$.

In summary, the ΔG_{H^*} index suggests that some sites in 4PP-protected ruthenium NPs could be very active regarding the HER. This is actually in full agreement with experiments [36], where such RuNPs were found to be highly active for this reaction in both acidic and basic media. They reached in 1MH_2SO_4 solution a current density of 10 mA cm^{-2} at only 25 mV overpotential, which was maintained for at least 12 hours without any sign of deactivation, outperforming commercial Pt/C and ruthenium black in short- and long-term stability tests as well as in terms of TOF and i_0. The corresponding exchange current density of $i_0 = 2 \times 10^{-4}$ A cm^{-2} is reported in Figure 15.6 as a gray horizontal line.

15.3.3 Optimal Ligands for the HER?

Owing to the previous discussion, the ΔG_{H^*} index confirms that it is a relevant descriptor of the activity of metal catalysts regarding the HER. It also confirms on the theoretical side, and in agreement with experimental measurements of exchange current densities [36], that surface ligands can significantly improve the activity of metal NPs. We shall now assess the possible influence of various surface ligands on the activity of small Ru clusters, on the basis of ΔG_{H^*}. The $Ru_{13}H_{14}$ cluster has been considered as a test case, where the Ru_{13} metal core has an icosahedral shape. With 1.2 H/$Ru_{surface}$, it is very close to the usual surface coverage at r.t.

The probing index ΔG_{H^*} has been calculated on three representative adsorption sites of $Ru_{13}H_{14}$ as well as in $Ru_{13}H_{14}L_x$ clusters, i.e. stabilized with x ligands L, with L_x = $(CO)_6$, $(CO)_{12}$, $(MeCOO)_3$, $(MeCOO)_6$, and $(n\text{OA})_6$, where nOA is the n-octylamine (see Figure 15.9). This is a preliminary study, aiming at showing that the capping ligands may modulate the HER activity of Ru clusters. These models are only test cases, some of them being compositions that are not realistic w.r.t. standard experimental conditions. Adsorption energies, ΔG_{H^*}, calculated for $Ru_{13}H_{14}$, are slightly stronger than those obtained for $Ru_{55}H_{53}$, with values in the range -10.8 to -16.6 kcal mol^{-1}. Based on this discussion, such a cluster would be a very inefficient catalyst for the HER.

The idealized volcano schematized in Figure 15.6 is reported in Figure 15.10 and serves as guideline to roughly predict the exchange current density i_0 as a function of calculated ΔG_{H^*}. This is what is going to be discussed now. i_0 for $Ru_{13}H_{14}$ is expected to be lower than 10^{-7} A cm^{-2} (white dots in Figure 15.10). When stabilized by three carboxylates ($Ru_{13}H_{14}(MeCOO)_3$), the H adsorption strength decreases on some sites (gray dots in Figure 15.10), thus making this cluster a pretty good catalyst for the HER, with expected values for i_0 close to 10^{-3} A cm^{-2}. The grafting of three additional MeCOO ligands stabilize too much the electronic structure of

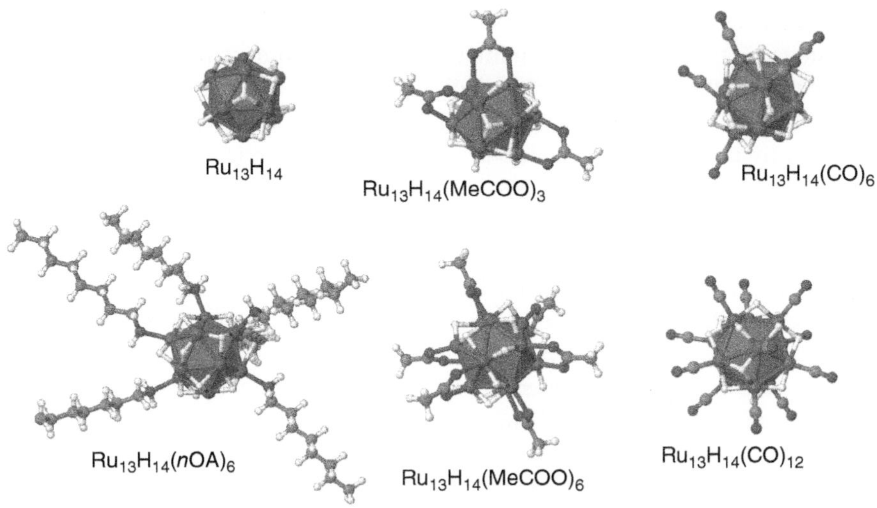

Figure 15.9 $Ru_{13}H_{14}L_x$ clusters, with L_x = $(CO)_6$, $(CO)_{12}$, $(MeCOO)_3$, $(MeCOO)_6$, and $(nOA)_6$. The unprotected reference $Ru_{13}H_{14}$ cluster is also shown.

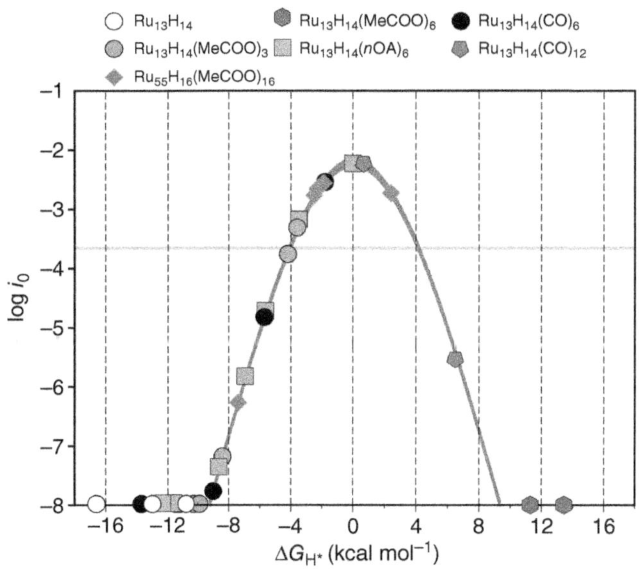

Figure 15.10 Evaluation of the possible HER activity of various $Ru_{13}H_{14}L_n$ clusters based on the volcano schematically plot in Figure 15.6 and on DFT dissociative adsorption Gibbs free energy ΔG_{H^*} (in kcal mol^{-1}) calculated for different sites. The predicted exchange current density i_0 is in A cm^{-2}. The horizontal gray line is the i_0 value experimentally measured for 4PP-capped RuNPs in acidic conditions (see also Figure 15.6). Source: Based on Creus et al. [36].

surface ruthenium atoms and involve positive, very unfavorable, Gibbs free energies (hexagons in Figure 15.10). Such surface composition must be considered as a borderline case, that is not representative of the likely compositions under usual experimental conditions [37]. The coadsorption of ethanoates and hydrides on a larger $[Ru_{55}]$ core was very recently studied [37]. The optimal surface composition determined by DFT calculations and *ab initio* thermodynamics appeared to be c. [0.4–0.6] H/Ru_{surf} and 0.4 ethanoate/Ru_{surf}, which was corroborated by experimental results. Moreover, for such a composition, namely, $Ru_{55}(CH_3COO)_{16}H_{16}$, several sites on the metal surface turned out to exhibit hydrogen adsorption Gibbs free energies in the range -3.0 to +2.0 kcal mol^{-1}. All ΔG_{H^*} values calculated in Ref. [37] are reported in Figure 15.10. According to this descriptor, ruthenium NPs stabilized by carboxylates are promising nanocatalysts for the HER in the electrolysis of water, with an expected exchange current density i_0 close to the 10^{-3} to 10^{-2} A cm^{-2} optimum. Regarding again $[Ru_{13}]$ clusters, an $Ru_{13}H_{14}(nOA)_6$ cluster stabilized by six *n*-octylamine ligands tend to lower $|\Delta G_{H^*}|$ on some sites. As a result, i_0 gets closer to the 10^{-3} to 10^{-2} A cm^{-2} optimum, in line with the catalytic performance of PP-capped RuNPs studied in Section 15.3.2. Other very good catalysts would be $Ru_{13}H_{14}(CO)_6$ or even $Ru_{13}H_{14}(CO)_{12}$ (dark dots and gray pentagons in Figure 15.10). However, it is again two borderline cases. According to stability diagrams calculated in Ref. [57], such surface compositions would be observed in harsh conditions only.

15.4 Summary

Although the presence of surface species is often neglected in computational calculations dealing with the catalytic properties of MNPs, we show in this chapter that they are not simple spectators and stabilizers of the metal core. Two reactions activated by ruthenium NPs or clusters were considered in the present study, in line with previously published works [35–37]: the isotropic H/D exchange at $C(sp^3)$ atoms and the HER.

In the first case, we have shown that the reaction profile is a function of the D/Ru_{surf} ratio, including in the vicinity of the active sites. It involves both a change in the local electronic structure and some steric hindrance. The motion of surface deuterides may also play a role, an assumption that deserves further investigation by *ab initio* molecular dynamics.

In both cases, the Sabatier principle has been used as a powerful framework. Regarding the HER, we have taken advantage of the so-called Balandin–Sabatier volcano plot, employed in several theoretical studies in heterogeneous catalysis, in order (i) to confirm the strong activity of 4-PP-protected RuNPs and (ii) to predict which ligands could involve the best activity of an Ru nanocatalyst. Ruthenium NPs stabilized by amines or carboxylates appear to be good catalysts for the HER, an assumption that still has to be experimentally checked in the case of carboxylates. In addition to local electronic and steric effects, the contribution of these ligands in the HER process could also stem from their rather moderate adsorption strength, which allows a large number of hydrogen atoms to be present on the Ru surface, in contrast with strongly adsorbed ligands, such as carbon monoxide. These results highlight, with a computational and theoretical perspective, that the nature and number of surface ligands can have a strong influence in the catalytic activity of small ruthenium NPs, a conclusion that can probably be generalized to other MNPs.

15.5 Computational Details

Periodic DFT calculations of metal clusters

Software: Vienna *ab initio* simulation package, VASP [63, 64]; spin-polarized DFT; exchange–correlation potential approximated by the generalized gradient approach proposed by Perdew, Burke, and Ernzerhof (PBE) [65]; projector augmented wave (PAW) full-potential reconstruction [66, 67]; PAW data sets for metal atoms treating the $(n-1)$p, $(n-1)$d, and ns states (14 valence electrons for Ru); kinetic energy cut-off: 500 eV [51]; Γ-centered calculations [68]; Gaussian smearing of 0.02 eV width; geometry optimization threshold: residual forces on any direction less than 0.02 eV Å^{-1}; supercell size: 27 × 27 × 28 Å for Ru_{13} clusters and 30 × 30 × 31 Å for Ru_{55} nanoclusters (set to ensure a vacuum space of c. 16 Å between periodic images of metal clusters).

NEB calculations

TSs were found by the climbing image nudge elastic band (CINEB) method [69–71]; three images; spring force between images: 5 eV; force tolerance of 0.02 eV Å^{-1}. The harmonic vibrational modes were systematically calculated in order to distinguish minima and saddle points. Reaction pathways around TSs were explored in forward and reverse directions after a slight alteration of the geometry of the TS according to the imaginary normal mode of vibration.

pDOS calculations

Software: Lobster package [72, 73]. It allows to compute reliable atom-projected density of states (pDOS) directly from plane-wave wave functions calculated with the VASP package. pbeVASPfit basis set of Lobster. Ru: {4p, 4d, 5s}; H: {1s}; C, N: {2s, 2p}. At least $117 + m + 4k$ bands are calculated for a $Ru_{13}H_m(X)_k$. The charge spilling, a criterion that assesses the quality of the projection, is systematically lower than 0.7%.

Calculation of d*-band centers*

d-Band centers, ε_d, were calculated with our home-made *tools4vasp* suite of utilities, which uses the DOS projected on a local basis set by the Lobster package (see also Ref. [62]). $\varepsilon_{d_{surf}}$ stands for the average *d*-band center of surface Ru atoms. The so-called coordination *d*-band center is calculated by averaging the atomic *d*-band center of the surface Ru atoms to which a given H atom is coordinated.

Atomic charges

Atomic charges were calculated by integrating the pDOS up to the Fermi energy. It is nothing else than a Mulliken population analysis (MPA) done after projection in an orthogonal basis set. An assessment of such charge analysis, called pMPA, is done in Ref. [74].

HER

The free energy of the adsorbed hydride, H^*, is usually calculated as

$$\Delta G_{H^*} = E(H^*) - E(\text{cluster}) - \frac{1}{2}E(H_2) + \Delta E_{ZPE} - T\Delta S_H$$

In practice, Nørskov et al. [60] have suggested that 5.5 kcal mol^{-1} can be considered as a representative of $\Delta E_{ZPE} - T\Delta S_H$ for all metals [60]. This means that $\Delta G_{H^*} = \Delta E_{H^*} + 5.5$ kcal mol^{-1}. This simple relationship has been used in Section 15.3.

The $Ru_{55}H_{53}$ model considered in Section 15.3 and in Figure 15.7 was obtained from a full geometry optimization of the $Ru_{55}H_{53}(\sigma\text{-PP})_9(\pi\text{-PP}_2)$ model cleared of the 11 4PP ligands. Normal mode calculations have confirmed that the six considered hydrides are true minima.

In order to avoid spurious steric effects after adsorption of an additional hydride on crowded surfaces, the ΔG_{H^*} index has actually systematically been calculated by removing an hydride from the surface.

Regarding Figure 15.8b, pMPA charges and *d*-band centers were calculated by considering the $Ru_{55}H_{47}$ and $Ru_{55}H_{47}(4PP)^{11}$ models, in order to evaluate the intrinsic adsorption capacity of a given coordination site, i.e. prior to the adsorption of an hydride on the surface of the nanocluster.

Acknowledgments

This article is dedicated to Professor Jean-François Halet on the occasion of his 60th birthday. We thank Drs Bruno Chaudret, Gregory Pieters, Jordi García-Antón, Karine Philippot, and Xavier Sala for helpful discussions at the frontier between theory and experiments and for stimulating scientific collaborations. We acknowledge the HPC CALcul en MIdi-Pyrénées (CALMIP-Olympe, grant P0611) for generous allocations of computer time. Université Paul Sabatier-Toulouse 3 and CNRS are also thanked for financial support.

References

1 Serp, Serp and Philippot, Philippot (eds.) (2013). *Nanomaterials in Catalysis*. Weinheim: Wiley-VCH.
2 Schmid, Schmid (ed.) (2010). *Nanoparticles. From Theory to Application*, 2e. Weinheim: Wiley-VCH.
3 Polshettiwar, V. and Varma, R.S. (2010). *Green Chem.* 12 (5): 743754.
4 Wilcoxon, J.P. and Abrams, B.L. (2006). *Chem. Soc. Rev.* 35 (11): 11621194.

5 Van Santen, R.A. (2009). *Acc. Chem. Res.* 42 (1): 5766.
6 Niu, S. and Hall, M.B. (2000). *Chem. Rev.* 100 (2): 353406.
7 Balcells, D., Clot, E., and Eisenstein, O. (2010). *Chem. Rev.* 110 (2): 749823.
8 Ziegler, T. and Autschbach, J. (2005). *Chem. Rev.* 105 (6): 26952722.
9 Sautet, P. and Delbecq, F. (2010). *Chem. Rev.* 110 (3): 17881806.
10 Neurock, M. (2003). *J. Catal.* 216: 7388.
11 Nørskov, J.K., Bligaard, T., Rossmeisl, J., and Christensen, C.H. (2009). *Nat. Chem.* 1 (1): 3746.
12 van Santen, R.A., Neurock, M., and Shetty, S.G. (2010). *Chem. Rev.* 110 (4): 20052048.
13 Nørskov, J.K., Abild-Pedersen, F., Studt, F., and Bligaard, T. (2011). *Proc. Natl. Acad. Sci. U.S.A.* 108 (3): 937943.
14 Calvo, F. and Carré, A. (2006). *Nanotechnology* 17: 12921299.
15 Gerber, I.C. and Poteau, R. (2013). *In silico* nanocatalysis with transition metal particles: where are we now? In: *Nanomaterials in Catalysis (eds. Ph. Serp and K. Philippot)*, 443471. Weinheim: Wiley-VCH.
16 Halperin, W.P. (1986). *Rev. Mod. Phys.* 58: 533606.
17 Kelsen, V., Wendt, B., Werkmeister, S. et al. (2013). *Chem. Commun.* 49: 34163418.
18 Novio, F., Monahan, D., Coppel, Y. et al. (2014). *Chem. Eur. J.* 20: 12871297.
19 Martínez-Prieto, L.M., Urbaneja, C., Palma, P. et al. (2015). *Chem. Commun.* 51: 46474650.
20 Philippot, K. and Chaudret, B. (2003). *C.R. Chim.* 6 (8–10): 10191034.
21 Baker, L.R., Kennedy, G., Krier, J.M. et al. (2012). *Catal. Lett.* 142 (11): 12861294.
22 Kwon, S.G., Krylova, G., Sumer, A. et al. (2012). *Nano Lett.* 12 (10): 53825388.
23 Chorkendorff, I. and Niemantsverdriet, J.W. (2003). *Concepts of Modern Catalysis and Kinetics*. Wiley-VCH.
24 Sabatier, P. (1911). *Ber. Dtsch. Chem. Ges.* 44 (3): 19842001.
25 Stoltze, P. and Nørskov, J.K. (1985). *Phys. Rev. Lett.* 55: 25022505.
26 Dahl, S., Logadottir, A., Jacobsen, C.J., and Nørskov, J.K. (2001). *Appl. Catal., A* 222 (1): 1929. Celebration Issue.
27 Filot, I.A.W., Shetty, S.G., Hensen, E.J.M., and van Santen, R.A. (2011). *J. Phys. Chem. C* 115 (29): 1420414212.
28 Medford, A.J., Vojvodic, A., Hummelshøj, J.S. et al. (2015). *J. Catal.* 328: 3642. Special Issue: The Impact of Haldor Topsøe on Catalysis.
29 Jacobsen, C., Dahl, S., Clausen, B. et al. (2001). *J. Am. Chem. Soc.* 123 (34): 84048405.
30 Abild-Pedersen, F., Greeley, J., Studt, F. et al. (2007). *Phys. Rev. Lett.* 99: 016105.
31 Balandin, A. (1958). *Adv. Catal.* 10: 96129.
32 Fahrenfort, H., van Reyen, L.L., and Sachtler, W.M.H. (1960). The decomposition of HCOOH on metal catalysts. In: *Mechanism of Heterogeneous Catalysis (ed. de Boer, J. H.)*, 2348. Amsterdam: Elsevier.
33 Tedsree, K., Chan, C.W.A., Jones, S. et al. (2011). *Science* 332 (6026): 224228.

34 Yoo, J.S., Abild-Pedersen, F., Nørskov, J.K., and Studt, F. (2014). *ACS Catal.* 4 (4): 12261233.

35 Taglang, C., Martínez-Prieto, L.M., del Rosal, I. et al. (2015). *Angew. Chem. Int. Ed.* 54 (36): 1047410477.

36 Creus, J., Drouet, S., Suri nach, S. et al. (2018). *ACS Catal.* 8: 1109411102.

37 González-Gómez, R., Cusinato, L., Bijani, C. et al. (2019). *Nanoscale* 11: 93929409.

38 Giri, R., Shi, B.F., Engle, K.M. et al. (2009). *Chem. Soc. Rev.* 38 (11): 32423272.

39 Junk, T. and Catallo, W.J. (1997). *Chem. Soc. Rev.* 26 (5): 401406.

40 Blanksby, S. and Ellison, G. (2003). *Acc. Chem. Res.* 36 (4): 255263.

41 Herron, J.A., Tonelli, S., and Mavrikakis, M. (2013). *Surf. Sci.* 614: 6474.

42 Jachimowski, T. and Weinberg, W. (1997). *Surf. Sci.* 372 (1–3): 145154.

43 Ciobîcă, I.M., Frechard, F., van Santen, R.A. et al. (2000). *J. Phys. Chem. B* 104: 33643369.

44 Filot, I.A.W., van Santen, R.A., and Hensen, E.J.M. (2014). *Catal. Sci. Technol.* 4 (9): 31293140.

45 Abbott, H.L. and Harrison, I. (2008). *J. Catal.* 254 (1): 2738.

46 Amiens, C., Ciuculescu-Pradines, D., and Philippot, K. (2016). *Coord. Chem. Rev.* 308: 409432.

47 Martínez-Prieto, L.M. and Chaudret, B. (2018). *Acc. Chem. Res.* 51: 376384.

48 Pieters, G., Taglang, C., Bonnefille, E. et al. (2014). *Angew. Chem. Int. Ed.* 53 (1): 230234.

49 Pery, T., Pelzer, K., Buntkowsky, G. et al. (2005). *ChemPhysChem* 6: 605607.

50 Truflandier, L.A., del Rosal, I., Chaudret, B. et al. (2009). *ChemPhysChem* 10: 29392942.

51 del Rosal, I., Truflandier, L., Poteau, R., and Gerber, I.C. (2011). *J. Phys. Chem. C* 115: 21692178.

52 Gutmann, T., del Rosal, I., Chaudret, B. et al. (2013). *ChemPhysChem* 14 (13): 30263033.

53 Mak, C.H., Brand, J.L., Deckert, A.A., and George, S.M. (1986). *J. Chem. Phys.* 85 (3): 16761680.

54 García-Antón, J., Axet, M.R., Jansat, S. et al. (2008). *Angew. Chem. Int. Ed.* 47: 20742078.

55 Hammer, B. and Nørskov, J.K. (1995). *Surf. Sci.* 343 (3): 211220.

56 del Rosal, I., Mercy, M., Gerber, I.C., and Poteau, R. (2013). *ACS Nano* 7 (7): 98239835.

57 Cusinato, L., Martinez-Prieto, L.M., Chaudret, B. et al. (2016). *Nanoscale* 8: 1097410992.

58 Walter, M.G., Warren, E.L., McKone, J.R. et al. (2010). *Chem. Rev.* 110 (11): 64466473.

59 Drouet, S., Creus, J., Collière, V. et al. (2017). *Chem. Commun.* 53 (85): 1171311716.

60 Nørskov, J.K., Bligaard, T., Logadottir, A. et al. (2005). *J. Electrochem. Soc.* 152 (3): J23J26.

61 Jaramillo, T.F., Jorgensen, K.P., Bonde, J. et al. (2007). *Science* 317 (5834): 100102.

62 Cusinato, L., del Rosal, I., and Poteau, R. (2017). *Dalton Trans.* 46: 378395.

63 Kresse, G. and Fürthmuller, J. (1996). *Phys. Rev. B* 54 (16): 1116911186.

64 Kresse, G. and Fürthmuller, J. (1996). *Comput. Mater. Sci.* 6 (1): 1550.

65 Perdew, J.P., Burke, K., and Ernzerhof, M. (1996). *Phys. Rev. Lett.* 77 (18): 38653868.

66 Blöchl, P. (1994). *Phys. Rev. B* 50 (24): 1795317979.

67 Kresse, G. and Joubert, D. (1999). *Phys. Rev. B* 59 (3): 17581775.

68 Monkhorst, H.J. and Pack, J.D. (1976). *Phys. Rev. B* 13 (12): 51885192.

69 Mills, G., Jónsson, H., and Schenter, G.K. (1995). *Surf. Sci.* 324 (2–3): 305337.

70 Henkelman, G., Uberuaga, B.P., and Jónsson, H. (2000). *J. Chem. Phys.* 113 (22): 99019904.

71 Henkelman, G. and Jónsson, H. (2000). *J. Chem. Phys.* 113 (22): 99789985.

72 Deringer, V.L., Tchougréeff, A.L., and Dronskowski, R. (2011). *J. Phys. Chem. A* 115 (21): 54615466.

73 Maintz, S., Deringer, V.L., Tchougréeff, A.L., and Dronskowski, R. (2013). *J. Comput. Chem.* 34 (29): 25572567.

74 Gerber, I.C. and Poteau, R. (2018). *Theor. Chem. Acc.* 137: 156.

Index

Nanoparticles in Catalysis: Advances in Synthesis and Applications, First Edition.
Edited by Karine Philippot and Alain Roucoux.

B

C

G

H

Q

R

S

X

Z